Dinosaur Tracks from Brazil

Life of the Past   James O. Farlow, editor

DINOSAUR TRACKS FROM BRAZIL

A Lost World of Gondwana

GIUSEPPE LEONARDI

AND

ISMAR DE SOUZA CARVALHO

Indiana University Press

This book is a publication of

Indiana University Press
Office of Scholarly Publishing
Herman B Wells Library 350
1320 East 10th Street
Bloomington, Indiana 47405 USA

iupress.org

Manufactured in the
United States of America

First printing 2021

Library of Congress
Cataloging-in-Publication Data

Names: Leonardi, Giuseppe, author. |
    Carvalho, Ismar de Souza, author.
Title: Dinosaur tracks from Brazil : a
    lost world of Gondwana / Giuseppe
    Leonardi and Ismar de Souza Carvalho.
Description: Bloomington, Indiana :
    Indiana University Press, [2021] |
    Series: Life of the past | Includes
    bibliographical references and index.
Identifiers: LCCN 2021008044 (print) |
    LCCN 2021008045 (ebook) | ISBN
    9780253057228 (hardback) |
    ISBN 9780253057242 (ebook)
Subjects: LCSH: Dinosaurs—Brazil.
    | Dinosaur tracks—Brazil.
Classification: LCC QE861.9.B6 L46
    2021 (print) | LCC QE861.9.B6
    (ebook) | DDC 567.90981—dc23
LC record available at https://
    lccn.loc.gov/2021008044
LC ebook record available at https://
    lccn.loc.gov/2021008045

Contents

Foreword

In 1912 Sir Arthur Conan Doyle, creator of the legendary Sherlock Holmes, published *The Lost World*, a novel featuring an equally influential literary creation: Maple White Land, a huge, steep-walled plateau in the wilds of tropical South America. This *tepui* was the home of a remarkably diverse antediluvian bestiary: nonavian dinosaurs, pterosaurs, plesiosaurs, phorusrhacids, notoungulates, and even basal hominins as well as anatomically modern, Stone Age humans. Conan Doyle's notion of a lost land still inhabited by prehistoric animals was hugely influential on popular culture, perhaps most notably in Edgar Rice Burroughs's Caspak and Pellucidar novels and the 1933 film *King Kong*.

The Lost World was my favorite childhood book. I read and reread it countless times. In my imagination I constantly invented new lost worlds of my own. But then I grew up and faced reality: tepuis are real geomorphological features in South America, and they do indeed house distinctive plants and animals, but the only dinosaurs living there are of the fluttery, feathery, avian kind.

Sir Arthur never knew it, and neither did my childhood self, but there really was a South American lost world waiting to be explored, one that allowed at least indirect glimpses of living dinosaurs, in a manner analogous to the Greek philosopher Plato's image of the wall of the material world's "cave" being the surface on which are cast the flickering "shadows" of metaphysical "reality." This real lost world—the equivalent of Plato's cave—comprises the Rio do Peixe Basins of Brazil, which preserve one of the world's great assemblages of fossil footprints and trackways (the "shadows") of dinosaurs and other Mesozoic land animals (the "reality"). Such trace fossils hold a special fascination for paleontologists like myself who so desperately want to know what living dinosaurs were like, because they record moments in the lives of the long-dead animals, revealing how they moved and interacted with each other. The authors of this book are the explorers who will describe for us the wonders of the Rio do Peixe lost world.

Giuseppe Leonardi was probably predestined (to make an ironic nod at John Calvin) to become a paleontologist. His great-great-grandfather Demetrio (1796–1881) was a geologist who researched metamorphic petrology and geohydrology. Giuseppe's father, Piero (1908–1998), was a geological polymath whose interests and publications spanned volcanism, structural geology, stratigraphy, planetary geology, archaeology, and, of course, paleontology. His paleontological research emphasized invertebrate and plant fossils, Cenozoic mammals, and vertebrate trace fossils.

Giuseppe Leonardi was born in 1939, one of Piero's four children. This was truly an academic family; Giuseppe's older sister became a teacher of natural science and eventually director of the Biblical School of the Patriarchate of Venice, and a younger brother became an archaeologist at the University of Padua. In 1948, at the age of nine, Giuseppe accompanied his father on a field trip to Bolzano Province of northeastern Italy, where they collected tetrapod footprints from Permian red beds. The lad became hooked on geology and paleoichnology. After high school he began attending the University of Padua (the family university for six generations!) with the intent of becoming a geologist.

However, perceiving a vocation to the priesthood, Leonardi left the university to become a seminarian in the Congregation of Schools of Charity or Institute Cavanis, a religious community in Venice, where he studied philosophy and theology and was ordained a priest in 1964. He then earned a master's degree in theology in Rome and pursued further studies at the Biblical Institute of Rome and in Jerusalem. Leonardi was invited to become a professor of Old Testament exegesis at the Pontifical Biblical Institute in Rome, but the superiors in his religious order asked him instead to become a teacher in Catholic schools. Giuseppe therefore studied natural science at the state University La Sapienza in Rome.

Throughout this time he never lost interest in geology, paleontology, and evolution. While in Rome Leonardi formed a paleoichnology study group with other students, some of who have gone on to become distinguished experts on vertebrate trace fossils. While on the mission field in South America and Africa, he carried out fieldwork as time permitted, eventually learning about the ichnological riches of the Rio do Peixe Basins that are the subject of this book.

Ismar Carvalho was born in 1962 in Resende, a small city in the interior of the Brazilian state of Rio de Janeiro. His is the third generation in a family of European immigrants. His grandfathers came from Portugal and Italy at the end of the nineteenth century. His forebears lived in agricultural colonies in the area around Resende, which was a truly multinational city, with inhabitants hailing from Italy, Portugal, Finland, and the Arab world.

Ismar's childhood studies were in the public schools of Resende. Everyone recognized the importance of learning as a way to improve one's life. As a child, Ismar wrote on his personal blackboard a phrase that has influenced him for his entire life: "*O estudo é a luz da vida*" — study is the light of life.

As a youth, the only contact Ismar and his peers had with the outside world was through the Sunday newspaper, which was generally read on Monday. His first contact with paleoichnology was at the age of fifteen, through a story in the paper: a notice that an Italian priest was looking for dinosaur footprints. Ismar was astonished by that news story and has kept the clipping in his collection of personal mementos.

The story revealed that the priest was not interested merely in the old bones of extinct animals. His research dealt with dinosaurs as living animals: how footprints and trackways demonstrate the way that dinosaurs walked and the environments in which they lived. By now you have surely guessed that the Italian priest was none other than Giuseppe Leonardi.

The story was decisive in Carvalho's decision to attend the university and study geology at a time when parents only wanted their sons to become physicians, engineers, or lawyers. During his first semester of classes, the university organized a week of geological studies, and Carvalho was delighted that his suggestion that Leonardi be invited to speak was accepted. So began the association that led to the present book.

During his second year of studies, Carvalho had the opportunity to transfer to Coimbra University (Portugal) to finish his undergraduate work. He then returned to Brazil and initiated graduate studies in geology and paleontology at Rio de Janeiro Federal University, where he works to the present day. His academic appointment is in the Geology Department of the Geoscience Institute of the university, and he was also the director of the Institute of Geoscience (which also includes the Geography and Meteorology Departments); presently he is director of the Casa da Ciência, the Science and Cultural Centre of Rio de Janeiro Federal University.

Carvalho has published technical papers describing several species of fossil dinosaurs and crocodyliforms as well as his work on dinosaur footprints. He is especially interested in the importance of microbial mats in preserving dinosaur tracks and invertebrate traces. Carvalho is editor of a textbook of paleontology that is widely used in Brazil and other Portuguese-language countries.

Leonardi and Carvalho will be our guides—our Professors Challenger and Summerlee—leading us through the lost world of the Rio do Peixe Basins. We will see many wonders, including the traces made by dinosaurs and other long-dead animals, with our physical eyes and in our mind's eye the fearfully great reptiles themselves. Prepare yourself for a scientific adventure!

James O. Farlow
Emeritus Professor of Geology
Department of Biology
Purdue University Fort Wayne

Acknowledgments

We are particularly grateful to James (Jim) O. Farlow (Department of Geosciences, Indiana–Purdue University at Fort Wayne, Indiana) for his wonderful foreword and for his excellent criticism, constructive suggestions, and complete revision of the manuscript, which resulted in substantial improvement.

We are very grateful to Richard T. McCrea (museum director and curator of paleontology at the Peace Region Palaeontology Research Centre, Tumbler Ridge, British Columbia, Canada), who, as a generous referee, gave us precious suggestions, criticism, and observations and his strong and hearty support. Finally, we thank Indiana University Press.

Marco Avanzini, of the Museo delle Scienze-MUSE of Trento, Italy, was very helpful in the production of this book, giving precious advice and discussing the data and their interpretation.

Maria de Fátima Cavalcante Ferreira dos Santos, Claude Luis Aguilar Santos, and Narendra Kumar Srivastava of the Federal University of Rio Grande do Norte at Natal (Rio Grande do Norte, Brazil); Diogenes de Almeida Campos, José Ferreira, Francisco Canuto de Araújo, and Sérgio Santa Rita de Queiróz of the Departamento Nacional de Produção Mineral-DNPM (the Geological Survey of the Federal Ministry of Mines and Energy of Brazil); the late Geraldo Barros da Costa Muniz and the late Jannes M. Mabesoone of the Federal University of Pernambuco at Recife; Luiz Carlos Godoy of the Universidade Estadual de Ponta Grossa, Paraná, Brazil; Francisco Henrique de Oliveira Lima of Petrobras; and Maria Dolores Wanderley of the Federal University of Rio de Janeiro were very helpful in the field.

Many colleagues and friends have generously helped us through these decades by providing us with bibliographic material, data, advice, and friendly support. Cordial thanks to all of them. We recall here the late Llewellyn I. Price (DNPM, Rio de Janeiro), the late Mário C. Barberena (Federal University of Rio Grande do Sul, Porto Alegre, Brazil), and the late José F. Bonaparte (Museo Argentino de Ciencias Naturales, Buenos Aires); Matteo Belvedere (University of Florence, Italy); Paolo Citton (National University of Río Negro, Argentina); Maria Alessandra Conti (Sapienza University of Rome); Pedro Proença Cunha and Maria Helena Henriques (University of Coimbra, Portugal); Cristiano Dal Sasso (Natural History Museum, Milan); Simone D'Orazi Porchetti (Italy); Antonio Carlos S. Fernandes (Federal University of Rio de Janeiro); Silvério Figueiredo (Polytechnic Institute of Tomar, Portugal); Peter L. Falkingham (Liverpool John Moores University, UK); Heitor Francischini

(Federal University of Rio Grande do Sul, Brazil); Marcelo A. Fernandes (Federal University of São Carlos, Brazil); José Carlos García Ramos (Jurassic Museum of Asturias, Spain); Martin G. Lockley (University of Colorado, Denver); Spencer Lucas (New Mexico Museum of Natural History and Science, Albuquerque); Lorenzo Marchetti (Museum for Natural History–Leibniz Institute for Research on Evolution and Biodiversity, Berlin); Christian A. Meyer (University of Basel, Switzerland); Paolo Mietto (University of Padua, Italy); Karen Moreno (Austral University of Chile, Valdivia); Umberto Nicosia (Sapienza University of Rome); In Sung Paik (Pukyong National University, Busan, Republic of Korea); Félix Pérez-Lorente (University of la Rioja, Spain); Fabio M. Petti (Italian Geological Society and Science Museum in Trento, Italy); Laura Piñuela (Museo del Jurásico de Asturias, Spain); José Henrique Popp and Marlene Popp (Federal University of Paraná, Curitiba, Brazil); Marco Romano (Sapienza University of Rome); Rafael C. da Silva (Geological Survey of Brazil); Fernando A. Sedor and Paulo Soares (Federal University of Paraná, Curitiba, Brazil); Mário Suárez R. (Noel Kempff Mercado Natural History Museum, Santa Cruz, Bolivia); Tony Thulborn and Susan Turner (University of Queensland at Brisbane, Australia); Alexander Wagensommer (Italy); and Fabiana Zandonai (Civic Museum of Rovereto, Italy).

Rodrigo Nascimento (Interciência, Rio de Janeiro) assisted us in the first phase of this volume, providing good suggestions on its structure and constant support. Thank you to David Miller, the Indiana University Press lead project manager/editor for this book, who did a wonderful job of reviewing and copyediting our book. Thank you to Leonardo Borghi (Federal University of Rio de Janeiro, head of the Sedimentary Geology Laboratory [Lagesed]) for laboratory support. Thank you, Deverson Silva (Pepi), Ariel Milani Martine, Franco Capone, Carlos Archanjo, Maria Judite Garcia, Wellington Francisco Sá dos Santos, and José Henrique Gonçalves de Melo for providing data and illustrations. Bruno Rafael de Carvalho Santos, Gilberto Raitz Jr., Diogo Lins Batista, and Jaime Joaquim Dias organized the illustrations and the final draft of this book.

The Museu Câmara Cascudo, of the Federal University of Rio Grande do Norte in Natal; the Departamento Nacional de Produção Mineral (DNPM; the Geological Survey of the Federal Ministry of Mines and Energy of Brazil); the Federal University of Pernambuco in Recife, Petrobras; and the Federal University of Rio de Janeiro gave logistical and technical support to our research in the field.

The successive mayors of Sousa—Gilberto de Sá Sarmento (1972–76), Clarence Pires de Sá (1977), Sinval Gonçalves (1978–82), Nicodemos de Paiva Gadelha (1983–88), and the late Salomão Benevides Gadelha (2002–08)—all in their respective times; José Nilton Fernandes, mayor of São João do Rio do Peixe (1983–89); Geraldo Nogueira de Almeida, mayor of Uiraúna (1983–88); and their administrations and collaborators helped very much with logistical support.

Luiz Carlos da Silva Gomes, president of the cultural group Movissauros; Miss Julieta Gadelha, news reporter; Tibério Felismino de Araújo, secretary of agriculture, then manager of Sousa; Neuricélia Teodoro Lima Moreira, then assistant attorney in the Sousa government; and Verniaud A. Breckenfeld Alexandre and Antônio Nogueira, both of São João do Rio do Peixe, and Lenice do Vale were in different ways very helpful during our expeditions to the Rio do Peixe Basins. Robson Araújo Marques contributed greatly to almost all of the expeditions as an agent of the Municipality of Sousa and was a most valuable guide during our fieldwork—besides being a very good friend.

We thank Anna da Schio Steiner of Venice for her helpful review of some chapters of the last draft of our manuscript.

João Carlos Moreira Rodrigues, sculptor and technician of the Museu Paraense Emílio Goeldi of Belém (Pará), created (1984–87) some dinosaur sculptures, with the valuable help of technicians Josenilton de S. Cavalcante and Francisco W. de S. Cavalcante, both from Sousa. Francisco Henrique de Oliveira Lima and Cláudia Valéria de Lima assisted in the statistical analysis of the footprint and trackway data (1985–87). Their three papers published with Leonardi (Leonardi et al. 1987a, 1987b, 1987c) are entirely reproduced herein. The engineer Mauro Augusto Modesto of Ponta Grossa, then a student, did the final drafts of a number of drawings.

The Conselho Nacional de Desenvolvimento Científico e Tecnológico-CNPq (the National Council for Research of Brazil) supported our research in the Rio do Peixe Basins, as well as in other Brazilian basins, over forty-six years (1975–2021); we are particularly grateful to Marcos Maciel Formiga of this agency (1984–89). The Fundação Carlos Chagas Filho de Amparo à Pesquisa do Estado do Rio de Janeiro (FAPERJ) supported field research in the area. The Departamento Nacional de Produção Mineral-DNPM (the Geological Survey of the Federal Ministry of Mines and Energy of Brazil) and its districts of Rio de Janeiro, Campina Grande (Paraíba), Fortaleza (Ceará), Recife (Pernambuco); and the Companhia de Pesquisa de Recursos Minerais-CPRM often provided concrete logistical and moral support. We are particularly grateful to Carlos Oití Berbert, formerly president of the CPRM and formerly director of the DNPM-DGM. The Centro Studi Ricerche Ligabue of Venice gave support to one of the expeditions and, in many other ways, helped the research.

Dinosaur Tracks from Brazil

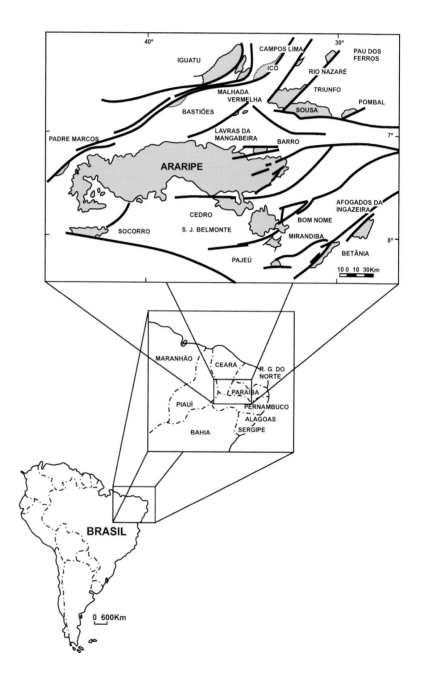

Figure 1.1. Location map of the Rio de Peixe basins, western Paraíba, Northeast Brazil (Carvalho et al. 2013a).

The Rio do Peixe basin comprises four sedimentary sub-basins: the Sousa, Triunfo (also called the Uiraúna-Brejo das Freiras basin), Pombal, and Vertentes (fig. 1.1). They are located in the western part of the Brazilian State of Paraíba, in the municipalities of Sousa, São João do Rio do Peixe (a small city formerly called Antenor Navarro), Aparecida (formerly a district of Sousa), Uiraúna, Poço, Brejo das Freiras, Triunfo, Santa Helena, and Pombal. In the first two basins—Sousa and Uiraúna-Brejo das Freiras (Triunfo), especially the Sousa basin—an abundant tetrapod ichnofauna have been preserved, consisting of footprints and trackways, mainly of large theropods, sauropods, and ornithopods. Invertebrate ichnofossils, such as traces and burrows produced by arthropods and annelids, are also common (Fernandes and Carvalho 2001). Along with the formation's dominant reddish color, typical of subaerial environments, there are some units of greenish shales, mudstones, and siltstones, where body fossils are present. These include ostracods, conchostracans, plant fragments, palynomorphs, fish scales, and bone fragments of crocodylomorph, dinosaurs, and probably pterosaurs.

Sousa (fig. 1.2) and Uiraúna-Brejo das Freiras (Triunfo) are intracratonic basins in Northeast Brazil that developed along preexisting structural

Figure 1.2. Location map of the Rio do Peixe basins, western Paraíba, Northeast Brazil, and the distribution of its local ichnofaunas (Leonardi and Santos 2006). The smaller maps are redrawn from Leonardi and Carvalho (2002).

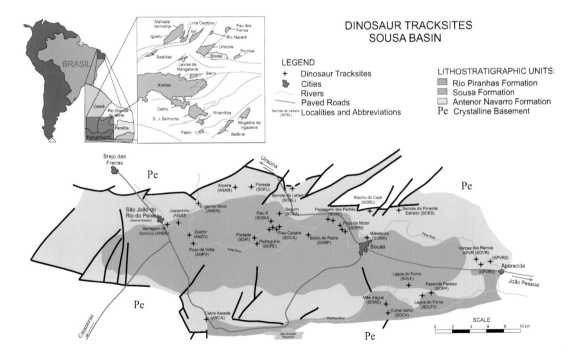

3

trends of the basement during the opening of the South Atlantic Ocean. The age of sedimentary deposits in these basins, based on ostracods (Sousa et al. 2019) and palynological material, is characteristic of the Rio da Serra (Berriasian to Hauterivian) and Aratu (early Barremian) local stages (Lima and Coelho 1987; Regali 1990).

Sedimentation in these basins was controlled by regional tectonic processes (Lima Filho 1991). During the Dom João time (Purbeckian Stage), due to crustal extension, sigmoidal basins were developed at the inflection of the NE–SW and E–W faults. During the Rio da Serra time (Berriasian to Hauterivian), under the same tectonic stress pattern, the basinal areas increased, and their shapes became rhomboidal. In the last stage, probably at the end of Aratu time (early Barremian Stage), there was a change in the tectonic pattern, and sediment accumulation began to diminish.

These deposits reflect direct control of sedimentation by tectonic activity. Deposition occurred along the faulted borders of the basins as alluvial fans, changing to an anastomosing fluvial system more distally. In the central region of the basins, a meandering fluvial system with a wide floodplain was established, where perennial and temporary lakes developed (Carvalho 2000a).

The paleontological-geological relevance of the Sousa and Uiraúna-Brejo das Freiras basins is the abundance of dinosaurian ichnofaunas that represent parts of an extensive Early Cretaceous megatracksite (Viana et al. 1993; Carvalho 2000a; Leonardi and Carvalho 2000, 2002) established during the early stages of the South Atlantic opening. In this area, 37 sites and approximately 96 individual stratigraphic levels preserve occurrences of more than 535 individual dinosaurian trackways as well as rare tracks and traces of the vertebrate mesofauna.

The Town of Sousa

The Brazilian town of Sousa (fig. 1.3) is indisputably the "capital" of the "valley of the dinosaurs" of western Paraíba. The Sousa municipality (Ferraz 2004) currently has a territory of 842 km² (square kilometers) and is situated in the region of the Alto Sertão da Paraíba (the inner semi-arid belt of Paraíba). It is bounded on the north by the municipalities of Vieirópolis, Lastro, and Santa Cruz; on the east by São Francisco and Aparecida; on the south by São José da Lagoa Tapada and Nazarezinho; on the southwest by Marizópolis; on the west by São João do Rio do Peixe (formerly Antenor Navarro); and on the northwest by Uiraúna.

The town of Sousa is located at 220 meters (m) above sea level, at (GPS datum: SIRGAS 2000) S 06 45.505, W 38 13.797 in the town center; at 420 kilometers west (as the crow flies) of João Pessoa, capital of the State of Paraíba. The town is crossed by the Peixe River (Rio do Peixe), a tributary of the Piranhas River (named Açu River in its lower reaches in Rio Grande do Norte); this latter river also crosses the territory of the municipality of Sousa.

The municipality of Sousa had (*Instituto Brasileiro de Geografia e Estatística*—IBGE, 2019) 69,444 inhabitants, 51,881 of whom live in the town of Sousa. The main access to the town is by car or bus from João Pessoa, along federal highway BR-220. The town of Sousa, through its mayors and administrations, is in many ways one of the most important supporters of the research that led to this scientific study.

The main occurrence and most readily accessible site of dinosaur footprints is at Passagem das Pedras (Ilha Farm) in the Sousa municipality; it is now a natural park. In December 1992, through a state act, the area was defined as a natural monument and named "Dinosaur Valley Natural Monument" (*Monumento Natural Vale dos Dinossauros*) (fig. 1.4).

Figure 1.4. Cretaceous Park at Sousa, with Carvalho, at the main gate.

History of the Paleontological Site

Figure 1.5. Luciano Jacques de Moraes (1896–1968), a Brazilian mining engineer, discoverer of the first 2 short trackways impressed in the rocky pavement forming the bed of the Peixe River, at Passagem das Pedras in the Ilha Farm, sometime before 1924. These were the first tetrapod trackways published from Brazil. Art by F. Fayez.

In the 1920s, Luciano Jacques de Moraes (1896–1968), a Brazilian mining engineer (fig. 1.5), was working for the DNOCS (*Departamento de Obras contra as Seccas*, the Department of Works against the Drought), surveying the then little-known Brazilian northeast. In the western area of the State of Paraíba, on the Ilha Farm, Moraes discovered 2 trackways impressed in the rocky pavement forming the riverbed of Rio do Peixe, a left tributary of the Piranhas River. They were 2 crisscrossing trackways of different size, made by very different animals. They were also the first fossil tetrapod trackways published from Brazil.

Moraes surveyed the tracks and later, with the help of Oliveira Roxo (*Divisão de Geologia e Mineralogia*, DGM), he correctly attributed the trackways to the Dinosauria. Moraes sent a slab containing a footprint (the sixth in the sequence) excavated from trackway SOPP 2 and a plaster cast of a footprint from trackway SOPP 1 to the United States to be studied by an unnamed paleontologist; he had not received an acknowledgment when he published his book in 1924, nor did he ever receive a reply. It seems probable that he had sent the tracks to the American Museum of Natural History in New York (this was a common opinion at the Departamento Nacional de Produção Mineral—DNPM—in the 1970s), but they were not found there by Leonardi in 1985. In his book, Moraes questionably attributed trackway SOPP 1, the largest, at that time 13 m long and with 15 footprints, either to a member of the Stegosauria or, as a less likely alternative, to the Ceratopsia; evidently, he interpreted this as the trackway of a quadruped. Probably he interpreted the true footprints themselves as impressions of the forefeet and the displacement rims, with their roundish collapsed mud cracks, as hind-foot prints. Moraes identified the maker of trackway SOPP 2 as a bipedal dinosaur without deciding between the Theropoda and the Ornithopoda. Moraes described the tracks with rare thoroughness, providing detailed drawings and good photographs. He also estimated the sizes of the presumed trackmakers, according to his different hypotheses about their nature. This was a good work for that time; however, Moraes was neither an ichnologist or a paleontologist, and his drawings were rather inaccurate. Among other things, he did not see the true fore prints of the semibipedal maker of SOPP 1.

Moraes eventually sent some photographs to Friedrich von Huene of Tübingen (1875–1969; fig. 1.6), who published, instead of the photos, some drawings from Moraes's (1924) book in his paper on new fossil tetrapods from South America (1931). He briefly described both tracks as digitigrade; wisely avoiding instituting new taxa, von Huene described SOPP 1 as the trackway of a quadruped with total overlap of manus and pes prints and the SOPP 2 as having been made by a biped. He tentatively attributed the first trackway to either a ceratopsid or a nodosaurid but preferred the latter; the second trackway he attributed to an ornithopod, more specifically to the trachodontids or kalodontids. His identifications were based largely on Moraes's unpretentious drawings.

Von Huene had worked in the field in Brazil in the 1920s but had never visited Paraíba. Brazilian geologists and paleontologists—among them the late Llevellyn Ivor Price (1905–1980; fig. 1.7), Diogenes de Almeida Campos, and Jannes M. Mabesoone—saw Moraes's tracks in the 1950s and 1960s. However, these tracks were not further studied or published, except for a brief mention (Price 1961) and one short compilation paper (Cavalcanti 1947). They were illustrated in a number of publications (Haubold 1971).

From a geological point of view, the Rio do Peixe basins were surveyed and published, along with other basins of Northeast Brazil, by Crandall (1910); Branner (1919); Sopper (1923); Oliveira and Leonardos (1943); Kegel (1965); Braun (1966, 1969, 1970); Costa (1964); Beurlen (1967a, 1967b, 1971); Beurlen and Mabesoone (1969); Costa (1969); Albuquerque (1970); Mabesoone (1972, 1975); Mabesoone and Campos e Silva (1972); Campos et al. (1974); Dantas (1974); Mabesoone and Campanha (1974); Brito (1975); Mabesoone et al. (1979); Vasconcelos (1980), Dantas and Caula (1982); Amaral (1983); Rand (1984); Barbosa et al. (1986); Lins (1987); Popoff (1988); Sénant and Popoff (1989); Lima (1990); Ponte et al. (1990); and Regali (1990).

Paleontological studies of invertebrates, plant megafossils, and pollen of the Rio do Peixe basins were published by Maury (1930, 1934), Mabesoone and Campanha (1974), Tinoco and Katoo (1975), Tinoco and Mabesoone (1975), Muniz (1985), Lima and Coelho (1987), and Carvalho (1993). However, in all of these geological and paleontological works, references to the dinosaur tracks, if present, were brief and incidental.

At the end of the twentieth century, a local legend arose in Sousa: the tracks had been discovered in 1897 by Anísio Fausto Silva, a local farmer. The centenary of this supposed discovery was celebrated in Sousa in 1997! The main tracks are quite conspicuous, and surely many farmers had seen them and, before them, ancient natives. However, the discovery, for scientific purposes, must be attributed to Luciano Jacques de Moraes.

Figure 1.6. The German paleontologist Friedrich von Huene of Tübingen (1875–1969) published Moraes's (1924) drawings of the Paraíban tracks in his 1931 paper on fossil tetrapods from South America. Von Huene used the drawings to interpret the nature of the trackmakers. Art by F. Fayez.

Figure 1.7. Llewellyn Ivor Price (1905–80), of the Departamento Nacional de Produção Mineral of Rio de Janeiro, was the main authority on dinosaurs in Brazil and deeply favored the expansion of the Brazilian paleoherpetology.

Figure 1.8. Leonardi crossing the Peixe River, a tributary of the Piranhas River, at Passagem das Pedras, in December 1975, when he was beginning his studies at the Rio do Peixe basins.

Figure 1.9. Paraíban "safari" scene at Piau-Caiçara farm. This was the first discovery at Piau in 1979: this well-preserved, deep footprint from either level 13 or—more probably—14 was collected from the stratum before the levels were assigned numbers and before the site was mapped. The print was made by an ornithopod.

Locally the tracks are known as the "ox tracks" (SOPP 1) and the "*ema* (south American ostrich [*Rhea*]) tracks" (SOPP 2).

In December 1975, Giuseppe Leonardi, who lived in Brazil from 1974 to 1989, began a series of thirty-three expeditions to the Rio do Peixe basins (1975–2016) to study the dinosaur tracks. He worked in the field alone during the first expeditions (fig. 1.8 and fig. 1.9) but later frequently with informal groups of researchers. These expeditions were mainly funded by the Brazilian *Conselho Nacional de Desenvolvimento Científico e Tecnológico* (the National Council for Research of Brazil). Ismar de Souza Carvalho of the Federal University of Rio de Janeiro led many other geological and ichnological expeditions, twice a year over the last twenty-three years. The result of this fieldwork has been a number of preliminary or partial publications, mainly in Brazilian journals (Leonardi 1979a, 1979b, 1979c, 1980a, 1980b, 1980c, 1980d, 1981a, 1981b, 1981c, 1982, 1984a, 1984b, 1985a, 1985b, 1985c, 1987a, 1987b, 1988, 1989a, 1989b, 1994; Leonardi et al. 1987a, 1987b, 1987c; Godoy and Leonardi 1985; Carvalho and Leonardi 1992; Santos and Santos 1987a, 1987b; Carvalho

Figure 1.10. Carvalho exploring the Pombal basin on 1988.

1989; Carvalho and Carvalho 1990; Azevedo 1993; Leonardi and Santos 2006; Antonelli et al. 2020).

The research on dinosaur ichnology was extended by Leonardi and others, with some success, from the Rio do Peixe basins to other Northeast Brazil basins (Leonardi 1980b, 1984c, 2008; Leonardi and Borgomanero 1981; Leonardi and Muniz 1985; Leonardi and Spezzamonte 1994; Leonardi et al. 1979; Assis et al. 2010). This research was later continued with great success by Carvalho (fig. 1.10) and others (Carvalho 1989, 1994a, 1994b; Carvalho and Fernandes 1992; Carvalho et al. 1993a, 1993b, 1994; Carvalho and Gonçalves 1994; Carvalho and Leonardi 2020; Dentzien-Dias et al. 2010; Santos and Santos, 1987a, 1987b, 1989; Viana et al. 1993).

Almost from the beginning of his fieldwork in the Rio do Peixe basins, Leonardi started a systematic campaign to protect the dinosaur tracks of the Paraíba by creating a natural park of the "Dinosaur Valley" (fig. 1.11), with the support of many friends and colleagues and the Brazilian press and television networks. Many public Brazilian agencies at the federal, state, and municipal levels were kept informed about these efforts, and their involvement was solicited.

Involved public agencies at the national (union) level included the Federal Ministry of Education, the Federal Ministry of Culture, the Senate of the Republic, the *Departamento Nacional de Produção Mineral*, the Federal Ministry of Mines and Energy, the *Companhia de Pesquisa de*

Figure 1.11. A map created for one of the projects for the four modules of the natural park "Vale dos Dinossauros," at Serrote do Letreiro (1984). The planned four modules were at Passagem das Pedras, Serrote do Letreiro (this map), Serrote do Pimenta at Estreito farm (all three in Sousa municipality), and at Engenho Novo (São João do Rio do Peixe). Only the Passagem das Pedras module was built, however.

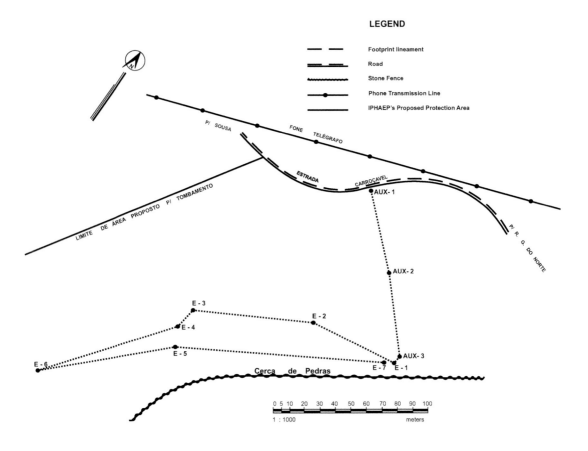

LEGEND

Footprint lineament
Road
Stone Fence
Phone Transmission Line
IPHAEP's Proposed Protection Area

Recursos Minerais, the Brazilian *Conselho Nacional de Desenvolvimento Científico e Tecnológico* (the National Council for Research of Brazil), the *Secretaria do Meio Ambiente* (Secretary of the Environment), the Federal University of the Paraíba, and the Federal University of the Rio Grande do Norte–Câmara Cascudo Museum of Natal. At the state level, there were the government and several successive governors of the State of Paraíba, the *Instituto Histórico e Geográfico Paraibano*, the *Instituto do Patrimônio Histórico e Arqueológico da Paraíba*, the Foundation "*Casa de José Américo*" at João Pessoa, and the *PBTour* (the Paraíba Agency for Tourism); and the municipalities of Sousa, São João do Rio do Peixe (than named Antenor Navarro); and Uiraúna.

Beginning in 1985, with grants from the Brazilian *Conselho Nacional de Desenvolvimento Científico e Tecnológico* (the National Council for Research of Brazil), fiberglass sculptures of dinosaurs were created for the park, in a laboratory in Sousa, by Leonardi and sculptor João Carlos Moreira Rodrigues of the Goeldi Museum of Belém (Pará, Brazil). Among these were a titanosaur, an abelisaurid, and some small theropods, all at full size.

With the collaboration of the Câmara Cascudo Museum of Natal (Rio Grande do Norte), a number of plaster casts of the more important and new tracks were made and deposited in that museum (fig. 1.12 and fig. 1.13; Santos et al. 1985). To avoid the danger of destruction by flood, vandalism, or theft, some original slabs with tracks were given to the Câmara Cascudo and to other museums and institutions. Details of where these specimens are deposited are provided in this book.

This memoir does not mark the end of ichnological research in the Rio do Peixe basins. The Sousa basin (1,250 km²) was largely explored by Leonardi over forty-two years and by Carvalho in the last thirty-four years, but it probably holds many new surprises. As many of the main ichnosites, especially those in the Sousa Formation, occur on bedding

Figure 1.12. Maria Dolores Wanderley, Maria de Fátima C. F. dos Santos, and Lenice do Vale preparing plaster casts at Passagem das Pedras in 1981.

Figure 1.13. Claude Luis de Aguilar Santos casting the first footprint of trackway SOPI 1 with plaster in 1981.

planes in a riverbed, natural erosion will uncover additional tracks in the old localities. Then new localities surely will be discovered. The Pombal basin was poorly explored, but it seems rather poor in outcrops, and the rocks are generally too coarse. The Brejo das Freiras-Uiraúna basin (480 km²) deserves further exploration.

The intent of the present work is to describe, illustrate, attribute, and, when possible, classify all the already discovered tracks and to extract from them all possible information. The tracks will be studied neither just like isolated biogenic structures nor just as a record of the passage of individual dinosaurs and other animals but rather as populations and associations in their whole paleobiological, behavioral, sedimentological, stratigraphic, paleoenvironmental, paleoclimatological, and paleogeographic context.

Methods

2

The main part of the ichnological fieldwork for this book was carried out over a period of fifteen years, from 1975 to 1989, before or at the beginning of the "dinosaur track renaissance" (Alcalá et al. 2016; Falkingham et al. 2016a, 3); however, the discovery and detailed study of most of the

Figure 2.1. Life in the *Sertão* (arid savanna) of Rio do Peixe. *A*, The crew from the Câmara Cascudo Museum of Natal, Rio Grande do Norte, casting tracks at Fazenda Piedade. *B*, Leonardi exploring astride the Serrote do Pimenta at Fazenda Estreito (Sousa) with vaqueiro hat and *facão de mato* (machete). Photo courtesy of Franco Capone. *C*, Surveying at Serrote do Letreiro (1988). Photo courtesy of Franco Capone. *D*, Praça dos dinossauros (Dinosaur Square) at Sousa, with a party of paleontologists on a field trip (1985). *E*, The 500-m-long bypass canal, excavated to provide an alternative path for the Peixe River and avoid flooding the main track surface in the bed of the river at Passagem das Pedras during the rainy season.

Figure 2.2. Working in the field in the Sertão of Rio do Peixe. *A*, The very first excavations at Passagem das Pedras (Sousa 1975), with workers and Fr. Mario Merotto. *B*, Excavations at Juazeirinho (São João do Rio do Peixe, formerly Antenor Navarro; 1984). *C*, Interpreting with chalk the large tracks of the sauropod herd at Serrote do Letreiro (1988). *D*, Searching for scattered tracks at Piau (Sousa, 1979). *E*, Drawing associations of tracks with the gridded quadrat, under the merciless equatorial sun, at Serrote do Letreiro 4 (December 1979). *F*, Locating the outcrops and all the tracks with the theodolite (Piau-Caiçara, level 15, 1980).

trackways and isolated tracks were done especially between 1975 and 1985 (figs. 2.1 and 2.2). After 1985, more emphasis was placed on the conservation and development of the sites' ichnological and environmental heritage, including selection of the areas to be protected, preparation of maps, creation of the modules (fig. 1.11) of the future park of "Dinosaur Valley," and the construction, in a special laboratory, of models of the local dinosaurs in fiberglass for display in the park. Moreover, Leonardi began visiting many ichnolocalities all over Latin America and then reported the remarkable number of local Latin American tetrapod ichnofaunas (no less than 115) in his *Annotated Atlas of South America Tetrapod*

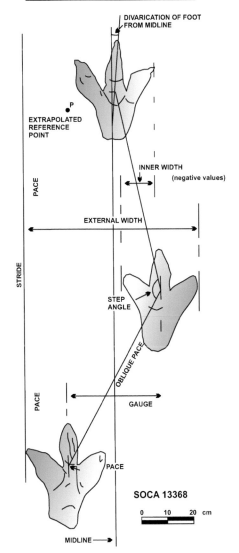

TRACKWAY PARAMETERS

DIVARICATION OF FOOT
FROM MIDLINE

EXTRAPOLATED
REFERENCE
POINT
• P

INNER WIDTH
(negative values)

PACE

EXTERNAL WIDTH

STRIDE

STEP
ANGLE

OBLIQUE PACE

PACE

GAUGE

PACE

SOCA 13368

0   10   20 cm

MIDLINE →

Footprint's Parameters

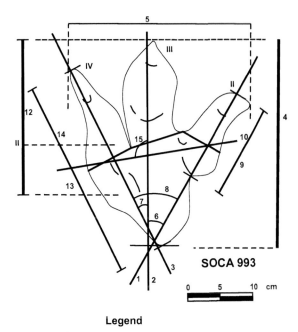

SOCA 993

0   5   10   cm

Legend

01 - Digiy axis II
02 - Digit axis III
03 - Digit axis IV
04 - Length of the footprint (FL)
05 - Width of the footprint (FW)
06 - Divarication between the digit II and III (DD II - III)
07 - Divarication between the digit III and IV (DD III-IV)
08 - Divarication between the digit II and IV (DD II-IV)
09 - length of the phalangeal portion of the digit II (LPD II)
10 - Free length of digit II (FLD II)
11 - Length of the phalangeal portion of the digit III (LPD III)
12 - Free length of digit III (FLD III)
13 - Length of the phalangeal portion of the digit IV (LPD IV)
14 - Free length of digit IV (FLD IV)
15 - Angle of the cross (AC)

Footprints (Devonian to Holocene) with an Appendix on Mexico and Central America (Leonardi 1994; cf. Alcalá et al. 2016, 102).

Many methods of fieldwork and lab work then employed are today considered rather vintage. For example, in the 1990s, one could reproduce trackways or even whole rocky pavements (fig. 2.5) with large PVC sheets or rolls. These did not exist in the years when we drew almost all the material illustrated in this book. Therefore, these subjects were drawn on graph paper. Sketches of the small or great rocky floors (up to 2,000 m²) were created with a compass, metric tape, and strings stretched taut in the main directions, above all and along the midlines of the trackways, and then with the gridded quadrant system (Leonardi 1977) using a rectangular wooden frame (40 × 80 cm) with nylon threads stretched at

Figure 2.3. *(left)* Trackway parameters, according to Leonardi (1979d, 1987a).

Figure 2.4. *(right)* Footprint parameters according to Leonardi (1979d, 1987a).

Figure 2.5. The main rock surface with tracks at Paraíso Farm contains a good-quality trackway of a large theropod and 8 isolated footprints, all of them probably attributable to theropods. The whole ichnofauna of the site, from this and other exposures, includes both theropod and sauropod tracks. From Leonardi and Santos (2006). Since the 1980s, the PVC plastic sheet helped reproducing the trackways or the whole track pavement.

Large Sauropoda

Large Theropoda

Figure 2.6. Working in the field in the Sertão of Rio do Peixe. *A*, Drawing long trackways with the gridded quadrat was endless work (Passagem das Pedras, trackway SOPP 1, January 1976). *B*, An iguanodontid trackway and an iguana (Passagem das Pedras, SOPP 1, 1983). *C*, Resting at noon in some rare shadow for a meal and cooking the *rubacão* (mixed rice and black beans) at Piau (December 1979).

Figure 2.7. Locality Piau-Caiçara (SOCA), level 13, sub-site 2 (SOCA 13/2). The rocky surface of this slope contains about 30 theropod tracks, in many cases collapsed, as well as numerous raindrop impressions. Graphic scale in centimeters. Photos and mosaic by Leonardi. Before the existence of digital cameras and image editing software, we used to take photos of the rock pavement and visualize them via photomosaics.

right angles forming squares with sides 5 cm long. A central stretch of a different color was used across the midline of the long side of the rectangular frame, matching the midline of the trackways or the longitudinal axis of a large footprint. The drawing of each 5 × 5-cm square was then reproduced on a suitable smaller scale (generally 1:5 or 1:10) on graph paper, following the trackway meter by meter (fig. 2.6; see also Lockley and Meyer 1999, 12–13). The angles among the strings were measured directly with a contact goniometer (Leonardi 1977).

Hendrickx et al. (2015, 2) state that the present-day state of the art on the relationship between theropod groups is deeply different from that of thirty years ago. The fieldwork and lab work done while preparing this book began earlier still, in 1975, when theropods were simply divided into coelurosaurs and carnosaurs, a division founded mainly on size (10). At that time theropod tracks were defined as coelurosauroid or carnosauroid. Since then, not only for theropods but for all of the dinosaurs, things have changed dramatically time and time again. The draft of this book has been updated many times; however, traces of those ancient terms and concepts can still be found here and there.

In the past, the most interesting tracksites were rocky pavements that were photographed in detail from above while the photographer stood on a ladder or a stool. All of these photographs were taken under the same lighting conditions, taking advantage of days with good weather. The photographs were printed and then glued in a photomosaic (see, e.g., fig. 2.7). At that time, there were no digital cameras or programs for merging photographs.

It would have been easier to use a drone to get a bird's-eye view (and images) of our trackways, but there were no drones when we did our field research (cf. Erin Parke, ABC News, April 24, 2015; Romilio et al. 2017; Xing et al. 2018). We made some short flights in a small plane and shot photographs over Sousa's Dinosaur Valley (fig. 2.8). In 1997 and 1999,

Figure 2.8. Aerial view of the Peixe River valley.

Figure 2.9. Working in the field in the Sertão of Rio do Peixe. *A*, Giuseppe Leonardi riding a small local horse *pé duro* (hard shoe) along the bottom of the dry bed of the Rio do Peixe at Piau in December 1979). *B*, Left to right: Maria Dolores Wanderley, Maria de Fátima C. F. dos Santos, and Lenice do Vale casting tracks with plaster at Fazenda Piedade (July 1981). *C*, Preparing with the theodolite the projects for four modules of the natural park Vale dos Dinossauros, here at Serrote do Pimenta (1984). *D*, Maria de Fátima C. F. dos Santos casting at Piau (July 1981). *E*, Drawing with the gridded quadrat at Piau (1979) on the bottom of the dry river. *F*, Measuring small tracks at Serrote do Letreiro (1977).

one of us used a radio-controlled model helicopter with cameras at the Dampier peninsula shores (Kimberley, Western Australia), with Tony Thulborn and Tim Hamley. However, the use of drones does not exempt an ichnologist from examining the tracks personally, nor is it a substitute when tracks lie in hard terrains, for example, in a karstic cave (Moreau et al. 2018) or on steep rocky walls, where climbing is necessary (Conti et al. 1975, 1977; Meyer et al. 1999; Lockley et al. 2002).

The main localities (Passagem das Pedras, Serrote do Letreiro, Serrote do Pimenta, and Piau-Caiçara) were later mapped by the land surveyors of the municipality of Sousa (fig. 2.9 and fig. 2.10), especially Sebastião Duarte Vieira (1980; scale 1:1,000) and Pedro L. de Lucena (1987; scale

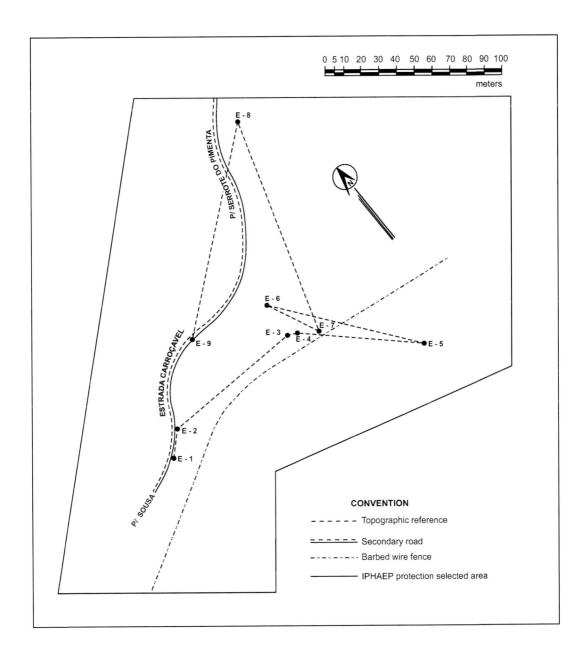

Figure 2.10. Serrote do Pimenta tracksite, in a map aimed to delimit the four main modules of the park. This map covers the whole area of the ichnosite, about 200 × 150 m, with an interesting set of dinosaur tracks exposed at three main levels. Many of the tracks are concentrated on the surface of a stratum with a surface exposure of 26 × 14 m (including the topographical points E-6 and E-7) on the right side (SE) of the road. Additional tracks occur in the rocky pavement of the dirt road in two contiguous layers: among them are trackways of 8 sauropods in a herd (immediately NE of the gate, topographical point-E-1–2), 1 theropod trackway (SOES 6, theodolite landmark E-9), and 1 hand-foot set of a quadrupedal dinosaur at the right (SE) of the road, 100 m uphill from the main pavement (SOES 7, theodolite landmark E-8), attributable to an ankylosaur. There is also 1 isolated fine trackway of a running large predator (SOES 1, theodolite landmark E-5) about 60 m SE of the main pavement. Altogether the site contains the tracks of 40 individuals: 27 large theropods, 3 ornithopods, 8 sauropods, 1 ankylosaur, and 1 lacertoid. There are 18 trackways (9 of them made by quadrupeds) and 22 isolated footprints. Paleontological survey and drawing by Leonardi.

1:500), aiming to delimit the main subdivisions of the park. These maps set the area and borders of the modules for eventual expropriation and fixed the exact location of the main tracks. Nowadays, more modern techniques using laser scanning allow imaging with higher detail level, registering with millimeter precision the shape and position of the tracks and footprints. With that technology, entire digital rock outcrops can be created and preserved for future studies (Medeiros et al. 2007; Romilio et al. 2017, a splendid work, indeed!).

In the first years of study, the chief purpose being comparison of our tracks with those of other places, tracks were manually drawn at a small scale, with numberless measurements, for a quantitative study (Falkingham et al. 2016a). The statistical analyses were carried out using a small, elementary pocket calculator, and graphs and diagrams were drawn by hand. Topographic maps of the studied region available for our use were at the low-resolution scale of 1:200,000. In addition, travels in northeastern Brazil were more difficult than today, with equatorial sunstroke, drought, torrential rains and sudden floods; scrubby and thorny vegetation; difficult and/or dangerous access to localities; political tension, financial discontinuity, and the chronic poverty of the "drought triangle" of northeastern Brazil; and high inflation of the country's currency (especially in the 1970s and 1980s), which made it difficult to raise money for research, to organize a budget, and to travel.

It is obviously difficult and tiring to work eight to ten hours per day in the field, measuring and manually drawing in the semiarid environment, at 6° of south latitude. Unfortunately, the best months for fieldwork in this region are the hottest and driest, especially November and December, when the Peixe River is almost completely dry and the outcrops on the bed of the river are almost totally exposed (fig. 2.11). The best years, from an ichnologist's point of view (absolutely not from the point of view of the poor farmers), were 1977 through 1981, during the locally famous great drought. "But this merely adds to the fun" (Farlow et al. 2015) for a tracker.

Methods of ichnological data analysis have become increasingly sophisticated over the years, but many of the newer methods were not available when our fieldwork was carried out, and we doubt how useful some of the new approaches would have been to us. For example, we did not employ a statistical technique often used in morphometrics—RFTRA (resistant-fit theta-rho analysis; Chapman 1990)—because it does not seem particularly applicable to the study of fossil footprints; nor did we employ airborne and handheld high-resolution LIDAR (light + radar, then an acronym for light detection and ranging)-based imaging, characterization and conservation of fossil tracks (Platt et al. 2018), a technique also useful to evaluate substrate consistency. Perhaps transformation grids (Thompson 1942; Haubold 1971) would have been more useful; however, these are mainly useful for quadrupedal and pentadactyl trackmakers, for which the reference points or landmarks in the pedes are more abundant. Not so in the bipedal and tridactyl dinosaurs, so abundant in our sample,

in which one can establish very few reference points, especially if the materials are not of very good quality.

The methods of geometric morphometric analysis proposed by Castanera et al. (2015) seem interesting and could be usefully tested for our rich ichnofaunas. However, we disagree with the viewpoint on the lesser utility of the hypexes as landmarks (186; in spite of Wings et al. 2016, 69; and following Hornung et al. 2016) but would possibly find good results in testing the dinosaurian ichnodiversity of the Rio do Peixe ichnofaunas. The tracks' morphologies, according to Paolo Citton et al. (2017), are also considered in order to analyze the trackmakers' behavioral patterns.

On the other hand, according to J. O. Farlow, geometric morphometrics is doubtfully much more useful than standard linear statistical methods when applied to footprints. A bone or a skull or a shell is a "real" morphological object, with true landmarks that can be quantified for analysis. Footprints are less "real" morphological objects because they nearly always are in part extramorphological. Consequently, there is greater subjectivity in identifying landmarks in footprints than in skeletons—which makes comparisons of landmarks from one footprint to another problematic (J. O. Farlow, personal communication, 2015).

Figure 2.11. The bed of the Rio do Peixe during the drought season.

However, also see, on this topic, Farlow (2018, 318–319), when he writes that "traditional and geometric morphometrics can both yield useful results."

Correct interpretation of a footprint or trackway is always a difficult and complex task, and to classify the trackmaker is also a tricky business. Understanding these facts takes us a step forward in a correct interpretation and in avoiding mistakes and biases (Falkingham et al. 2016a). There are also new perspectives in studying the fossil tracks of the Rio do Peixe basins (and generally all current and fossil footprints), using some new methods for the documentation of footprints, such as laser scanning, aerial and close-range photogrammetry, three-dimensional (3-D) models, biplanar X-ray 3-D motion analysis, and particle-based perspective and analysis (Belvedere et al. 2012; Breithaupt and Matthews 2012; Petti et al. 2018; Costa-Pérez et al. 2019; Gatesy and Ellis 2012, 2016; Matthews et al. 2016; Romilio et al. 2017; Wings et al. 2016). Especially important is the concrete proposal of a standard protocol on these issues put forward by Falkingham et al. (2019).

We decided to record, describe, and illustrate with photographs and drawings virtually all the tracks of the Rio do Peixe basins and to publish in tables virtually all the measurements of their parameters, without excluding bad-quality material. In the statistics section, it probably would have been better to consider only the good-quality tracks. In this way, we would have followed in advance the suggestions very recently proposed by Belvedere and Farlow (2016) for quantifying the quality of the ichnological material on a numerical scale and following Hornung et al.'s (2016) example.

Only a few new names of ichnotaxa were instituted for the Sousa ichnofaunas during our forty-two years of work, and this happened in the very first years in the Rio do Peixe basins, when Leonardi (1979a) instituted three taxa for some tridactyl trackways of the main outcrop at Passagem das Pedras (SOPP). We did not do so again later, with the exception of naming the very productive *Caririchnium* trackway (Leonardi 1984a). As a matter of fact, we believe that in the literature there is an excessive inflation of ichnotaxa, especially for the theropods, most of which are absolutely *nomina vana* or *nomina dubia* (see the same idea in Haubold 2012). We agree, then, with Farlow et al. (2015) on the advice that a trackway and its footprints—especially tridactyl tracks—are extremely variable objects as regarding their aspect, form, outline, volume, relationship with the substratum, and behavior of the maker and with the different situations of emplacement, preservation, and erosion (see, e.g., Sciscio et al. 2016). As a consequence, Farlow et al. (2015) suggest, the ichnotaxonomy of tridactyl dinosaur may seem, in general, rather futile. On this kind of judgment, see also Pérez-Lorente (2015, 26–27) and Xing et al. (2018, Conclusions).

It is futile, if we want to apply to fossil tracks what Dodson (1990) said about the fossil bones of dinosaurs. This happens, as Dodson says, because of the archaic taxonomy, the imprecise biostratigraphy, the preservation of poor or modest qualities that affect our understanding of diversity. We can add to this problem also the rarity of the good and necessary revisions of the tetrapod ichnological material.

On the other hand, we agree with Rich T. McCrea (personal communication, November 20, 2016) when he defends the practical usefulness of a prudent and temperate ichnotaxonomy:

> Classifications are an aid to the memory because it helps us remember the shared characteristics of a particular classification group. Classifications improve our predictive and explanatory powers about the relationships among things. This is very helpful in getting a good and accurate picture of the diversity of things as well as possible evolutionary or relationship pathways. Classification also provides generally stable, unique and unequivocal names for things which dramatically reduces the degree of provinciality in names (i.e., common names). All of the above help with scientific communication and help us all move forward and improve the classification system since we have the power to remove junior taxonomic names and as well as group different ichnotaxa where there are obvious commonalities under a larger grouping such as an ichnofamily. If I tell the authors of this book that I have found a *Caririchnium*-like trace in western Canada, they, and others in our field, will understand immediately that the prints I am working with have a number of characters in common with that ichnotaxon and it links to any work that has been done on ascribing the prints to a particular body fossil group (e.g., large ornithopods).

On this subject, from now on, an emerging and sound tool could be used for naming tracks more correctly and for instituting new taxa: the statistical-based virtual tracks recently proposed by Belvedere et al. (2018) and a concept of morphological preservation (MP) and the corresponding and useful scale of MP, recently suggested by Lorenzo Marchetti et al. (2019). We have not followed the trend of studying the ichnological material of the basins of the Rio do Peixe under the aspect of the land-vertebrate ichnofacies, introduced by Lockley et al. (1994, 2007) and a concept in which we do not blindly believe (see also Santi and Nicosia 2008).

The vertebrate paleoichnological terminology and methods used here mainly follow Leonardi (1987a). For a graphic representation of the parameters of footprints and trackways, see figures 2.3 and 2.4 in this chapter. To avoid repetition in the systematics, the authors and year of institution and publication of the ichnotaxa will be listed only at their first mention.

Figure 3.1. Geological context of the Rio do Peixe basins in northeastern Brazil (Carvalho 2004b).

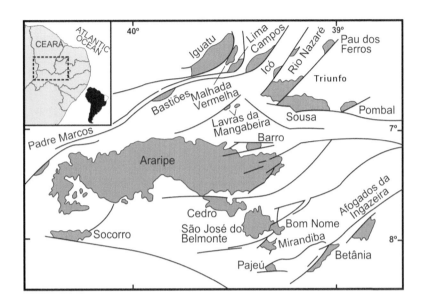

Geological Context of the Footprints

3

The intracratonic basins of Northeast Brazil (fig. 3.1) are sites of Creta-ceous sedimentation, the origin and evolution of which were controlled by the reactivation of preexisting tectonic structures in the basement (Precambrian rocks) during the Jurassic and Cretaceous. This reactiva-tion was closely related to the opening of the South Atlantic Ocean, and normal and transcurrent faults were the main structural style for the open-ing of grabens and half-grabens in the central region of Northeast Brazil during the Early Cretaceous. These sedimentary basins are known in the literature as the intracontinental basins of Northeast Brazil (Ponte 1992) and show great similarities in their origin and evolution. They originated as a consequence of tectonic movements that resulted in the separation of South America and Africa (Mabesoone 1994; Valença et al. 2003). This low-lying trend has been called the Araripe-Potiguar depression (Mabe-soone 1994; Valença et al. 2003).

The Precambrian basement was characterized by thermal and tec-tonic-magmatic processes that took place during the Meso- and Neo-Pro-terozoic periods, continuing into the Cambrian-Ordovician (Santos and Brito Neves 1984). The province shows various zones of supracrustal rocks embedded among gneissic-migmatitic terrains with diverse structural trends that are grouped into fold systems, resulting in the superposition of diverse tectonic, metamorphic, and magmatic events upon the sedi-mentary and volcanic rocks that accumulated after the Meso-Proterozoic (Almeida and Hasui 1984). According to Trompette et al. (1993), the Borborema province belonged to a larger Precambrian paleocontinent extending into Africa. The region was periodically affected by the forma-tion of intracontinental rifts, for the last time from the Callovian onward (Matos 1992). As a consequence, several sedimentary basins resulted from differential reactivated fault movements within the ancient Precambrian belt zone (fig. 3.2). Crustal extension was then generated along the pre-existing Precambrian fault lines, SW–NE oriented, called "half-grabens" (Ponte 1992; Valença et al. 2003).

The basins lie in the western part of Paraíba, in Rio Grande do Norte, Piauí, Pernambuco, and in the southern part of the State of Ceará, Northeast Brazil (fig. 3.3). They present a wide variety of vertebrate and invertebrate ichnofossils, especially Early Cretaceous dinosaur footprints (Leonardi 1979a, 1989a; Carvalho 1989; Carvalho et al. 1993a, 1993b, 1994; Leonardi 1994; Leonardi and Spezzamonte 1994), which demonstrates that in spite of the paucity of skeletal remains, the dinosaur faunas of Brazil were abundant and diverse.

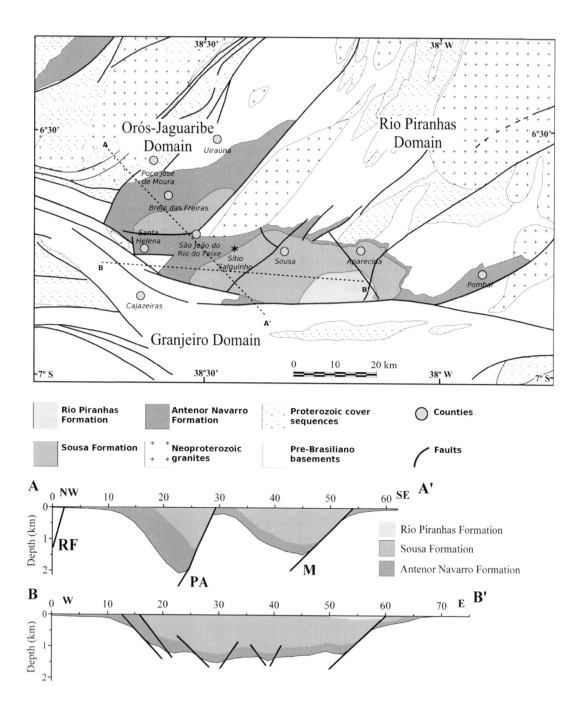

Figure 3.2. Geological framework of the region adjacent to the Rio do Peixe basins, and geological sections (A-A' and B-B') through the Triunfo and Sousa basins (modified from Castro et al. 2007).

The Rio do Peixe basins (fig. 3.3) are related to a synrift context. Córdoba et al. (2008) considered these, like other northeast Brazilian interior basins, as erosional remnants of a series of basins located in the south of the Potiguar basin, overlying the crystalline basement of the Precambrian Borborema Province. Seismic sections presented by Córdoba et al. (2008) allow an improved view of the 3-D architecture of the Rio do Peixe basins. The combination of the current erosion level with the geometry of the main faults highlights the existence of different

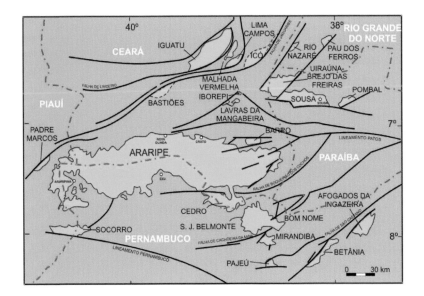

Figure 3.3. Vertentes, Uiraúna-Brejo das Freiras (Triunfo), Sousa, and Pombal are grouped in the Rio do Peixe basins (Carvalho 2004a).

half-grabens (Pombal, Sousa, Brejo das Freiras) whose sedimentary filling, apart from Cenozoic deposits, defines the Rio do Peixe Group, comprising the Antenor Navarro (alluvial fans/braided channels), Sousa (shallow lacustrine/floodplain), and Rio Piranhas (alluvial fans/braided channels) Formations. Based on the structural style and petrographic-diagenetic features, Córdoba et al. (2008) inferred some larger original dimensions for this basin and similar counterparts in the region, which were reduced (with the exposure of the crystalline highs) by significant erosion that occurred in late to post-rift and subsequent evolutionary stages. Two alternative hypotheses try to explain the evolution of these rift basins, either by a model considering northwestern extension during the Berriasian-Barremian ages or by a model involving reactivation of E–W and N–E strike-slip Precambrian lineaments but also involving northwestern extension.

The Atlantic Proto-Ocean and the Origin of the Rio do Peixe Basins

After the end of the Permian, sediment accumulation decreased in many Brazilian sedimentary basins, restricted mainly to portions of the Amazonas, Paraná, and Parnaíba intracontinental basins. In the Jurassic, the stability of the Brazilian Platform was broken. Intense tectonic activity related to the beginning of the rupturing of the Gondwana crust created many small sedimentary basins and led to rapid accumulation of continental sediment. Many of the individual half-grabens that resulted from this tectonism contained small lakes that captured the drainage network and had an eventual physical linkage (Machado et al. 1990).

According to Popoff (1988), at its beginning, the South Atlantic was divided into three tectono-sedimentary domains: austral (southern), tropical (midlatitude), and equatorial (northern). Their tectonic history was diachronous and did not coincide during the Early Cretaceous. The Brazilian northeastern interior basins were included in the midlatitude tropical

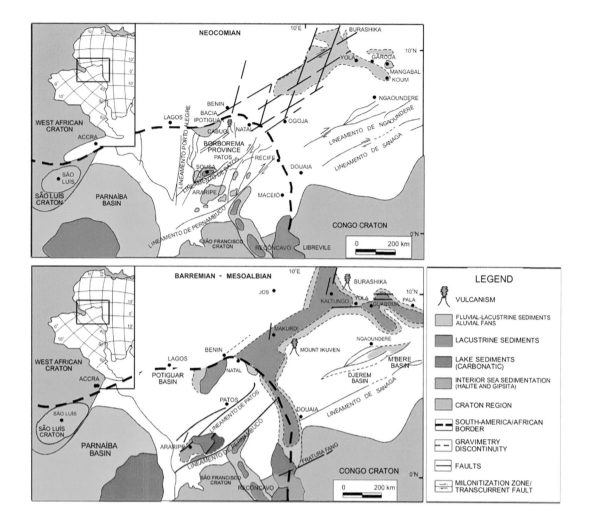

Figure 3.4. Geological evolution of northeastern Brazil and west Africa during the Early Cretaceous (modified from Popoff 1988).

domain and were limited by the fracture zones of Pernambuco-Birao-Khartoum (N of the Congo Craton), Portalegre, and Paraíba (fig. 3.4).

Gravimetric data (Rand 1984) showed that the depocenter of the basins is located nearside the faults of Portalegre and Malta. Lima Filho (1991) defined the tectono-sedimentary evolution of these sedimentary areas as occurring in the following phases:

Phase I: reactivation of previous regional faults of NE–SW and E–W direction inside the craton and the posterior origin of sigmoidal basins with rapid syntectonic siliciclastic deposition in grabens or half-grabens. This phase is represented by the Antenor Navarro Formation and probably ranges from the Purbeckian (known locally as the Dom João stage) through the Berriasian (locally, the Rio da Serra stage).

Phase II: progressive diminishing of tectonic activity, with long intervals of tectonic stability reflected by fluvial and lacustrine environments with sedimentation of fine siliciclastic and marls. During this interval, dinosaur tracks, probably from the Valanginian age (locally, the Rio da Serra stage) were easily preserved in the Sousa Formation.

Phase III: drastic change of tectonic conditions due to reactivation of the fault system. Here we see the deposition of immature coarse silici-clastic sediments in a braided fluvial system (Rio Piranhas Formation) of the Hauterivian-Barremian age (local stage Aratu).

Phase IV: reactivation of postdepositional faults that induced fractures and folds in the sediments of the Rio do Peixe Group.

The structural and geophysical data presented by Córdoba et al. (2008) indicate that the sediment column may be more than 2 km thick in the deep portion of depocenters, or even 2.5–3 km thick in the case of the Brejo das Freiras half-graben. In each half-graben, the layers are tilted to the faulted borders (fig. 3.2)—an area in which syntectonic conglomerates may occur.

The seismic stratigraphic interpretation, based on sedimentological information and relations with the faults, shows that the strata of the three units occur in an interfingering manner and therefore constitute approximately chrono-equivalents units in a synrift tectonosequence (Córdoba et al. 2008). The lower limit of this sequence is represented by an unconformity surface that separates it from the crystalline basement below. Its upper limit is characterized by a discordance of strong erosive character whose hiatus tends to increase toward the flexural margins of the half-grabens. These unconformities are progressively uplifted and eroded while the hanging wall blocks subside and accumulate greater thicknesses of sediments. The lateral relationships between the facies show that the sedimentary filling of half-grabens was controlled by the relationship between the rates of creation of accommodation space and of sediment supply, which is ultimately a direct function of the intensity of tectonic activity.

Paleoclimate

During the early Mesozoic, a hot and arid climate prevailed in the Southern Hemisphere (Lima 1983). This is clearly evidenced by the widespread occurrence of aeolian deposits along the Brazilian and African intracratonic basins. The connection of South America and Africa as a single, large continental block did not permit higher humidity in what was at that time the continental interior. Higher humidity came with the breakup of this continent and the establishment of a lacustrine and fluvial system among the new rift basins, suggesting a link to those tectonic events that drove the separation of South America and Africa and led to the origin of the equatorial Atlantic Ocean.

Topographic barriers along the borders of the Afro-Brazilian Depression played an important role in reducing wind velocity from the southeast (Golonka et al. 1994; Da Rosa 1996)—a factor that likely affected the distribution of the flora and associated fauna. Rainfall distribution was also affected by the winds and topography. Moisture carried by winds faced northern, northwestern, and western margins as a final obstacle. Although low, this moisture contributed to higher precipitation locally and led to the growth of abundant vegetation, mainly along the northern margin. Climatic differentiation of the Afro-Brazilian Depression into two regions led to the establishment of a Berriasian coniferous forest on the northern margin, as suggested by the presence of silicified logs, foliage, and palynomorphs in fluvial deposits of the prerift sequence. Diagenetic studies point to generally semiarid conditions in the region (García and Wilbert 1994; Da Rosa 1996; Da Rosa and Garcia 2000).

Throughout the Early Cretaceous, hot climatic conditions were widespread, although there was probably a wide range of humidity (Skelton 2003). According to Petri (1983, 1998) and Lima (1983), in the earliest Cretaceous the climate was more humid in regions located to the south of the tropical domain (the Recôncavo-Tucano-Jatobá basins). Despite the tropical domain's hotter and drier climate, interpretations of depositional environments and fossils suggest the existence of lakes, which, during the Neocomian, locally provided more humid conditions. The presence of the large conchostracans *Palaeolimnadiopsis reali* in some lacustrine facies of the Sousa basin shows there were optimum conditions for this group, with abundant freshwater and a warm, wet climate (Carvalho 1989). At that time, South America was still connected to Africa, and the Atlantic Ocean was in its initial developing phase. In what is now northeastern Brazil, across an area of hundreds of square kilometers, ephemeral rivers and shallow lakes constituted important environments for an abundant endemic biota in many basins limited by faults (Lima Filho et al.1999; Mabesoone et al. 1979, 2000).

Petrological Aspects of the Sandstones of the Rio do Peixe Basins

The Rio do Peixe basins contain mature sandstone at the base of the section; its content, both quartz and lithoclasts, indicates an equilibrium between metamorphic and plutonic sources. As sedimentation continued, there was a relative increase in plutonic quartz and lithoclasts and a much greater input of feldspar. This can be linked to a source-area evolution: the metamorphic cover of the granitic Rio Piranhas Massif was eroded until its roots were exposed by continuous uplift and erosion. Da Rosa (1996) suggested that the source area shift and uplift represent the disconnection of the Rio do Peixe basins and their correlatives from the Afro-Brazilian Depression. Plots of petrographic data in provenance diagrams identify a stable continental block being continuously uplifted (Da Rosa 1996; Da Rosa and Garcia 1993).

The progressive uplift of the Jaguaribe Lineament began to release sediments to the northern margin of the Afro-Brazilian Depression via the erosion of a Precambrian metamorphic source area and the formation of alluvial fans and braided streams. These deposits are represented by the basal parts of the Antenor Navarro and Sousa Formations in the Rio do Peixe basins. The uplift of source areas, indicating the separation of South America from Africa, was an important event that allowed rejuvenation of southward-flowing streams and led to the filling of shallow lakes (Da Rosa and Garcia 2000).

The Sousa basin comprises an area of 1,250 km² in the western part of the State of Paraíba, in the counties of Aparecida, São João do Rio do Peixe, and Sousa (fig. 3.5). The basement is composed of highly metamorphosed Precambrian rocks aligned structurally in a NE–SW direction. The predominant rocks are migmatites, granites, gabbros, and amphibolites. The main lithologies in the Sousa basin are clastic rocks, including breccias and conglomerates, sandstones, siltstones, mudstones, and shales. In some cases, the carbonate content is high in the form of marls and thin (centimeter-thick) limestones.

A formal lithostratigraphic subdivision of the Cretaceous in the Sousa basin and the neighboring Uiraúna Brejo das Freiras (Triunfo) and Pombal basins was erected by Mabesoone (1972) and Mabesoone and Campanha (1973/1974). These authors identified the Rio do Peixe Group, with a total thickness of 2,870 m, and subdivided it into the Antenor Navarro, Sousa, and Rio Piranhas Formations (fig. 3.6). The Antenor Navarro (figs. 3.7, 3.8) and Rio Piranhas Formations are composed of immature sediments, including breccias and conglomerates, with pebbles of metamorphic and magmatic rocks in a coarse arkose matrix. These lithology types are located near the faulted margins of the basin (fig. 3.9). Toward

Figure 3.5. Geological map of the Sousa and Triunfo basins (modified from Carvalho 1996a).

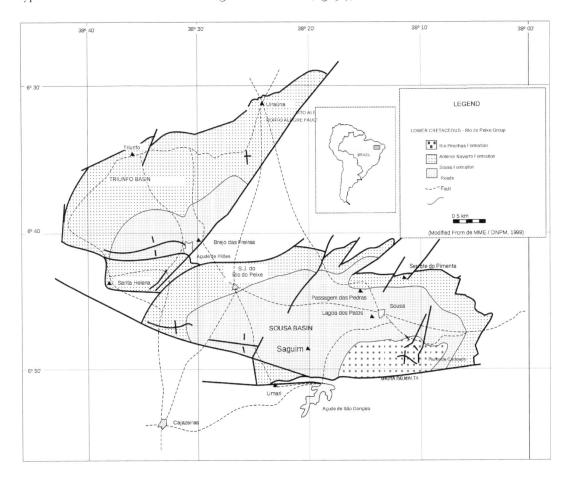

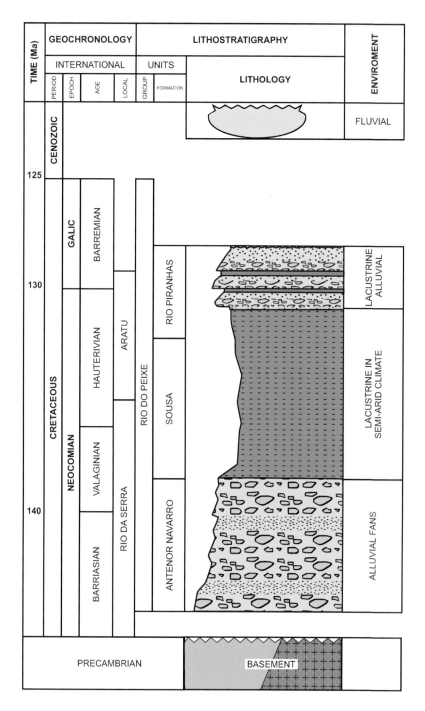

Figure 3.6. Mesozoic and Cenozoic stratigraphic section of the Rio do Peixe basins (modified from Srivastava and Carvalho 2004).

the basin's depocenter, there are conglomeratic and fine sandstones, sometimes interbedded with siltstones and shales. Cross-channel and tabular stratification, climbing ripples, and ripple marks are the main sedimentary structures. The Sousa Formation is composed of reddish sandstones, siltstones, mudstones, and carbonate nodules; marls also

Figure 3.7. Antenor Navarro Formation, Sousa basin. Locality of Serrote do Letreiro. *A*, Tabular geometry of the sandstone beds. *B*, Tabular cross-stratification. *C*, "Elephant skin" structures. *D*, Sandstone intraclasts.

occur. Common sedimentary structures include mud cracks, convolute structures, ripple marks, climbing ripples, rain prints, and evidence of bioturbation (fig. 3.10).

During the Dom João time (the latest Jurassic "Purbeckian" stage), because of crustal extension, a sigmoidal basin developed at the inflection of the NE–SW and E–W faults. Sediments of this time are not preserved in the Rio do Peixe basins (Arai 2006). During the Rio da Serra time (the Berriasian–Hauterivian), under the same tectonic stress pattern, the basinal area increased, and its shape became rhomboidal. Eventually, probably at the end of the Aratu time (the lower Barremian Stage), there was a change in the tectonic pattern, and the rate of sediment accumulation began to decline. The deposits reflect direct control of sedimentation by tectonic activity. Along the faulted borders of the basin, deposition consisted of alluvial fans (Mabesoone et al. 1979). This fan system passes distally into an anastomosing fluvial system. In the central region of the basin, a meandering fluvial system with a wide floodplain was established, where perennial and temporary lakes developed.

Paleontology and Age of Deposits

An abundant ichnofauna of vertebrates—footprints and trackways of theropods, sauropods, and ornithopods—is one of the primary characteristics

Figure 3.8. Antenor Navarro Formation, Sousa basin. Locality of Serrote do Letreiro. *A*, Tabular cross-stratification in medium-grain-size sandstones. *B*, Interbedding of fine-grain sandstones and siltstones. *C*, Exhumed surface of a sandbar with dinosaur footprints. *D*, Rounded sauropod footprint in a coarse-grain sandstone surface. Note also the mat wrinkles on the surfaces (cf. Dai et al. 2015).

of the Sousa basin. Invertebrate ichnofossils, such as tracks and burrows produced by arthropods and annelids, are also common. Despite the sediments' strong reddish color, typical of sediments that accumulated in subaerial environments, there are some levels of greenish shales, mudstones, and siltstones where fossils are common. These consist of ostracods, conchostracans, plant fragments, palynomorphs, and fish scales.

The sediments of the Sousa basin were first dated by Moraes (1924), who proposed a Comanchean age (Early Cretaceous) for dinosaur tracks at Passagem das Pedras. From an analysis of ostracods, Braun (1966, 1969, 1970) and Mabesoone and Campanha (1973/1974) suggested an

age between the Berriasian and Hauterivian. The palynological assemblages are characteristic of the Rio da Serra (Berriasian–Hauterivian) and Aratu (lower Barremian) local stages (Lima and Coelho 1987; Regali 1990).

INVERTEBRATE FOSSILS

Invertebrate fossils from the Sousa basin are restricted to ostracods and conchostracans (figs. 3.11, 3.12). Although Maury (1934) identified the Bivalvia *Diplodon lucianoi*, Beurlen and Mabesoone (1969) considered

Figure 3.9. Antenor Navarro Formation, Sousa basin. Locality of Serrote do Letreiro. *A*, Exhumed surface of a sandbar with dinosaur footprints. *B*, Fractured sandstone, tectonically controlled. *C*, Theropod footprint with a reddish color that makes it distinct relative to the surrounding sandstone. *D*, Theropod footprint in coarse-grain sandstone.

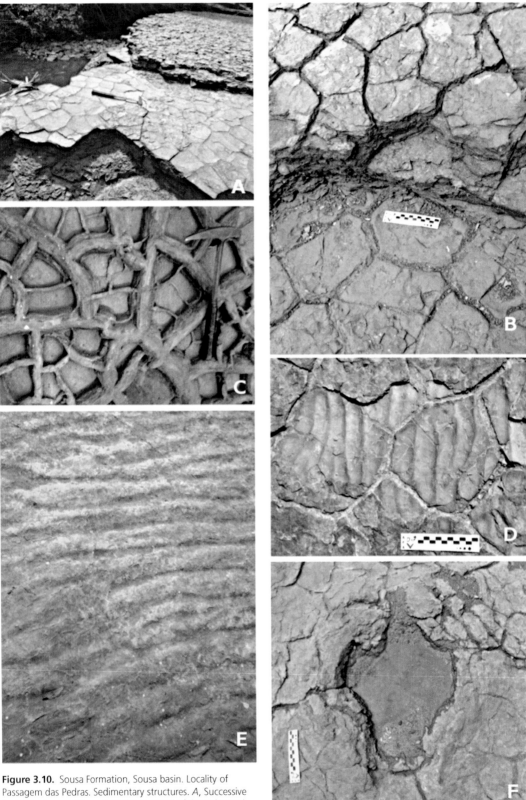

Figure 3.10. Sousa Formation, Sousa basin. Locality of Passagem das Pedras. Sedimentary structures. *A*, Successive bedding planes with mud cracks. *B*, Detail of the polygonal mud cracks. *C*, Large molds of mud cracks. *D*, Ripple marks and mud cracks in successive layers. *E*, Asymmetrical ripple marks. *F*, Footprint with a displacement rim.

this species an estheriellid conchostracan. Carvalho (1993) identified it as the conchostracan species *Estheriella lualabensis* (fig. 3.13).

Tinoco and Katoo (1975) identified four species in the Sousa basin of the order Conchostraca: *Petriograpta* cf. *reali*, *Palaeolimnadiopsis freybergi*, *Graptoestheriella fernandoi*, and *?Pseudograpta barbosai*. Braun (1969) recognized in the Uiraúna-Brejo das Freiras basin *Pseudograpta* sp., *Pseudoestheria* ? sp., *Palaeolimnadiopsis* sp., and *Graptoestheriella* aff. *G. brasiliensis*. Tasch (1987) and Carvalho (1993) revised this classification. Nowadays the valid species names for Sousa Basin are *Cyzicus brauni*, *Cyzicus cassambensis*, *Estheriella brasiliensis*, *Estheriella lualabensis*, *Palaeolimnadiopsis freybergi*, and *Palaeolimnadiopsis reali* (Carvalho 1993).

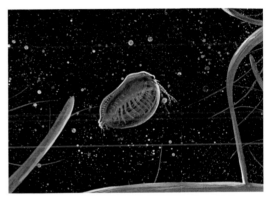

Figure 3.11. *(top left)* Ostracods were common organisms in the lakes of the Sousa basin during Early Cretaceous. Art by Ariel Milani Martine.

Figure 3.12. *(top right)* In addition to ostracods, conchostracans were very common crustaceans in the temporary lakes of the Sousa basin. Art by Ariel Milani Martine.

Figure 3.13. *(left) Estheriella lualabensis* is a small-size conchostracan with a wide distribution. It is also found in the Lualaba Series of the Congo basin, Africa.

The following species are considered to be valid and are included in the families Palaeolimnadiopsidae, Esthereliidae, and Cyzicidae:

Palaeolimnadiopsis reali Teixeira, 1958 (from the locality of Pedregulho, Sousa basin)

Palaeolimnadioposis freybergi Cardoso, 1971 (from the Sousa–Aparecida road, Sousa basin (fig. 3.14)

Estheriella brasiliensis Oliveira, 1953 (from the Lagoa do Forno drilling F.EST.LF.PB-1, Sousa basin; the Sousa–Aparecida road, Sousa basin; and Fazenda Santa Rita, São João do Rio do Peixe, Uiraúna-Brejo das Freiras basin)

Estheriella lualabensis Leriche, 1913 (from Passagem das Pedras, Sousa basin)

Cyzicus codoensis (Cardoso, 1962) from the São João do Rio do Peixe–Cajazeiras road (Sousa basin)

As for the Triunfo basin (Uiraúna-Brejo das Freiras), the species are *Cyzicus abaetensis, Cyzicus brauni, Cyzicus cassambensis,* and *Cyzicus pricei,* based on Carvalho (1993).

Figure 3.14. *Palaeolimnadiopsis reali,* a gigantic conchostracan from the Sousa basin. Some specimens reach lengths of 3.5 cm, indicative of nutrient-rich freshwater lakes with an alkaline pH.

Invertebrate microfossils, specifically ostracods, were recognized in the Sousa basin by Maury (1934) Braun (1966, 1969, 1970), Mabesoone and Campanha (1973/1974), Tinoco and Katoo (1975), Tinoco and Mabesoone (1975), and Barbosa et al. (1986). These authors identified isolated valves of the freshwater genera *Darwinula, Cypridea, Ilyocypris,* and *Candona.*

Almost all the ostracod species and genera from the Sousa basin are from the Sousa Formation. The identified taxa are

Bisulcocypris Pinto and Sanguinetti, 1958
Brasacypris Krömmelbein, 1965
Brasacypris ovum Krömmelbein, 1965
Candona cf. *C. condensa* Krömmelbein, 1962
Candona cf. *C. redunca* Krömmelbein, 1962
Candona subreniformis Jones, 1893
Cypridea cf. *C. ambigua* Krömmelbein, 1962
Cypridea lunula Krömmelbein, 1962
Cypridea vulgaris Krömmelbein, 1962
Darwinula Brady and Robertson, 1885
Darwinula cf. *D. leguminella* (Forbes, 1855)
Darwinula cf. *D. oblonga* (Roemer, 1839)
Ilyocypris Brady and Norman, 1889
Paraschuleridea sp. 5 (in Braun, 1966)
Paraschuleridea micropunctata nom. nud.

Mabesoone and Campanha (1973/1974), based in Braun (1966, 1969, 1970) and Silva (1970, in Mabesoone and Campanha 1973/1974), established biostratigraphy zones for the Sousa Formation correlated to the biozones of the Recôncavo basin: *Darwinula* cf. *oblonga* (RT-001), *Cypridea* cf. *ambigua* (RT-003/004/005/006), and *Cypridea vulgaris* (RT-003, RT-004, RT-005). The age of the strata where these biozones are distributed are related to the Dom João (Purbeckian) and Aratu stages (Neocomian).

PALEOBOTANY

Plant fossils from the Sousa basin generally do not allow identification to the species level. Carbonized wood fragments are particularly common (Moraes 1924; Braun 1969, 1970; Dantas and Caula 1982; see also fig. 3.15), as are permineralized logs (fig. 3.16). Braun (1969) identified some plant remains from the Sousa Formation ("Unit B" of Braun 1969) as leaves and stems of *Podozamites* F. Braun, 1843 (Coniferophyta, fig. 3.17) and a genus of *Cycadoidea*, *Otozamites* (figs. 3.18, 3.19).

Permineralized logs have been discovered in the locality of Bênção de Deus, Sousa County (Sousa basin). They are up to 1.5 m long and are reworked fragments of larger logs deposited in coarse sandstones of the Antenor Navarro Formation. These plant remains do not show detailed anatomical structures and have generally been assigned to the genus *Dadoxylon*—though according to Mussa (2001) and Stewart (1983), this is a problematic designation. *Dadoxylon* was first used for structurally preserved fragments of cordaitalean secondary wood, but anatomically *Dadoxylon* is within the same organization as the Mesozoic and Tertiary form genus *Araucarioxylon* (with affinities to the recent genus *Araucaria*). As Stewart (1983) pointed out, to resolve this incongruity, some authors use *Dadoxylon* as a generic name for Paleozoic wood fragments and *Araucarioxylon* for those that have the same characteristics but are

Figure 3.15. Riachão dos Oliveira (SORO), Antenor Navarro Formation, Sousa. *A,* Overall view of the site. *B, C,* Fossil plants (probably conifers) in the sandstones. *D,* A poor-quality ornithopod footprint from the same locality and level as the fossil plants.

from the Mesozoic. Mussa (2001) followed this approach; Mussa also considered that due to the secondary stellar pattern of *Dadoxylon benderi* (Mussa, 1959), this species should be designated as *Araucarioxylon benderi* (Mussa) Mussa, a Brazilian Araucariaceae.

Members of the Araucariaceae are now restricted to the South American and southwest Asia–western Pacific regions despite their extensive distribution in both hemispheres during the Mesozoic. Despite this unusual present-day distribution pattern, there is some consistency in representation in relation to climatic conditions, with a concentration in lower midlatitudes. This is well illustrated by the present latitudinal distributions of the major continental *Araucaria* species (*A. angustifolia* in southeastern South America and *A. cunninghamii* and *A. bidwillii* in Australia), which are remarkably similar despite a long period of continental separation of these taxa and their systematic placement in different subgroups of the genus. All extant species have a minimum requirement of about 10°C in the winter months, which, perhaps together with a requirement for significant summer rainfall, could explain the southern extent of the species. Major exceptions to this distribution are the *Agathis* species that occur in equatorial peat swamps, where substrates with low nutrients and poor drainage perhaps allow them to be competitive with angiosperms, and

Figure 3.16. *A*, Permineralized log of *Araucarioxylon* from the locality Bençâo de Deus, Sousa basin. *B*, Detail of the *Araucarioxylon* log showing destruction of the wood anatomy by silica recrystallization. Visitor Center of the Dinosaur Valley Natural Monument.

Figure 3.17. Reconstruction of *Podozamites*, a coniferophyte. Art by Ariel Milani Martine.

Figure 3.18. *Otozamites*, a genus of Cycadoidea found in the Sousa Formation. Art by Ariel Milani Martine.

Araucaria araucana in Chile, which extends to the tree line. However, *A. araucana* does not extend into the cool, temperate forests of South America that support the majority of southern conifers on this continent. The araucarias also appear to have maintained a preference for subtropical or mesothermal conditions (Kershaw and Wagstaff 2001).

The fossil Araucariaceae *Araucarioxylon* is also present in the other basins near the Sousa, including the Araripe (Missão Velha Formation), Coronel João Pessoa (Antenor Navarro Formation), and Jatobá basins (Sergi Formation). The great number of these silicified logs, along with their wide distribution and geological context, indicates the growth of luxurious forests in mountains adjacent to river plains, in subtropical

Figure 3.19. The Cycadoidea and Coniferophyta were the most common plants during the Early Cretaceous in the Sousa basin. Art by Ariel Milani Martine.

conditions. This genus probably includes many distinct species of Coniferales (Araucariaceae), as conifers reached their maximum diversification during the Mesozoic. If so, the Brazilian *Araucarioxylon* is likely a mosaic of "organ-taxa" and names. Further work on this group will require well-preserved material to allow observation of the variability of wood characteristics and, in this way, identification of the diversified wood types in these Mesozoic floras (fig. 3.20). In addition, a detailed level-by-level study of the dinosaur-plant interactions within the ecosystems (Slater et al. 2017) in the approximately ninety-six track-bearing stratigraphic levels of the Rio do Peixe basins would be very interesting.

PALYNOMORPHS

Palynological data of Cretaceous sediments (figs. 3.21, 3.22, 3.23) were obtained through a stratigraphic borehole (LF-01-PB) drilled in the Sousa

Figure 3.20. Reconstruction of the Early Cretaceous vegetation and landscape of the Sousa basin. Araucariacean forests cover the highlands surrounding the main depositional area. Art by Ariel Milani Martine.

basin, which is thought to cut through the entire sequence. The studied assemblage from this borehole is very rich, allowing the characterization of 99 taxa. The species *Dicheiropollis etruscus* is present at all fertile levels, as are other stratigraphically important species. This fact suggests that the three stratigraphic units recognized in the basin (Antenor Navarro, Sousa, and Rio Piranhas Formations), all present in the borehole, were deposited within a continuous, relatively short time interval, here interpreted as the Aratu Age (Neocomian). The depositional environment was under hot climatic conditions, somewhat drier in the lower part of the section than at the top (Lima and Coelho 1987). The identified palynomorph taxa (Lima and Coelho 1987) include spores, pollens, scolecodonts, and algae.

Spores

Stereisporites psilatus (Ross) Manum
Leitotriletes sp. cf. *L. breviradiatus* Döring
Todisporites major Couper
Deltoidospora juncta (Kara Murza) Singh
Deltoidospora hallii Miner
Biretisporites pontoniaei Delcourt and Sprumont
Undulatisporites pannuceus (Brenner) Singh
Undulatisporites undulapolus Brenner
Concavissimisporites variverrucatus (Couper) Singh
Leptolepidites major Couper
Leptolepidites verrucatus Couper
Leptolepidites crassibalteus Filatoff
Mathesisporites tumulosus Döring
Gemmatriletes sp. cf. *G. clavatus* Brenner
Acanthotriletes varispinosus Pocock
Pilosisporites sp. cf. *P. semicapillosus* Dörhofer
Echinatisporis sp.
Cicatricosisporites microstriatus Jardiné and Magloire
Cicatricosisporites stoveri Pocock
Cicatricosisporites minutaestriatus (Bolkhovitina) Pocock
Cicatricosisporites recticicatricosus Döring
Cicatricosisporites swardi Delcourt and Sprumont
Cicatricosisporites crassistriatus Burger
Cicatricosisporites exilioides (Malyavkina) Dörhofer
Cicatricosisporites subrotundus Brenner
Cicatricosisporites sp. cf. *C. augustus* Singh
Appendicisporites sellingii Pocock
Appendicisporites parviangulatus Döring
Ischyosporites variegatus (Couper) Schultz
Concavisporites obtusangulus (Potonié) Pflug
Dictyophyllidites harrisi Couper
Densoisporites microrugulatus Brenner
Bullasporis aequatorialis Krutzsch
Perotrilites pseudoreticulatus Couper

Pollen

Zanallapollenites dampieri Balme
Zanallapollenites trilobatus Balme
Zanallapollenites microvelatus Schultz
Inaperturopollenites sp. cf. *I. dubius* (Potonié and Venitz) Thomsom and Pflug
Inaperturopollenites turbatus Balme
Inaperturopollenites sp. cf. *I. patellaeformis* Pflug and Thomsom
Coptospora kutchensis Venkatachala
Coptospora aequalis Tralau
Araucariacites limbatus (Balme) Habib

cf. *Exesipollenites tumulus* Balme
Peltrandipites sp.
Sergipea sp. cf. *S. naviformis* Regali et al.
Sergipea sp. cf. *S. variverrucata* Regali et al.
Sergipea sp. cf. *S. simplex* Regali et al.
Cycadopites glottus (Brenner) Wingate
Cycadopites giganteus Stanley
Cycadopites carpentieri (Delcourt and Sprumont) Singh
Cycadopites minimus (Cookson) Pocock
Cycadopites follicularis Wilson and Webster
Cycadopites sp. 1
Cycadopites sp. 2
Cycadopites sp. cf. *C. dijkstrae* Jansonius
Bennettitaepollenites minimus Singh
cf. *Bennettitaecuminella* sp.
Alisporites bilateralis Rouse
Cedripites cretaceus Pocock
Podocarpidites biformis Rouse
Podocarpidites typicus Sah and Jain
Podocarpidites ellipticus Coockson
Podocarpidites epistratus Brenner
Podocarpidites alareticulosus Sah and Jain
Podocarpidites sp. cf. *P. fangii* Pocock
Podocarpidites sp. cf. *P. selowiformis* (Zaklinskaya) Drugg
Podocarpidites sp. cf. *P. multesimus* (Bolkhovitina) Filatoff
Podocarpidites sp. cf. *P. arcticus* Pocock
Pinuspollenites sp.
Phyllocladidites inchoatus (Pierce) Norris
Gamerroites sp. 1.
Gamerroites sp. 2.
cf. *Rugubivesiculites* sp.
Vitreisporites pallidus Reissinger (Nilsson)
Vitreisporites sp. cf. *V. itunensis* Pocock
Classopollis simplex (Danze-Corsin and Laveine) Reiser and Williams
Classopollis torosus (Reissinger) Couper
Classopollis minor Pocock and Jansonius
Dicheiropollis etruscus Trevisan
Circulina parva Brenner
Equisetosporites ovatus (Pierce) Singh
Equisetosporites virginiaensis Brenner
Equisetosporites cancellatus Paden, Phillips and Felix
Gnetaceaepollenites boltenhageni Dejax
Gnetaceaepollenites uesuguii Lima
Gnetaceaepollenites lajwantis Srivastava
Gnetaceaepollenites chlatratus Stover
Gnetaceaepollenites sp. cf. *G. mollis* Srivastava
Gnetaceaepollenites sp.

Steevesipollenites sp.
Eucommiidites tredssonii (Erdtman) Hugues
Eucommiidites minor Groot and Penny
Eucommiidites sp. 1
Eucommiidites sp. 2

Scolecodonts

cf. *Marlenites* sp.
cf. *Pronereites* sp.
Scolecodont, indet.
Algae
Tasmanites sp.

VERTEBRATE FOSSILS

As a rule, formations rich in tracks are poor in body fossils and vice versa, because of the different environmental conditions necessary for the preservation of these two different kinds of fossils. Localities with rich bone and track remains together are very rare throughout the world. There is one rare instance where both skeletal fossils and footprints of dinosaurs from the Upper Cretaceous are extraordinarily abundant in several sites in southern Mongolia (Ishigaki and Matsumoto 2009). Another case, more recently published, is that of the Camarillas Formation (~130.6–128.4 million years ago [mya], Barremian age) in Teruel, Spain; along with large numbers of sauropod, ornithopod, and theropod tracks, vertebrate body fossils from this unit are abundant and include bones and teeth of Chondrichthyes, Osteichthyes, Amphibia, Testudines, Lepidosauria, Crocodylomorpha, Pterosauria, and Dinosauria (Ornithopoda, Theropoda, Sauropoda) (Navarrete et al. 2014).

The Sousa basin, although rich in dinosaur tracks, shows few skeletal fossils. There are, however, a few instances of vertebrate fossils in the counties of Aparecida (Acauã) and Sousa. Unfortunately, the material in these localities is extremely fragmented, with a severely limited distribution in the Sousa and Rio Piranhas Formations.

Maury (1934) revealed the presence of fish scales in the Acauã shale of the Sousa basin. Braun (1970) also showed the existence of fish bone fragments in Braun's B and C units, 1969, which correspond to the Sousa and Rio Piranhas Formations, respectively. Tinoco and Katoo (1975) identified fragments of fish bones, teeth, and scales in core F.EST.LF. PB-1 in Lagoa do Forno (Sousa County). Beurlen and Mabesoone (1969) stated that the scales found in the Rio do Peixe region belong to the genus *Lepidotes* Agassiz, 1832. In Passagem das Pedras (Sousa County), there are some unidentified bones in very fine sandstone (fig. 3.25B), probably belonging to the Crocodyliformes. Ghilardi et al. (2014, 2016) recognized a sauropod fibula at the Lagoa do Forno site in the Rio Piranhas Formation, in the vicinity of Sousa County. The specimen (UFPE-7517) is 45 cm in length and has a general sigmoidal shape that identifies it as a titanosaur.

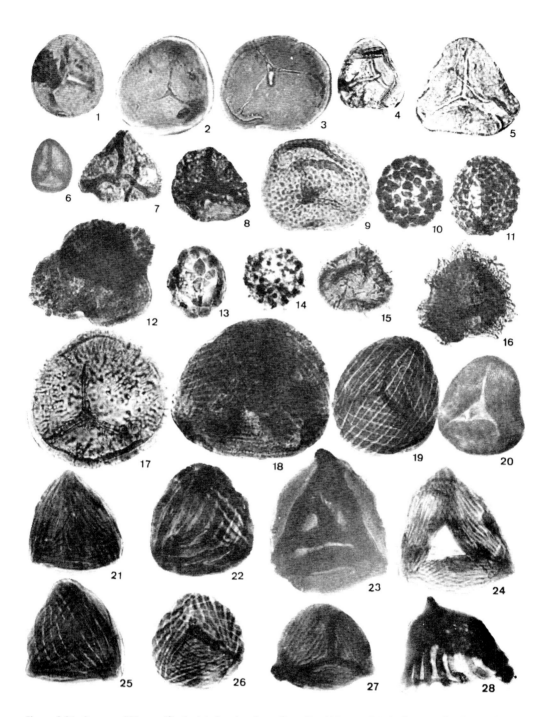

Figure 3.21. (Images ×800 magnification) *1, Stereisporites psilatus* (Ross) Manum. *2, Leiotriletes* sp. cf. *L. breviraditus* Döring. *3, Todisporites major* Couper. *4, Deltoidospora juncta* (Kara Murza) Singh. *5, Deltoidospora hallii* Miner. *6, Biretisporites pontoniaei* Delcourt and Sprumont. *7, Undulatisporites undulapolus* Brenner. *8, Undulatisporites pannuceus* (Brenner) Singh. *9, Concavissimisporites variverrucatus* (Couper) Singh. *10, Leptolipidites major* Couper. *11, Leptolepidites verrucatus* Couper. *12, Mathesisporites tumulosus* Döring. *13, Leptolepidites crassibalteus* Filatoff. *14, Gemmatriletes* sp. cf. *G. clavatus* Brenner. *15, Acanthotriletes varispinosus* Pocock. *16, Pilosisporites* sp. cf. *P. semicapillosus* Dörhöfer. *17, Echinatisporis* sp. *18, Cicatricosisporites microstriatus* Jardiné and Magloire. *19, Cicatricosisporites stoveri* Pocock. *20, Cicatricosisporites minutaestriatus* (Bolkhovitina) Pocock. *21, Cicatricosisporites recticicatricosus* Döring. *22, Cicatricosisporites sewardi* Delcourt and Sprumont. *23, Cicatricosisporites crassistriatus* Burger. *24, Cicatricosisporites* sp. cf. *C. augustus* Singh. *25, Cicatricosisporites exilioides* (Malyavkina) Dörhöfer. *26, Cicatricosisporites subrotundus* Brenner. *27, Appendicisporites sellingii* Pocock. *28, Appendicisporites parviangulatus* Döring (figure modified from Lima and Coelho 1987).

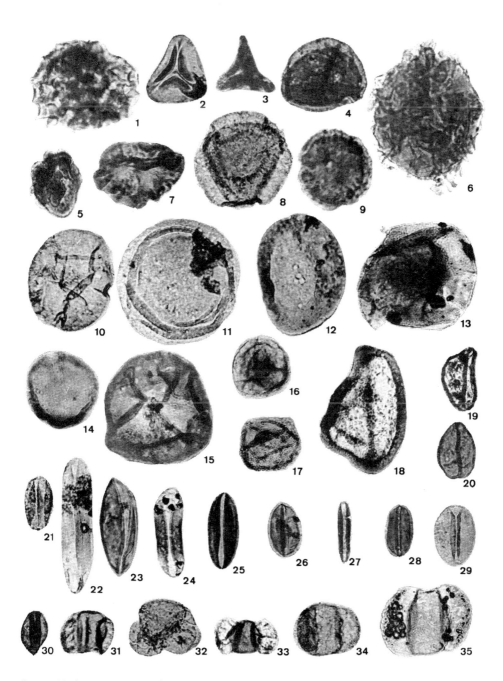

Figure 3.22. (Images ×800 magnification) *1, Ischyosporites variegatus* (Couper) Schultz. *2, Concavisporites obtusangulus* (Potonie) Pflug. *3, Dictyophyllidites harrisi* Couper. *4, Densoisporites microrugulatus* Brenner. *5, Bullasporis aequatorialis* Krutzsch. *6, Perotrilites pseudoreticulatus* Couper. *7, Zonallapollenites dampieri* Balme. *8, Zonallapollenites trilobatus* Balme. *9, Zonallapollenites microvelatus* Schultz. *10, Inaperturopollenites* sp. cf. *I. dubius* (Potonie and Venitz) Thomsom and Pflug. *11, Inaperturopollenites* sp. cf. *I. patellaeformis* Pflug and Thomsom. *12, Coptospora kutchensis* Venkatachala. *13, Inaperturopollenites turbatus* Balme. *14, Coptospora aequalis* Tralau. *15, Araucariacites limbatus* (Balme) Habib. *16,* cf. *Exesipollenites tumulus* Balme. *17, Peltrandipites* sp. *18, Sergipea* sp. cf. *S. naviformis* Regali et al. *19, Sergipea* sp; cf. *S. variverrucata* Regali et al. *20, Sergipea* sp. cf. *S. simplex* Regali et al. *21, Cycadopites glottus* (Brenner) Wingate. *22, Cycadopites giganteus* Stanley. *23, Cycadopites carpentieri* (Delcourt and Sprumont) Singh. *24, Cycadopites minimus* (Cookson) Pocock. *25, Cycadopites follicularis* Wilson and Webster. *26, Cycadopites* sp. *1. 27, Cycadopites* sp. *2. 28, Cycadopites* sp. cf. *C. dijkstrae* Jansonius. *29, Bennettitapollenites minimus* Singh. *30,* cf. *Bennettitaecuminella* sp. *31, Alisporites bilateralis* Rouse. *32, Cedripites cretaceus* Pocock. *33, Podocarpidites biformis* Rouse. *34, Podocarpidites* sp. cf. *P. fangii* Pocock. *35, Podocarpidites* sp. cf. *P. selowiformis* (Zaklinskaya) Drugg (figure modified from Lima and Coelho 1987).

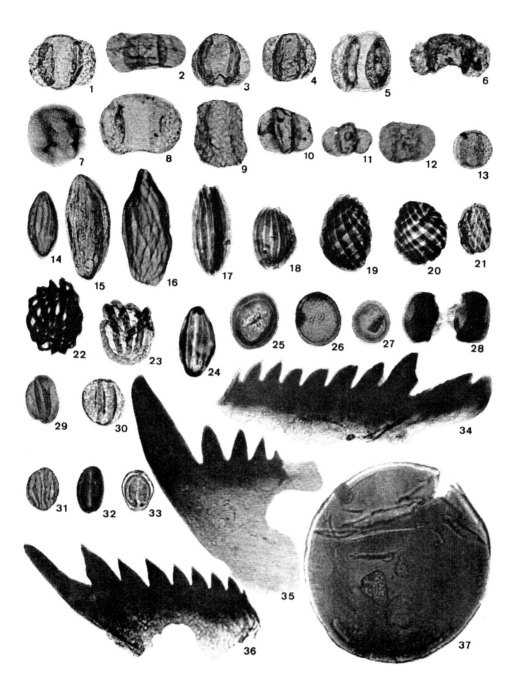

Figure 3.23. (Images ×800 magnification) *1, Podocarpidites typicus* Sah and Jain. *2, Podocarpidites* sp. cf. *P. multesimus* (Bolkhovitina) Filatoff. *3, Podocarpidites ellipticus* Coockson. *4, Podocarpidites epistratus* Brenner. *5, Podocarpidites alareticulosus* Sah and Jain. *6, Pinuspollenites* sp. *7, Phyllocladidites inchoatus* (Pierce) Norris. *8, Vitreisporites pallidus* Reissinger. *9, Podocarpidites* sp. cf. *P. arcticus* Pocock. *10, Gamerroites* sp. 1. *11, Gamerroites* sp. 2. *12,* cf. *Rugubivesiculites* sp. *13,* cf. *Vitreisporites itunensis* Pocock. *14, Equisetorporites ovatus* (Pierce) Singh. *15, Equisetosporites virginiaensis* Brenner. *16, Equisetosporites cancellatus* Paden Phillips and Felix. *17, Gnetaceaepollenites boltenhageni* Dejax. *18, Gnetaceaepollenites uesuguii* Lima. *19, Gnetaceaepollenites lajwantis* Srivastava. *20, Gnetaceaepollenites chlatratus* Stover. *21, Gnetaceaepollenites* sp. *22, Gnetaceaepollenites* sp. cf. *G. mollis* Srivastava. *23,* Gnetaceaepollenites sp. *24, Steevesipollenites* sp. *25, Classopollis torosus* (Reissinger) Couper. *26, Classopollis simplex* (Danze-Corsin and Laveine) Reiser and Williams. *27, Classopollis minor* Pocock and Jansonius. *28, Dicheiropollis etruscus* Trevisan. *29, Circulina parva* Brenner. *30, Eucommiidites troedssonii* Erdtman. *31, Eucommiidites minor* Groot and Penny. *32, Eucommiidites* sp.1. *33, Eucommiidites* sp.2. *34,* cf. *Marlenites* sp. *35,* cf. *Pronereites* sp. *36,* Scolecodont indet. *37, Tasmanites* sp. (figure modified from Lima and Coelho 1987).

The dinosaur footprints and trackways in the Sousa basin occur with a high paleodiversity in at least 37 localities, in clastic deposits within all the formations of the Rio do Peixe Group. The tracks represent at least 535 individual dinosaurs (Leonardi et al. 1987a, 1987b, 1987c; Carvalho and Leonardi 1992). They are part of a widely distributed set of track localities that Viana et al. (1993) named the Borborema Megatracksite, including the Icó, Iguatu, Malhada Vermelha, Lima Campos (Leonardi and Spezzamonte 1994), Cedro (Carvalho 1993a), and Araripe (Carvalho et al. 1994, 2019, 2020) basins.

Footprints are rare in the Antenor Navarro and Rio Piranhas Formations. The lithofacies, sedimentary structures, and geometry of the beds point to sedimentation in fan delta, alluvial fan, and anastomosing fluvial environments. Footprints are preserved only in fine sediments that accumulated as subaerial sandy bars in alluvial fans and anastomosing rivers close to the basin margins. In the Sousa Formation, the sediments' generally finer grain size rendered them more suitable for track preservation. The essentially microclastic sequence points to lacustrine, swampy, and meandering braided fluvial paleoenvironments.

The Triunfo (Uiraúna-Brejo das Freiras) Basin

Located in the west of the State of Paraíba in the counties of Uiraúna, Poço, Brejo das Freiras, Triunfo, and Santa Helena, this 480-km² basin is an asymmetric graben controlled by a NE transcurrent fault system. The Precambrian basement is composed of igneous (granites, gabbros, and diorites) and metamorphic (migmatites, gneisses, quartzites, and marbles) rocks. In the basin, the main lithologies are clastic rocks: breccias, conglomerates, sandstones, siltstones, shales, and mudstones. Limestone is rare, occurring as nodules or as centimeter-thick levels in marls. The lithostratigraphic terms are the same as for the Sousa basin. The total thickness of these deposits is unknown. Besides the breccias and conglomerates near the faulted margins, there are coarse arkosic sandstones and medium-fine quartzose sandstones with an argillaceous matrix or siliceous cement in the Antenor Navarro Formation. The main sedimentary structures are cross-channel and planar stratification. The basin is an asymmetric graben, and the finer-grained lithologies are distributed in the south-southeast region, where tilting was greater than elsewhere. Such deposits are referred to the Sousa Formation, which comprises shales and mudstones interbedded with sandstones and siltstones. The main sedimentary structures are ripple marks, climbing ripples, mud cracks, convolute lamination, and features indicating liquefaction.

Like the Sousa basin and others, the origin of this basin was a result of reactivation of basement transcurrent faults. Deposition of coarse-grained sediments on the margins occurred under the strong influence of tectonic activity. The tilted blocks created a pronounced rupture in

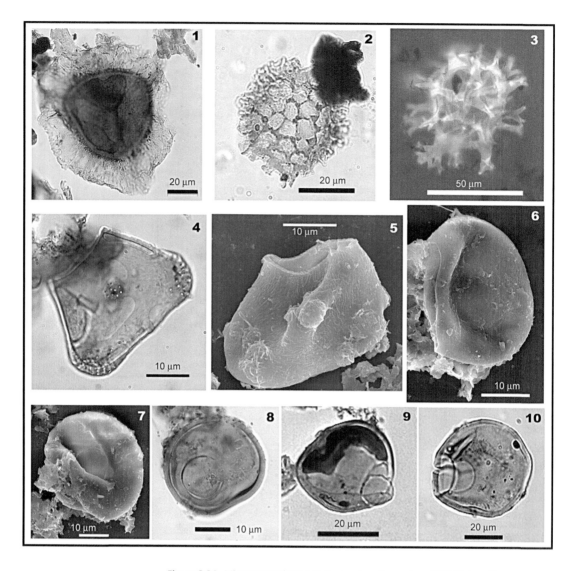

Figure 3.24. Miospores and microphytoplancton of organic wall (MPO) from the eo-devonian section (Lochkovian–Praguian?) from the Rio do Peixe basins. Specimens from samples of drill 1-PIL-1-PB, with exception of specimen 3, from a sample of drill 1-TRF-1-PB. *1, Zonotriletes* sp. 1. *2, Dictyotriletes emsiensis. 3,* Acritarc acantomorph indeterminate (ultraviolet light). *4, Pulvinosphaeridium brasiliensis. 5, Pulvinosphaeridium brasiliensis* (MEV image). *6* and *7, Nanocyclopia* sp. (MEV images, specimens with opercule in situ). *8–10, Nanocyclopia* sp. (*8* and *9*—specimens with opercule in situ, *10*—specimen without oper-cule) (Roesner et al. 2011).

terrain topography, and in the southern part of the basin the decrease in gradient favored the establishment of meandering fluvial and lacustrine environments.

Paleontology and Age of the Deposits

Conchostracans, bone fragments, and both invertebrate and vertebrate ichnofossils occur in the Triunfo basin. Except for the dinosaur footprints,

Figure 3.25. Notosuchian fossils from the Antenor Navarro Formation, Uiraúna-Brejo das Freiras (Triunfo) basin. Locality of Poço. *A*, Femur. *B*, Bone fragments from Passagem das Pedras, Sousa Formation.

5 cm

which are located close to the northern margin, the fossiliferous localities are in the central-south region of the basin. There are no micropaleontological or palynological records for the sediments, which inhibits precise determination of their age. However, as noted by Carvalho et al. (1993a, 1993b), the geological setting for the ichnofossils on the basin margins is the same as that of the Sousa basin. The paleoenvironmental interpretation indicates coalescing alluvial fans and an anastomosing fluvial system. By analogy with the sediments dated by palynology in the Sousa basin, and considering the similarities among the ichnofaunas, the main depositional phase in the Uiraúna (Triunfo) basin probably dates from between the Rio da Serra and Aratu stages (Berriasian–Lower Barremian). Roesner et al. (2011) also identified Early Devonian rocks through palynological analysis from boreholes drilled by Petrobras (fig. 3.24).

There are a few instances of vertebrate fossils in the Uiraúna-Brejo das Freiras basin, in the counties of Poço and Triunfo, where the material was extremely fragmentary, with a severely limited distribution in the Antenor Navarro and Rio Piranhas Formations. At the Poço locality (Poço County), bone fragments belonging to Crocodylomorpha, possibly notosuchian, were identified in a succession of fine sandstones with clay intraclasts (fig. 3.25). This is a rare occurrence in the geological context of the Rio do Peixe basins, given the rarity of other tetrapod skeletal elements. Only a right femur was identified amid the bone fragments. It displays an expanded and relatively flat proximal epiphysis, and the distal extremity is fragmented. The diaphysis is straight and cylindrical, expanding and curving in its distal portion. The femur is 8 cm long (fig. 3.25). This bone material differs from that found in other small crocodiles from the Lower Cretaceous (Aptian) of northeastern Brazil, such as *Araripesuchus* Price, 1959 (Araripe basin) and *Candidodon* Carvalho and Campos, 1988 (Parnaíba basin). Despite the morphological similarity of the proximal portion of the femur in these two genera, the distal region is much straighter and more cylindrical (Carvalho and Campos 1988; Carvalho, 1994c). The identification of a notosuchid crocodile in rocks considered as originating in the Rio da Serra and Aratu stages (Neocomian) expands the temporal distribution of this fossil group, since previously only isolated teeth attributed to *Araripesuchus* (Koum basin, Cameroon) were known from the Neocomian (Carvalho and Nobre 2001).

Another important occurrence comes from the Rio Piranhas Formation (Lower Cretaceous). Skeletal elements assigned to Dinosauria (Sauropoda) were recognized in the Triunfo region (Uiraúna-Brejo das Freiras or Triunfo basin, State of Paraíba): a set of three articulated caudal vertebrae, isolated chevrons, and ischium whose characteristics indicate attribution as a Titanosauriformes, named *Triunfosaurus leonardii* (Carvalho et al. 2017). These fossils are aligned in an SSE direction, in the context of gravel bars in the same direction of channeled flows. This is one of the oldest Titanosauriformes, common in the Upper Cretaceous rocks of Brazil and Argentina (Carvalho et al. 2014).

The Dinosaurian Ichnofauna of Triunfo Basin and Its Stratigraphic Context

There are few footprints in the Triunfo basin. To date, only 4 isolated footprints and 2 incomplete trackways have been identified (Carvalho 1989, 1996a; Leonardi 1994). Among the isolated footprints, 3 probably belong to theropods. The others' poor preservation does not allow the identification of their makers. One incomplete trackway consists of just 2 digitigrade, tridactyl, and mesaxonic footprints (on the concept of "axony," see Roman et al. 2018). There is a protuberance at the proximal outline of the footprints that would correspond to digit I or to a more basal pad of digit IV. The rounded extremities of the digits, without clear claw marks, suggest they were made by a small ornithopod. These footprints are preserved in a fine sandstone and do not show morphological details.

The stratigraphic succession where the footprints occur in the Antenor Navarro Formation begins with conglomeratic sandstone that is overlain by coarse and immature sandstones. The tracks were encountered in a sequence of interbedded medium- to fine-grained sandstones and siltstones. The main sedimentary structures are cross-channel and planar stratification. The beds show a tabular geometry. The succession has been interpreted as an alluvial fan deposit. The footprints would have been made during periodic breaks of sedimentation when channel sandbars became subaerial owing to discharge fluctuations (Carvalho 2000a, 2000b).

The Megatracksite of the Rio do Peixe Basins

The Rio do Peixe Group outcrop region can be regarded as a mega-tracksite (Viana et al. 1993, sensu Lockley 1991, modified) composed of the 37 tracksites of the Rio do Peixe basins. It crops out over an area of about 1,730 km² (or more, if one includes the thus far nearly sterile Pombal basin and perhaps other small basins nearby). The Rio do Peixe megatracksite is an instance in which fossil footprints rather than skeletal assemblages provide information about the composition of the dinosaur fauna of the region, corresponding in practice to category 1 of Martin Lockley's classification of fossil vertebrate sites: "1. Footprints are the only evidence of dinosaurs" (Lockley 1991, 85).

The way a track is preserved is directly related to its geological context, and the northeastern Cretaceous Brazilian basins provide a wide variety of examples of fossil track preservation. It is possible to recognize the following environmental settings of the track-bearing strata in these basins: alluvial fan-braided fluvial system, floodplain of meandering fluvial system, and lacustrine margin (Carvalho 2004a, 2004b).

Alluvial Fan—Braided Fluvial System

Footprints are rare in the Antenor Navarro and Rio Piranhas Formations. The coarse lithologies of these units, such as conglomerates, coarse sandstones, and sandstones interbedded with siltstones, were certainly a restrictive factor for fossil track preservation. The lithofacies, sedimentary structures, and geometry of the beds point to sedimentation in fan delta, alluvial fan, and braided/anastomosing fluvial environments (Carvalho 2000a, 2000b).

The alluvial fans are characterized by coarse, immature, poorly sorted, detrital sediments built up by a mountain stream at the base of a mountain front (fig. 3.26). The deposits are coarsest and thickest near the fan head area and decrease rapidly toward the base of the alluvial fan deposit.

Figure 3.26. An alluvial fan with a sparse dinosaurian fauna. Depositional context of the Antenor Navarro Formation, Triunfo basin (art by José Henrique Gonçalves de Melo).

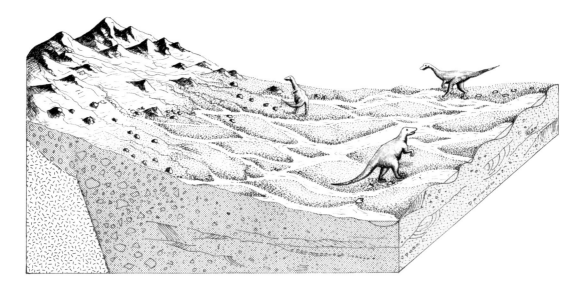

These may be found associated with braided river deposits, depending on topographical and climatic conditions. The braided river facies comprises braid bars of coarse material developed in mountainous streams and finer-grained material developed in rivers with large seasonal discharge and sediment load in the lower reaches of the river (Reineck and Singh 1986). These kinds of environments present extremely oxidizing conditions, and organic matter is rare. The coarse-grained facies assemblages that are susceptible to reworking and scour presumably reflect a low footprint preservation potential in these areas (Lockley and Conrad 1991; Kozu et al. 2017; Xing et al. 2017b).

Despite the low preservation potential of alluvial fans and braided fluvial systems, dinosaur tracks and trackways are found in these contexts in the Sousa and Triunfo basins (Carvalho 1989, 1996a; Carvalho and Leonardi 1992; Carvalho et al. 1993a, 1993b). The grain size of the sediments that formed the substrate over which the dinosaurs traveled was probably the main factor that limited footprint preservation. In the Sousa basin, footprints are found in the Antenor Navarro and Rio Piranhas Formations, preserved only in fine sediments that were accumulated as subaerial sandy bars. They are also present in the Uiraúna-Brejo das Freiras basin, in the Antenor Navarro Formation, in a sequence of interbedded medium- to fine-grained sandstones and siltstones. A conspicuous aspect of the footprints from these areas is that often they seemingly are preserved in apparent convex epirelief; however, they are concave hyporeliefs filled with harder material from the contiguous upper layer, with a coarser grain size than the surrounding matrix, and with red color pigmented by iron oxides. The Antenor Navarro Formation footprints at Serrote do Letreiro (Sousa basin) are a good example of this preservation mode. Small theropod footprints are reddish in color and sometimes filled with coarse sandstones, contrasting with the surrounding substrate. Such aspects have already been analyzed by Kuban (1991a) in his study of footprints from the Glen Rose Formation (Lower Cretaceous, south central–north central Texas). Kuban interpreted the color differences on the footprints and matrix as resulting from secondary sediment infilling of the original track depression and differential oxidation of iron on the surface of the infilling material. Although some Sousa basin tracks do not show any textural differences from the surrounding substrate, they are reddish in color, contrasting with the pavement where they occur. This could be related to a differential packing of the sediment where footprints are present—a factor that allowed distinct iron concentrations due to fluid percolation during diagenesis.

Meandering Fluvial Floodplain—Lacustrine Environments

The floodplain of meandering fluvial rivers presents the finest grain size of all the alluvial sediments—generally fine silt and finely laminated clay, interrupted by some sandy intercalation. Perennial and temporary lakes may develop in the floodplain area. In hot and dry climates, little or no organic matter is incorporated into flood basin deposits, and if the water flow is sluggish, it is possible for saline lakes to develop. Mud cracks, rain-drop impressions, bioturbation, and other surface features are widespread because of repeated exposure (Reineck and Singh 1986). This is a favorable environment for fossil track preservation. Such floodplains cover a wide distribution area, forming many repeated track-bearing strata, and it is possible to have detailed footprint morphology impressions, such as those found in the Sousa Formation (Sousa basin).

The Sousa Formation is composed of reddish mudstones, siltstones, and fine-grained sandstones; carbonate nodules and marls also occur. Common sedimentary structures include mud cracks, convolute structures, ripple marks, climbing ripples, raindrop impressions, and both

Figure 3.27. Floodplain and temporary lakes with extensive biogenic reworking of the substrate by a diversity of animals. Depositional context of the Sousa Formation, Sousa basin (art by José Henrique Gonçalves de Melo).

Figure 3.28. Alluvial fan with sauropods and theropods at the border of the Sousa basin, Antenor Navarro Formation. Footprints are rare but occur in successive levels of sandbars (art by José Henrique Gonçalves de Melo).

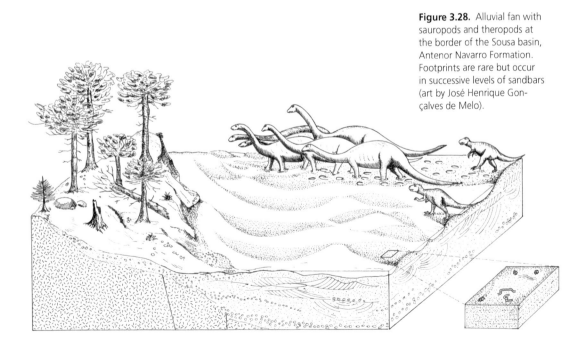

Figure 3.29. The dinosaurs of the Rio do Peixe basins lived near the shores of temporary lakes, as reconstructed in this scene. Art by Ariel Milani Martine.

vertebrate and invertebrate bioturbation. The dominance of theropod footprints in this setting is attributed to an ecological zonation of the dinosaurian biota and a taphonomic artifact. The theropods probably had a preference for the low floodplain areas, which happened to be where the chances of footprint preservation were highest, due to superb preservational potential of the fine-grained sediments (fig. 3.27). The sauropods, in contrast, did not routinely pass through or occupy these regions. They lived in the upland areas of the basin near its borders, where there was a lower potential for track preservation due to the coarse nature of the substrate sediments (fig. 3.28).

The footprints are preserved as concave epireliefs, sometimes with detailed morphology such as claws, wrinkles, digital pads, and sole pads. The essentially microclastic sequence allowed such preservation. Lockley and Conrad (1991) presented many examples from distal fluvial floodplain environments and lake borders, where there is a diversity, abundance, and wide distribution of dinosaur footprints. The lakes in the Sousa basin developed in the sedimentary depocenters of those areas and are characterized by clastic sediment. They resulted from fluvial systems that graded into fluvio-lacustrine complexes. Because the climate during the main interval of deposition was hot, with a high evaporation rate, it was possible for saline lakes rich in calcium carbonate to develop. Dinoturbation was significant along the margins of these lakes, as discussed by Carvalho and Carvalho (1990) and Carvalho (1993). The lakes were mostly small and temporary, hot and shallow, and their water chemistry had an alkaline character (pH between 7 and 9). The lakes held large amounts of nutrients and chemical ions such as calcium and phosphorous—optimal environmental conditions for flourishing life.

The cyclic succession of mudstones, siltstones and fine-grained sandstones in fluvial-lacustrine environments, a product of periodic flooding and a lake's changing shoreline, allowed the establishment of many successive surfaces adequate for track preservation (fig. 3.29). This can be observed in recent environments, as described by Cohen et al. (1991) regarding Lake Manyara (Tanzania). They observed that a strong shoreline-parallel zonation of substrate conditions correlates with differences in track preservation style (Lockley and Xing 2015; see also Falkingham et al. 2014a; Melchor et al. 2019).

Texture, composition, and moisture regulate the probability of initial track registration and depth of penetration. Furthermore, in paleolakes a cyclic pattern of deposition occurs, as described by Prince and Lockley (1991). They mapped bedding plane exposures of the Morrison Formation (Jurassic, Colorado) that revealed hundreds of dinosaur footprints in a package of cyclic lacustrine sediments at the top of the shoreline facies beds. In the Sousa basin, Leonardi (1989a, 1994) recognized twenty-five levels with dinosaurian ichnopopulations in the Sousa Formation (Caiçara-Piau locality). These different levels represent cyclic deposition on a Cretaceous lake's shoreline. Similarly, Paik et al. (2001) analyzed the dinosaur track–bearing deposits of the Jindong Formation (Upper Cretaceous, Korea), which were interpreted as the result of repeated deposition by sheet floods on a mudflat associated with a perennial lake,

Figure 3.30. Medium-size to large theropod footprint at the raindrop-marked level 13, subsite 2 of the Piau-Caiçara farm (SOCA 13217). Graphic scale in centimeters. The arrow indicates the direction of the light.

Figure 3.31. Dinosaur footprint with displacement rims. Such structures occur around the margins of footprints made in thick, plastic, moist to water-unsaturated microbial mats on top of moist to water-unsaturated sediment. Sousa Formation, Passagem das Pedras locality.

utilized by dinosaurs as a persistent water source during drought in an arid climate. For their part, Matteo Belvedere et al. (2010) illustrate a huge number of dinosaur fossil tracks present in at least 21 trampled levels with diverse successive ichnofaunas in floodplain deposits in the Late Jurassic Iouaridène tracksite, Central High Atlas, Morocco.

The footprints and undertracks in the basins of northeastern Brazil were produced in both subaerial and subaqueous settings. It is possible to identify footprints with well-defined morphologies (fig. 3.30), as well as prints that progressively lose their clarity due to alteration by mud cracks, fluidization, and convolute and radial structures (fig. 3.31). Well-preserved footprints with impressions of claws, nails, and soft tissues like sole and phalangeal pads are considered to have been produced in muddy sediments with high plasticity and low water content, probably in a subaerial setting of floodplains and lake margins. This setting is easily recognized by the association of the footprints with raindrop impressions (fig. 3.30) and mud cracks, the latter of which sometimes originate at the margin of the track or as distal extensions from the digits. The dehydration of muddy sediments produces structures similar to those described by Lockley et al. (1989). In contrast, if the geological setting of the footprint is alluvial fan sediments, even though such prints were also made in a subaerial setting, their morphology is often restricted to the gross outline of the footprint. In this case the track is more disruptive of the depositional surface and can be considered a dinoturbation structure. There are some exceptions, as in the case of Antenor Navarro Formation footprint tracks from Serrote do Letreiro that show well-defined contours due to differential iron oxidation on the surface of the infilling material. Finally, in subaqueous environments, there is a decrease in the morphological clarity of the footprints that are preserved, with poorly defined nails, claws, pads, and sole marks.

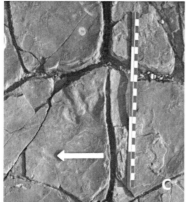

Figure 3.32. Passagem das Pedras (SOPP). Tracks in the siltstones and shales of the Passagem das Pedras locality, Sousa Formation. *A*, The bank of the rocky pavement seen from the river, in vertical section. *B*, The savanna near Passagem das Pedras. *C*, The smallest (5.6 cm long) dinosaur footprint (SOPP 18) presently known from the Rio do Peixe basins, probably attributable to a hatchling ornithopod or, less likely, theropod. *D*, The holotype of *Sousaichnium pricei* during the second phase of its excavation in 1975, after a power shovel had removed about 2 m of overlying sand and gravel. *E*, Another view of *Sousaichnium pricei*. *F*, The trench dug in 1977 to uncover a portion of trackway SOPP 5, paratype of *Moraesichnium barberenae*.

Lockley and Meyer (1999, 73) strongly recommend an integrated study taking into account tracks, substrates, and paleoenvironments, and then a holistic approach to both the maker of the tracks and the environment in which it was living. For a geologist, this means studying not just the tracks but the tracks in their sedimentological context. An analogous recommendation is found in Pérez-Lorente (2015); Lallensack et al. (2016, 2019); and Gatesy and Ellis (2016). Differences in substrate conditions, such as variable water content, sediment plasticity, and the presence or absence of microbial mats (Marty 2012; Carvalho et al. 2013; see also Xing

Tracks and Substrate

et al. 2016d) have a biasing influence on the kinds of animals that could imprint their footprints. For example, with just one exception (SOPP 18; fig. 3.32C), in the Rio do Peixe basins we have no footprints of very small dinosaurian individuals, with hind-foot prints less than 12 cm of length, and almost no footprints of the vertebrate mesofauna. It is possible that hatchlings or juveniles, and adult dinosaurs of very small size, could not leave footprints because of the relative compactness and/or dryness (also superficial) of the substrate (Leonardi 1984a). The same is probably true for lizards and for Early Cretaceous mammals and for other groups of small tetrapods, as well as crawling traces (Repichnia) of invertebrates (mainly arthropods), which are also very rare.

Different substrate conditions from place to place, even over short distances (microdifferences or microfacies), have a strong influence on the variability of footprint morphology, both within individual trackways as well as from one individual track to another among trackways and isolated footprints. Several trackways show large differences among their footprints when it is certain that those prints pertain to the same trackway and then to the same individual trackmaker. Examples include trackways ANEN 1, ANJU 4, SOMA 1, SOCA 1336, SOCA 48, and many others.

This fact was similarly demonstrated by Razzolini et al. (2014) for the El Frontal dinosaur tracksite (Early Cretaceous, Cameros basin, Soria, Spain), where one can frequently observe a trackway, evidently made by just 1 trackmaker, showing great variability among the geometry of its footprints and among its parameters. Multidata analyses of this ichnosite demonstrate that the differences among the depth and other characteristics of the footprints of the same trackway depend on the different consistencies of substrate of the sediment where the maker was proceeding. So, the high variability of the footprints in the trackways of that ichnosite depends on the high variability of the substrate, due to frequent and systematic lateral changes in microfacies. Similar phenomena were described in a very interesting paper by Ishigaki (2010), who reported a trampled bedding surface from the Upper Cretaceous Abdrant Nuru fossil site in Mongolia, across which a number of theropods laboriously slogged their way, and also in a paper by Ősi et al. (2011) regarding some Hungarian Hettangian tracks. The same phenomenon of footprint variation in a given trackway is demonstrated by Castanera et al. (2018) for the theropod-dominated ichnofauna of the Reuchenette Formation (Kimmeridgian) in the Swiss Jura Mountains, northwest Switzerland.

The weight of large animals surely compacted the sediment, expelling air and water from spaces between sediment grains and rearranging their fabric (Lockley and Meyer 1999, 232; Thulborn 2012, 5–8). These changes sometimes affected the way future erosion impacted the rocks created from those sediments; most notably, the stone material surrounding the footprint might be abraded more rapidly than the material filling the footprint itself, leaving the track in fake-positive relief, as happened with footprint ANEN 14. Both lab and field experiments in neoichnology, with living animals moving on different substrata, were performed with the purpose of better understanding the relationship between the trackmaker and the substrate, including the microtectonic of the tracks, although only a small part of the results was published (Leonardi 1975).

A noteworthy feature of the Rio do Peixe footprint assemblages is the absence of trackways with very short steps, the kind that would have been made by large herbivores walking slowly while browsing on vegetation. This probably reflects substrate conditions at the time the footprints were made. Presumably the surfaces of plastic mud where the tracks were impressed, which became the rocks of the Sousa Formation (in which the tracks are found today, in the bed of the Peixe River), correspond to the silty bottoms of ancient temporary lakes. Ordinarily the lakes would have had no water, and so their beds would have been dry and hard, likely with little plant cover. Under temporary wet conditions after heavy rains, when lake levels were high, there would have been insufficient time to grow enough aquatic plant material in the lake bed to attract large plant eaters. As the water progressively dried up, successively more of the lake bottoms would have been exposed. Once again, there would not have been enough time for the newly emergent silt bottom to develop plant cover before the dinosaurs trod across them. There likely would have been permanent vegetation on terrains situated above the maximum level of the water of the lakes. In such places the ground was probably too hard to record many footprints, and it may have been covered by brush—additional factors making it difficult for footprints to be impressed in the soil. On the alluvial fans that are represented today by the Antenor Navarro and Rio Piranhas Formations, the wet zones corresponding to both sides of the temporary and anastomosing creeks, which were sufficiently plastic to receive the impression of the footprints in the wet sand, probably had little or no vegetation.

Another phenomenon tied to the substrate is development of displacement rims of footprints, which are quite variable in size and form (figs. 3.33, 3.34). Some of them are wide and thick, sometimes with the aspect of true bulges of mud, now evidently lithified (e.g., ANEN 10 and SOCA 138). Such conspicuous mud bulges are especially well developed in sauropod footprints of both the fore- and hind-limbs. In these cases, the very heavy animal impressed its feet into a surface of very plastic and/or waterlogged mud. Under these circumstances the displacement rim of the hind-foot will frequently fill in and/or squash the horseshoe-shaped footprint of the forefoot from behind so that it becomes very narrow or little more than a crescent-shaped slit (fig. 3.34).

In other cases, and frequently at our sites, the displacement rim around a footprint is very low and narrow, indicating a compact and firm mud (in the trackway SOES 9, *Caririchnium magnificum*). In other instances, it seems as though the weight of successive overlying layers deposited over the tracks compressed and squashed the footprints and their displacement rims (Lockley and Xing 2015).

Other kinds of displacement rims are:

- Elliptical displacement rims jointly around a pair or set of hand-foot prints in the trackway of a quadrupedal animal, mainly seen in sauropod tracks (e.g., at Floresta dos Borba or in those of the herd of the Riacho do Pique at Serrote do Letreiro)
- Displacement rims that control the development of mud cracks (e.g., in SOPP 1 at Passagem das Pedras, figs 3.35; cf. Schanz et al. 2016, fig. 19.4, 376).

In addition to complete displacement rims, there are other analogous structures of expulsion:

- Compressed sediment between two toes, especially in footprints made by running theropods (e.g., SOPP 7)
- Crescentic displacement rims immediately in front of footprints in trackways made by running dinosaurs (e.g., SOPP 5)
- Crescent-shaped convexities at the rear margins of each of the three incomplete digit impressions in footprints made by half-swimming trackmakers, as in many of the theropod footprints from level 16 at SOCA (SOPP 16). We use the term "half-swimming" gait or tracks, first described by Coombs (1980), instead of "swim tracks": "In this kind of progression, the trackmakers swam in rather shallow water in temporary lakes, perhaps engaged in 'fishing,' and they impressed in the sediment only the distal portions of the toes—the claws or, in one case, the hooves. The morphology of theropods footprints interpreted as showing this manner of locomotion is very characteristic: the tip of the claw mark of digit III is triangular and straight; in contrast, the tips of the claws of digits II and IV are curved, because impressed with an outward and rearward rotational movement" (Leonardi 1987a, 34–35, 51, and plate IX, L–O). For a more recent interpretation and description of this kind of tracks, see Milner and Lockley (2016).

For a very interesting discussion, accompanied by good graphs and illustrations, on the genesis of the displacement rims and the movements of sediment during their production, see Falkingham et al. (2014b).

To say that the positive (convex) volume of the displacement rim is almost the same as the negative (concave or void) volume of the corresponding footprint on the sediment surface is probably a tautology. However, it is convenient here to stress this concept. Very deep footprints are indeed frequently associated with high and wide displacement rims. In contrast, the dinosaur footprints of the Rio do Peixe basins are rarely very deep, and so their displacement rims generally have only modest dimensions, often even with tracks of sauropods and large ornithopods.

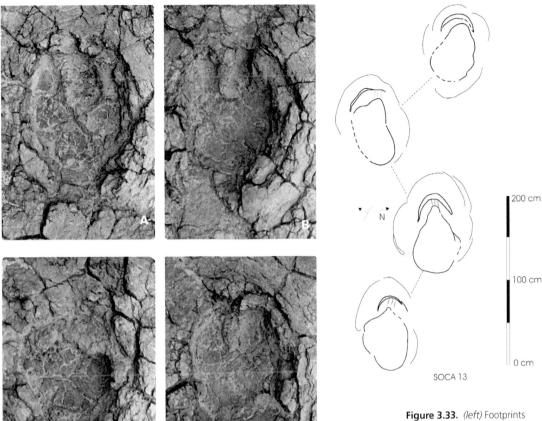

SOCA 13

In these formations, one gets the impression that while the superficial sediments were sufficiently plastic to receive the impression of the footprints, the sediment layers beneath the track-bearing stratum were often already hard at the time tracks were impressed, perhaps due to partial and early lithification. As a result, the feet of the trackmakers sank just through the thickness of the more recent and superficial layer. A good example of this phenomenon is the holotype of *Caririchnium magnificum* Leonardi, 1984, in the Antenor Navarro Formation at Serrote do Letreiro (SOES 9). The holotype was produced by a massive, heavy ornithopod whose footprints are nonetheless very shallow, with very low and thin displacement rims (see also Lockley and Xing 2015).

Although the positive volume of the displacement rim is almost the same as the negative volume of the corresponding footprint, the key word here is *almost*, because one part of the volume of the sediment is not displaced but rather compacted on the "sole" of the footprint. Consequently, there are not many very deep footprints at the Sousa basin. The deepest tracks are those of the main trackway of *Sousaichnium pricei* Leonardi

Figure 3.33. *(left)* Footprints of trackway *Sousaichnium pricei* SOPP 1, Passagem das Pedras (SOPP). *A*, SOPP 1/10 (left pes). *B*, SOPP 1/12 (left pes), one of the best footprints in the trackway. *C*, SOPP 1/13 (right pes). *D*, SOPP 1/14 (left pes). Note how development of the mud cracks was controlled by the displacement rims of the prints. The red color of the footprint is due to the shale that filled in the footprints. In SOPP 1/13 (*C*) note that a mass of mud collapsed into the footprint after extraction of the foot, covering the impression of the toes. Before the footprint was withdrawn from the sediment, the mud had squeezed against the metapodium of the trackmaker.

Figure 3.34. *(right)* Diagram of very well-preserved sauropod trackway (SOCA 138) at the Piau site. The excavated portion of the trackway includes 4 hand-foot sets. From Leonardi and Santos (2006).

Figure 3.35. *(above)* This long ornithopod trackway (SOPP 1), in the Dinosaur Valley Natural Park at Passagem das Pedras, Ilha Farm, is the holotype of *Sousaichnium pricei*. Note the high displacement rims and how these are circularly and radially controlling the mud cracking. The ichnofauna of the park includes many tracks of large and small theropods, this large ornithopod trackway, and a short trackway of a small ankylosaurs.

Figure 3.36. *(left)* Another view of the holotype trackway *Sousaichnium pricei*, Passagem das Pedras (SOPP). Some 52 hind-foot prints were excavated, along with about 25 forefoot prints, but only of the right manus. The trackway is attributed to a Gondwanan semibipedal ornithopod. Note the sinuous and serpentine walking gait and the conspicuous displacement rims around the footprints. The surrounding rock surface also shows ripple marks and mud cracks. Graphic scale = 1 m.

1979 at Passagem das Pedras (SOPP 1, figs. 3.35, 3.36) and especially some sauropod undertracks at Piau (SOCA level 13/3).

In our basins, one never sees tracks so deep that the superficial layers where the dinosaurs walked were broken and the trackmakers' feet punched through them so that the footprints are composite, like they are, for example, in the Morrison Formation at the Purgatory River site in southeast Colorado, in a sauropod trackway photographed by Martin Lockley (published in Leonardi 1987a, plate XVI, fig. D). This kind of deep sauropod track that penetrated through a thick package of sediment

layers or laminae was reported by Browne (2009), from the Upper Cretaceous (Maastrichtian) North Cape Formation at northwest Nelson, from
the South Island of New Zealand.

Such deep composite footprints also occur at the Lavini di Marco
tracksite at Trento, Italy (Calcari Grigi di Noriglio Formation, Hettangian-Sinemurian; Leonardi and Mietto 2000; Petti et al. 2020b). In contrast, the footprints in the Rio do Peixe basins have a continuous surface,
generally smooth and rather shallow, and the laminae are parallel to the
outer surface. This also depends on the fact that, due to the flat landscape
of the Rio do Peixe basins, one almost never sees the rocks in sections. As
a result, as in the case of the huge number of tracks in Altamura, Puglia,
Italy (Nicosia et al. 1999a, 1999b; Petti et al. 2200a), you can practically
never examine the tracks in cross section, right in the outcrop. One can
do this instead very well, just for example, in Lee Ness Sandstone, Ashdown Formation, southern England (Shillito and Davies 2019), in the
North Horn Formation of the Wasatch Plateau, central Utah (Difley and
Ekdale 2019) and in the Aptian of the Chapada do Araripe in northeastern
Brazil (Carvalho et al. 2020). In trackway SOPP 1, when the trackmaker,
a large iguanodontid, withdrew its foot from the very plastic but firm
mud with a backward movement, the mud sticking to the front of the
metatarsal portion of the foot sometimes collapsed into the large and
deep footprint (SOPP 1, footprints 2, 6, and 13).

At our sites, impressions of scales were not found in the footprints, as
they were in other ichnosites (e.g., Paik et al. 2017). This may be due less
to bad impression or modern erosion than to the fact that the feet of the
trackmakers were dirty—covered by a film of fresh mud—especially in
the case of the Sousa Formation. Some convexities, perhaps corns, can
be observed in some hind-foot prints of *Caririchnium magnificum* at the
Serrote do Pimenta (hind-foot print 4). In contrast, for some good examples of the impression of pedal scales in large theropods, see McCrea
et al. (2014, especially fig. 10 and its comment); and for splendid skin and
scale impression of sauropod and diminutive feet from the Cretaceous of
Korea, see Kim et al. (2010, fig. 2; and 2019, fig. 4; respectively).

Another phenomenon of the interaction between track and substrate is
that of sticky mud and the sucker effect, with consequent counter-relief of
the footprints. There are many examples of this among pedal prints of sauropods of the Riacho do Pique at Serrote do Letreiro (association SOSL
2, 11–39, 50, 400–489; fig. 3.37A): in this ichnopopulation the hind-foot
print often presents a kind of reverse relief, which might be interpreted
as a suction structure, produced as the flat or concave hind-foot sole was
lifted off the sticky mud/sand mixture of the substrate.

Another example of counter-relief or, in this case, apparent counter-
relief is that of theropod footprints at Serrote do Letreiro (SOSL, fig.
3.37B–D), at Curral Velho (SOCV), and at Fazenda Paraíso (SOFP),
where the footprints, which are quite normally concave epireliefs, are
nevertheless filled in by the coarser sediment of the underside of the
overlying layer and so present an apparent counter-relief.

Figure 3.37. Serrote do Letreiro (SOSL), Antenor Navarro Formation, Sousa. This was the first locality discovered outside the classic site of Passagem das Pedras (1977). *A*, The dry bed of the Riacho do Pique rivulet; the environment is savannah and parkland, with pereiro trees (*Aspidosperma pyrifolium*). The streambed is crossed (from right to left) by the shallow trackways of a herd of at least 15 sauropods as well as trackways of some small to medium theropods. Photo courtesy of Franco Capone. *B*, Typical theropod footprint (SOSL 44/5; right pes) from Serrote do Letreiro, subsite 4, filled in by coarser sand and so in apparent reverse relief. *C*, A similar track, belonging to trackway SOSL 41, left pes. *D*, the same footprint as in *B*, associated with another track (SOSL 41/3) of the same kind and an ancient native petroglyph.

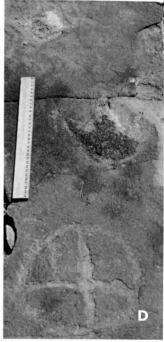

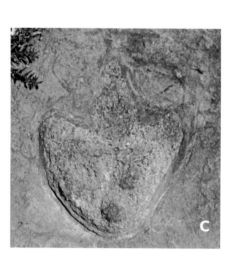

In general, it is important to note that a footprint is not a two-dimensional structure, marked just on the surface of the air-soil interface; in contrast, it is a very complex 3-D structure that involves to varying degrees the depth of the substrate. This can be studied with a computerized axial tomography sensu stricto, by mechanically cutting a specimen of track in parallel slices, horizontally and/or vertically. However, there are also more sophisticated methods, like those illustrated by Falkingham and Gatesy (2014): a combination of XROMM (X-ray reconstruction of moving morphology) and DEM (discrete element method) methodologies, which those authors describe (1, 3–5) as a temporal sequence they term "track ontogeny." This combination of methodologies fosters a synthesis between the surface-layer-based perspective of footprint shape prevalent in paleontology and the particle volume–based perspective essential for a mechanistic understanding of sediment redistribution during track formation. In some lucky cases, the 3-D character of footprints can be directly observed in the field, such as at the Lavini di Marco site (Hettangian-Sinemurian Calcari Grigi di Noriglio Formation, of the Italian Southern Alps), where a track or a trackway, because of erosion, allows one to directly examine its inner and deep structure (Leonardi and Mietto 2000). Note that in rare cases, the undertrack can be better preserved (more distinct, more complete, deeper, and more readily classifiable) compared with the actual trampled surface, where the tracks can be poorly preserved, that is, shallower, indistinct, and unclassifiable (Marchetti 2019).

Depending on the substrate, Falkingham et al. (2010, 356) suggest using tracks (not only of the dinosaurs but the fossil tracks in general) as if they were paleopenetrometers. In geotechnics and pedology, a penetrometer is a device utilized to gauge the resistance, consistency, and structure of soil, mainly in house building. The depth of a track could be seen and utilized in an analogous way (and with some due reservations) to study these characteristics of the paleosoil where the track was printed. This study would help analyze the paleoenvironment. On this topic, see also Demathieu's chapter in Leonardi (1987a) and Platt et al. (2018).

Fossil tracks are preferentially found in fine-grained sediments, such as carbonates and mudstones. This might lead to the interpretation that footprints can be preserved only in such fine-grained sediments. In reality, however, fossil tracks can be found in sedimentary rocks with many substrate grain sizes, including, rather unexpectedly, coarse-grained deposits (Hasiotis et al. 2007; Dai et al. 2015). Microbial mats covering sediment surfaces can stabilize those surfaces because such biofilms allow the surface to lithify by precipitation of calcium carbonate (Chafetz and Buczynski 1992; Dupraz et al. 2004; Dupraz and Visscher 2005), consequently enhancing the preservation potential of primary structures like ripple marks and mud cracks (Dai et al. 2015)—and footprints.

Microbial mats are organosedimentary deposits of benthic microbial communities, a multilayered sheet of microorganisms, mainly bacteria and cyanophytes, that grow at interfaces among different types of material, mostly on submerged or moist surfaces. The products of benthic microbial communities are called "biolaminites" or "biolaminations" for the flat laminated type of stromatolites, or "biolaminoids" for less significantly laminated sediment that accumulated through the activity of microbial communities (Marty 2005; Marty et al. 2009).

The preservation of animal footprints in the fossil record is strongly dependent on taphonomic processes, although it is the sedimentation regime that determines whether preservation will take place, enabling a footprint to be incorporated into the sedimentary record. The possibility

Preservation of Dinosaur Tracks Induced by Microbial Mats

Figure 3.38. Photomicrographs of the (*A*) microclastic microbialite and (*B*) microbialite facies of the Sousa Formation. *A*, Very thin (centimeter scale) graded terrigenous microclastic bed (flooding event), interbedded with microbial laminites, infilling the tracks. Note biofilms/thin microbial mats (dark laminae) and their wrinkles (cf. Dai et al. 2015). *B*, disrupted carbonate microbial laminites, interbedded with very thin terrigenous microclastic beds, associated with the footprints from Passagem das Pedras, Ilha Farm. Note the dark laminae (reddish color), related to organic matter replaced by siderite (Carvalho et al. 2013a).

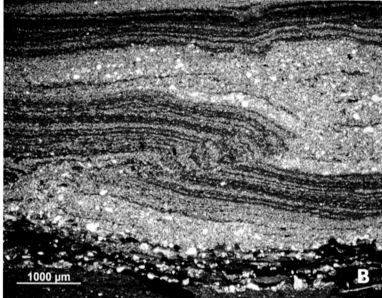

of preservation is minimal during long-lasting periods of exposure without any sedimentation, and preservation is favored by rapid and significant preservation events. Consequently, footprints are most commonly preserved in environments that undergo cyclic sedimentation. Therefore, the final preservation of fossil vertebrate tracks in laminated sediments has been explained by the stabilization process of the sediment surface by microbial mats that would cover the tracks and protect them from erosion (Thulborn 1990; Avanzini 1998; Conti et al. 2005; Marty 2005).

Marty et al. (2009) showed that microbial mats play a major role in the preservation of footprints, in some cases through consolidation by desiccation or lithification of the mat, and sometimes by continuing growth of the mat itself. The formation and morphology of footprints in microbial mats depend on the nature of the mat but also on the characteristics (water content, grain size, lamination, degree of consolidation, presence of a lithified horizon, yield strength) of the underlying sediment. These authors also observed that footprint morphology is a function of microbial mat thickness and water content of the mat and the underlying sediment. Generally poorly defined or no footprints are produced in dry mats while in water-saturated microbial mats the imprints are well defined, sometimes with well-defined displacement rims. The formation of these rims around prints of large dinosaurs occurs in thick, plastic, moist to water-unsaturated microbial mats on top of moist to water-saturated sediment. These features are commonly observed in the tracks of Passagem das Pedras, Sousa basin (figs. 3.32E–F; 3.33, 3.35, 3.36).

The occurrence of microbial structures (fig. 3.38) in the Sousa basin was described by Silva Filho (2009), who recognized distinct sedimentary facies of microbial and siliciclastic origin. The lithologies are marls, sandstones, siltstones and mudstones, interpreted as deposits of microbial mats and decantation of fine-grained sediments in which dinosaur tracks are found.

The large number of dinosaur tracks and associated sedimentary structures in the Sousa basin (Sousa Formation, Lower Cretaceous) is certainly related to the role of biofilms in their consolidation. Footprint consolidation and its early lithification probably occurred due to the existence of algal mats that allowed a more cohesive substrate that was more resistant to disaggregation. The sediments were initially stabilized by early cementation and by the network of biofilms over the tracks. This happened repeatedly, resulting in a large number of sediment layers with preserved dinosaur tracks and sedimentary structures (Carvalho et al. 2013).

The proliferation of the abundant conchostracan fauna found in the inner basins of the Brazilian northeast requires that the large bodies of water where these organisms lived had specific physical-chemical characteristics and available nutrients. Present-day conchostracans inhabit fresh, well-oxygenated, alkaline waters (pH between 7 and 9) in aquatic settings with muddy bottoms. The chemistry of the water in small lakes where conchostracans occur fundamentally affects morphological features of their shells, especially concerning their size and calcification of the valves. Through the study of conchostracans, Carvalho and Carvalho (1990) and Carvalho (1993) inferred that they lived in small temporary lakes, warm and shallow, in which water chemistry was alkaline (pH between 7 and 9). The dimensions of some of the conchostracans (up to 3.5 cm in length) suggest that an ecological optimum must have existed in which large amounts of nutrients and chemical ions, such as calcium and phosphorus, were present (figs. 3.12, 3.13, 3.14).

Ecological Aspects of the Cretaceous Lakes

Figure 3.39. During the Early Cretaceous some perennial lakes occurred in the Sousa basin, as indicated by specimens of the big fossil teleostean fish *Lepidotes*. Art by Ariel Milani Martine.

Chemical analyses of samples of sedimentary microclastic and chemical rocks (mudstone, marl, and limestone) allow evaluation of the geochemical conditions of the environment in which the Cretaceous conchostracans lived (Carvalho 2009), since there is a relation between sediments that would have served as a substrate and a source of nutrition for the conchostracofauna, and the chemical conditions of the surrounding microclimate. Despite the possibility that diagenetic factors might have significantly modified the initial concentration of the analyzed chemical elements, this quantitative evaluation allows one to infer some ecological parameters through the comparison with current environments, the better to distinguish among paleoecological interpretations.

Hydrothermal activity is very intense in areas with wrench faults, indicating a region with a high heat-transfer rate. Hydrothermal alterations in these regions are likely reflections of the mechanical importance of fluid pressure on the fault mechanism. Thus, tectonic activity in northeastern Brazil during the Early Cretaceous could have remobilized chemical elements such as calcium, sodium, iron, fluorine, phosphorus, manganese, magnesium, potassium, and sulfur through hydrothermal solutions, which would rise through the reactivated faults to affect the chemical character of the waters inhabited by the conchostracans (Carvalho 1993). At the present time, in the Uiraúna-Brejo das Freiras basin,

one can observe continuing hydrothermal phenomena caused by this tectonic activity. According to Boa Nova (1940), the water temperature at hydrothermal springs in Brejo das Freiras varies from 33.5°C to 38°C; its chemical composition is dominated by calcium (0.0028 mg/l), magnesium (0.0078 mg/l), sodium (0.1787 mg/l), potassium (0.039 mg/l), chlorine (0.1237 mg/l), sulfate (0.0263 mg/l), and silicon (0.0328 mg/l) ions as well as iron and aluminum oxides (0.0064 mg/l). The pH of these waters is 8.9. The existence of hydrothermal springs that fed into the lakes at the beginning of the Cretaceous period might have been responsible for nutrient enrichment and the maintenance of the alkaline pH—factors that propitiated the blooming of the conchostracofauna.

Another possible source of nutrients in the water in which the Cretaceous conchostracans lived is the chemical weathering of the rocks that comprise the Precambrian basement as well as the basic igneous rock intrusions, such as the one at Lavras da Mangabeira. Alteration of the minerals in those rocks (syenite, granite, granodiorite, gabbro, phyllite,

Figure 3.40. Endorheic drainage characterized the Rio do Peixe basins, establishing both temporary and permanent lakes. Art by Ariel Milani Martine.

quartzite, marble, and diabase) would provide a rich ionic variety (calcium, magnesium, sodium, iron, phosphorus, aluminum, manganese, silicon, potassium, and fluorine) that would be progressively concentrated within the pull-apart basins due to the predominantly endorheic drainage (figs. 3.39, 3.40). At the present time, the local ponds and lakes almost always reflect the geological conditions of the region, with their ions originating from weathering of rocks and soils.

The main ions found in continental lakes are calcium, magnesium, sodium, potassium, iron, manganese, chlorine, sulfate, carbonate, and bicarbonate. The water's ionic composition will be influenced by the geology of the surrounding drainage area and the lake's accumulation basin and by the rain regime. The vertical distribution of dissolved ions in shallow waters with no permanent thermal stratification is homogenous. According to Esteves (1988), continental bodies of water usually have a pH between 6 and 8. Those with lower values have high concentrations of dissolved organic acids that result from the metabolic activity of their aquatic organisms or as inputs from allochthonous sources. Aquatic ecosystems with high pH values are usually found in regions with a negative hydric balance, regions influenced by the sea, or karstic regions.

Limnic sediments can be considered as the products of the integration of all of the processes that occur in an aquatic ecosystem. Taking into consideration the cycling of nutrients and the energy flux, sediment is one of the basic compartments of continental aquatic ecosystems because this is where the biological, physical, and/or chemical processes that influence the metabolism of the whole system occur (Esteves 1988). As limnic sediments are generated during the different phases of the history of a lake, they constitute one of the sources of ions available to organisms and thus serve as indicators of the lake's changing environmental conditions and thus as proxies for changes in lentic faunas. Although diagenesis results in several modifications in the mineralogical (and thus also chemical) composition of the sediments during and after deposition, chemical analysis of samples of sedimentary rocks from the Sousa basin where the Cretaceous conchostracans are found (table 3.1) provided significant data about the ionic content and its relation to the dimensions and species compositions of the conchostracan assemblages found in the samples.

Table 3.1 Concentration of major chemical elements in samples of sedimentary rocks with conchostracans from the inner basins of the Northeastern Brazil.

sample	species	chemical elements – TOTAL SAMPLE (%)									
		Si	Fe	Mn	Ca	Mg	Na	K	P	S	C total
UFRJ-DG 18-Co Fm. Sousa, Serrote do Pimenta	*Cyzicus* sp.	23.3	6.1	0.05	1.0	1.6	0.80	3.4	0.08	0.01	0.14
UFRJ-DG 26-Co Fm. Sousa, Pedregulho	*Palaeolimnadiopsis reali*	21.7	3.2	0.07	6.4	1.8	2.6	2.8	0.17	0.02	2.0
UFRJ-DG 32-Co Fm. Sousa, Lagoa dos Patos	*Palaeolimnadiopsis reali*	31.7	4.0	0.50	1.0	1.6	2.4	1.7	0.10	0.02	0.21

Oil exudes from the ground at the Saguim Farm, Sousa County (fig. 3.41), Sousa basin. Although the region is surrounded by rocks mapped as the Sousa Formation, the oil actually occurs in metamorphic rock from a structural dome in the basement rock. An expressive system of normal faults on the southeastern border of the basin caused uplift of the basement. This discovery prompted new economic interest in this sedimentary area and investigations into the basin's origin.

Although the source rocks of the Saguim oil are unknown, ANP (Agência Nacional de Petróleo) data (2008) indicated that the petroleum systems of the Sousa basin have their source in black shales from the Sousa Formation and, as reservoir rocks, the sandstones of this same lithostratigraphic unit, and also from the Antenor Navarro Formation. The sealing rocks are shales and limestones from the Sousa Formation, and the traps are of structural, stratigraphic, and paleogeomorphic origin.

Mineral Resources and the Oil of the Sousa Basin

Figure 3.41. Oil exuding from the ground at Sítio Saguim, Sousa basin.

Considering the geological framework of the Sousa basin, the sediments overlie a tonalitic gneiss basement that has undergone severe changes in mineralogy due to retrometamorphic and alteration phenomena. In addition, expressive fracturing of the rock is noteworthy, as it relates to the normal fault system that uplifted this basement. These modifications possibly caused disturbances in physical parameters of the gneiss, such as porosity and permeability (Carvalho et al. 2013). The formation of secondary porosity should be related mainly to the development of cracks and microcracks, possibly enhanced by rearrangement of the minerals from retrometamorphic and alteration processes. In the Sousa basin basement, the secondary porosity and permeability of the rock could be markedly related to fracturing, but the fabric of chlorite and biotite, defining a lepidoblastic texture, should have improved this parameter in order to make easy the migration of the alteration fluid and, at last extension, the above mentioned oil.

The oil sample from the well located at Saguim Farm was analyzed through chromatography. It presents characteristics of a light oil, with 81.1% of saturated compounds and a predominance of C_{17} and C_{23} n-paraffins. The analysis of biomarkers indicates the presence of tricyclic and tetracyclic terpanes; the dominance of the $17\alpha(H)$; $21\beta(H)$; 30-Hopane (C_{30}); gammacerane/C_{30} $17\alpha(H)$; $21\beta(H)$; a 30-Hopane (C_{30}) ratio of 0.23; and a C_{30} $\alpha\beta$-Hopane/C_{30} $\beta\alpha$-Hopane (moretane) ratio reaching 80%. These aspects point to a nonbiodegraded mature oil from a lacustrine freshwater environment (Mendonça Filho et al. 2006) that would probably be in a context similar to the environments interpreted for the Sousa Formation.

In relation to the qualitative and quantitative analyses of particulate organic matter (kerogen) through microscopic techniques (palynofacies), in the geological context in which oil was found, it was observed that the organic matter assemblage presents a predominance of phytoclasts (woody material), a moderate percentage of amorphous organic matter, and a low to moderate percentage of palynomorphs. According to the geochemical parameters, the TOC (total organic carbon) and C/S ratios suggest predominantly oxic and brackish water conditions during sedimentation of the analyzed interval of the Sousa Formation. Rockeval pyrolysis verified the predominance of type II and III kerogen in the studied sedimentary section. The palynofacies and organic facies parameters indicate paleoenvironmental variations in the studied sections from oxic shallow lacustrine to anoxic-dysoxic brackish lacustrine. The thermal maturity parameters (spore color index, vitrinite reflectance, spectral fluorescence, and Tmax) show that the analyzed samples present an initial stage of thermal maturity. Integration of the geochemical and petrographical data also contributes to an assessment of potential for the generation of hydrocarbons. These results suggest a low liquid hydrocarbon generation potential but a moderate gas potential for the studied sedimentary section (Iemini 2009). Geochemical patterns of the oil samples recovered by rock solvent washing and the oil exudation sample present a good correlation and show a characteristic freshwater lacustrine depositional environment with different levels of biodegradation. Oil-rock correlation reveals similarities of geochemical parameters related to the depositional environment. Nevertheless, the maturation parameters point to an early stage of oil generation for the rock extracts in comparison with the oils (Sant'Anna 2009).

The siliciclastic rocks of the Rio do Peixe basins are mainly exploited for building and ceramics. Clays of the Sousa Formation are used in ceramics, and the conglomerate and sandstone of the Antenor Navarro and Rio Piranhas Formations serve as reservoirs for water. The availability of water in these aquifers allows the economic activity, both industrial and agricultural, that is very important for all the regions where these sedimentary deposits are found. Locally, the slabs of sandstones are used as flagstones and curbs for street pavement, for buildings, and for stone fences.

ABBREVIATIONS

Code number of a track: for example, SOSA 1: SO = municipality of Sousa;
SA = locality Saguim; 1 = track 1 of the locality

Ank	Ankylosauria
Bis	twice
CI	95% confidence interval
Croc	Crocodylia
CV%	coefficient of variation
FL	footprint length (cm)
GOr	graviportal Ornithopoda
h	skeletal hip height, i.e., height of the hip joint (cm)
h = 4FL	trackmaker hip height (cm) estimated by multiplying mean footprint length by 4
h = yFL	trackmaker hip height (cm) estimated by multiplying footprint length by y, where y is an allometric coefficient specific to each group or form of dinosaur (Thulborn 1990, tables 10.1–10.6).
Hs	half-swimming tracks
II^III	divergence between digits II and III
III^IV	divergence between digits III and IV
II^IV	(total) divergence between digits II and IV
Is	incertae sedis
km/h	speed entered in kilometers per hour
L/W ratio	length/width of a footprint
Lac	lacertoid
Length of digit	length of the free digit (measured proximodistally from the hypexes)
Lower	lower endpoint
LTh	large Theropoda
M	mean value
max	maximum value of the range
min	minimum value of the range
ML	midline
MTh	medium-size Theropoda
N	number of measured values
Q	quadrupedal
Sau	Sauropoda
SL/FL	ratio of mean stride length to mean footprint length
SL/FL ratio	stride length/footprint length
SL/h	ratio between mean stride length and height of the hip (relative stride length; Thulborn 1990, 259)
SL/h ratio	stride length/height of the hip joint
SOr	small Ornithopoda
STh	small Theropoda with long (and sinuous) III digit
Upper	upper endpoint
V	speed or velocity, in km/h
V_c	calculated trackmaker speed (km/h), following Thulborn (1990, 287–308)
V_t	theoretical or predicted trackmaker speed (km/h), obtained following Thulborn (1990, 294–95)
xFL	mean footprint length in the trackway (cm)
σ^{n-1}	standard deviation

The Ichnofaunas of the Rio Do Peixe Basins and Their Trackmakers

4

Collectively, the 37 ichnosites and the about 96 ichnofossiliferous levels of the Rio do Peixe basins contain the following counts of trackways assigned to different categories of dinosaurs: 329 large theropods; 31 smaller theropods having a third toe substantially longer than the other two toes; 5 additional, different kinds of small theropods; 16 medium-size theropods from Serrote do Letreiro (for a total of 381 individual theropods); 59 sauropods; 38 graviportal ornithopods (among them 4 quadrupedal and 1 sub-quadrupedal trackways, along with some isolated footprints, probably pertaining also to quadrupedal animals); 1 ankylosaur; 1 small quadrupedal ornithischian; 2 small ornithopods (altogether 42 ornithischians); and at least 53 indeterminate dinosaurian tracks. In total, the number of identifiable individual dinosaurs is 482, and the number of individual dinosaurs including the indeterminate tracks is at least 535. There are also 4 possible dinosaurian tail impressions.

There are 101 trackways assigned to herbivorous dinosaurs (21% of the identifiable individual trackways) and 381 trackways attributed to theropods (79% of the identifiable individual trackways); the ratio of herbivorous to theropod individual trackways in this dinosaurian ichnofauna is 1:3.77. The implications of this rather low ratio will be discussed below.

There are at least 65 trackways of quadrupeds (13.48% of the identifiable individual trackways) and 417 trackways of bipeds (86.51% of the identifiable individual trackways). The ratio of quadrupedal to bipedal trackways is 1:6.42.

The meso-ichnofauna, very rare in these basins, is represented by just one set of batrachopodid prints; crocodilian traces (tracks and body imprints in the mudstone); 1 lacertoid footprint; and a very large number of small chelonian swimming tracks (Leonardi and Carvalho 2000).

Sousa Formation

Altogether, the 21 sites from the Sousa Formation represent at least 66 stratigraphic levels that preserve an overall ichnofauna that is numerically dominated by isolated footprints and trackways of carnivorous dinosaurs. There are individual tracks of 296 theropods, 16 sauropods, 20 ornithopods, 38 bipedal dinosaurs of uncertain identity, and some tracks representing the mesofauna.

| Abreu (SOAB) | At least one large theropod track, with curved digit marks (fig. 4.1), occurs in the bed of the Peixe River between Lagoa do Canto and Abreu, or in the Sítio Abreu, in the Sousa municipality, about 8.5 km WNW from Sousa, as the crow flies. The rocks here pertain to the Sousa Formation and show mud cracks and a black-varnished surface. Information and photo were kindly provided by Carlos Antônio Leme and Luiz Carlos da Silva Gomes. We didn't observe it. |

| Araçá de Cima (ANAC) | Several theropod tracks (fig. 4.1) occur at the Araçá de Cima farm, in the municipality of São João do Rio do Peixe, on a greenish surface of the Sousa Formation (S 06 44.995, W 038 24.673). (Here and for the other ichnosites of the Rio do Peixe basins, the datum is WGS 84.) |

| Barragem do Domício (ANBD) | This site takes its name from a small dam nearby, Barragem do Domício, which itself is named after a former farm owner (fig. 4.2). This locality is 1 km E of the center of the small town of São João do Rio do Peixe, at the outskirts of town, some 50 m downstream from the small dam (S 06 44.165, W 038 26.288). At this site the outcrop is a wide siltstone exposure of the Sousa Formation, which here is a brick color, and it gently slopes to the bed of the Peixe River. The strike of the pavement is N88°E, and the dip is 10° to the south. This locality was discovered by Leonardi in 1984.

References: Leonardi 1985b, 1989a, 1989b, 1994; Leonardi et al. 1987a, 1987b, 1987c. |

ANBD 1

This is a short trackway of 3 footprints, the first and second of which are incompletely preserved. The footprints are tridactyl, mesaxonic, longer than wide; the toes are thin, and the third footprint presents some digital pads (see Conti et al. 1977, 6–7, and Leonardi 1987a for a discussion of the terms *mesaxonic, paraxonic, ectaxonic,* and *entaxonic* and of the axony of a footprint). Claw marks are sometimes present and, in one case (digit IV of the third footprint), bend outward from the fleshy portion of the digit impression. The pace angulation is relatively high (158°). The foot axis is almost parallel to the midline. The trackway is directed to N24°E. It is attributed to a medium-size theropod. It was left in situ (fig. 4.2).

ANBD 2

This is an incomplete trackway with 2 preserved footprints. The first is very incomplete and shows only the outline of a broad "heel"; the second footprint is missing; the third is complete but shallow and shows just the outline. It is tridactyl and mesaxonic and has strong digits; digit IV shows a claw impression. This trackway can be attributed to a medium-size theropod. The animal was going approximately N15°W. The trackway was left in situ.

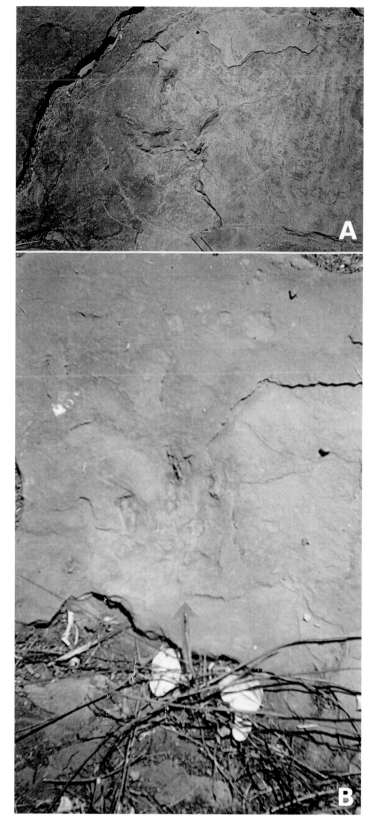

Figure 4.1. *A*, Abreu (SOAB; municipality of Sousa), and *B*, Araçá de Cima (ANAC; municipality of São João do Rio do Peixe). Sousa Formation. Two small localities, each with just 1 medium-size isolated theropod footprint. Courtesy of Luiz Carlos da Silva Gomes.

Figure 4.2. Barragem do Domício (ANBD). Municipality of São João do Rio do Peixe. Sousa Formation. *A,* Outcrop of red siltstones near the Domício dam and its small açude (artificial lake), with 3 theropod trackways showing a quick walking gait. *B,* The third footprint of the first trackway, ANDB 1/3. Graphic scale = 10 cm.

This is a trackway of 5 footprints. The first and the fifth footprints are incomplete; the others are complete but shallow and poorly preserved (probably undertracks). The footprints are tridactyl, mesaxonic, and symmetrical; the digits are broad. One claw mark is preserved, in digit III of the first footprint. The "heel" is broad. The pace angulation is moderately high (average = 163°). The footprint axes are parallel to the midline or directed a little inward with respect to the animal's direction of travel. The maker was directed N24°W. This trackway can be attributed to a large theropod. It was left in situ.

The 3 trackways consist of subtracks and are impressed in different but contiguous levels. Consequently, there is not a true association. All the trackmakers were medium-size to large theropods. All of the trackways are directed more or less northward.

Engenho Novo (ANEN)

This locality is situated on the Engenho Novo farm, on the dirt road São João do Rio do Peixe-Baixio and Catolé, close to where the road crosses Rancho Creek, some 5.5 km as the crow flies ENE of São João do Rio do Peixe (S 06 42.870, W 038 24.737), in the homonymous municipality. It was discovered by Leonardi in December 1984 and studied later by him and by Santos and Santos (1987a). The principal outcrop is a rocky pavement of about 18 × 7 m of surface on a gentle slope (13° to S; strike N78°W) crossed by the unimproved road. The rock is siltstone to fine sandstone of light chocolate color and a little crumbly. The upper surface of the bed presents some ripple marks. The lithology is not very typical of the Sousa Formation, and indeed it is situated not very far south of the boundary between the Sousa and the Antenor Navarro Formations.

Road traffic has abraded the surface of the pavement and damaged many of the footprints.

Another outcrop can be seen 100 m SSSW from this main outcrop, downstream, along the bed of the Rancho Creek. Beneath a concrete bridge over this creek and a little south of it a few tracks were found (dip 14° to the S; strike N75°W). This outcrop stratigraphically lies some 15 m above the main one.

All the footprints are shallow except for the sauropod hand-foot pair (ANEN 10). This specimen was unfortunately destroyed by vandals after discovery and study of the site.

All of the tracks of the main pavement had been left in situ, with the intention of protecting and stabilizing them in a small outdoor museum. The project we presented never came to fruition. Some plaster casts from the site are kept in the Câmara Cascudo Museum of Natal (Rio Grande do Norte, Brazil). A plaster cast of ANEN 1 is kept at the collections of the visitor center of the Dinosaur Valley Natural Monument in Sousa. Both ANEN 40 and ANEN 41 were given to the pro tempore mayor in 1985 to be kept in the town hall of São João do Rio do Peixe.

References: Leonardi 1985b, 1989a, 167, 1994; Leonardi et al. 1987a, 1987b, 1987c; Santos and Santos 1987a.

ANEN 1

This is a trackway of 6 tridactyl, mesaxonic footprints that are shallow but clear (figs. 4.3, 4.4A, 4.5, 4.6, 4.7). The footprints are small. The three toes are generally joined together at their proximal ends, but in the sixth footprint they are completely separated. The impression of digit III is the longest; the impressions of digits II and IV are more or less the same length. The interdigital angle between digits II and IV varies, but it is generally low. The second, third, and fifth footprints show digital pads. All footprints have claws marks except the first, which is incomplete. The trackway midline shows a curved path, and because of this the pace angulation varies. The stride length increases through the end of the trackway.

The trackway was attributed to the ichnogenus *Coelurosaurichnus* Huene, 1941 (now synonymized with *Grallator*, cf. Leonardi and Lockley, 1995), from the Triassic (Carnian) of Tuscany (Italy). Despite the shortness of the trackway, it is very interesting because in only 6 footprints it shows diverse enough characteristics that the individual prints would be classified as different "ichnospecies" of that ichnogenus if they were found separately; in reality, they are different due to different conditions of the substrate and the position of the foot in the footfall as the animal was cornering. So, for example, ANEN 1/2 is reminiscent of *Coelurosaurichnus* according to Baird (1957, in Haubold 1971, fig. 42/6); ANEN 1/3 of *Coelurosaurichnus sassandorfensis* Kuhn, 1958; ANEN 1/4 of *Coelurosaurichnus kehli* Beurlen, 1950 or *Coelurosaurichnus moeni* Beurlen, 1950; and ANEN 1/6 of *Coelurosaurichnus toscanus* Huene, 1941 itself. The digit III impression is sometimes completely straight (print 1/3), sometimes slightly curved (1/2), and still other times more or less strongly

Figure 4.3. Map of the track-bearing pavement of the Sousa Formation, at the first subsite (first ichnofauna) of the Engenho Novo farm (ANEN), at São João do Rio do Peixe. The ichnofauna contains numerous theropod tracks, 1 sauropod hand-foot set, and a long trackway of an apparently bipedal dinosaur, attributed to a large ornithopod or, more probably, to a sauropod whose hind-foot tracks overprint its forefoot tracks. Draft of a map by Santos and Santos, in preparation for their 1987 publication about this locality. See also figures 4.4–4.8. Graphic scale = 1 m.

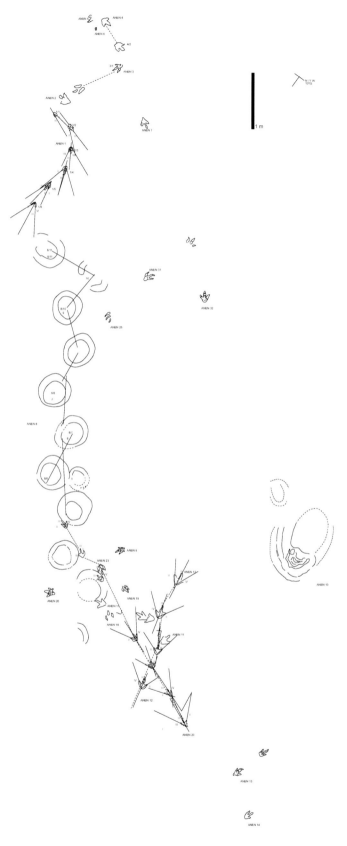

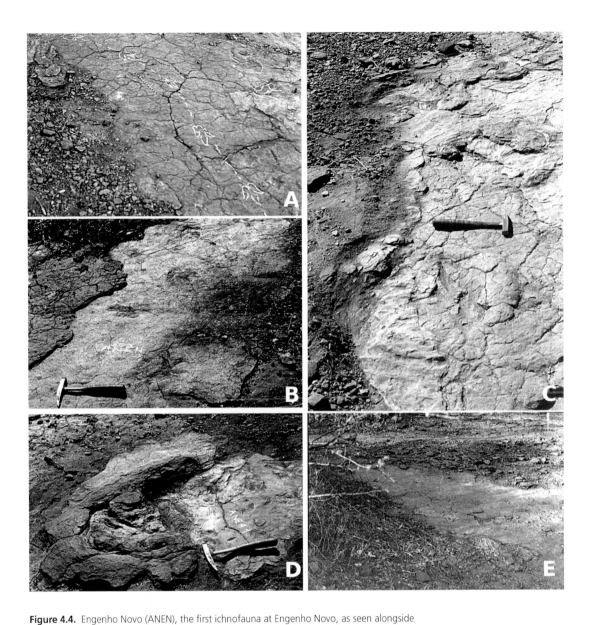

Figure 4.4. Engenho Novo (ANEN), the first ichnofauna at Engenho Novo, as seen alongside the road. *A*, A curvilinear trackway of a small theropod, ANEN 1. *B*, The rocky pavement at Engenho Novo; *C*, A large and long trackway, ANEN 8, probably made by a seemingly bipedal (or maybe semibipedal) large herbivore. This trackway was initially attributed to an ornithopod, albeit with reservations because the footprints lacked morphological details; on the other hand, the trackway showed some similarities with other trackways that had been attributed, with good reason, to ornithopods. However, the trackway is now thought more probably to have been made by a sauropod. *D*, An important, rare (in the whole Rio do Peixe basins ichnofaunas) and unfortunately incomplete hand-foot set of a large sauropod (ANEN 10). The footprints were already broken at the moment of discovery in 1984 and are now destroyed. In the Rio do Peixe region, sauropod footprints are generally very badly impressed and difficult to identify as such. In this case, to the contrary, the ichnite was a very typical set that was easy to classify as sauropod. The hand in particular shows the typical crescentic outline. The displacement rim is broad and high; *E*, Another group of isolated theropod footprints from the same locality.

Figure 4.5. *(above)* Engenho Novo (ANEN). The second ichnofauna at Engenho Novo, the latter of which was excavated near a small bridge crossing Rancho Creek. *A,* Carrying the heavy slab containing trackway ANEN 40. *B,* Leonardi working with chisel and hammer to trim specimen ANEN 40/1. *C,* ANEN 41, an incomplete theropod footprint at the Rancho Creek bridge. *D,* ANEN 40/1, the first footprint of a short trackway attributed to a large theropod, at the Rancho Creek bridge. Natural cast, left pes. Note the very long claws on toes, with a long scratch mark extending backward from the tip of toe III.

Figure 4.6. *(below)* A sauropod and its hand-foot set. Art by Ariel Milani Martine.

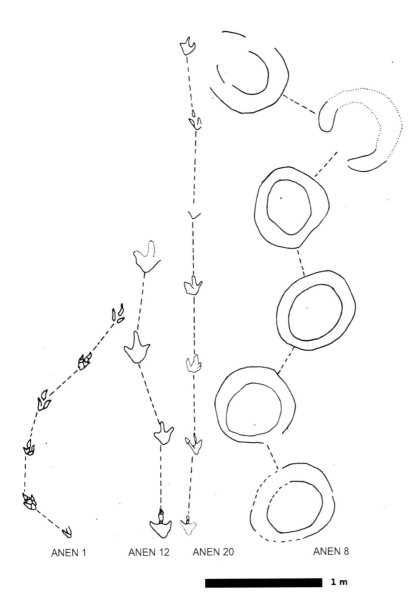

Figure 4.7. Trackways from Engenho Novo (ANEN). ANEN 1: a curvilinear trackway of a small theropod. ANEN 8: the trackway of a large-size herbivorous dinosaur, probably a sauropod, or possibly an ornithopod. ANEN 12 and 20: 2 theropod trackways along the dirt road. All figures are redrawn from Santos and Santos (1987).

ANEN 1   ANEN 12   ANEN 20   ANEN 8

1 m

crooked (1/4, 1/5, 1/6) as the animal increased its speed. The angle between the toes also varies greatly among the prints in this trackway. This highlights the futility of naming taxa on the basis of isolated footprints (Razzolini et al. 2014; also see chap. 3, Tracks and Substrate).

This trackway, like several specimens of this or similar kind in other Peixe River outcrops and especially at the Piau-Caiçara (SOCA) site, level 13, pertains to the *Grallator-Eubrontes* plexus. This fact contributes to confirming the wide temporal and geographic distribution of this plexus (Xing et al. 2015c).

ANEN 2

This is an isolated tridactyl footprint, probably attributable to a theropod (fig. 4.8). The toes are very short compared with the "heel" (the region

Figure 4.8. Theropod footprints and trackways from Engenho Novo (ANEN). ANEN 18 was made by a half-swimming dinosaur. Redrawn from Santos and Santos (1987a).

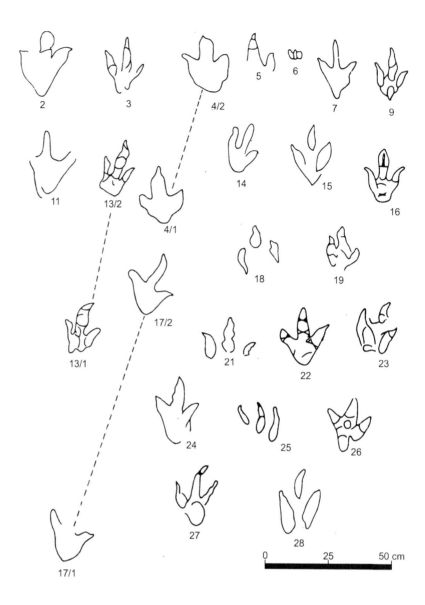

of the print proximal to the individual toe marks), which is long and wedge-shaped.

ANEN 3

This trackway features a single tridactyl footprint of a small theropod (fig. 4.8), with wide interdigital divergence; the digit III impression is longer than the outer ones. There is a single digital pad in one of the outer toes and two such interdigital pads in the third digit.

ANEN 4

This is a short trackway of 2 tridactyl footprints (fig. 4.8), with stout toe marks with distinct claw marks. The track can be assigned to a large theropod.

ANEN 5

An incomplete, probable theropod footprint (fig. 4.8).

ANEN 6

A small, unclassifiable print lacking morphological details (fig. 4.8).

ANEN 7

This is a small, tridactyl, probably theropod footprint, probably a left, with a long, thin, slightly curved digit III impression (fig. 4.8); the mark of digit IV toe is relatively short and blunt. It is probable that the mud collapsed inside the digit III impression as well as the distal part of the digit II impression.

ANEN 8

This is a long, badly impressed trackway, by appearances made by a large biped—perhaps an ornithopod or, more probably, a sauropod (figs. 4.4C and 4.7). There are 11 circular footprints, without morphological details. The prints have large but shallow displacement rims. The values of the pace angulation and of the stride fluctuate, and the pace angulation drops between footprints 10 and 11, indicating a probable change of direction. Between footprints 4 and 5 there is a small footprint of the same general style, which seems to belong to the same trackway. If the trackway is attributed to an ornithopod, this could be a case of semibipedality; if it is and, likely more correctly, attributed to the Sauropoda, it could be a case of total overlap of the foot on the hand, except in the case of just this footprint. The putative manus print seems relatively too large to be a handprint of an ornithopod. In the diagrams (fig. 5.6B; Santos and Santos 1987a), this trackway seems to fit better among the sauropodian tracks.

One cannot determine the trackmaker's direction of travel but only the trackway's linear orientation. In this case, the difficulty of distinguishing between sauropods and ornithopods derives specifically from the absolute absence of morphological details. In other cases of such mistaken identity, the confusion arises from a wrong observation of the material in the field, as in the case of the original classification of the seven large parallel trackways of the Corda Formation (Leonardi 1980e, 1994; Lopes et al. 2019). An analogous and particularly striking case of misidentification was that of *Iguanodonichnus frenki* from the Upper Jurassic at the Baños del Flaco Formation, in the Baños del Flaco locality, Chile; the trackways were first classified as iguanodontid trackways by Casamiquela and Fasola (1968). However, more recently they were redescribed, reclassified, and quite correctly attributed to sauropods by Karen Moreno and Michael J. Benton (2005).

ANEN 9

A single theropod footprint, probably a right, with long and thin toes, and with one digital pad in the digit III impression (fig. 4.8). In the proximal portion of the "heel," one pad is probably the first pad of digit IV. The digital divergence II^IV is low.

ANEN 10

An important, locally rare, unfortunately incomplete (broken), and, to-day, destroyed hand-foot pair of a large sauropod (figs. 4.3, 4.4, 4.6). In the Peixe River region, sauropod footprints are generally very badly impressed and difficult to identify as such. This specimen, in contrast, is a very typical pair and very easy to classify as sauropod. The manus print is particularly typical with its crescent outline and was almost completely obliterated by the pressure of the displacement rim of its foot, which is large and high. It is impossible to attribute this pair to a trackmaker at the genus or even family level.

ANEN 11

An incomplete, perhaps theropod footprint (fig. 4.8). The "heel" is long and wedge shaped. The toes are short.

ANEN 12

This is a theropod trackway with 4 footprints (fig. 4.7). In the first foot-print, one can see two digital pads and a claw in the digit III impression. The total divergence between digits II^IV is wide. The trackmaker was slightly decelerating.

ANEN 13

This is a short trackway of 2 footprints (fig. 4.8). The shape of the "heel" differs between the first and second footprints. The trackway is attribut-able to a theropod and is similar to *Grallator*, cf. *Grallator variabilis* Lapparent and Montenat, 1967.

ANEN 14

A tridactyl footprint, perhaps theropod, with the impression of digit III toe as long as one of the outer toes, but longer than the other (fig. 4.7). The total divergence between digits II^IV is low. The footprint is in apparent convex epirelief because of an inversion of relief, probably caused by unequal compaction of the substrate and subsequent erosion.

ANEN 15

A tridactyl, wedge-shaped footprint attributable to a theropod, with com-pletely separated and only partially impressed toe marks (fig. 4.8). One of the toes has higher relief than the surrounding rock, as described for ANEN 14. The total interdigital divergence is high.

ANEN 16

A tridactyl footprint attributable to a theropod, with pointed and curved impressions of digits II and IV (fig. 4.8). The distal portion of digit III shows a medial, antero-posterior ridge, which can be interpreted as the impression of a furrow or wrinkle of the toe.

ANEN 17

A trackway of 2 consecutive footprints of a theropod, very similar to ANEN 12 (fig. 4.8). The first footprint was partially overlapped by ANEN 24. The second overlaps the displacement rim of footprint ANEN 8/2.

ANEN 18

This trackway comprises three associated elongated marks, similar to scratches, that must be interpreted as a footprint made by a half-swimming theropod (fig. 4.8). If so, this would have been the first footprint impressed in this pavement, before the water level in the temporary lake dropped. The depth of the water at the time the print was made would have been about 1.5 m.

ANEN 19

A small, tridactyl, theropod footprint (fig. 4.8).

ANEN 20

This is a rectilinear, probably theropod trackway with 8 footprints, with high interdigital divergence (fig. 4.7). The divarication of the footprints from the midline is negative. From the morphologic point of view, the trackway is similar to those of the *Eubrontes-Grallator* plexus. The track-maker maintained a constant cruising speed.

ANEN 21

An incomplete footprint, attributable to a large theropod, with wide digital divergence (fig. 4.8).

ANEN 22

A theropod footprint reminiscent of *Irenesauripus* Sternberg, 1932 and/or *Columbosauripus* Sternberg, 1932 (fig. 4.8).

ANEN 23

An incomplete footprint of poor quality, probably theropod (fig. 4.8).

ANEN 24

A footprint attributable to a large theropod (fig. 4.8).

ANEN 25

A footprint attributable to a half-swimming theropod (fig. 4.8).

ANEN 26

This is a left footprint of a theropod (fig. 4.8). The base (proximal end) of the digit III impression shows a rounded feature that can be interpreted as a callosity. This footprint is similar to the theropod prints of Serrote do Letreiro (SOSL) attributed to *Grallator* or *Eubrontes*.

ANEN 27

A thin-toed theropod footprint, probably a right (fig. 4.8). The distal end of the digit III mark is bent slightly outward.

ANEN 28

A footprint attributable to a theropod with separated toes (fig. 4.8).

ANEN 40 and ANEN 41

About 100 m S of the new bridge over Rancho Creek, the following material was found:

ANEN 40 (fig. 4.5) is a short trackway of 2 large footprints, attributable to a very large theropod. The first footprint was found, as a natural cast, on a large slab (about 240 × 50 × 14 cm) of medium to coarse sandstone. Such preservation is rare in the Sousa Formation, but, as previously noted, this site is at the boundary with the Antenor Navarro Formation.

The slab had been detached and overturned but left almost in place by a flood during the rainy season. The other footprint was uncovered in situ. The first footprint is probably a left footprint, 40 cm long and 27 cm wide; it is tridactyl, with stout toes and long, sharp claws, which have probably been extended by scratches. Digit IV is narrower than the others, with a shorter claw. It shows only one digital pad. Digit III is long and distally broad, with one well-developed pad. Another small pad or callus can be seen between the hypexes. The digit II impression is short and broad. It shows one stout distal pad, a medial callosity, and a large, well-developed pad in the "plantar" region. The "heel" is rounded. The divarication of digits II^III (15°) is larger than that of digits III^IV (11°); the total divarication is very low (25°). Digits III and IV are closer together than are digits II and III. Both hypexes are rounded. Medially, laterally, and posteriorly, the footprint is surrounded by a shallow displacement rim. The second footprint is an incomplete right, very similar to the first print. The oblique pace is very high (1.70 m). The direction of motion of the trackmaker of this very narrow trackway was about S40°W.

This trackway is not easily comparable to any known and named form. The very low digital divergence, the relative shortness of the digit III, and the presence of the medial callosity are reminiscent of "*Bückerburgichnus*" *maximus* Kuhn, 1958 from the Lower Cretaceous of Germany.

ANEN 41 (fig. 4.5) is a single incomplete theropod footprint with a stout digit III toe impression and one other toe mark, found 140 cm S of ANEN 40, heading in the same direction.

The first ichnofauna (ANEN 1–28; figs. 4.3–4.8), found on the road rocky pavement, is richer and older than the second one (to be discussed below). Summing up, it includes:

- 24 tracks attributable to Theropoda, 5 of which are attributable to small theropods with a long III digit, 10 of which pertain to large theropods
- 1 hand-foot pair attributable to Sauropoda
- 1 trackway attributable to Sauropoda or, less probably, to Ornithopoda
- 2 indeterminate dinosaurian footprints

The difficulty of more exactly classifying many theropod tracks is due to the poor quality of the impressions and the damage they suffered from road traffic. Whatever the exact systematic affinities of the trackmakers, the theropod footprint assemblage was made by rather small-size animals compared with the other assemblages of the Rio do Peixe basins. The average length of the footprints at this site is 22 cm; the standard deviation is low (average = 3.18); no footprints longer than 27 cm are recorded. The digit III impression is frequently long and curved or sinuous.

Some of the tracks were impressed when the surface was flooded (the water was ≈1 m–1.5 m deep) and were made by dinosaurs that were half-swimming (ANEN 18, 25, ?28). Others were recorded later, when the water had subsided but the mudflat remained wet, with the mud consistently yielding, so the reptiles produced continuous trackways (ANEN 1, 8, 12, 20). Other tracks were recorded still later, when the mudflat was partially dry, and the tracks occur as short trackways (ANEN 4, 13, 17) or isolated footprints recorded in the last puddles or soft spots of mud.

The sauropod hand-foot pair was impressed when the mud was deep and yielding, producing a wide and deep displacement rim. One would expect to find a complete trackway instead of an isolated pair. Very probably the other footprints were destroyed by human activity, just as the one recently found was vandalized after its discovery.

A remarkable feature of this footprint association is the large number of theropod (24; 92.3%) individual footprints and trackways and the very small number of tracks of herbivorous dinosaurs (2; 27.7%). This is, however, a rather common situation in the Rio do Peixe basins, as it is for dinosaur footprint assemblages throughout South America.

The directions of the tracks of this ichnosite are randomically scattered in many ways, occupying almost all of the 20° sectors in a rose diagram. Ten of them are concentrated in both ways of a direction NNE–SSW; other modes are in the sectors 260°–280° (4 individuals) and 320°–340° (3 individuals). Six individual directions are comprised in the I quadrant, 5 in the II, 9 in the III, and 7 in the IV. No social behavior can be demonstrated.

Both theropod tracks of the second ichnofauna, found downstream (ANEN 40 and 41; fig. 4.5A–D), are similar between them and go approximately the same way (about northward).

Juazeirinho (ANJU)

The Juazeirinho farm locality is in the municipality of São João do Rio do Peixe, at the side of an old, unimproved road from Sousa to São João do Rio do Peixe, 2.3 km E of São João do Rio do Peixe, at a place locally known as Rio dos Tanques (S 06 44.685, W 038 25.144). The Peixe River has here exposed a 2-m-thick section of the Sousa Formation. In 1977 Leonardi found an isolated footprint in situ at this site as well as a short trackway of 2 footprints in a large, detached slab, both of which are attributable to large theropods. Leonardi carried out excavations in 1984, resulting in the discovery of 2 more trackways at a different stratigraphic level.

The Sousa Formation is here, as always, represented by alternating siltstone and massive mudstone or marl beds. Some mud cracks and ripple marks are present on the siltstone layers. The 3 trackways and the isolated footprint come from three different levels, so their occurrence at this locality is not a true association.

References: Leonardi 1979c, 1980b, 1981b, 1985b, 1989a, 1989b, 1994; Leonardi et al. 1987a, 1987b, 1987c.

ANJU 1

An isolated tridactyl, mesaxonic footprint with stout toes, clear digital pads, and distinct triangular claws (figs. 4.9A–4.10). The total digital divergence (between digits II^IV) is relatively high (56°). The divergence between digits III^IV is less than that between digits II^III, unlike the case of SOES 11. The footprint is longer than it is wide. Digit III is separated from the other two toes.

The slab bearing this in situ footprint was cut away from the layer and displayed in front of the headquarters of the municipality of the town of Sousa, along with the slab containing ANJU 2 (which had become detached from the rock by natural processes), at the initiative of a former mayor of Sousa, which led to a controversy between the municipalities of São João do Rio do Peixe and Sousa from 1980 through 1984. Both slabs containing ANJU 1 and ANJU 2 are now kept in front of the museum at the Dinosaur Valley Natural Monument.

ANJU 2

A short but fine trackway of 2 footprints, the first preserved as a subtrack, the second well impressed and well preserved on a mud-cracked bed surface (figs. 4.9B–4.10). The footprints are tridactyl and mesaxonic, with stout toes and clear digital pads: three on both digits II and III and three or four on digit IV. The claw marks are clear, deep, and trapezoid-shaped; those of the outer toes bend slightly outward. The total interdigital divergence (between II^IV) is high (average = 69°). The posterior margin of the footprint is bilobate, with an indentation clearly separating digit II from digit IV. Digit III, as always, projects well beyond the tips of the other two toe marks and is closed between the outer toes. The average ratio of footprint length/width is about 1.

The dinosaur was taking relatively long steps. If the pace from the second footprint to the (not preserved) following print was comparable to

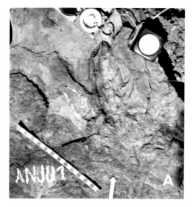

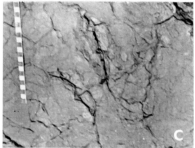

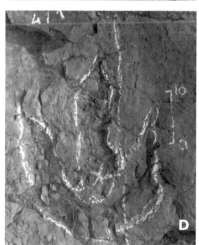

Figure 4.9. Juazeirinho (ANJU). In 1988, 2 tracks were discovered on two different levels at Juazeirinho (municipality of São João do Rio do Peixe, Sousa Formation); other prints were excavated in 1984. *A*, ANJU 1, an isolated theropod track with wide digital divarication, preserved as a natural cast. *B*, The short trackway ANJU 2, with the same morphological features as ANJU 1, on the upper surface of a detached slab. Both specimens are now displayed in the museum of the park in Sousa. *C*, ANJU 2/1, the first footprint of trackway ANJU 2. *D*, ANJU 4/1, the first footprint of trackway ANJU 4. *E*, The excavated upper level at Juazeirinho in 1984, with the shallow theropod trackway ANJU 4. Graphic scale in centimeters.

that between the preserved first and second footprints, the stride length/ footprint length (SL/FL) ratio of the trackway would be about 6; this value suggests a rather quick-walking trackmaker. Given the relatively long step, one can assume that the pace angulation of this trackway would be high (~170°) and the width of the trackway, as a consequence, low. The footprints turn slightly inward (-9°30').

The trackway was found on a detached slab, and so the animal's direction of travel is not exactly known but was probably approximately N60°W. The specimen is reposited with ANJU 1, as already described.

Both the ANJU 1 and ANJU 2 tracks can be attributed to large theropods of the same kind. They are similar to 2 trackways, also of 2 footprints,

Figure 4.10. Tracks from the localities of Matadouro (SOMA; figs. 4.12–4.15), Piedade (SOPI; figs. 4.58–4.60), and Pedregulho (SOPE; figs. 4.17 and 4.59F) in the municipality of Sousa, and from the site of Juazeirinho (ANJU) in the municipality of São João do Rio do Peixe, all from the Sousa Formation. The tracks from Matadouro (SOMA 1, 2, 3) and from Juazeirinho (ANJU 1, 2) pertain only to theropods; the tracks from Piedade were made by both theropods (SOPI 1, 2, 3) and ornithopods (SOPI 3, 4); those from Pedregulho (SOPE 1, 3) pertain to theropods. SOPE 1 is one of the largest theropod tracks from the Sousa basin. From Leonardi 1979b. Graphic scale 10 cm.

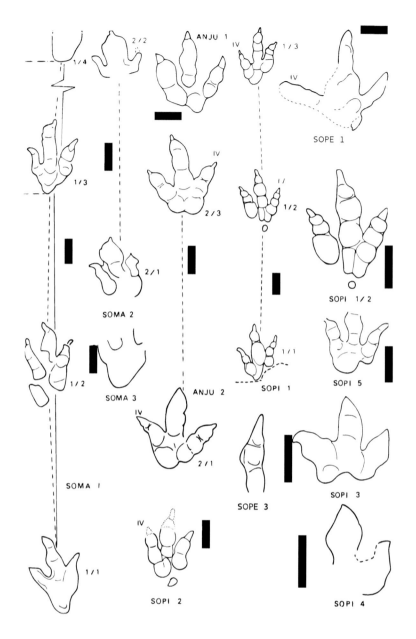

classified as *Irenesauripus cf. mclearni*, catalog numbers 8558 and 8567 in the collections of the Paleobiology Division of the Canadian Museum of Nature in Ottawa, Canada, from the Lower Cretaceous of the Peace River, British Columbia (Canada). However, our tracks can be better classified as *Eubrontes* (*Brontozoum*) *divaricatus* (E. Hitchcock, 1865).

A large metatarsophalangeal area is commonly observed in large theropod footprints from post–Lower Jurassic strata. The metatarsophalangeal pads of some Jurassic *Eubrontes*-type tracks are aligned with the axis of digit III, and this feature appears to be common in Lower Cretaceous theropod (*Eubrontes*-type) tracks from China (Xing et al. 2014b).

ANJU 3 and ANJU 4

These 2 trackways occur in the same siltstone level, 16 m apart, and were excavated in 1984. ANJU 3 heads S44°E and ANJU 4 N17°E. Both trackways are poorly impressed and very probably are subtracks. They were left in situ.

ANJU 3 is a poorly impressed and badly preserved trackway of 4 extant footprints; the second footprint in the sequence is missing. The first and third footprints are incomplete. The footprints are tridactyl, mesaxonic, and relatively wide. The digital divergence is rather high. The digits are broad (probably because these are subtracks). Claw marks can be seen in digit III of the fifth footprint and in digit II of the fourth. No claw marks are preserved in the other footprints, or any other morphological details. The pace angulation is relatively low (average 141°). The gait was generally pigeon-toed. The trackway is attributed to a large theropod.

ANJU 4 is a short trackway of 3 large footprints, shallowly impressed (probably subtracks) and badly preserved (figure 4.9D–E). The third footprint is incomplete; the first two are better preserved but are very different from each other. Both show impressions of sharp claws, but the first footprint has relatively long toes and a short and broad proximal region, behind the hypexes, while the second print seems to have digits with relatively short free lengths and a long and wedge-shaped "heel." These differences are surely due to different substrate consistencies where the two feet trod. The pace angulation is moderately high (159°). This trackway is attributed to a large theropod.

All the dinosaurs of this small ichnofauna, distributed in three levels, were theropods of medium to large size. No group behavior is recorded. The directions of travel of the dinosaurs are in the I and II quadrants (respectively 0°–90° and 91°–180°).

Baixio do Padre (SOBP)

In the municipality of Sousa, at the locality Baixio do Padre, very near the hamlet of Riachão dos Anísios, alongside the old tracks of an abandoned narrow-gauge railway (S 06 45.113, W 38 19.993), an ichnolocality was discovered by Leonardi, Maria de Fátima C. F. dos Santos, and Tibério Felismino de Araújo in the bed of Riachão dos Anísios, a creek tributary of the Peixe River. The discovery was made in terrains of the Sousa Formation at the end of a fruitful expedition to the Potiguar basin (Rio Grande do Norte, Brazil) in January 2008. One may get to this site by turning left, at S 06 45.113, W 38 19.993, from the old, unimproved Sousa-São João do Rio do Peixe road, soon after leaving the hamlet of Caiçara.

The ichnofauna consists of a theropod trackway (SOBP 2, not studied because it was under water) and a trackway (SOBP 1; fig. 4.11) of a large, graviportal ornithopod, classifiable as *Caririchnium magnificum*, with four excavated hind-foot prints. Forefoot prints are not imprinted—or, more probably, are not preserved in this trackway because of the poor condition of the rocky surface. The main trackway has the following

Figure 4.11. Baixio do Padre
(SOBP), Sousa, Sousa Formation. *A*, A trackway classifiable
as *Caririchnium magnificum*
and attributed to a quadrupedal ornithopod. *B*, A single
large footprint of the same. A
theropod trackway is associated with it. Graphic scale in
centimeters.

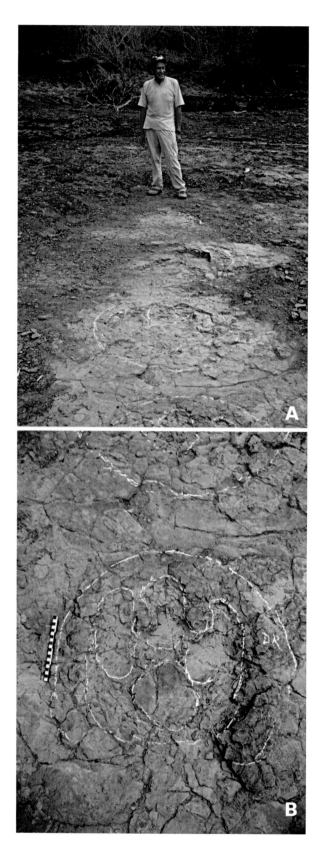

mean parameters (in cm): footprint length 54; footprint width 50; length of the footprint including the displacement rim 70; width of the footprint including the displacement rim 71; free length of digit III 20; "sole" length 34; "sole" width 30; length of pace 104; stride 192; pace angulation ≈145°. The trackway is otherwise unpublished.

Matadouro (SOMA)

This locality is situated in the bed of the Peixe River, at the northwestern margin of the town of Sousa, immediately behind the municipal slaughterhouse (S 06 45.113, W 038 13.716). At this site, the rocky pavement is frequently covered by heaps of manure, so it is not so easy to examine the trackways, as was most definitely the case in 1979, the year of the site's discovery by Leonardi. Three contiguous layers of chocolate-colored siltstones, with black surfaces, crop out here. The strike is parallel to the present river direction (N82°W; dip 7° to the south). The track-bearing level has primary and secondary mud cracks. The contiguous lower layer shows ripple marks with crests oriented N–S.

References: Leonardi 1980b, 1981b, 1989a, 1994; Leonardi et al. 1987a, 1987b, 1987c.

The main layer crops out on a 40 × 2.5-m surface, presenting a small ichnoassociation of bipedal dinosaurs, very probably all of them large theropod tracks (figs. 4.12–4.16).

SOMA 1

This is the best-preserved trackway in the outcrop and the longest, with 4 uncovered footprints, in a length of 3.76 m (figs. 4.10, 4.12, 4.13, 4.15). The first 3 footprints were probably impressed in the upper layer and are very well impressed and preserved; the fourth print is a shallow, incomplete undertrack in the lower layer.

The footprints are longer than they are wide, with long, stoutly constructed toes that end in long but blunt (truncated) claws. The proximal portion of the fourth toe forms a "heel," sometimes divided from the other digits, so the proximal outline of the footprint is bilobate. The total digital divergence between digits II and IV is low (average 58°15'), but the distal tips of the the outer digits bend outward. In the first and fourth footprints the toes join at their proximal ends, in the second print they are totally separated, and in the third print they are partially united. This reflects the changing consistency of the mud as it resulted in different footprint depths along the trackway. The digits show three, three or four, and five pads on the second, third, and fourth digits, respectively.

Although the footprints in this trackway are generally similar in morphology, there are interesting minor shape variations from one to the next:

1. The tip of the claw mark of digit III points inward in the first and second footprints (SOMA 1/1 and 1/2) and outward in the third foorprint (SOMA 1/3).

2. The tip of the claw mark of digit IV is directed forward in footprints 1/1 and 1/2 and outward in footprint 1/3. These features point to the mobility of the toes.
3. The impression of a large proximal digital pad occurs in footprint 1/2 but is absent in the other footprints.
4. The claw marks in footprints 1/2 and 1/3 are completely separated from the distal pads, whereas the boundaries between the claws and the pads are indistinct in print 1/1.

All of these differences are due to different conditions of the substrate at different spots along the trackway surface.

This is the trackway of a bipedal dinosaur, a theropod of medium-large size, with very high pace angulation (178°) and a very narrow trackway width, a relatively short stride (cm 162), and relatively low SL/FL ratio (5.3). The data indicate a dinosaur moving in a rather fast walk. The divarication of foot from the midline is variable but angles slightly outward. The direction of travel of the trackmaker was N6°W. This trackway is strongly reminiscent of a trackway (catalog number 8563, Canadian Museum of Nature) assigned to the ichnogenus *Irenesauripus*.

SOMA 2

This is a trackway of two excavated footprints from a bipedal dinosaur (figs. 4.10, 4.12, 4.14A–C). The footprints are stoutly constructed and relatively short (average footprint L/W ratio = 1.07). Digit III is much longer than digits II and IV, which are short and slender. However, the considerable difference in appearance between digit III and the others is probably due more to the way they were impressed than to any morphological differences in the trackmaker's toes. All digits show claw marks.

The SL/FL ratio of the trackway would be rather high (approximately 7.5), suggesting that the trackmaker was walking rather quickly. The animal proceeded in the direction N12°E. The prints are attributable to a medium-size theropod.

SOMA 3

Another trackway of a bipedal dinosaur, represented by 2 excavated footprints (figs. 4.10, 4.12, 4.13D). The prints were deformed by mud collapse and are poorly preserved but nevertheless can be attributed to a large theropod. The animal was walking quickly in a S4°E direction.

SOMA 4

An isolated, incomplete, shallow footprint, probably an undertrack (fig. 4.12). The long axis points S51°W. Because of the relatively long outline of the print and its association with other trackways of carnivorous dinosaurs, it could also have been made by a large theropod.

The SOMA trackways comprise a small ichnoassemblage of bipedal dinosaurs, all of them theropods (fig. 4.16). The tracks belong to

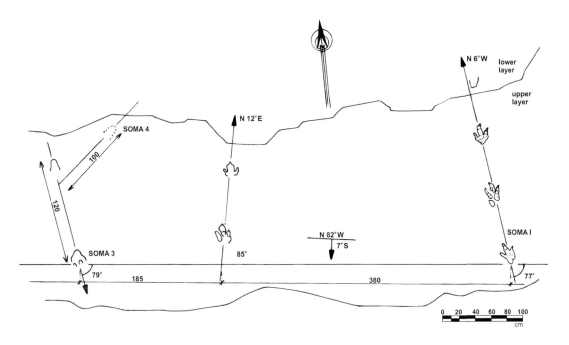

Figure 4.12. *(above)* Locality of Matadouro (SOMA; figs. 4.10, 4.13–4.16), municipality of Sousa, Sousa Formation. The outcrop is in the town of Sousa, alongside the municipal slaughterhouse, in the bed of the Peixe River. All 4 trackways at this site are attributable to theropods. Survey and drawing by Leonardi.

Figure 4.13. *(left)* A fine theropod trackway from Matadouro (SOMA 1). It was discovered underneath 2 m of manure from the slaughterhouse.

103

Figure 4.14. Additional photographs of footprints and trackways from Matadouro (SOMA). *A,* View of the trackway surface alongside the slaughterhouse. *B,* The first footprint of trackway SOMA 2 (SOMA 2/1). *C,* The second footprint of SOMA 2 (SOMA 2/2). *D,* Shallow trackway SOMA 3. Graphic scale in centimeters.

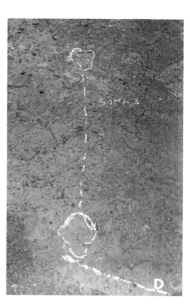

4 theropods, attributable to different ichnogenera, that moved as single individuals (no gregarious behavior) across the site. Three of the 4 trackways are oriented approximately N–S, parallel to the crests of the ripple marks in the underlying contiguous layer. Two trackways (SOMA 1, 2) are directed to the north and the third (SOMA 3) to the south. All of them belong to different species and probably to different genera. The maker of SOMA 4, the fourth trackway, proceeded diagonally with respect to the other 3 trackways. We have here 4 individual dinosaurs that walked across the site one by one. The distance separating the tracks is small, with a concentration of 4 trackways in approximately 15 m² — that is, 1 individual per 3.75 m², or 0.27 individual per square meter.

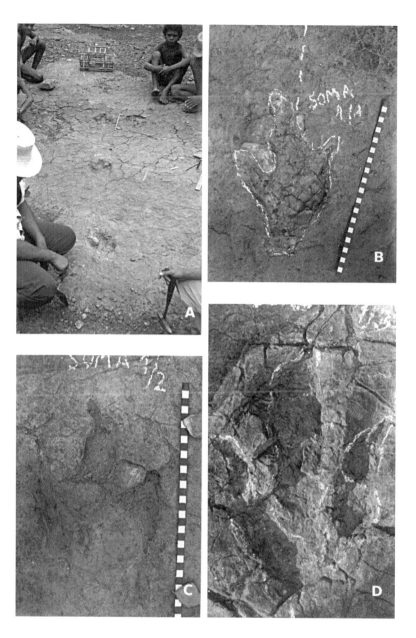

Figure 4.15. More photographs of the Matadouro (SOMA) tracksite. *A*, Trackway SOMA 1. *B*, The first footprint of the same trackway (SOMA 1/1, left pes). *C*, The second footprint of the same trackway (SOMA 1/2, right pes). *D*, The third, lighted laterally using a mirror (SOMA 1/3, left pes). See also figure 4.13. Graphic scale in centimeters.

These outcrops are situated at the border between the municipalities of Sousa and São João do Rio do Peixe, at the place where the old Sousa-São João do Rio do Peixe dirt road fords the Peixe River (S 06 45.371, W 038 20.911; fig. 4.17A). At this site a bank of layers of the Sousa Formation forces the river to flow over a modest waterfall or, better, a small system of rapids (fig. 4.17B). The rock layers are alternating siltstone and mudstone, red (brick) color, with beds containing abundant and gigantic conchostracans. This locality was discovered by Leonardi in 1977.

References: Leonardi 1979c, 1980f, 1981a, 1989a, 1989b, 1994; Leonardi et al. 1987a, 1987b, 1987c.

Pedregulho (SOPE)

Figure 4.16. Environmental reconstruction of the Matadouro (SOMA) tracksite, by Ariel Milani Martine.

SOPE 1

This is a very large footprint—tridactyl, mesaxonic, wider than long, badly eroded but clearly attributable to a very large theropod (figs. 4.10, 4.17C). It has a very high digital divergence II∧IV (105°). The toes are very long and stoutly constructed. Because of the poor state of preservation, neither claw marks nor digital pads are visible. There is a large and deep concavity at the base of digit III, where the sediment experienced the maximum stress of the animal's weight. However, this depression could be just a random irregularity in the rock eroded surface. The footprint L/W ratio (0.88) is extremely low for an archosaur, especially for a theropod. This is the second largest theropod footprint in the Sousa basin (and in Brazil), exceeded in size only by the very large footprint SOCA 81; generally, Sousa basin theropod footprints are between 20 and 35 cm in length. The footprint points to N44°E. This track could not be found in 2008 and was very probably completely eroded by the river.

SOPE 2

A small theropod footprint.

SOPE 3

This is an isolated and incomplete toe mark with three pads and a claw; the preserved portion is 15 cm long. The print can perhaps be attributed to a large theropod (fig. 4.10).

Altogether, this site shows, or at least once showed, a very large theropod footprint worn by the river and two smaller theropod footprints.

Figure 4.17. Pedregulho Ranch (SOPE; see also fig. 4.10). The outcrops are situated in the bed of the Peixe River, at the border between the municipalities of Sousa and Sao João do Rio do Peixe, Sousa Formation. *A*, The site and some local children (1977). *B*, The crossing of the Rio do Peixe, where a sequence of layers of the Sousa Formation gives rise to a modest waterfall or, more accurately, a small system of rapids in the rainy season. Photo courtesy of Franco Capone. *C*, A very large theropod track, worn by the river when first studied (1977) and now completely eroded away. Graphic scale in centimeters.

Nothing else was found in the adjacent wide, rocky pavements to the north and northeast. The footprints were left in situ.

This is a very interesting locality because there are at least twenty-five levels with dinosaurian ichnoassociations (figs. 4.18–4.57) in a rather small area. A large number of dinosaur track-bearing levels at particular sites have also been found elsewhere, most notably South Korea (Paik et al. 2001, 2006).

Piau-Caiçara is situated at the border between the Piau and Caiçara farms, in the municipality of Sousa, nearly 10 km WNW from Sousa. Access is by the old dirt road that connects the town of Sousa with the town of São João do Rio do Peixe, and then from the Caiçara

Piau-Caiçara (SOCA)

village by the small dirt road to the bed of the Peixe River near the Caiçara dam.

The discontinuous outcrop consists of about 2 km of the rocky bed of the Peixe River (coordinates of the dam: S 06 44.413, W 038 19.908) (fig. 4.18).

When fieldwork began at Piau-Caiçara in 1979, the twenty-five track-bearing strata unfortunately were numbered beginning at the Caiçara dam, which provides the main access to the outcrops; consequently, they were numbered starting from the uppermost stratum and counting downward. A further problem is that, due to the shifting of subaquatic sand dunes during the rainy season and of small temporary residual pools in the drought season, the numbering was not continuous, and, in successive years, bis-numbers (= twice numbers) were entered. In the field, we used to number new ichnofossiliferous levels (and then the tracks they bore) by adding to the serial number the Latin and also English adverb "bis," meaning "twice," for newly discovered levels, for example, "5 bis" for a level discovered (under a subaquatic sand dune grown during the rainy season, or on the bottom of small temporary residual pools in the drought season on the riverbed) and located between the previously discovered levels 5 and 6 of a locality. So, the numbers of the levels are sometimes similar to "5 bis," and the code number of a track discovered on this level could be, for example, SOCA 51 bis (= municipality of Sousa, locality Caiçara, track 51 bis [of the level 5 bis]). It does not now seem useful to assign different and more logical continuous numbering, starting from the lowermost stratum and working upward, and, in consequence, to assign new codes to every track of this site, because of the confusion this could introduce in comparisons with the numbering employed in previous preliminary papers (especially Godoy and Leonardi 1985) and in vintage photographs and maps.

This site was discovered by Leonardi, with Robson Araujo Marques, in 1979.

References: Leonardi 1981a, 1981e, 1982, 1984a, 1984d, 1985a, 1989a, 1989b, 1994; Leonardi et al. 1987a, 1987b, 1987c; Carvalho 1989; Carvalho and Leonardi 1992; Lockley 1991; Thulborn and Wade 1984; Leonardi and Santos 2006.

In addition to at least twenty-five levels with dinosaurian ichnoassociations (fig 4.18; figs. 4.19, 4.20 in sequence; fig. 4.21), there are without a doubt other ichnofossiliferous levels in the nearby strata upstream of the dam, on the Caiçara farm, but these are under the waters of a reservoir that very rarely dries up. A detailed study of the ichnofauna at Piau-Caiçara farms reveals:

> I. A great predominance of theropod trackways and isolated footprints, among which 100 were made by large theropods, about 25 similar to the *Grallator* kind, imprinted by small theropods.

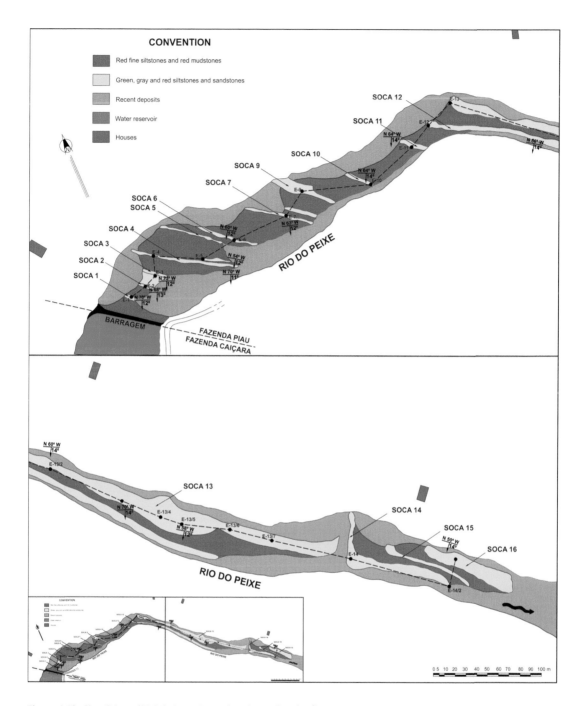

Figure 4.18. Piau-Caiçara (SOCA), Sousa Formation, Sousa. See also figs. 4.19–4.57. Map of the bed of the Peixe River from the dam (left) on the border between Caiçara, south of the dam, and Piau farms, north and east of the dam, as well as downstream, from level 1 at left to level 16 at right. Geological and paleontological survey and drawing by Leonardi.

II. Eleven large undertracks of sauropod individuals, 7 of which were probably in a herd (gregarious behavior; about 30 footprints altogether) (figs. 4.32–4.34, 4.36, 4.37, 4.39, 4.55, 4.56).

III. Six or 7 trackways or isolated footprints of ornithopods (figs. 4.24–4.26, 4.32, 4.45, 4.46, 4.48, 4.49).

IV. One crocodiloid hand-foot set, attributable to the morpho-family Batrachopodidae (tracks SOCA 13334–13335; figs. 4.36, 4.41, 4.43).

V. Hundreds of small footprints (around 1–5 cm long) of half-swimming reptiles. A few of them were made by small-size dinosaurs, but many of them were probably produced by turtles (figs. 4.27–4.30, 4.55).

VI. Some unidentifiable bipedal dinosaur footprints. Thirty others identified as indeterminate theropod prints (figs. 4.22–4.24, 4.26–4.30, 4.31, 4.32, 4.34–4.40, 4.42–4.44, 4.48, 4.49, 4.51, 4.53–4.55, 4.56).

VII. In addition to tetrapod tracks and traces, trails and burrows attributable to invertebrates as well as fish trails (Muniz 1985) also occur.

The outcrop shows red, dark gray, or dark green siltstone with mud cracks and ripple marks, alternating with reddish mudstone and shale. The stratigraphic column consists of 62 m of sediments (fig. 4.21). The strike and dip of the layers change from N68°W, 13° to the S to N50°W, 14° to the S. Almost all of the tracks were left in situ, but the majority of the better-quality prints were scheduled to be collected. Some of the pieces are kept at the Federal University of Pernambuco, Department of Geology, at Recife; the Federal University of Rio de Janeiro, Department of Geology; the Federal University of Sergipe, in Aracaju; the Departamento Nacional de Produção Mineral, Paleontology Section, in Rio de Janeiro; and the visitor center at the natural park of Sousa.

This locality was discovered by Leonardi in 1979. Several expeditions for excavation, surveying, and study have been made with his collaborators—mainly Luis Carlos Godoy, Maria de Fátima F. C. dos Santos, Claude L. Aguilar Santos, Carvalho, and others.

Because of the large number of individual animals recorded by the ichnofauna, paleontological description of the dinosaur tracks will herein be done first by groups of track types and then by level association.

Theropod Tracks

TRACKS OF SMALL-SIZE THEROPODS WITH LONG AND SINUOUS III DIGIT

About 26 individual tracks are attributed to this morphogroup (~13% of the Piau-Caiçara tracks). The best footprint of this kind is SOCA 1321 bis (fig. 4.38A), from level 13, locality 2 (13/2). This is a very typical right footprint of this kind, 10 cm long, with a very narrow, sinuous, and relatively long digit III and divergent, short digits II and IV. The slab found in situ was collected and given to the mayor of Sousa pro tempore with a small collection of other footprints in order to start a small municipal museum; unfortunately, the tracks were mislaid by the next municipal administration. In fact, the lot were used to construct the foundations of a building! This footprint was eventually recovered and displayed in the small museum of the Dinosaur Valley Natural Monument at Sousa. From the morphologic point of view, it can be attributed to *Grallator cf. Grallator olonensis* Lapparent and Montenat, 1967, from the Infralias of Veillon (Vendée, France). This belongs to the ichnofamily Grallatoridae Lull, 1904; it can be attributed to an unnamed South American group very similar to the Noasauridae.

Footprints SOCA 13319, 13325, 13329, 13345, 13351, 13354, 13378, and 13399, all come from the dinosaur track cluster of level 13, subsite 3 (13/3, fig. 4.36). All are longer than 20 cm and are relatively large footprints attributable, especially on the basis of their large size but also because of their morphological structure (particularly the long digit III and the relatively short digits II and IV), to *Grallator cf. G. maximus* Lapparent and Montenat, 1967 from the same age and locality as the above-mentioned *G. olonensis*. These footprints are also likely to have been made by noasaurids. SOCA 33 (fig. 4.22) from level 3 can be attributed to the same taxa.

SOCA 13319 (fig. 4.36, bottom left), in contrast, is also reminiscent of *Ornithomimipus* Sternberg, 1926 and could perhaps be attributed to Ornithomimidae Marsh, 1890, but this dinosaurian family has not been found in the Gondwanan continents.

The small footprints SOCA 13317, 13321, 13346, 13346, 13355, from the same level and subsite (fig. 4.36), can also be attributed to *Grallator*, perhaps to *Grallator cf. G. variabilis* Lapparent and Montenat, another ichnospecies from the Lower Jurassic of France.

Trackway SOCA 1331 (5 footprints; fig. 4.36, top left; figs. 4.42B, 4.42C, 4.42E; fig. 4.44, second from the right), from the same level and locality (13/3) as the prints thus far described, is very similar to *Grallator*, especially to *Grallator variabilis* and *Grallator maximus*. However, the impression of the *claw* of digit II is extraordinarily long, and its impression in footprint 2 is not extramorphological but rather morphological—that is, the long claw mark is not a claw drag. The impression of digit IV is widely separated from the other toe marks. This trackway could be attributed to a very large trackmaker of the Noasauridae or another similar family. Despite the fact that this trackway is very fine and something different from other theropod tracks, it did not seem useful to institute a new taxon for it. The trackway was left in situ, but some casts are kept in the Câmara Cascudo Museum of Natal (Rio Grande do Norte, Brazil).

The four excavated footprints in trackway SOCA 99 (fig. 4.28E), from level 9, also present some features of *Grallator*, but digit III is not quite long enough, and the general dimensions (>30 cm long) point rather to a probable abelisaurian rather than a noasaurid trackway.

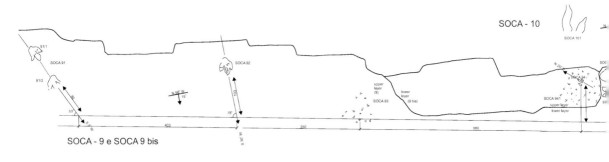

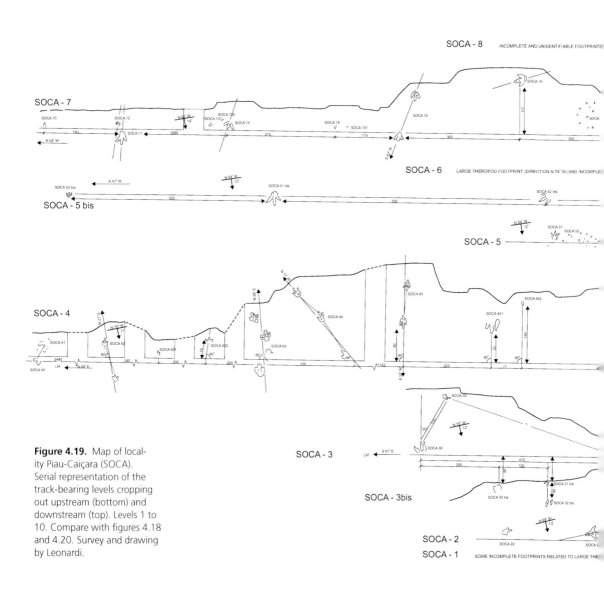

Figure 4.19. Map of locality Piau-Caiçara (SOCA). Serial representation of the track-bearing levels cropping out upstream (bottom) and downstream (top). Levels 1 to 10. Compare with figures 4.18 and 4.20. Survey and drawing by Leonardi.

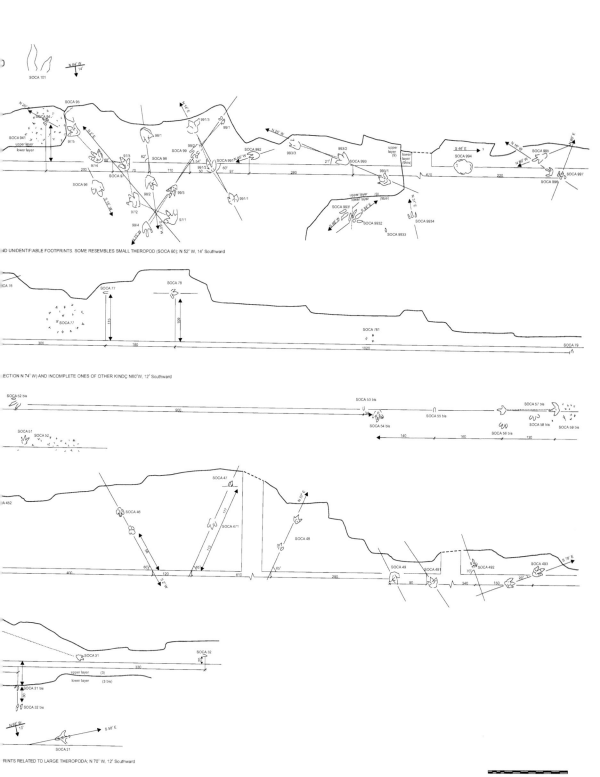

AND UNIDENTIFIABLE FOOTPRINTS. SOME RESEMBLES SMALL THEROPOD (SOCA 80); N 52° W, 14° Southward

ECTION N 74° W) AND INCOMPLETE ONES OF OTHER KIND; N60°W, 12° Southward

RINTS RELATED TO LARGE THEROPODA; N 70° W, 12° Southward

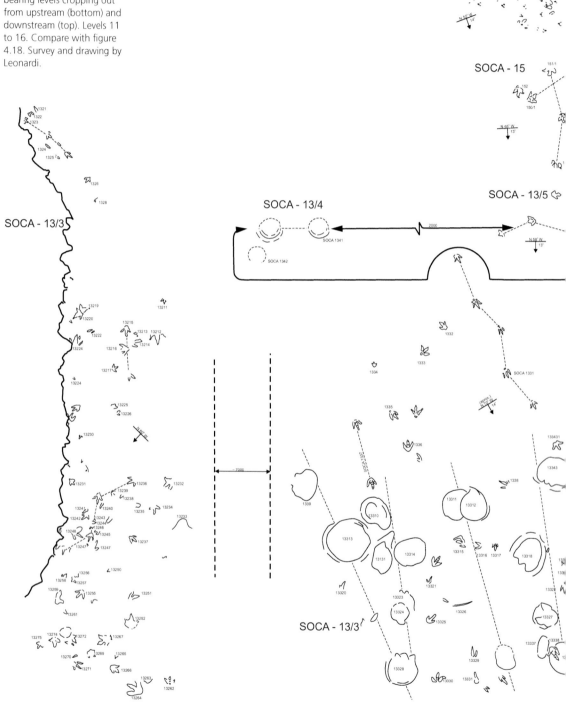

Figure 4.20. Map of locality Piau-Caiçara (SOCA), showing track levels that follow those shown in figure 4.19. Serial representation of the track-bearing levels cropping out from upstream (bottom) and downstream (top). Levels 11 to 16. Compare with figure 4.18. Survey and drawing by Leonardi.

SOCA - 16

SOCA - 15

SOCA - 13/5

SOCA - 13/4

SOCA - 13/3

SOCA 1341

SOCA 1342

SOCA 1331

SOCA - 13/3

SOCA - 12   UNIDENTIFIABLE FOOTPRINT

SOCA - 11   UNIDENTIFIABLE

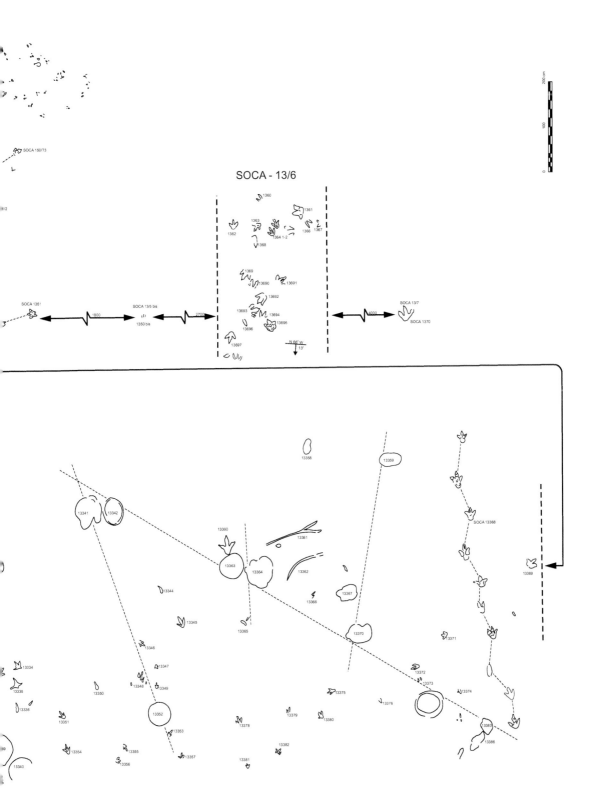

SOCA - 13/6

Figure 4.21. Stratigraphic section of the Piau-Caiçara (SOCA) tracksite compiled over the years 1983–1985 by Leonardi and Luiz Carlos Godoy, created mainly to locate the track-bearing levels at the site, along the bed of the Peixe River. This took years because sometimes some of the levels were under water, and at other times other levels were buried beneath a sand dune. The thin graphic scale on the left of the figure represents 15 m of height of the stratigraphic column.

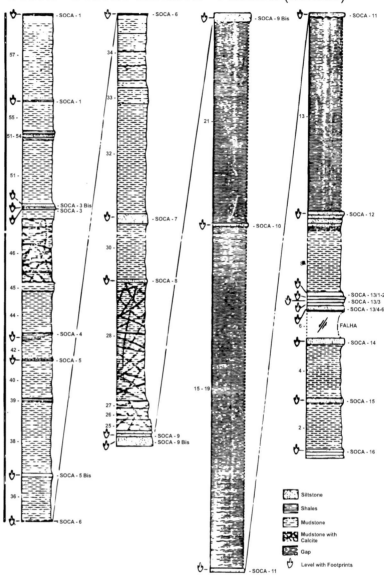

Piau outcrop profile:
Sousa - Sousa Formation - Paraíba (Braszil)

A mixed lot of footprints in SOCA 1323, 13230, and 13276, coming from level 13 at exposure 2 (13/2; fig. 4.35), along with SOCA 1332, 13322, 13333, 13356, 13357, 13375, and 13382 from level 13, exposure 3 (13/3, fig. 4.36), pertain also to this group and look like *Grallator*, but no precise ichnological attribution can be made. However, SOCA 13322, 1333, and 13382 are not very typical in morphology and can with uncertainty be attributed to small and/or juvenile theropods.

An isolated tetradactyl footprint, SOCA 13227 (fig. 4.35), with a digit I impression backward directed as a crooked spur, is very similar to *Satapliasaurus tschabouchianii* Gabounija, 1951, attributed then by Haubold

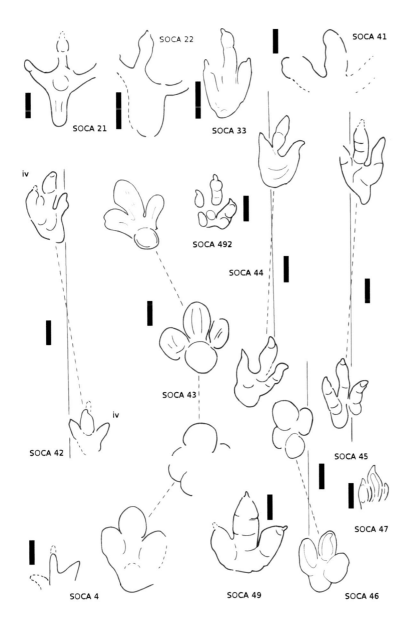

Figure 4.22. Tracks from the Piau-Caiçara (SOCA) locality. Sousa Formation, levels 2 to 4. SOCA 21 and 22 are named *Moraesichnium barberenae* and presumed to have been made by long-heeled theropods. *SOCA 33, 4, 492, 42, 44, 45,* and *47* are theropod tracks. *SOCA 49* is a typical "fat-toed theropod". *SOCA 43* and *46* are ornithopod trackways very similar to *Staurichnium diogenis* (probably a nomen dubium). *SOCA 4* is an unclassifiable bipedal footprint. Graphic scale = 10 cm. Drawing by Leonardi.

(1971) to Coelurosauria. We feel this could be so, or the putative mark of digit I might be alternatively attributed to the overstepping of this print with another small footprint of the same kind of predatory dinosaur.

Level SOCA 13 is the only level in the whole Sousa basin in which tracks of this kind, similar to tracks of the *Eubrontes-Grallator* plexus, are relatively abundant: 24 specimens, with 3 specimens at exposure 2 and 21 at exposure 3. The only other site where this kind of tracks is found is Serrote do Letreiro.

TRACKS OF LARGE THEROPODS

At the Piau site, there are about 120 probable individual large theropod tracks (~60% of the Piau-Caiçara tracks). There is frequently, especially

Figure 4.23. Theropod footprints from levels 3 and 4, Piau-Caiçara. *A*, SOCA 31. *B*, SOCA 44/1. *C*, SOCA 33. *D*, SOCA 45/2. *E*, SOCA 45/1. Graphic scale in centimeters.

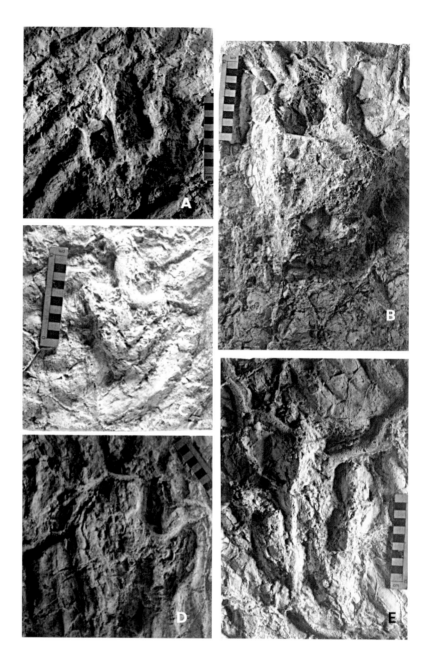

at Piau-Caiçara but also in other Peixe River localities, a striking similarity between many of their theropod tracks and those from the mainly contemporaneous ichnofaunas of La Rioja, Spain (Moratalla 2012; Moratalla et al. 1988a, 1988b; Pérez-Lorente 2001, 2002, 2003, 2015). The largest theropod track at Piau-Caiçara is one isolated footprint (SOCA 101), longer than 75 cm, found at level 10. This is also the largest theropod footprint in this basin (fig. 4.19).

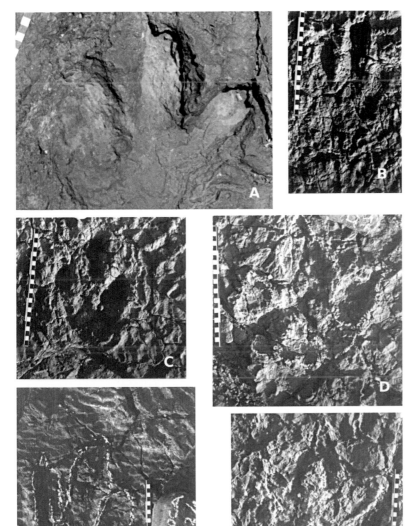

Figure 4.24. Piau-Caiçara, the bed of the Peixe River at Piau and Caiçara farms, is the richest locality in terms of the abundance of tracks. Tracks in this figure all come from level SOCA 4. *A*, A fine ornithopod right footprint (SOCA 172, found in a detached block). *B–F*, Probable theropod tracks: *B*, SOCA 48 bis; *C*, SOCA 492/2; *D*, SOCA 49; *E*, SOCA 492/1; *F*, SOCA 48. Graphic scale in centimeters.

Gigandipodidae-like Tracks (Ichno- or Morphofamily Gigandipodidae)

A trackway with 4 footprints (SOCA 1351; figs. 4.39A, 4.39C, 4.44, first trackway at right), from level 13, exposure 5 (13/5), can be attributed to this morphofamily and be compared with *Gigandipus caudatus* E. Hitch-cock, 1855, and better with *Hyphepus fieldi* E. Hitchcock, 1858, both from the Portland Arkose of Connecticut and Massachusetts, Lower Jurassic (see also Xing et al. 2018). This is the trackway of a bipedal dinosaur with a relatively long stride, as indicated by a high SL/FL ratio (mean = 7.48). The feet are turned inward (negative rotation) with respect to the

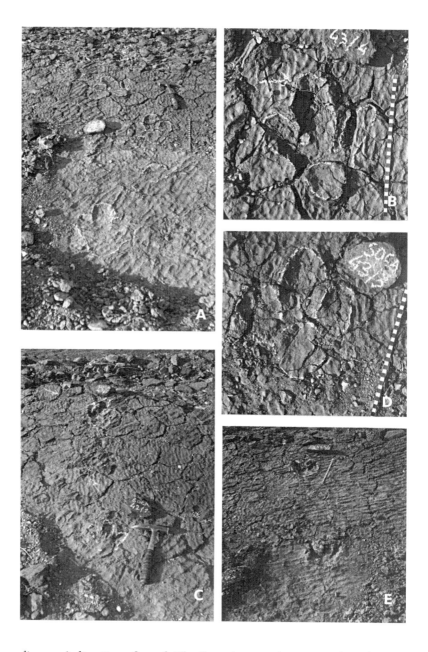

Figure 4.25. Piau-Caiçara. Level 4. *A*, The ornithopod trackway SOCA 43 on a surface with multiple ripple marks. *B*, SOCA 43/4. *C*, Theropod trackway SOCA 42. *D*, SOCA 43/3. *E*, Theropod trackway SOCA 44. Graphic scale in centimeters.

dinosaur's direction of travel. The footprints are almost as wide as they are long, present a high digital divergence, and are tetradactyl due to the presence of a well-developed digit I impression in footprints 1 and 2, directed backward as a spur. This spur is very delicate in the third footprint. Digits II–IV are broad and wedge-shaped at the "heel." In one footprint, they present a nail or claw, but generally they have blunt distal extremities.

This trackway is distinct enough to justify a new generic and specific name, as it is different from both *Gigandipus* and *Hyphepus*, because the trackway lacks "tail drags." However, a new taxon will be not instituted because the trackway remains in situ, frequently covered by sand dunes or water and so not easily found, and unfortunately no casts were made of it.

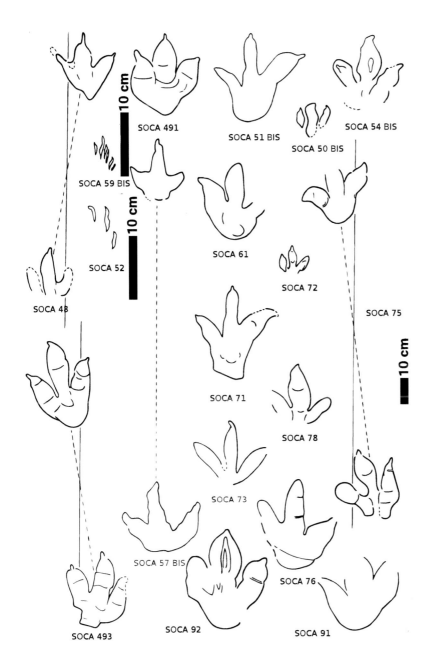

Figure 4.26. Tracks from locality Piau-Caiçara, derived from levels 4 to 9. Short trackways SOCA 48 and 493; 51 bis 1 and 71 (with a very asymmetrical footprint because of a very large digit IV mark); 50 bis, a print made by a half-swimming dinosaur whose body was largely supported by the water; 54 bis, 75, 76, 92 (with fat toes but with scratches from claws), are attributable to theropods. SOCA 72 was made by a very small biped and could be theropod or ornithopod; SOCA 491 is attributable to an ornithopod; and SOCA 52 is interpreted as the trace of a half-swimming turtle. SOCA 59 bis could be the hand-foot set of a very small, half-swimming tridactyl dinosaur. Graphic scale = 10 cm.

Eubrontidae-like Tracks (Ichno- or Morphofamily Eubrontidae)

Some tracks are very *Eubrontes*-like:

SOCA 1364/1–2 (figs. 4.45 [top], 4.46), a pair of footprints found close together, probably belonging to the same individual, which stopped on both feet. These tracks are very similar to *Eubrontes divaricatus* (E. Hitchcock, 1865) or to *Eubrontes platypus* Lull, 1904. Both footprints show slipping marks of the "heel." These Lower Cretaceous *Eubrontes*-type tracks also show a large metatarsophalangeal area that is aligned with the axis of digit III (Xing et al. 2014b).

Figure 4.27. Piau-Caiçara, levels 5 and 5 bis. *A*, SOCA 51/1, the collapsed left footprint of a theropod. *B*, SOCA 59 bis, an unusual half-swimming hand-foot set of a very small tridactyl dinosaur. *C*, Footprint SOCA 5 bis 3 has the appearance of an ornithopod print and shows scratches impressed by claws, along the main axes of toes II and III, produced during the retraction of the foot. *D*, SOCA 5 bis 2, a theropodian footprint. *E*, SOCA 51 bis 1, a footprint of a theropod. *F*, SOCA 52 bis 2, a probable theropod footprint. *G*, SOCA 5 bis 6, a theropod footprint. Graphic scale in centimeters.

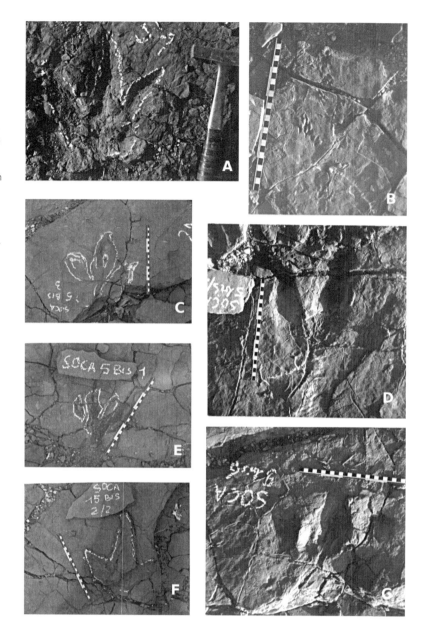

The slab bearing these two fine tracks was stolen from the field and illegally kept in a private collection in the small town of Santa Rita (Paraíba), at the Museu Comunitário Jeová B. De Azevedo. In 2014, this slab was recovered by means of a lawsuit and was brought back to Sousa, where it is now kept at the visitor center and museum of the Dinosaur Valley Natural Monument.

Footprints SOCA 492/1 and 2 (figs. 4.24C, 4.24E) are similar to both *Eubrontes tuberatus* (E. Hitchcock, 1858) and *Eubrontes divaricatus* (E. Hitchcock, 1865).

Footprints SOCA 13237 (fig. 4.35), 1369 (figs. 4.45, 4.46), 13697 (figs. 4.45, 4.46, bottom left), 1363 (figs. 4.45–46), and 13338 (fig. 4.36, overstepping

Figure 4.28. Piau-Caiçara, levels 7 and 9. *A–C*, Three photos of tracks impressed by small animals, probably turtles that were probably half-swimming, from the level SOCA 7. *D*, The outcrop of the level 9, extending across the Peixe River. This is the way layers crop out across the river from level 1 to level 10 at Piau-Caiçara. As seen in this view, the Peixe River flows from left (west) to right (east). The succeeding levels (11–16) crop out, in contrast, more or less parallel to the riverbed. *E*, SOCA 99, a theropod track associated with small chelonian tracks like those in *A–C*. See also figure 4.29. Graphic scale in centimeters.

a large sauropod track) can also be compared with some forms of the genus *Eubrontes*. SOCA 1369 is kept in the collections of the Laboratory of Paleontology of the Federal University of Sergipe, in Aracajú (Brazil).

Columbosauripus-like Tracks

Some footprints have a general outline comparable to that of *Columbosauripus ungulatus* Sternberg, 1932 from the Lower Cretaceous of the Peace River (British Columbia, Canada). These include tracks SOCA 9931 (figs. 4.31G, 4.55), 13215 (fig. 4.35), 13217 (figs. 4.35 detail, 4.47D), 13255 (fig. 4.35), 1326 (fig. 4.35), 13271 (fig. 4.35), 1336 (fig. 4.36, top left),

Figure 4.29. Locality Piau-Caiçara (SOCA), level 9, Sousa Formation, at Sousa. The rocky surface of this slope shows several tracks made by small, half-swimming chelonians (SOCA 94 bis) along with some isolated theropod tracks. See also figure 4.30. Graphic scale in centimeters. Photomosaic by Leonardi.

and 1337 (fig. 4.36, top left). *Columbosauripus* is probably not very different from *Irenesauripus* Sternberg, 1932.

Moraesichnium-like Footprints

Isolated footprints SOCA 21 and 22 (fig. 4.22, top left) from the second level of this site, SOCA 995 (figure 4.31 B) from its ninth level, and a

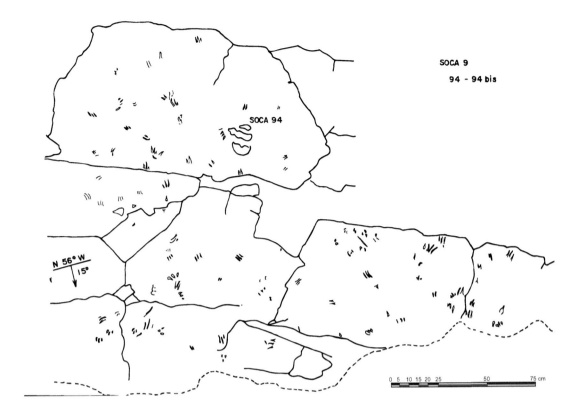

SOCA 9
94 - 94 bis

SOCA 94

N 56° W
15°

0 5 10 15 20 25 50 75 cm

Figure 4.30. Locality Piau-Caiçara (SOCA), municipality of Sousa, Sousa Formation, level 9. The map shows a portion of the large surface of this level; this portion contains the track of a half-swimming, medium-size theropod (SOCA 94) as well as some 70 small tracks of half-swimming chelonians (SOCA 94 bis). Compare with the corresponding photo-mosaic (fig. 4.29). Survey and drawing by Leonardi. Graphic scale in centimeters.

single incomplete print from the twelfth level are classifiable as *Moraesichnium barberenae* Leonardi, 1979 because of the long heel or metatarsus pad, the high digital divergence, and the small claw marks. SOCA 21 and 22 are rarely examinable, being almost always submerged in the river or in the remaining pools in the drought season. SOCA 995 is a rare case in the Sousa basin in which mud cracks do not follow the outline or the morphological features of the print, by which they are generally controlled, but rather cut cross the surface of the sole of the footprint. As for paleontological discussion on the trackmaker of *Moraesichnium*, see the description of locality SOPP, from which the holotype comes.

"Bückeburgichnus"-like Footprints

Some footprints are reminiscent of "*Bückeburgichnus maximus*" Kuhn, 1958 from the Lower Cretaceous of Bückeburg, Niedersachsen, Germany (Kuhn 1958; Hornung et al. 2012a). Although the Brazilian prints do not always have the long "heel" or the impression of the first toe, they do show the high digital divergence and the general features of this taxon—as if a "*Bückeburgichnus*" maker had proceeded in a completely digitigrade way instead of in a semiplantigrade fashion.

Among the footprints that could be characterized as "*Bückeburgichnus*"-like is SOCA 173, a print with a very long and stout digit IV (fig. 4.56A). Another is SOCA 9932, whose left external digit was compressed and deformed by another large theropod that crossed the mudflat in the opposite

Figure 4.31. Piau-Caiçara, levels 9 and 15. *A*, Layer SOCA 9, with theropod and turtle tracks. *B*, A probable theropod footprint (SOCA 995) disrupted internally by mud cracks, a rare case at Sousa. *C*, Layer SOCA 9, with abundant tracks. *D*, A short trackway (151, from the level SOCA 15), attributable to a theropod or, less probably, an ornithopod. *E*, SOCA 151/1, first footprint of the same trackway as in figure 4.31*D*. *F*, The maker of footprint 9931 (at left) passed by after the maker of SOCA 9932 (at right), laterally compressing the latter's footprint. *G*, SOCA 9991, a collapsed theropod footprint with associated tracks of turtles (at right side of photograph). Graphic scale in centimeters.

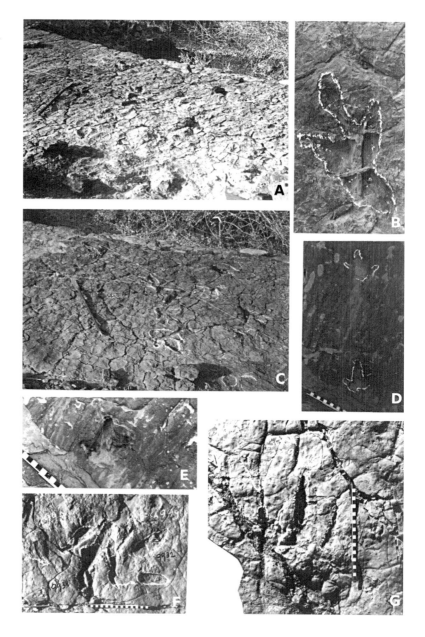

direction, putting its right foot very near footprint SOCA 9932 when the mud was still plastic (figs. 4.31G, 4.55). A particularly striking candidate is trackway SOCA 13368 (10 footprints) from level 13, exposure 3 (13/3) (figs. 4.36, 4.40, 4.44, third trackway from the left). The general outline of the footprints in this trackway is very similar to "*Bückeburgichnus*," although SOCA 13368 lacks the digit I impression of the German ichnotaxon, and its "heel" is relatively a little shorter and its toes are fatter, sometimes with hooflike tips. It is noteworthy that in this trackway the shape variability of the footprints is very high, surely due to different local conditions of the substrate and also to the way the trackmaker put its feet on the sediment. There is a longitudinal (sagittal) furrow along the axis of digit III in a

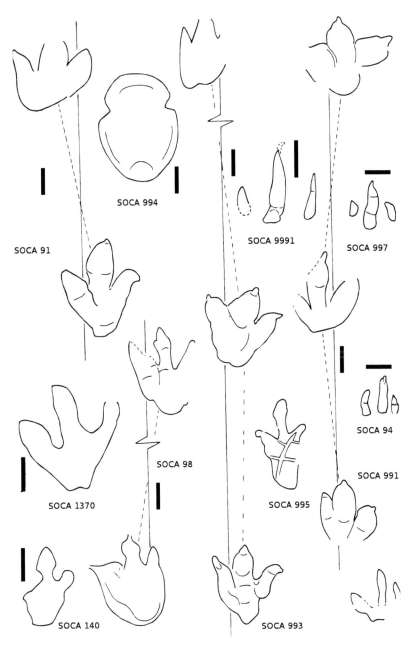

Figure 4.32. Tracks from locality Piau-Caiçara. SOCA 91, 98, 99, and 993 are trackways of "fat-toed" theropods, from level 9. SOCA 995 is a probable theropod track broken only internally by mud cracks. SOCA 9991, 94, and 997 are prints made by half-swimming theropods. SOCA 994 is probably a hand-foot set of a medium-size sauropod. SOCA 140, an incomplete footprint, is perhaps attributable to an ornithopod. Graphic scale = 10 cm. Drawing by Leonardi.

number of footprints of this trackway; it was probably dug by the long claw during the retraction of the feet.

The very large incomplete and isolated footprint SOCA 101 (75 cm long; fig. 4.19, level 10) and also SOCA 13360 (fig. 4.36) could perhaps also be attributed to this group.

"Fat-Toed" Theropod Footprints

At the Piau-Caiçara locality, there are several tracks that show true claw marks; although they are generally small, they can be attributed to large

Figure 4.33. A recently discovered sauropod trackway at the Caiçara-Piau Farm, level 13/site 7. Note the very wide and high displacement rims of the prints. The Piau ichnofauna is one of the richest and most diverse of the basin; it contains large numbers of tracks belonging to several forms of large and small theropods, large ornithopods, small bipedal ornithischians, sauropods, turtles, and crocodiles. Photograph from Leonardi and Santos 2006. On the right, a photo of twin footprints (SOCA 1364/1–2 from level 13/6 of the site Caiçara-Piau) imprinted by a theropod that stopped on both feet (see also figs. 4.45, 4.70C).

theropods. However, they present very fat and wide toes, and, because of this, they sometimes have a general footprint outline very similar to that of some Iguanodontidae footprints, apart from the claw marks. Moreover, when these footprints occur in a trackway, the pace angulation is high and the trackway is narrow.

These footprints sometimes are reminiscent of the general features and outline of *Tyrannosauropus petersoni* Haubold, 1971; however, this ichnogenus and species does not present claws. In some cases, it is very possible that these tracks are so fat-toed because they are under-tracks. Alternatively, this phenomenon may depend on compression exerted by the successive "upper" layers (Lockley and Xing 2015).

To this group, we assign the short trackways SOCA 493 (2 footprints; fig. 4.26), 91 (2 footprints; figs. 4.26, 4.32), 97 (5 footprints), 98 (2 footprints; fig. 4.32), and 991 (3 footprints; fig. 4.32) and the isolated footprints SOCA 49 (fig. 4.22), 76 (fig. 4.26), 92 (fig. 4.26), 1361 (figs. 4.45, 4.46), 13692 (figs. 4.45, 4.46), 176, and perhaps 171 (fig. 4.56F). This last footprint, with a large, callused digit III, can alternatively be attributed to an ornithopod.

A number of these fat-toed theropod tracks, especially SOCA 49, 493, (91?), (92?), 97, 98, and 993, could be attributed to the ichnoform *Corpulentapus lilasia* Li et al., 2011 from the Cretaceous Longwangzhuang Formation at a site in the Zhucheng region of China. We note, in passing, that this specific name is particularly fine and poetic.

Asymmetrical Large Theropod Tracks

Some large theropod footprints show strong asymmetry, with a very long digit IV impression—for example, SOCA 51 bis (fig. 4.26), 54 bis, 61, and 71 (fig. 4.26). This same kind of asymmetry is seen in footprints as *Columbosauripus amouraensis* Bellair and Lapparent, 1948 and *Shensipus tungchuanensis* Young, 1966, respectively from the Cenomanian of Amoura, Algeria, and the Middle Jurassic of Tungchuan, Shensi, China.

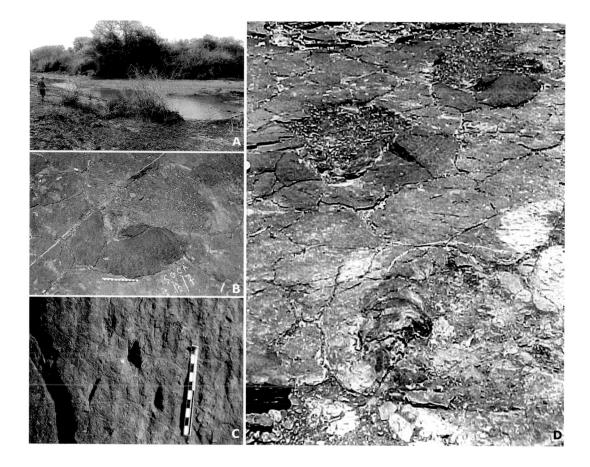

Other Large Theropoda

There are 24 tracks belonging to this group (~12% of the Piau-Caiçara tracks):

- The short trackways SOCA 75 (fig. 4.26) and 151 (figs. 4.37D, 4.37E)
- The isolated footprints SOCA 47 (fig. 4.22), 1322, 1324, 13212, 13216, 13219, 13228, 13232, 13234, 13251, 13259, 13264, 13274, and 13275 (all of them in figs. 2.2, 4.35); tracks 1360, 1366, 13690, 13691, 13693, 13694 (all of them in figs. 4.45, 4.46); and track 1370
- Footprint SOCA 1351 (fig. 4.39A).
- The short trackway SOCA 48 (figs. 4.24B, 4.24F, 4.26), whose 2 footprints are very different from each other. One of them is very similar to *Satapliasaurus kandelarii* Gabounija, 1951.

FOOTPRINTS OF HALF-SWIMMING THEROPODS

Almost all of a set of about 40 footprints and short trackways from level 16 (figs. 4.34C, 4.39B, 4.39D, 4.47B, 4.47E, 4.51–4.55) belong to this group, attributable to medium and large theropods. These tracks are very similar to those illustrated by Coombs (1980) from the Lower Jurassic of Rocky

Figure 4.34. *A*, The bed of Rio do Peixe at Piau-Caiçara at level 13. photo by Wellington Francisco Sá dos Santos. *B* and *D*: photographs of the sauropod trackway at the Piau site. *B*, Hind-foot prints of the sauropod trackway are oval-shaped. The horseshoe-shaped forefoot prints are almost obliterated by the high anterior displacement rim of the hind-foot print. *C*, A fine track of a half-swimming theropod at Piau-Caiçara, level 16, print SOCA 1619.

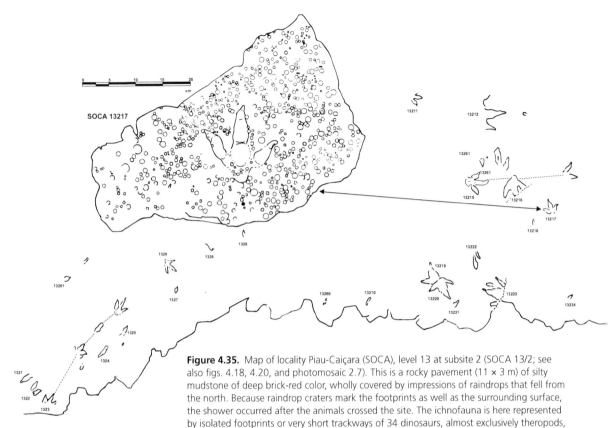

Figure 4.35. Map of locality Piau-Caiçara (SOCA), level 13 at subsite 2 (SOCA 13/2; see also figs. 4.18, 4.20, and photomosaic 2.7). This is a rocky pavement (11 × 3 m) of silty mudstone of deep brick-red color, wholly covered by impressions of raindrops that fell from the north. Because raindrop craters mark the footprints as well as the surrounding surface, the shower occurred after the animals crossed the site. The ichnofauna is here represented by isolated footprints or very short trackways of 34 dinosaurs, almost exclusively theropods, as well as a number of isolated digit impressions. The tracks are generally shallow and rarely well preserved and in general have very narrow toe impressions because of the mud that collapsed after the animals lifted their feet. The track surface was generally almost dry, and the trackmakers impressed their footprints only in wet patches where some puddles were still present. The track SOCA 13217 is highlighted in detail (see also figs. 3.30, 4.47C). Survey and drawing by Leonardi.

Hill, Connecticut. In addition, SOCA 59 bis (very small, mesaxonic; figs. 4.26–4.27B) and SOCA 94 (figs. 4.29, 4.30), 941, 997 and 9991 (fig. 4.32), 13270 (figs. 3.30, 4.35), and 175 (fig. 4.55) are examples of this kind of footprint. The tracks of half-swimming theropods constitute about 23% of the Piau-Caiçara dinosaur tracks.

As mentioned above, half-swimming is the progression of animals that float in shallow waters and make progress by setting the tips of their feet against the bottom (the submerged substrate). The corresponding trackways are frequently incomplete and irregular. Prints of this type could be made also by animals fully immersed, that is, "over its head" in water, as hippos kick along bottom in this way, fully submerged. The footprints consist principally of scratches or indentations left by the claws, digits or hooves. Fossil trackways of this type are common in red-beds of the Permian and Triassic ages. Trackways of half-swimming theropods were found and reported from the Connecticut Valley (Coombs, 1980; and are now discussed; see Milner and Lockley, 2016) and from Paraíba, Brazil (Godoy and Leonardi, 1985; Leonardi, 1994a); hadrosaur trackways

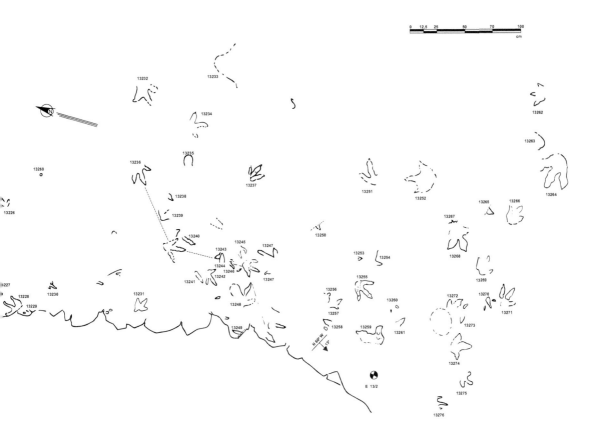

of this type occur in the Middle Cretaceous of the Peace River Valley, British Columbia (Sarjeant 1981). The term "semi-natation," from which corresponding terms in other languages were derived, was proposed by D. Heyler (oral communication). (Leonardi 1987a, 34–35, 51, plate IX, L–O)

On this topic, see also Xing et al. 2015f; Milner et al. 2006; Xing et al. 2013; Lockley et al. 2014a; Romano and Whyte 2015; Segura et al. 2016; and especially Milner and Lockley 2016).

Many other footprints at this site also seem to have been made by half-swimming reptiles. However, these tetradactyl and pentadactyl footprints were not made by dinosaurs but instead very probably by turtles (as discussed previously), and less frequently by small, lizardlike reptiles.

Sauropod Tracks

In level 13, locality 3 (13/3), there are 26 large to very large deep, rounded, or elliptical footprints (SOCA 1339, 13310, 13311, 13312, 13313, 13318, 13323, 13324, 13327, 13328, 13334, 13340, 13341, 13342, 13343, 13352, 13359, 13363, 13364, 13367, 13370, 13377, 13383, and 13384; figs. 4.35, 4.38D), which are perhaps arranged in 7 short and discontinuous trackways (fig. 4.20, level 13/3) and are associated with many theropod tracks. These sauropod tracks are large and deep but of poor quality and do not show any morphological details. However, because of their very large dimensions and their

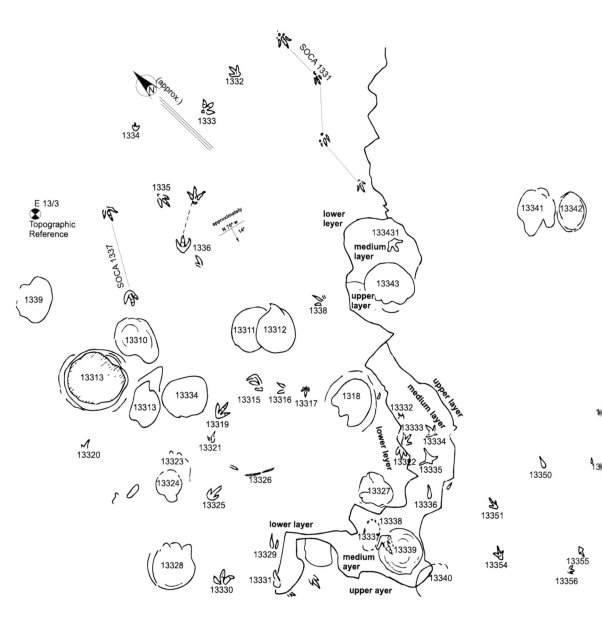

Figure 4.36. Piau-Caiçara (SOCA). Detailed map of level 13, subsite 3 (SOCA 13/3; figs. 4.18, 4.20; see also figs. 4.36, 4.37, 8.6, 4.38D, 4.40–4.44). This is a large pavement (20 × 10 m) of deep gray siltstone, which weathers to a deep brick-red or pink color. The upper surface is sometimes covered by a black varnish. The level consists of three strata, in all of which a number of dinosaur tracks are seen. Upper: the sample comprises 1 trackway and 20 isolated footprints, representing 21 individual theropods. Middle: the sample includes 7 isolated footprints (5 theropod tracks and 2 crocodiloid footprints; figs. 4.41, 4.43). Lower: the sample comprises 18 individual theropods represented by 3 trackways and 15 isolated footprints. In addition to the tracks already described above, there are 26 large, round, and very deep undertracks (diameter up to 120 cm; minimum diameter ~50 cm) that can be observed in one of the three strata at the subsite 13/3, and that probably were impressed, as true tracks, in a higher stratum now destroyed by erosion. These footprints, attributable to sauropods, are perhaps arranged in 7 short and discontinuous trackways whose direction of travel would lay in the sector 25°–60° (or the opposite direction, 205°–240°), a spread of only 35°, which could indicate the crossing of a herd of large sauropods (figs. 4.36, 8.6). Survey and drawing by Leonardi.

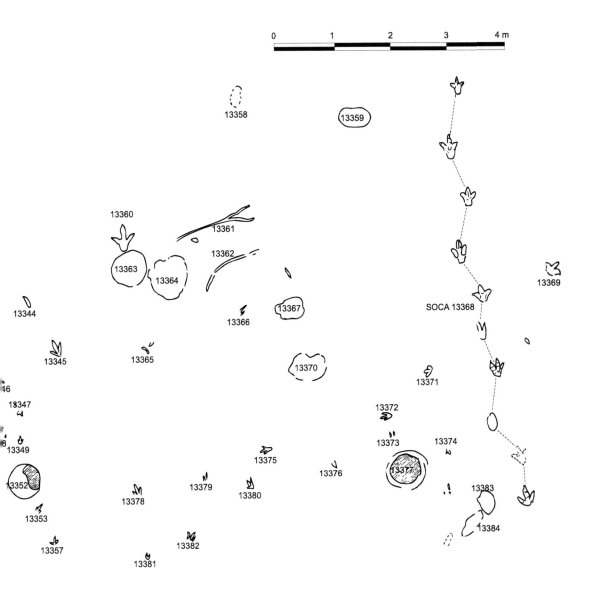

generally rounded or elliptical outline, they are attributed to the Sauropoda. The largest footprint among them has a diameter of about 120 cm.

The large dimensions of these footprints pose a problem, though they are not the largest sauropod tracks in the world; see, for example, the very large sauropod hind-foot prints (up to about 170 cm long) at Toothpick Beach, near Minyirr (Riddell Point), Broome, Western Australia (Thulborn 2012, figs. 13–15), visited twice in the field by Leonardi with Tony Thulborn, in 1997 and 1999 (and cf. Salisbury et al. 2017). They pose a problem, as do similar-size tracks the entire world over, because some of them seem to be much larger than the largest known feet skeletons of sauropods. (To date, the largest theropod tracks discovered in the world appear to be those described by Boutakiout et al. [2009].)

Figure 4.37. Paleoenvironmental reconstruction scenario of the Piau site in the Sousa basin, Sousa Formation. Four large theropods follow and prepare to attack a small herd of 3 titanosaurids along a drying valley at the bottom of the Sousa basin. Compare with figures 4.36 and 8.6. Redrawn and modified by Leonardi in India ink, from an original color picture by Renzo Zanetti.

In any case, the size of these prints, which is greater than the size of most known skeletal specimens, is not troubling. There is a greater chance for any particular dinosaur to be recorded in the geologic record by footprints rather than by a skeleton, simply because it could have taken many steps during its life, but it died and left a skeleton only once. That being the case, one would expect the largest dinosaurs to be more likely represented by tracks than by bones (J. O. Farlow, personal communication, 2015).

The 120-cm diameter of the largest footprint at Caiçara-Piau is partially explained by the fact that this footprint and the others of the group are probably undertracks. In fact, it has been noted (Thulborn 1990, 2012; Lockley 1991; Xing et al. 2015e; Gatesy and Ellis 2016) that undertracks are larger and larger, with increasing depth, than the original tracks because the pressure of the trackmaker's feet on the substratum is exerted downward and outward. So, the ideal surface joining the external outlines of the original footprint and its undertracks in one or more "lower" layers is subconical, and the footprint so seen is not really the impression of a pes on a surface but a "track volume" (Gatesy and Ellis 2016).

We can also attribute the short trackway of 2 high-rimmed, rounded footprints, partially damaged by fluvial erosion, to sauropods, coded SOCA 1341 (fig. 4.20, level 13/4, fig. 4.39E–F). Same goes for the isolated footprint 1342, from level 13, exposure 4 (13/4, fig. 4.20, level 13/4), simply because of its large dimensions, and also for the strange footprint SOCA 994 from level 9 (fig. 4.32).

In December 2003, a relatively well-preserved narrow-gauge sauropod trackway (SOCA 138) was discovered and recorded (Leonardi and Santos 2006). It had been naturally exposed by river erosion during the "winter" (= rainy season) of 2002 at the bottom of the bed at level 13/site 8 (figs. 4.33, 4.34). Currently, the trackway includes 4 hand-foot sets, with very wide and high displacement rims.

The footprints are oval, with a mean length of ~75.5 cm and a mean width of ~42.5 cm; no morphological details are evident. The horseshoe-shaped handprints are relatively small and almost completely filled in by the anterior part of the high (up to 20 cm) displacement rim of the hind-foot print. The mean length of the fore footprints is 21.25 cm; their mean width is 42.5 cm. The mean length of the stride is 236 cm; the mean pace angulation is 119°; and the mean length of the oblique pace is 129 cm. The trackway is narrow gauge and of the *Parabrontopodus* type. The footprints do not overlap the midline but are very close to it. The animal was moving in a westerly direction.

Altogether, there are about 11 individual sauropods recorded at the Caiçara-Piau site (5.3% of the individual tracks of the site), and the total number of their footprints is 38.

Ornithopod Tracks

Ornithopod tracks are not common at the Piau-Caiçara farm; there are about 6–7 individuals (3.6% of the individual tracks of the site) and, with some doubt, possibly 11 more.

SOCA 43 (fig. 4.25), a curved trackway of 4 footprints, and SOCA 46, a short trackway of 2 prints (fig. 4.22), are assigned to the form *Staurichnium cf. Staurichnium diogenis* Leonardi, 1979, whose holotype is at Passagem das Pedras (SOPP). Both show rounded, hooflike terminal ends in a typical clover-shaped pattern and a fleshy "heel" pad. The stride is short, and the SL/FL ratio is very low.

OTHER ORNITHOPODS

An isolated but deep and well-preserved footprint, SOCA 172 (figs. 1.9, 4.24A), from level 13 or more probably 14 (it was collected from the stratum before the levels were given numbers and before mapping of the site), belongs to ornithopods and is something similar to the tracks *Iguanodontipus* Sarjeant, Delair and Lockley, 1998. It is kept in the collections of the School of Geology of the University of Pernambuco in Recife, Brazil.

Footprints SOCA 30, 41, 140 (fig. 4.32), 134991, and 13499 (fig. 4.55) may also belong to the ornithopods. Some isolated tracks from level 13, exposure 6, such as SOCA 1361 and 13695 (fig. 4.45), could perhaps be compared with *Stegopodus* sensu Gierlinski and Sabath (2008). See also Gierlinski et al. (2009; fig. 5).

Figure 4.38. Piau-Caiçara (SOCA). *A*, Track 1321 bis, a fine theropod track (about 9 cm long), found in situ in a rock floor exposed along a path between the localities SOCA 13/1 and SOCA 13/2. *B*, SOCA 54 bis, an asymmetrical right theropod footprint with very large outer toe (digit IV). *C*, Raindrop impressions and 2 collapsed theropod tracks at SOCA 13/2. *D*, Overall view of the very rich exposure 3 of level SOCA 13, gridded with chalk for detailed mapping, with Robson de Araujo Marques (1979). *E*, Short trackway SOCA 1311, attributable to a theropod.

TRACKS ATTRIBUTABLE TO ORNITHOPODS
OR TO LARGE THEROPODS

Some footprints or short trackways show the general outline typical of some ornithopod footprints like *Gypsichnites* Sternberg, 1932 or *Iguanodontipus* Sarjeant et al., 1998, respectively from the Lower Cretaceous of the Peace River, British Columbia, Canada, and from the European Wealden. However, they sometimes also show small claws or nails, as do some *Iguanodontipus* footprints. They are reminiscent as well of some footprints attributed to theropods by Haubold (1971)—for example, the unnamed form of Marsh, 1899 (Haubold 1971, fig. 49/4) and those of

Figure 4.39. Piau-Caiçara (SOCA), levels 13 to 16 (figs. 4.18, 4.20). *A*, The 4 footprints of trackway SOCA 1351, level 13, subsite 5. The identifying code written on the stone in chalk was only provisional and later replaced by the label used here (fig. 4.44). *B* and *D*, Footprints of half-swimming theropods from level SOCA 16, the highest and youngest of the levels at Piau-Caiçara (figs. 4.51–4.54). *C*, The first footprint (right pes) of the trackway from *A*, attributed to a "fat-toed" theropod, with the impression of the first toe like a sickle-shaped (falciform) spur. This feature can be compared with *Gigandipus caudatus* E. Hitchcock, 1855, or even better with *Hyphepus fieldi* E. Hitchcock, 1858 of the morphofamily Gigandipodidae. *E* and *F*, Two photographs of a sauropod trackway (SOCA 1341), with rounded print outlines and high, regular displacement rims, from level 13, subsite 4.

Brodrick, 1909 (Haubold 1971, fig. 49/7). As a matter of fact, Lockley (2000), McCrea (2000), Gierlinski et al. (2008), and Díaz-Martínez et al. (2015) all suspected *Gypsichnites* was a theropod track.

However, on May 9, 1985, Leonardi had between his hands the specimens NMNH 8553 and 8554 of *Gypsichnites pacensis* in the National Museum of Natural History at Ottawa and had the clear impression that they were true ornithopod tracks. These footprints from Piau can be alternatively classified, case by case, as ornithopod, or they might be broad tracks or also undertracks of theropods. Among them are SOCA 57 bis and 78 (fig. 4.26); tracks 13266 and 13268 (figs. 4.35); the group of tracks 1333, 13330, 13369, and 13498 (fig. 4.36); and 13695 (figs. 4.45,

Figure 4.40. Piau-Caiçara (SOCA). *A–E*, Footprints from theropod trackway SOCA 13368 and their occurrence on the pink, sloping rock surface of level 13, subsite 3 (*F*; fig. 4.44). The identifying codes written on the stone in chalk were only provisional and later replaced by the labels used here. *A*, SOCA 13368/1 (right pes). *B*, 13368/3 (right pes). *C*, 13368/5 (right pes). *D*, 13368/6 (left pes). *E*, 13368/8 (left pes). *F*, The sloping surface containing this trackway. Graphic scale in centimeters. This trackway shows some noteworthy features. There is considerable shape variability among the footprints in the trackway. Claw marks are apparent only in footprints 1 and 8. In footprint 6 the claw mark of toe III was obliterated by primary mud cracks, but a long scratch made by that claw runs along the main axis of toe III and was made when the trackmaker was closing its toes together before elevating its foot.

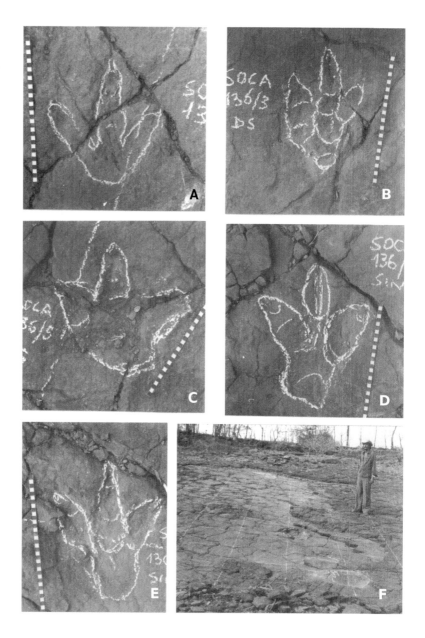

4.46), 150, and 153 (fig. 4.55). Together they comprise about 5.5% of the Piau-Caiçara tracks.

Crocodiloid Footprints

The crocodiloid set of footprints SOCA 13334–13335 from level 13, exposure 3 (13/3; see the map in fig. 4.36, middle layer) is a rare case of nondinosaurian footprints in the Sousa basin; in fact, these are rare in any of the basins of Northeast Brazil. They are of poor quality but can be compared to some forms of the morphofamily Batrachopodidae Lull, 1904, especially, for the manus, to *Cheirotheroides pilulatus* E. Hitchcock, 1858 and to *Batrachopus gracilior* (E. Hitchcock, 1858), both from the

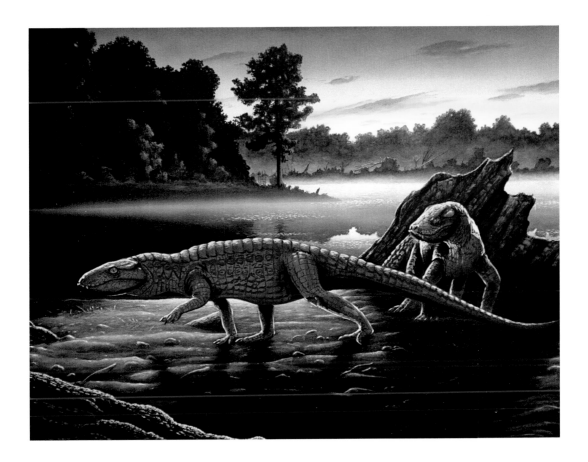

Portland Arkose, Liassic of Massachusetts; however, the pes is different (Lockley et al. 2017). There is some chance that the trackmaker of this pair may be attributed to the order Crocodylia Gmelin, 1788.

Turtle Footprints

Hundreds of small footprints (~1–5 cm long), attributable to half-swimming turtles, occur at the Piau-Caiçara farm (Leonardi and Carvalho 2000). They could be attributed to the ichnogenus *Chelonipus* (H. Ruhle von Lilienstern, 1939; see also Milner and Lockley 2016). They are especially abundant in levels SOCA 5, 7, 9, 16; see, for instance, the rocky pavement of level 9 (figs. 4.29, 4.30) and tracks SOCA 59 bis, 94 (fig. 4.32), 1359, and 174 (fig. 4.55). See also figures 4.19 and 4.20. The Araripemydae family are the main candidates for the trackmakers. It is puzzling that we did not find full walking tracks of turtles or lizardlike reptiles corresponding to the half-swimming tracks.

All in all, to date, a total of about 210 individual identifiable dinosaurian trackways and isolated footprints have been uncovered at the Piau-Caiçara site, along with hundreds of footprints of very small reptiles, probably turtles, that appear to have been half-swimming. A study of the ichnofauna reveals a great predominance of theropod tracks (170 individuals), among which 129 were made by large theropods and 27 by small

Figure 4.41. Piau-Caiçara (SOCA). Paleoenvironmental reconstruction of level 13, subsite 3. This outcrop shows the only occurrence of crocodiloid footprints in the Sousa basin, apart from the possible impression of crocodilian bodies in the mudstone at Tapera. The footprints are crocodiloid hand-foot set SOCA 13334–13335 (see the map in fig. 4.36, middle layer, and also fig. 4.43), attributable to the Morphofamily Batrachopodidae. The trackmaker was probably a notosuchian (cf. figs. 4.20 at level 13, subsite 3, and 4.36). Painting by Ariel Milani Martine.

Figure 4.42. Piau-Caiçara (SOCA), level 13, subsites 1 and 3. *A* and *D*: Two footprints from theropod trackway SOCA 1311 from level 13, subsite 1. *A,* SOCA 1311/1. *D,* SOCA 1311/2. *B, C,* and *E*: Three plaster casts (courtesy of the Câmara Cascudo Museum of Natal, RN) of the fine theropod footprints from trackway SOCA 1331 from the great sloping surface at Piau, level 13, subsite 3 (cf. figs. 4.20, 4.36). *B,* SOCA 1331/4 (right pes). *C,* SOCA 1331/2 (right pes). *E,* SOCA 1331/5 (left pes).

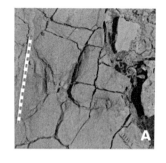

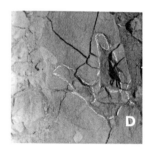

Figure 4.43. A Cretaceous crocodile with its hand-foot set of footprints (Piau-Caiçara level 13, subsite 3). Painting by Ariel Milani Martine.

theropods; at least 6 or 7 trackways and isolated footprints of ornithopods; a cluster of very large, rounded subtracks, attributed to sauropods, 7 of them probably in a herd, as well as 2 more trackways and an isolated footprint; some unclassified dinosaurian isolated footprints or trackways—in this group we allot, among others, SOCA 422 and the group of tracks SOCA 13315, 1334, 13371, and 13372 (fig. 4.36); and 1 crocodiloid hand-foot set, attributable to the Batrachopodidae. With the exception of the sauropods, the tracks do not indicate group behavior.

The Associations and Behavior of the Dinosaurs at Piau-Caiçara

The same tracks that have been previously described, discussed, and classified are herein examined from the point of view of their associations or faunas, level by level, and of their behavior, especially their directions of travel (fig. 4.57 and Godoy and Leonardi 1985).

LEVEL SOCA 1

This is a small rocky surface immediately downstream from the dam of the Caiçara reservoir. It is always under water and could never be really examined for behavior.

LEVEL SOCA 2

Two isolated footprints of 2 different individual large theropods, attributable to *Moraesichnium barberenae*. Both trackmakers were traveling in an ~S58°E direction (fig. 4.22).

LEVEL SOCA 3 BIS

A small exposure of a layer immediately above the underlying level SOCA 3. There is 1 incomplete and badly preserved footprint, perhaps attributable to an ornithopod, and 2 incomplete footprints attributable to large theropods. The possible ornithopod print was directed in a ~NNE direction; both large theropods were directed to ~S67°W (fig. 4.22).

LEVEL SOCA 3

An incomplete small theropod trackway directed to ~S50°E, and an isolated theropod footprint directed to about N20°E, as well as some isolated theropod toes (figs. 4.22, 4.23).

LEVEL SOCA 4

This sample consists of 15 tracks (trackways or isolated footprints; figs. 4.22–4.23). Two of them are ornithopod, and the remainder are theropods. A rose diagram of the trackmakers' directions of travel (with sectors marked off at 10° intervals; fig. 4.57) presents two modes. One of them (N = 3) has an azimuth of 215°; the other (N = 2) points to the NNE. The orientations of almost all of the tracks (93.33%) fall into two circular sectors: 340°–39° (counting clockwise) and 180°–219°, which together

Figure 4.44. The main theropod trackways from site Piau-Caiçara (SOCA), level 13, subsite 3, Sousa, Sousa Formation. Graphic scale = 1 m. Drawing by Leonardi.

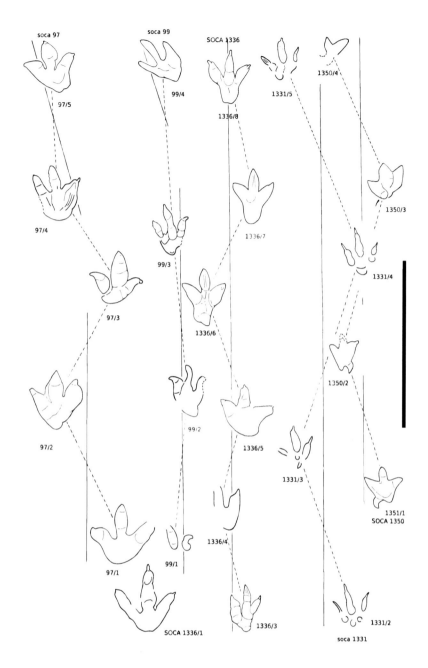

comprise a sector of 120° of arc (33.33% out of 360°); 12 tracks (80%) lie in the two sectors 10°–59° and 180°–219° (a combined sector of 90°; 25%). We thus note that overall the sample shows two opposite and preferred directions, with 9 and 5 tracks, respectively. The difference in the count between the two main directions of travel is 4 (26.66% of the total) in favor of the direction ~NNE. There is another track heading S78°E (azimuth 102°) and so crossing the general pattern shown by the ichnofauna of this level. This general course, with its two opposing directions, probably corresponds to a more frequently traveled lane, analogous to present-day game paths or corridors. This corridor was surely influenced, here as well

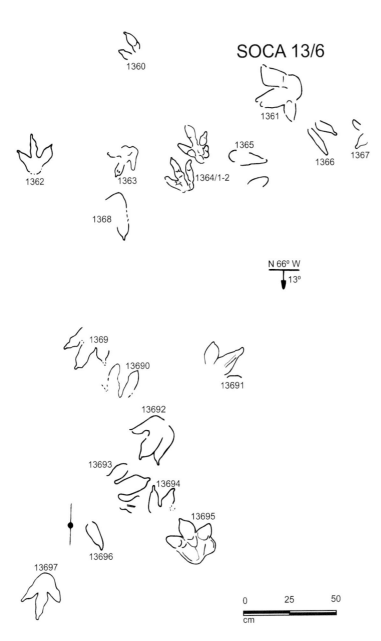

SOCA 13/6

N 66° W
13°

0   25   50
cm

Figure 4.45. Tracks from site Piau-Caiçara, level 13, subsite 6 (cf. the corresponding photomosaic, figs. 4.20, 4.46). The sample includes 19 isolated but fine footprints. Many of them are from theropods, but two of them can be attributed to ornithopods (SOCA 1361?, 13695). SOCA 1364/1–2 is a pair of adjacent theropod footprints, probably belonging to the same individual, which stopped on both feet (figs. 4.33, 4.70C). Graphic scale = 10 cm. Drawing by Leonardi.

as at many other levels at Piau-Caiçara, by the shoreline of the temporary lakes, as indicated by the orientation of the wave-ripple crests. These, in this stratum, are aligned N17°E (17° and 197°) and N75°E (75° and 255°). That corresponds almost perfectly to the preferred dinosaur travel directions. The trackmakers therefore moved along the shore, perhaps because the shorefront was free of dense vegetation or because it was a good place for feeding and drinking. Of course, it should be noted that there is an evident bias because the lake beaches and mudflats, with their moist sediment, could better receive and preserve the impressions of the animals' feet.

There is no difference between the direction of travel of the theropods and the ornithopods: the latter's tracks split into almost opposite directions (N26°E and S2°W), as do some theropods (15 individuals).

LEVEL SOCA 5

The sample consists of a single isolated theropod footprint, directed to ~N45°E, along with many small tracks, very probably attributable to half-swimming turtles (figs. 4.26, 4.27).

LEVEL SOCA 5 BIS

This sample consists of a trackway of 2 footprints and 6 isolated footprints (figs. 4.26, 4.27). The distribution of the directions of travel differs from that of all the other levels (fig. 4.57). All track directions at SOCA 5 bis fall into the two western quadrants (III and IV), within a preferred sector of 110° (210°–319°; 30% out of 360°). The distribution is unimodal. The direction of the mode (215°) corresponds satisfactorily with the two alignment directions of the ripple crests (196° and 212°). The main group (N = 5, 71.4% of the tracks) occurs in a sector of only 40° (210°–249°; 11.11% of 360°) and represents animals that moved parallel to the shoreline, southwestward; however, two other individuals moved in directions more or less orthogonal to the shoreline, indicated by the two series of ripple crests. The dinosaurs walking in directions contained in the III and IV quadrants were not matched by other animals heading in opposing directions of the other quadrants.

LEVEL SOCA 6

One large footprint attributed to a large theropod, directed to ~N74°W (286°), as well as some incomplete footprints.

LEVEL SOCA 7

The dinosaurian ichnofauna is limited at this level, with a single trackway and 5 isolated footprints (N = 6; fig. 4.28). However, there are many tiny tracks made by half-swimming animals, very probably attributable to turtles and perhaps a few very small dinosaurs that swam in very shallow water (probably 5–10 cm). The rose diagram (fig. 4.57) has a bimodal trend, with groups of 3 and 2 individuals, respectively, heading in directions similar to those of sample SOCA 5 bis, and a single individual that heads NE (44°). This last individual and those moving southwestward (mode: 235°) were paralleling the shore, as indicated by the wave ripple crests (50°–230°). Here also, similar to what was observed at other levels, 2 individuals were going off the shore, in directions almost orthogonal to the waterline.

At this level, almost all the individuals can be attributed to theropods, other than 1 footprint questionably attributed to the ornithopods. Here, too, no difference is seen in the behavior of the theropods and the

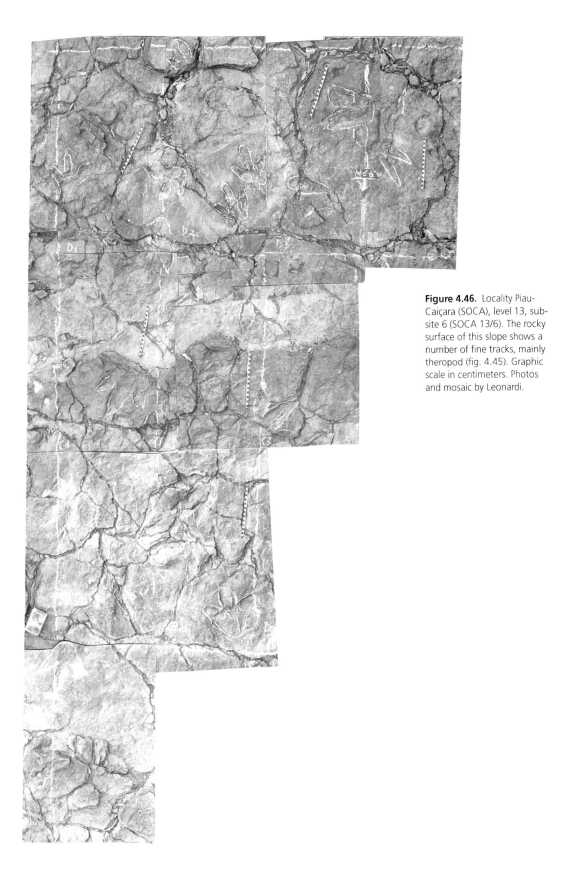

Figure 4.46. Locality Piau-Caiçara (SOCA), level 13, sub-site 6 (SOCA 13/6). The rocky surface of this slope shows a number of fine tracks, mainly theropod (fig. 4.45). Graphic scale in centimeters. Photos and mosaic by Leonardi.

Figure 4.47. Piau-Caiçara (SOCA), levels 13 and 16. *A*, View of the almost completely dry bed of the Peixe River at level 16 (gray rocks in foreground). *B*, Footprint of a half-swimming theropod from level 16. *C*, Medium-size theropod footprint SOCA 1321, which occurs with more 30 other theropod tracks on the same surface, and also raindrop impressions, level SOCA 13/2. This rocky pavement was completely destroyed by equatorial floods. *D*, a large theropod footprint from level SOCA 13. *E*, Another print of a half-swimming theropod from level 16. Graphic scales in centimeters.

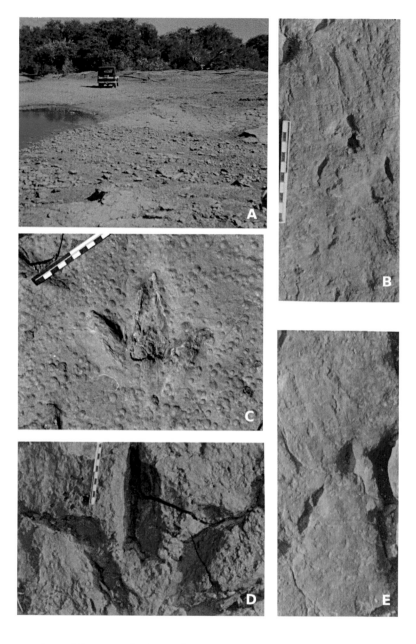

ornithopod because the latter footprint is almost parallel to 1 theropod footprint, and both of these prints correspond to the mode of the sample.

LEVEL SOCA 8

At this level are several incomplete, unidentifiable footprints, apart from one that may be attributed to a small theropod.

LEVEL SOCA 9

This stratum crops out in an exposure that is wider and at a higher elevation than most of the preceding levels, along a narrow stretch of the

Figure 4.48. Piau-Caiçara (SOCA), levels 15 and 16. *A*, Two isolated theropod footprints, SOCA 150/1 (deep) and SOCA 152 (shallow) from level 15. *B*, Possible ornithopod print, the only one attributed to this group from level 16 (SOCA 1621). *C*, Another view of the same footprints from *A*. *D*, Footprint of a half-swimming theropod track (SOCA 1656), level 16. *E*. A theropod footprint from level 15.

river, and it is positioned immediately above the underlying level 9 bis. Its ichnofauna consists of 9 trackways and isolated footprints (figs. 4.28–4.32). The distribution of the directions of travel of the sample (fig. 4.57) occurs almost totally in the western quadrants (III and IV) and is completely included in the sector 180°–19° (counting clockwise). Some concentrations of individual travel directions are found in the sector 1°–19° (N = 2) and in the group of sectors 180°–219° (N = 4). Together, these two sectors account for 60° of arc (16.66% of 360°) and include 6 individuals (66.66% of all the tracks). The concentration of travel directions in the west quadrants was also noted in the strata 5 bis and 7, thus showing a certain continuity of landscape and/or animal behavior between the times associated with levels 5 bis–9.

Figure 4.49. Locality Piau-Caiçara (SOCA), level 15 (SOCA 15, as in fig. 4.50). The rocky surface of this eroded slope shows a number of theropod and ornithopod footprints. Graphic scale in centimeters. Photos and mosaic by Leonardi.

LEVEL SOCA 9 BIS

The sample includes 8 isolated individual footprints. The directions of travel of the 9 bis dinosaurs (fig. 4.57) show an arrangement over a sector of 240° of arc (260° to 139°, clockwise)—a rather different situation from that of levels 5 bis–9. The other sector (140° to 259°) is wholly empty. A

SOCA 15

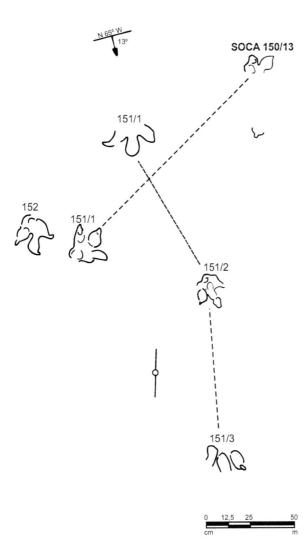

Figure 4.50. Locality Piau-Caiçara (SOCA), level 15. The drawing shows 3 sets of tracks: SOCA 150 is probably an ornithopod trackway; 151 and 152 are probably collapsed and partially deformed theropod tracks of the same kind. The arrow points to the direction of the ripple marks. Drawing by Leonardi.

noteworthy difference between this and the preceding three levels is that in the former the quadrants devoid of trackmaker travel directions were the eastern ones, unlike SOCA 9 bis. This points to a sudden change of landscape or local paleogeography, with its effects on behavior of the dinosaurs.

LEVEL SOCA 10

This bed displays just 1 isolated and incomplete footprint of a theropod of very large size (SOCA 101).

LEVEL SOCA 11

Some incomplete and unidentifiable footprints.

Again, some incomplete and unidentifiable footprints.

LEVEL SOCA 13

This level is actually a bundle of strata cropping out over a distance of 400 m along the northern side (left side) of the river Peixe, opposite level 12 (which is almost devoid of trace fossils), which crops out along the southern (right) side of the river (figs. 4.33–4.37, 4.57). This is controlled along this stretch by the outcrops of these two parallel strata sets and flows in its normal course between them (mean strike: N65°W; mean dip: 14° to the south). Eight track-bearing exposures were discovered along the outcrop of level 13 and were numbered from W to E as 13/1 to 13/8. At exposure 13/3, a very rich rocky pavement, three different levels were delimited as the lower, middle, and upper strata. It was not possible to obtain a good lateral correlation of the different strata of this thirteenth level among its different exposures, and so each site is described separately.

LEVEL SOCA 13/1

This is a trackless rocky pavement where we marked a permanent topographical point. It is part of the same level as SOCA 13/2.

LEVEL SOCA 13/2

This is a rocky pavement (11 × 3 m) of silty mudstone of deep brick-red color, wholly covered by the impressions of raindrops fallen (from north- and southward) after the crossing of the fauna: in fact, one notes that the raindrop craters are marked on the footprints as well as on the surrounding surface (figs. 2.7, 3.30, 4.35). The ichnofauna is here represented by the isolated footprints or very short trackways of 34 dinosaurs, almost exclusively theropods, along with a number of isolated digit impressions. The tracks are generally shallow and rarely well preserved, usually with very narrow toes, because of the collapse of the mud after the trackmakers' feet were lifted off the substrate (Farlow et al. 2012b, 52–54, figs. 17 and 18; Abbassi and Madanipour 2014, fig. 8). The surface was probably almost dry, so the trackmakers marked their footprints only where some puddles were left.

The orientations of the tracks (fig. 4.57) occur in sixteen out of thirty-six sectors of 10° width (61.5%). The most frequented sectors are 110°–129° (N = 7); 160°–179° (N = 4); 190°–239° (N = 10), and 340°–19° (clockwise, N = 8). No evident preferred directions are recognizable; no marks of wave ripples are present, either.

LEVEL SOCA 13/3

This is a large pavement (20 × 10 m) of deep gray siltstone, turning here and there to a deep brick-red or pink color due to the weathering.

Figure 4.51. *(facing left)* Locality Piau-Caiçara (SOCA), level 16. The dark-gray rocky surface of level 16, with a great number of prints made by half-swimming theropods. The surface is gridded in square meters, as in other mosaics, for purpose of mapping (figs. 4.34C, 4.47A, 4.47B, 4.47E, 4.48B, 4.51–55). Graphic scale in centimeters. Photo and mosaic by Leonardi.

Figure 4.52. *(facing right)* Map of the same locality and level as in figure 4.51. Survey and drawing by Leonardi.

The upper surface is sometimes covered by a black varnish. The level consists of three strata; a number of dinosaur tracks are seen in each (figs. 4.36–4.44):

> Upper stratum: the sample comprises 1 trackway and 20 isolated footprints, representing 21 individual theropods (fig. 4.57). The sector 210°–299° is empty. There is some concentration in the sector 0°–109° (30.55%), mainly corresponding to the NE quadrant, with 13 individuals (61.9%).
>
> Middle stratum: the sample includes 7 isolated footprints (5 theropod tracks and 2 crocodiloid footprints). These tracks are predominantly oriented (fig. 4.57) in two opposite sectors (20°–29° and 180°–209°), together amounting to a sector of 40° (11.11%) and comprising 85.7% of the ichnofauna. Only 1 footprint points to the east.
>
> Lower stratum: the sample comprises 18 individual theropods represented by 3 trackways and 15 isolated footprints. The pattern of trackmaker travel directions (fig. 4.57) is clearly bimodal; the modes are in the sectors 40°–49° and 230°–239°. This N and NE sector of 140° (330°–109°, counting clockwise; 38.88% out of 360°) includes 13 individual dinosaurs (72.22%).

The direction of the mode (45°) and that of the mean direction (58°36') are not so different from the orientation of the shoreline (76°) indicated by the wave-ripple crests in an adjacent thin layer. The group of tracks of the SW quadrant (N = 5) has travel directions in a sector of 60° (180°–239°; 16.66% out of 360°) but comprises 27.77% of the ichnofauna. The mode (235°) is not far from the direction of the crests of the ripple marks (256°). The sectors 330°–109° and 180°–239° together form a sector of 200° (55.55% of 360°) that includes all of the ichnofauna (100%). The two sectors seem to represent a preferred corridor or path nearly parallel to the shoreline. The difference in the counts of trackmaker directions between the two dominant sectors is 8 (44.44%); the trackmakers were mainly moving southwestward.

In addition to the theropod tracks previously described are 26 large, rounded, and very deep subtracks (diameter up to about 120 cm; minimum diameter about 50 cm) that occur in one or another of the three strata at subsite 13/3 and that probably were marked, as original tracks, on a higher stratum that has been destroyed by erosion. These footprints, attributable to sauropods because of their general shape and very large size, are perhaps arranged in 7 short and discontinuous trackways (figs. 4.36 and especially 4.20) whose direction of travel would lie in the sector 25°–60° (or, alternatively, 205°–240°), a circumscribed arc of only 35°, which suggests the crossing of a herd of large sauropods, some of which would have been true giants, weighing scores of tons. They impressed their footprints not only on the superficial stratum of mud but in several deeper layers, among them those three we are dealing with. The direction

Figure 4.53. Paleoenvironmental reconstruction of the circumstances under which footprints and trackways of half-swimming dinosaurs (lower right insert and figs. 4.34C; 4.47A, 4.47B, 4.47E, 4.48B, 4.51–55) are thought to have been made. Most such prints are interpreted as having been made by theropods. Dinosaurs progressing in this manner half-swam in rather shallow water in the temporary lakes, leaving in the sediment marks of just the distal tips of the toes, claws, or, in one case, hooves (of a single half-swimming ornithopod present at this level). Perhaps the carnivorous dinosaurs were "fishing." Drawing by Leonardi.

data of these 7 trackways was not recorded on the rose diagrams because their actual directions of travel are not certain, and because they, as true tracks, are pertinent to another higher level that no longer exists. However, their directions correspond well to the general trends of site 13/3.

LEVEL SOCA 13/5.

A theropod trackway attributed, from a morphological point of view, to Gigandipodidae, pointing to N58°W (figs. 4.20, 4.39A, 4.39C, 4.44, last at right).

LEVEL SOCA 13/5 BIS.

Here is a single tridactyl footprint, possibly made by a half-swimming theropod, but different from all examples of this kind (SOCA 1350 bis;

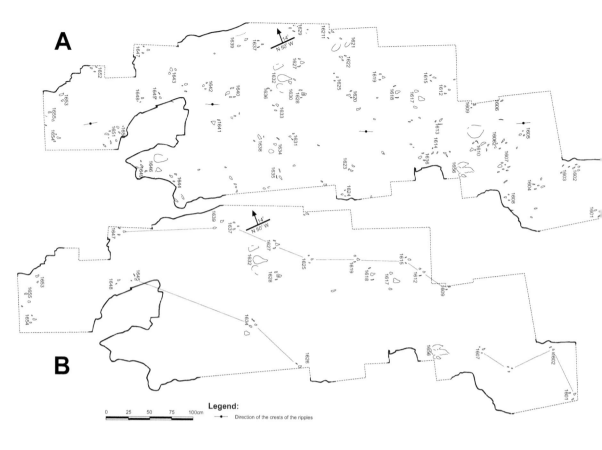

A

B

Legend:
0   25   50   75   100cm
→— Direction of the crests of the ripples

Figure 4.54. *(above and facing)* Locality Piau-Caiçara, level 16 (SOCA 16). The same rocky pavement with numerous footprints and trackways of half-swimming theropods illustrated in figures 4.51–4.52. A, Overall map of the site, with short line segments indicating the local orientation of ripple mark crests, which are parallel to the direction of almost all the tracks; B, Tracks heading more or less southward, parallel to the long axis of the outcrop and also to the orientation of the ripple crests (presumably parallel to the ancient beach); C, Tracks crossing the outcrop along its lesser axis, more or less eastward or westward; D, Tracks again oriented parallel to the long axis of the site (parallel to the beach), but in this case headed in the opposite direction (more or less northward) from those shown in B. Survey and drawing by Leonardi.

fig. 4.20). It could instead be identified as a pes print of a half-swimming crocodile (Vila et al. 2015; fig. 2 and comment).

LEVEL SOCA 13/6

This site presents many tracks on a small surface (3 × 2 m; figs. 4.45, 4.46). The sample includes 19 isolated footprints, 13 of which are complete and are herein examined. Most of them are assigned to theropods, but two or three of them can be attributed to ornithopods. There are two preferred directions of travel (fig. 4.57), in opposite sectors. One group of tracks occurs in the sector 340°–118° (clockwise) and includes 1 short trackway and 6 isolated footprints (N = 7); the other group occupies the sector 210°–279° with 6 isolated footprints. Together, these two sectors are a sector of 190° (55.77% out of 360°) and contain all the fauna (100%). Sensu lato, we can speak of two preferential corridors, mainly in the I (NE) and III (SW) quadrants.

The direction of the shoreline, as shown by the ridges of the ripple marks, oriented 25°/205°, corresponds not too badly to the median of the first sector (30°), but the mode is more northerly. Almost all of the individuals therefore proceeded parallel to the shoreline. The probable ornithopod track directions are in the NW quadrant.

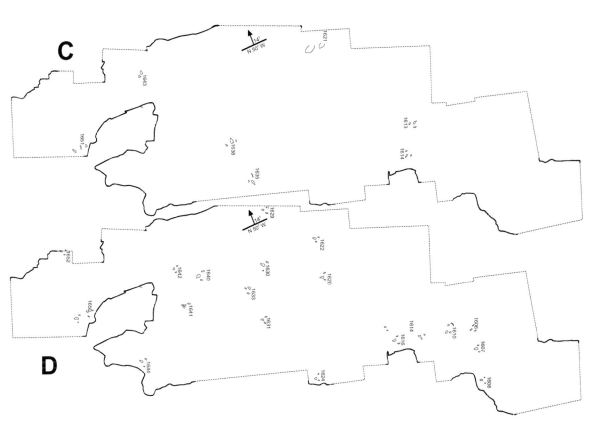

LEVEL SOCA 14

Here we find 1 isolated and poorly preserved footprint (SOCA 140) attributed to an ornithopod (fig. 4.32).

LEVEL SOCA 15

The very small sample (2 short trackways and one isolated footprint) does not allow firm conclusions (figs. 4.49, 4.50, 4.57). However, trackway SOCA 151 and footprint 152, both theropod, follow the same general NE and SW (average = 216°) trend, one that is not very different from that of the ripple crests (41°/221°); trackway SOCA 150, probably made by an ornithopod, has a very different azimuth (85°). SOCA 153 is a short trackway attributable to a small theropod or ornithopod; SOCA 154 was made by a very small theropod.

LEVEL SOCA 16

This stratum crops out in a pavement of 7 × 2 m and shows a large sample of 40 trackways and isolated footprints (figs. 4.50–4.55). All of them seem to have been made by small- and medium-size theropods swimming in shallow water (1–1.5 m) and thrusting themselves along the bottom with the tips of their hind-feet. These footprints are very similar to those published by Coombs (1980) from the Lower Jurassic of Connecticut,

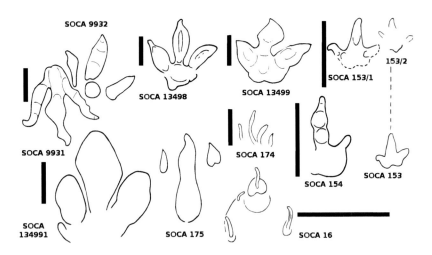

Figure 4.55. Mostly isolated footprints from the upper levels of locality Piau-Caiçara (SOCA). SOCA 9932 and 9931 are tracks of 2 large theropods (see also fig. 4.31*F*); the maker of footprint 9931 (at left) passed by after the maker of SOCA 9932 (at right), laterally compressing its footprint. SOCA 13498 is a theropod print from level 13, subsite 4; the blunt toe impressions are marked by long furrows produced by the claws during retraction of the foot. SOCA 134991 could be a blunt-toed theropod or an ornithopod. SOCA 153 is a short trackway attributable to a small theropod or ornithopod. SOCA 154 was made by a very small theropod. SOCA 174 is a track of a half-swimming small reptile, probably a turtle or a lacertoid, rotating its foot. SOCA 175 is a print of a half-swimming theropod, and SOCA 16 is an even better example (one of the best such tracks) of a print made by a half-swimming dinosaur at level SOCA 16 (see also figs. 4.51–4.54). In this footprint, the dynamic of the motion of the foot is particularly clear. Graphic scale = 10 cm. Drawing by Leonardi.

and their trackmakers were surely similar to the *Eubrontes* maker, as demonstrated by Coombs. The distribution of the trackmakers' directions of travel is clearly bimodal, with modes at 75° and 340° (fig. 4.57).

The sample of the SE quadrant (N = 19) is itself bimodal (modes in the sector 140°–149° [4 individuals] and 170°–179° [5 individuals]). The quadrant contains 47.5% of all of the ichnofauna of this level. The sector 140°–189° contains 42.5% of the sample and 89.47% of the SE quadrant. It evidently indicates a preferred corridor. The mean of the track directions in the II quadrant is 153°; the mode is 175°. This matches fairly well the orientation of the wave-ripple crests (average = 151°/331°).

The sample of the NW (IV) quadrant (N = 15) shows a unimodal, strongly peaked curve. It includes 37.5% of the ichnofauna; the sector 330°–349° (N = 10) includes 25% of it. This is another preferred corridor. The direction of travel of the dinosaurs of this sector corresponds very well with the orientation of the ripples (average = 151°/331°, as previously noted); the mode is 340°.

The NE and SW quadrants (I and III) are almost devoid of tracks, containing only 6 footprints (15% of the ichnofauna). These are concentrated in the sectors adjacent to the two preferred corridors. The sector 210°–229° may indicate a secondary corridor, representing individuals swimming transversely to the ripple mark orientations. So, they swam either away from the shore or toward it.

Evidently, the movements of the theropods of this ichnofauna were strongly bipolar, with two opposite mean directions along an axis positioned at about 160° or 339° of the rose diagram. This is very similar to the orientation of the ripple marks (151°/331°). Some 67.5% of the sample is concentrated in the opposite sectors 140°–189° and 330°–349° (together, 19.44% out of 360°). There is a slight preponderance of dinosaurs trending NNW as opposed to SSE (10%).

It seems probable that as a rule the theropods of this fauna swam in shallow freshwater parallel to the beach; occasionally some of them went away from the shore or moved toward it. One has the impression that, after reaching their position in the lake, they coasted for a long time, perhaps searching for fish or other aquatic animals like turtles, before coming back to the shore.

We will now examine all the ichnological material, identifiable and measurable, from the Piau-Caiçara locality, excluding the innumerable small footprints attributed mainly to half-swimming turtles. It is a large sample of 194 individuals (data from 1985, by which time almost all the best specimens from this site had been discovered and recorded), almost all of them dinosaurs.

The directions of travel shown by the trackways were specifically studied (Godoy and Leonardi 1985). If we organize these data in a general rose diagram (fig. 4.57) or histogram, the observed distribution is essentially bimodal, with modes in the SW and NW quadrants. However, this summary distribution is of limited value because it combines data for the first through fifteenth levels on the one hand, with data for the sixteenth level (the lowest and oldest level), the latter of which is a large sample (40 individuals; 20.6% of the Piau-Caiçara ichnofauna) with a different directional trend than that of levels 1–15. Restricting ourselves to the sample from levels 1–15 (N = 154), the distribution is clearly tetramodal, with two dominant modes in the NE and SW quadrants, and two secondary modes in the other two quadrants. Considered in this way, the distribution of trackmaker directions of travel falls into four groups, which are separated by three sectors of 10° arc (140°–149°; 270°–279°; 320°–329°) that lack any footprints, and by a sector with just 1 individual (70°–79°).

The mode of the group of NE-trending tracks (and adjacent sectors) is 9 at N15°E (azimuth 15°); this group occupies a sector of 100° (330°–69° clockwise; 27.77% out of 360°), presents a sample of 58 individuals (37.66%), and is no doubt a preferred corridor.

The group of SW-trending tracks (and adjacent sectors) has mode 15 at S35°W (azimuth 215°), occupies a sector of 120° (150°–269°; 33.33%), and presents a sample of 60 individuals (38.96%). If one considers just the sector 180°–239°, of 60° (16.66%), the sample consists of 49 individuals (31.81%). This is therefore the main concentration of animals and thus the main travel direction corridor.

The group of SE-trending tracks, whose mode is 5 at S75°E (azimuth 105°), occupies a sector of 60° (16.66%) and has a sample of 24 individuals (15.58%).

The group of NW-trending tracks, the one with the smallest count, has its mode 5 at N75°W (azimuth 285°), occupies a sector of 40° (11.11%), and presents a sample of 12 individuals (7.79%).

Thus, there seems to have been a main direction of travel or corridor, oriented NE–SW, along which dinosaurs moved in both directions. There were slightly more animals proceeding southwestward, with their paths more concentrated in a narrow sector with a clearer orientation.

Figure 4.56. A small collection of loose, scattered pieces of slabs containing footprints, collected along the bed of the Peixe River at Piau, where they were deposited during the rainy season. *A,* SOCA 173, a theropod footprint. *B,* SOCA 172, an ornithopod footprint. *C,* SOCA 174, a footprint recording rotational movement of the foot of a half-swimming small reptile, probably a turtle or lacertoid. *D,* SOCA 175, a footprint of a half-swimming theropod. *E,* SOCA 171, a group of small prints probably attributable to half-swimming turtles. *F,* SOCA 176, a print of the same kind as in *D. G,* SOCA 170, a possible ornithopod footprint. The caliper is open to 5 cm in all photographs.

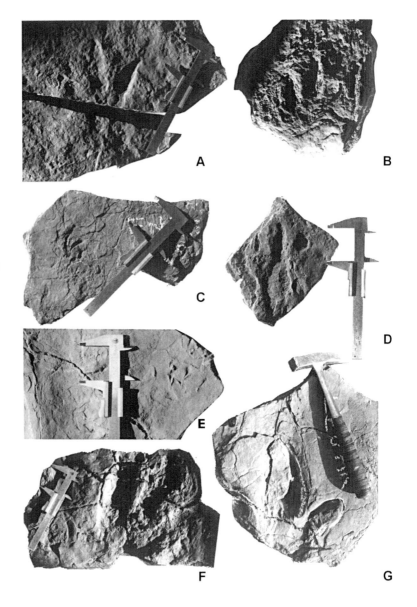

The very high figure of the mode (N = 15) stresses this last point. Together, the two NE and SW groups make a sector of 220° (61.11%) that includes 118 individuals (76.62%).

In addition to this dominant set of two opposite directions, there is another set of directions, secondary but also well defined, distributing its data around the axis 75°–255°. Here one also finds two (lesser) groups, moving in opposite directions, with a clear predominance of the group heading SE. The southeast and northwest sectors together constitute 100° of arc (27.78%) and have a sample of 36 individuals (23.38%).

Let us now examine successively and in graph the directions of movement of the animals of the Piau-Caiçara farm, from the oldest to the youngest levels, and observe the changes. The lowest level, SOCA 16, shows a main corridor corresponding to an axis running 160°–340°.

POLAR HISTOGRAMS

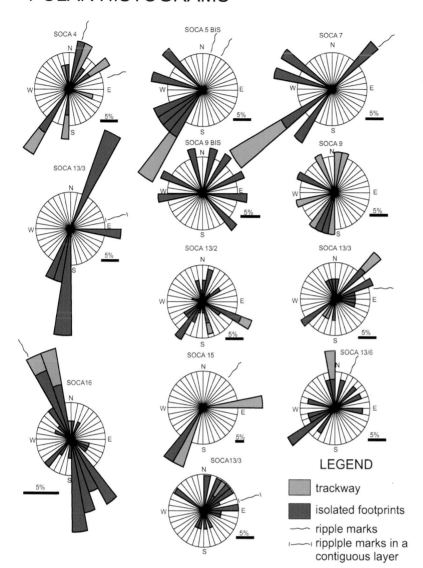

Figure 4.57. Rose diagrams of the directions taken by the trackmakers in several levels of the Piau-Caiçara farm site. Redrawn from Godoy and Leonardi 1985. Compare with figures 4.18–4.21, 4.35, 4.36, 4.45, and 4.51–4.54.

LEGEND

▨	trackway
▬	isolated footprints
～	ripple marks
⌐―⌐	rippple marks in a contiguous layer

Starting from the next level (15), we note a substantial shift of the pattern, corresponding to a clockwise rotation of about 55°. The ichnofaunas of levels 15 up to 4 present a nearly constant pattern, with a dominant orientation (in both ways) ~NE–SW, in which the data place themselves around an axis at about 35°–215°. However, one must note that in levels (and sites) 15 and 13/6 the situation is not so clear. Levels 1, 2, 3, and 3 bis present very small samples with diverse patterns.

The patterns of orientation of the wave-ripple crests (when ripple marks are preserved) and the corresponding presumed shorelines of the temporary lakes in this locality show corresponding changes. In level 16 their orientation is ~151°–331°; but from level 15 to level 4 their orientation lies always in the sector 16°–75° (and 196°–255°), with a mean direction

41°–221°. Within this sector one notes a counterclockwise shift from level 15 to 13/6; a clockwise shift from 13/6 to 13/3; and a long counterclockwise shift from level 13/3 to level 4. The clockwise shift between the main direction for level 16 and the main direction of levels 15 to 4 is of 70°—a figure not very different from 55°, pertinent to the main shift of the animals' travel directions from the sixteenth level to the other levels (a difference of 15°).

It seems evident that there was a major change in the landscape between the time associated with level 16 and the time associated with the following level 15 and that later there was a remarkable constancy to the landscape's structure. On the other hand, the dominant directions of the shorelines, as indicated by the crests of the ripple marks, are frequently parallel to some regional faults (rejuvenated in the Early Cretaceous) that gave origin to the Sousa basin, in association with the opening of the South Atlantic Ocean. It is evident, therefore, that these dinosaur populations and associations were in general strongly influenced in their movement by the local and regional landscape, in particular by the bodies of water (waterlines) (Moratalla 2012) and, indirectly, by the regional tectonic texture, traveling along preferential corridors (Godoy and Leonardi 1985; Leonardi 1989a, 1989b).

Piedade (SOPI)

The Piedade farm lies mostly in the municipality of São João do Rio do Peixe, but along the Peixe River it borders the Pedregulho farm inside the Sousa municipality. Access to the tracksite is through a fenced road ("*corredor*") leading to the main farmhouse, in the territory of the São João do Rio do Peixe municipality. However, the small rocky surface where the tracks occur is situated across the river, in the Sousa territory, so one must enter the farm from the São João do Rio do Peixe side and wade across the river to reach the outcrop. This is situated on the Peixe River bed, in the Piedade farm, a few hundred meters from the main farmhouse, 13 km as the crow flies WNW of Sousa (S 06 44.927, W 038 20.954). This locality was discovered by Leonardi in 1977.

The outcrop comprises a thick sedimentary section and a corresponding rocky pavement in the Sousa Formation. The tracks are concentrated on the right of the river, in three contiguous layers of gray siltstones (N58°W; 11° to south), with an outcrop of 14.5 × 2 m, corresponding to about 29 m² (figs. 4.58, 4.59A, 4.59B, 4.60B, 4.61). It contains tracks of 7 individuals, with a concentration of 1 individual track per 4 m², or 0.25 per m².

References: Leonardi, 1979c, 1980b, 1981a, 1989a, 1989b, 1994; Leonardi et al. 1987a, 1987b, 1987c.

This is a diverse ichnofauna, dominated by large theropods. It includes 1 short trackway and some isolated prints attributed to the ichnogenus *Eubrontes* E. Hitchcock 1845, and isolated footprints of ornithopods.

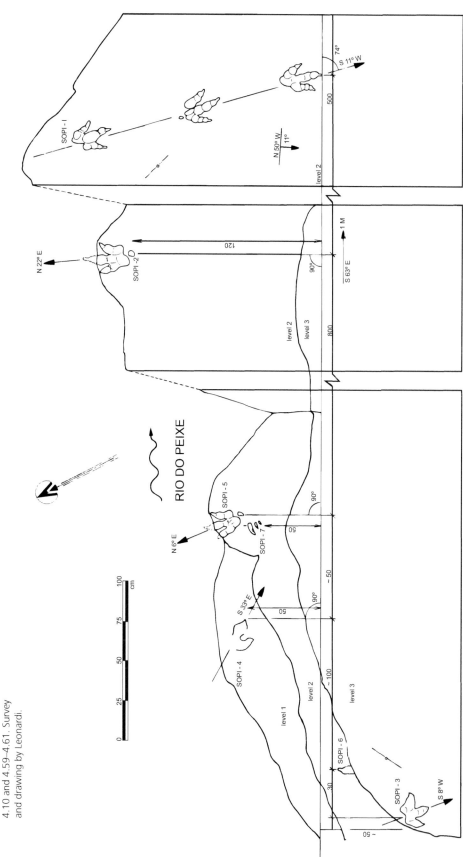

Figure 4.58. Map of the ichnofossiliferous outcrop at Piedade Farm, Sousa, Sousa Formation. Theropod and ornithopod tracks were discovered here. See also figures 4.10 and 4.59–4.61. Survey and drawing by Leonardi.

Figure 4.59. Representative footprints from the ichnofaunas of Fazenda Piedade (SOPI) and Pedregulho (SOPE), Sousa. See also figures 4.10 and 4.58. *A*, The fine theropod trackway SOPI 1 soon after excavation in 1979. *B*, The small outcrop with its levels and ichnofauna. *C*, SOPI 5, a shallow left theropod footprint at the margin of the exposure, associated with an asymmetrical print probably made by a half-swimming theropod. *D*, SOPI 3, a probable ornithopod left footprint. *E*, SOPI 4, a small ornithopod footprint. *F*, SOPE 3, an isolated theropod digit at Pedregulho site (fig. 4.10). Graphic scale in centimeters.

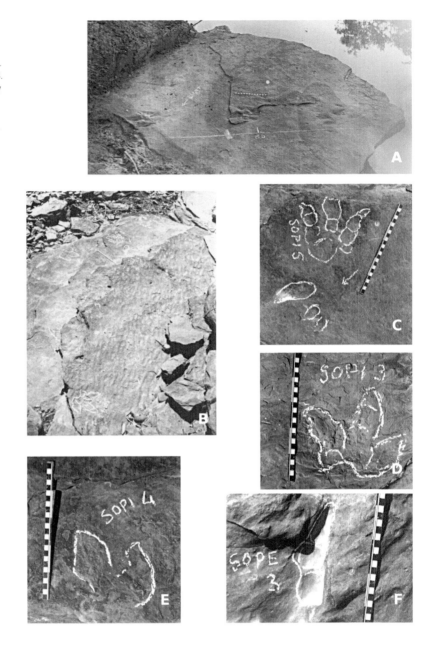

SOPI 1

This is the only complete trackway at this outcrop, with 3 footprints of a theropod, shallow but well impressed and preserved, crossing the narrow slab in situ (figs. 4.10, 4.58, 4.59A, 4.60A, 4.60C, 4.60D). The footprints are tridactyl, subdigitigrade, mesaxonic, and moderately long (L/W ratio averages 1.35). The toes are slender. Digit II is almost completely separated from the other two toes; digits III and IV are joined, with III much longer than IV. Digit II is clearly shorter than digit IV and proximally shows an indentation that would correspond in the living animal to a distinct parasagittal furrow on the "heel." The claws are long and acuminate

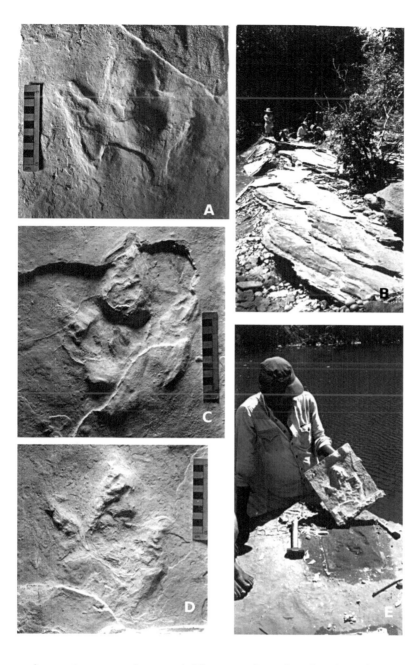

Figure 4.60. Fazenda Piedade (SOPI; see also figs. 4.10, 4.58). *A*, *C*, and *D*. Plaster casts of the footprints of the trackway SOPI 1, courtesy of the Câmara Cascudo Museum of Natal, Rio Grande do Norte. *A*, SOPI 1/3 (left pes). *C*, SOPI 1/2 (right pes). *D*, SOPI 1/1 (left pes). *B*, the little ichnosite at Piedade farm, with the staff of the Câmara Cascudo Museum of Natal at work. *E*, Claude Luis Aguilar Santos casting SOPI 1/1.

and sometimes turned outward. There are three digital pads in digit II, four in III, and five in IV. As also seen in SOPI 2, the proximal pad of digit IV is clearly separated from the other, more distal digital pads in the second footprint of SOPI 1; this proximal feature is absent in the other 2 footprints.

The trackway is rectilinear, with very high pace angulation (176°30') and very low width (negative breadth between tracks: -20 cm). However, the stride is relatively short (SL/FL ratio = 5.4), indicating walking at a low speed, in which the trackmaker was slowing down. This was a medium-size theropod. The divarication of the foot from the midline is negative

Figure 4.61. Paleoenvironmental reconstruction of the sites of the Sousa Formation, with ripple marks in the substrate and a running, medium-size theropod. Drawing by Leonardi.

(inward rotation with respect to the direction of travel): average -10°40'. The animal moved S11°W. A plaster cast of this track is kept at the Câmara Cascudo Museum of Natal-Rio Grande do Norte, Brazil.

SOPI 2

This is an isolated and incomplete left footprint, heading N22°E, in the opposite direction from SOPI 1, which probably had the same kind of maker (figs. 4.10, 4.58). However, SOPI 2 was a different, larger individual.

SOPI 3

This is another isolated footprint, of poor quality (figs. 4.10, 4.58, 4.59D). It is perhaps a left and is tridactyl, digitigrade, and mesaxonic, and quite unlike the 2 previously described tracks. It comes from the upper layer, so it is from a different association. The footprint is wider than it is long (width/length ratio = 0.86) and of small to medium size (18.5 × 21.5 cm).

The digits are broad and spatulate, with blunt and hooflike rather than acuminate tips. Digit III seems to show three pads. The proximal margin of the footprint is concave, with a wide indentation, corresponding to the advanced forward position of digit III and separating digit II (the shortest) from digit IV. The trackmaker was likely an ornithopod and was heading S8°W.

SOPI 4

This is another incomplete and poor-quality footprint, from the lowermost layer (figs. 4.10, 4.58, 4.59E). It seems similar to SOPI 3 but could be a deformed footprint of a theropod. It is tridactyl, digitigrade, and broad-toed, with a spatulate digit III; all three toes are blunt. The trackmaker was heading approximately S33°E.

SOPI 5

An isolated left footprint (figs. 4.10, 4.58, 4.59C), heading N6°E, of the same form and size as SOPI 1.

SOPI 6

An isolated digit III impression, perhaps of the same form as SOPI 5 (fig. 4.58). The foot was probably directed approximately N26°E.

SOPI 7

An isolated footprint of a half-swimming animal (fig. 4.58). Its three digits are separated and slightly divergent; the rightmost digit is the longest, the others diminishing in length in sequence leftward. It seems that the foot of the trackmaker was lacertoid—that is, strongly asymmetrical and ectaxonic. Identification seems impossible.

This ichnofauna is, in fact, distributed in three levels: the lowermost stratum (1) shows just SOPI 4; the main, middle stratum (2) includes SOPI 1, 2, 5, 6, and 7; the upper stratum (3) shows only SOPI 3. Stratum 2 also shows mud cracks. The ichnofauna of the middle stratum (2) is completely different from the other two. Both the lowermost (1) and uppermost (3) strata each show a single footprint of a bipedal dinosaur of small to medium size, probably an ornithopod. In contrast, the middle level records a small assemblage of 3 or 4 individual theropods, all of them of the same form and nearly similar size (although SOPI 2 is 15–20% larger than the others), quite probably the same Linnaean species. SOPI 2, 5, and 6 had similar travel directions (respectively N22°E; N6°E; N26°E), an arc of only 20°; but SOPI 1 headed in the opposite direction.

All the tracks at this site are of comparable depth and show displacement rims of the same size. The individuals of this site proceeded in two directions, back and forth along a corridor oriented approximately NNE to SSW (N6°E–N26°E to S8°W–S11°W). The dinosaurs seem not to have been moving in groups. This orientation is nearly the same as that of the crests of the ripple marks (and thus of the shoreline) in the middle (and main) layer (N16°W). However, the orientation of the ripple crests in the uppermost layer, where the single track SOPI 3 is found, is different: N61°E.

Riverbed of Rio do Peixe between Passagem das Pedras and Poço do Motor

This locality corresponds to almost 1 km of the rocky bed of the Peixe River, containing ten levels with fossil tracks. Passagem das Pedras is the locality discovered by Luciano Jacques de Moraes in the 1920s. Collectively among the richest sites, in terms of the number of fossil tracks, the portions of the bed of the Peixe River that are locally known as Passagem das Pedras and Poço do Motor are in our basin, as well as the stretch between them and the adjacent sectors upstream and downstream. These add up to almost a kilometer of exposed rocky surface, which is only rarely covered by sand dunes, except for the main pavement. The surface is dry in the drought season (June to January) and frequently flooded in the rainy season (January to May). The stratigraphic sequence here consists mainly of brick-colored, poorly stratified mudstones and siltstones, alternating with rare harder and darker ichnofossiliferous siltstones or fine sandstone strata, frequently dark gray or black-varnished on the surface (fig. 4.62).

The main site of Passagem das Pedras, with its large rocky pavement of the ford (fig. 4.66), shows two contiguous ichnofossiliferous strata (levels 3–4); one of the other levels (6) is sterile. The sterile exposures among the track-bearing levels, reckoning from upstream to downstream (measured horizontally along the riverbed) are 200, 100, 140, 40, 90, 45, and 140 m. The mean dip is about 6° to the south in the first half (upstream) and about 12° to the south downstream. The overall thickness of the whole sequence (or section) is about 100 m.

The ichnofossiliferous levels contain tracks of about 30 dinosaurs, among which 1 was quadrupedal, 1 was semibipedal, and 28 were bipedal. Adjacent stretches of the river bed, upstream as far as the Jangada farm, and downstream as far as the Lagoa dos Patos site, were thoroughly explored, but no good tracks were discovered. The concentration of the tracks does not show evident patterns of abundance but seems to be random, depending especially on the area of the exposed surface.

References: Moraes 1924; Cavalcanti 1947; Price 1961; Haubold 1971; Leonardi 1979b, 1979c, 1979d, 1980b, 1981a, 1984a, 1984b, 1985a, 1989a, 1989b, 1994; Leonardi et al. 1987a, 1987b, 1987c; Lockley 1991; Thulborn 1990; Carvalho et al. 2013; Santos et al. 2015, 2019; other authors.

On the whole, the main pavement includes the large iguanodontid trackway *Sousaichnium pricei* Leonardi, 1979, the ornithopod trackway *Staurichnium diogenis* Leonardi, 1979, and at least 5 trackways of the long-heeled theropod ichnotaxon *Moraesichnium barberenae* Leonardi, 1979. In the other levels, there are trackways and isolated footprints of 8 large theropods, 1 small theropod footprint, 1 ornithopod track, and a trackway (undertrack) of a small quadrupedal dinosaur that is difficult to identify. The directions of travel of all the dinosaur tracks in the riverbed between

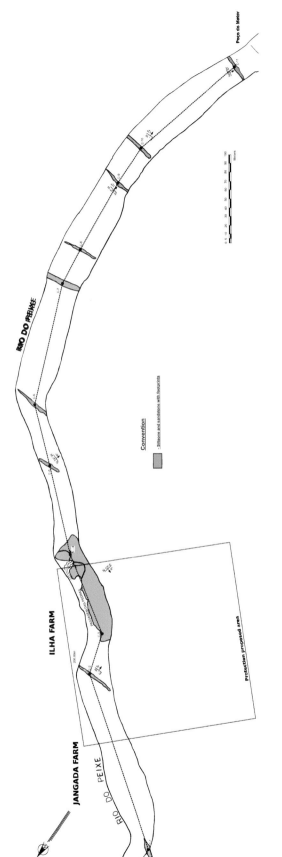

Figure 4.62. Map of the localities Passagem das Pedras (SOPP) and Poço do Motor (SOPM), Sousa Formation, Sousa. The map shows the bed of the Peixe River from the Jangada farm locality at the upstream end (left) through the Poço do Motor locality at the downstream end (right), from level 1 at the left to level 11 at the right. See also figs. 4.63–4.80. Survey and drawing by Leonardi.

Passagem das Pedras and Poço do Motor lie mainly in the II (SE) and III (SW) quadrants of the compass in the lower levels, with a tendency to shift from east to west as one moves upward through the levels. However, upstream of the large pavement of Passagem das Pedras, the trend passes to the IV (NW) and I (NE) quadrants. The overall distribution of the SOPM-SOPP site is almost random. However, some concentrations do exist (10 tracks [29.4%] in the sector 135°–199° [17.7%]).

Poço do Motor (SOPM)

This locality is in the riverbed of the Rio do Peixe, a short distance downstream from Passagem das Pedras and upstream of Lagoa dos Patos (S 06 45.688, W 038 14.751), and of a paved ford (*lajão*; recently replaced by a bridge), of the old unimproved Sousa-São João do Rio do Peixe road. Above the six older ichnofossiliferous strata of Passagem das Pedras, there are three siltstone layers with tracks in the riverbed.

Beginning with the uppermost level and working downward, there are (fig. 4.63, bottom) a stratum of dark-gray siltstone (level 11) cropping out from the river with a transverse exposure of 40×3 m across the river bed, showing mud cracks and ripple marks, overlying other dark siltstone beds. This surface shows one theropod trackway (SOPM 1) heading N58°W and two isolated footprints (SOPM 2 and 3) directed to N4°W. All three tracks have the same form. The strike of the stratum is N88°W, the dip 12° to the south. The concentration of tracks here is of 1 dinosaur per 40 m² or 0.025 per m². Trigonometric point or theodolite landmark E-11.

A siltstone layer (level 10) with outcrop dimensions of 40×1.5 m, again shows mud cracks and ripple marks, bearing 3 short theropod trackways of three different forms, heading in three different directions (SOPM 4 to S67°W; SOPM 5 to S19°W; SOPM 6 to ~N44°E), and thus displaying no gregarious behavior. The concentration of tracks is 3 dinosaurs per 60 m² or 0.05 per m². Strike E–W; dip 12° to the south. Theodolite landmark E-10.

A stratum of gray-greenish fine siltstone or mudstone (level 9) crops out 40×1.5 m, with 1 trackway (SOPM 7) heading S32°E and 1 isolated footprint (SOPM 8) heading S45°E. The 2 trackways belong to two different forms attributable to theropods. The concentration of tracks is low, with 1 dinosaur per 30 m² or 0.033 per m². Strike N82°W; dip 12° to the south. Theodolite landmark E-9.

All the tracks at this locality were left in situ.

SOPM 1

This is a 4.5–m-long trackway of 6 footprints, documented over the course of 10 years of successive expeditions (1977–1987; figs. 4.63, 4.64A, 4.78). During this time erosion by the river progressively eroded away the more exterior part of the exposure, destroying a very fine footprint but also excavating new footprints at the base of the sloping head of the stratum, during the annual flood season (January–May). The footprints are large and wider than they are long (L/W ratio = 0.91) due to the high digital divergence (II^IV = 83°) and abduction of the distal part of the extreme digits, as well as because of the unusually large distances separating the bases of the toes (very wide hypexes)—something rarely seen in theropod footprints. The toes were stout but flexible and mobile, with a strong callosity covering the pads on the "plantar" surface, and show long triangular claws. Digit III is spatulate.

SOPM 1 is an interesting theropod trackway, straight, whose maker was reducing speed (fig. 5.1). Thus, the pace angulation progressively diminishes from 158° to 145° (the last value rather low for a theropod), and the stride from 191 to 133 cm. The trackway widths are uniformly high for a theropod; negative values of the breadths between tracks, for example, are low, pointing to a heavyset animal walking at a slow speed: 5 km/h at the beginning of the track and 2.6 km/h at the end of it (mean = 4.2 km/h). During the quicker phase, the footprint axis was directed inward (negative rotation), but during the phases of slower walking the print axis was parallel to the midline. The displacement rim is higher in correspondence of one side or another of digit III, indicating the high mobility of this digit. The maker was directed N58°W.

SOPM 2

(figs. 4.63, 4.65). An isolated footprint, incomplete, very similar to SOPM 1, with a spatulate digit III. Stoutly constructed claws. The print is the same morphotype as SOPM 1. The trackmaker headed northward.

SOPM 3

An isolated, incomplete footprint, very similar to SOPM 1, with a strongly curved digit II (figs. 4.63, 4.64B, 4.65). It seems that the toe was truly laterally flexible because the curved impression of it is not a drag mark but a true morphological impression.

SOPM 4

A theropod trackway of 4 footprints (figs. 4.63, 4.64E, 4.65). These are typical of a large theropod; however, they vary greatly in shape along the trackway due to their poor quality, particularly in the preservation of the hypexes and the claws. For example, in the second footprint the digit III impression is very broad distally, but this is an extramorphological feature; the true shape of the trackmaker was probably more like that seen in the first footprint.

This trackway has pace angulation of ~180°; the external trackway breadth is equal to the mean footprint width. The SL/FL ratio is 6.2, and the inferred speed is 5 km/h. The divergence of the footprint axis from the midline is variable. The trackway heads S67°W.

SOPM 5

This is a short trackway of 2 footprints belonging to a small theropod (figs. 4.63, 4.64C, 4.65). The SL/FL ratio was probably about 7. The footprints are small, subdigitigrade, and deepest at the tips of the toes, which are

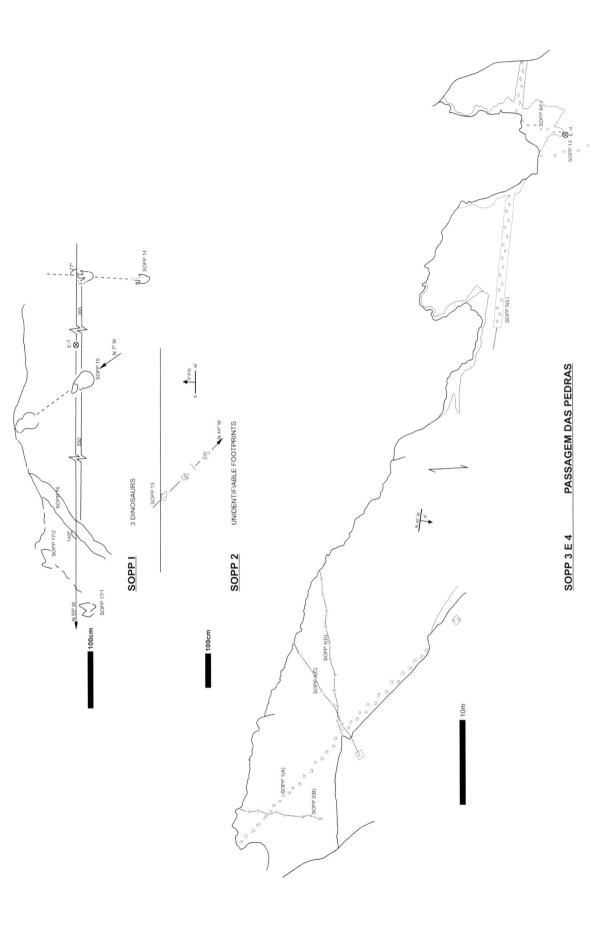

SOPP I

3 DINOSAURS

100cm

SOPP 2

UNIDENTIFIABLE FOOTPRINTS

100cm

SOPP 3 E 4   PASSAGEM DAS PEDRAS

10m

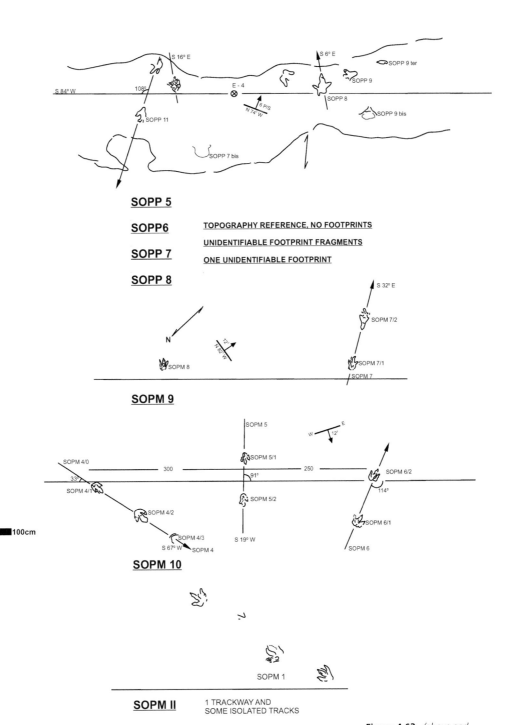

SOPP 5

SOPP6   TOPOGRAPHY REFERENCE, NO FOOTPRINTS

                UNIDENTIFIABLE FOOTPRINT FRAGMENTS

SOPP 7   ONE UNIDENTIFIABLE FOOTPRINT

SOPP 8

SOPM 9

SOPM 10

SOPM II   1 TRACKWAY AND
SOME ISOLATED TRACKS

Figure 4.63. *(above and facing)* Localities of Passagem das Pedras (SOPP) and Poço do Motor (SOPM). Serial representation of the track-bearing levels that crop out from upstream (top) to downstream (bottom). See also figures 4.63–4.80. Survey and drawing by Leonardi.

Figure 4.64. A number of theropod footprints were found in several layers of the siltstone and shale of the Sousa Formation at Poço do Motor (SOPM; see also figs. 4.62–4.63), downstream from the Passagem das Pedras (SOPP), at the locality Poço do Motor (SOPM). The SOPM exposures are younger than those of SOPP. *A*, Footprint SOPM 1/4, part of a trackway of 4 footprints. *B*, Isolated footprint SOPM 3. *C*, Footprint SOPM 5/1, from a short track-way of 2 footprints. *D*, SOPM 7/2, the collapsed and eroded second print of a short track-way. *E*, Footprint SOPM 4/2, from a short trackway of 4 footprints. *F*, Trackway SOPM 7 in the process of excavation. Graphic scale in centimeters.

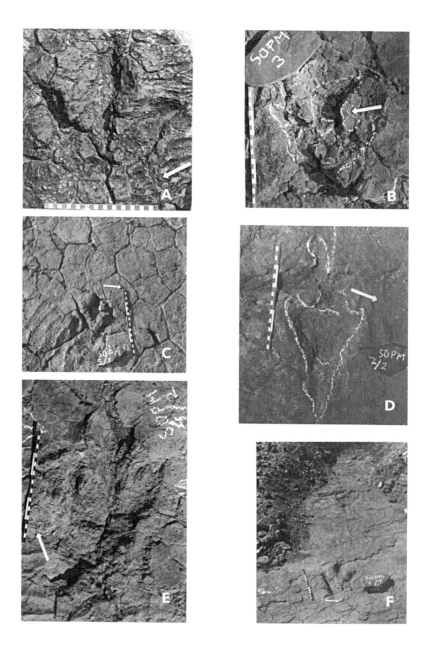

clearly separated, and show a low digital divergence (average = 40°) and a small "plantar" pad. The trackmaker moved S19°W.

SOPM 6

This is another short trackway of 2 footprints, made by a speedily pacing theropod of small–medium size, heading N44°E (figs. 4.63, 4.65). The 2 footprints differ somewhat in shape, but both are truncated proximally, have short, broad, clearly separated toes, low interdigital divergence (mean = 43°30'), and small, pointed claws.

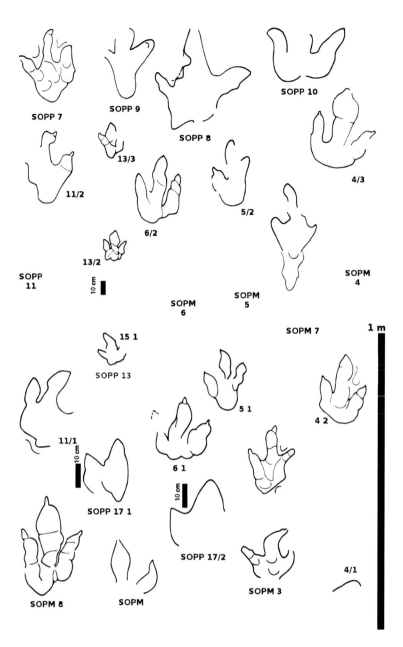

Figure 4.65. Footprints from localities Passagem das Pedras (SOPP) and Poço do Motor (SOPM), Sousa Formation, Sousa (see also figs. 4.62–4.63). Almost all the tracks are attributable to theropods. SOPP 17 is probably attributable to an ornithopod, but it is of poor quality. Graphic scale = 100 cm. Drawing by Leonardi.

SOPM 7

A short trackway of 2 footprints made by a quickly pacing theropod of small-medium size (figs. 4.63, 4.65D, 4.65F).

SOPM 8

An isolated, large theropod footprint, probably a left, showing good morphological details, heading NW, and accompanied by some small footprints of swimming turtles (figs. 4.63, 4.65).

| Passagem das Pedras (SOPP) | Passagem das Pedras, incorrectly called Passagem da Pedra or Passagem do Pedro on some maps, is the name of a large rocky pavement (100 × 20 m; surface of ~2,000 m²) in the bed of the Peixe River, 4.5 km NW of the town of Sousa, in the Ilha farm (S 06 44.031, W 038 15.657). Sometimes the whole stretch of about 1 km from the Jangada farm to the Poço do Motor is called Passagem das Pedras by extension. In the stretch here examined, seven ichnofossiliferous levels were found: one (level 3) by Luciano Jacques de Moraes in the 1920s, another (level 4) by Leonardi in the 1970s, like the levels of Poço do Motor. Among them the river-bed consists in sequences of brick- or chocolate-color or gray-greenish mudstones and silty mudstones; the stretch between the levels 5 and 6 is covered by recent coarse sand deposits. The levels of this site (fig. 4.63) will be shortly described starting from the more ancient, following the older, previously described three levels of SOPM: |

Level 8, with 1 incomplete, isolated footprint of a bipedal dinosaur, crops out with a surface of 40 × 3 m. Theodolite landmark E-8.

Level 7 consists of two strata of silty mudstone, red color, with mud cracks, both with some incomplete and unclassifiable footprints, cropping out with a surface of 40 × 7 m. Theodolite landmark E-7.

Level 6, without ichnological material, is just like a landmark of survey (E-6).

Level 5 is a bundle of four gray and greenish silty sandstone and fine sandstone strata, with mud cracks and ripple marks. The outcrop surface has an area of 25 × 4 m. Strike N74°W, dip 6° to the south. Theodolite landmark E-5. It shows a short trackway and 7 isolated footprints, constituting an association of 8 dinosaurian individuals (SOPP 7, 7 bis, 8, 9, 9 bis, 9 ter, 10, 11; figs. 4.65, 4.79A–B) among which 5 or 6 are theropod and 2 or 3 are unclassifiable individuals. These tracks do not present a pattern of social behavior, and their directions are scattered in the I, II, and IV quadrants. The concentration of individuals is 1 track per 12.5 m² or 0.08 track per m².

Level 4 is the silty (chocolate changing to greenish color) lower stratum of the large rocky pavement of Passagem das Pedras sensu stricto, which crops out at the northeastern margin of that pavement and by means of two artificial trenches. It shows 1 trackway of 7 excavated footprints *Staurichnium diogenis* (SOPP 6) and 1 trackway *Moraesichnium barberenae* (SOPP 5), interrupted by the course of the river, with 13 excavated footprints on the left side of the river and at least 34 excavated footprints on the right side of the same; and a very small isolated tridactyl footprint, perhaps attributable to a very small theropod (SOPP 18). No social pattern is seen. The theodolite landmark E-4 is situated at the SE margin of this level. Strike about N40°W; dip 6° to south.

Level 3 is the upper and main stratum of the large pavement of
Passagem das Pedras, in which the first 2 tracks of this basin
and of Brazil were discovered by Luciano Jacques de Moraes in
the 1920s (presently SOPP 1 and 2). It has an area of about 100
× 20 m and theodolite landmark E-3 at the northern margin
of the outcrop. This level shows 1 quadrupedal (or, better,
semiquadrupedal) trackway *Sousaichnium pricei* (SOPP 1), with
52 excavated footprints and ~25 (just right) hand-prints, with a
total length of 48 m of excavated trackway (2008); 4 trackways of
respectively 10 (SOPP 2), 7 (SOPP 3), 10 (SOPP 4) and 5 (SOPP
12) footprints of *Moraesichnium barberenae* and some incomplete
and unclassifiable footprints. More theropod trackways have
appeared here because of the natural erosion by floods, rain, the
sun, and thermal expansion after the overall surface was dug out
and exposed in the first activities of fieldwork by Leonardi (1975–
79). These last were not yet completely surveyed.
The concentration of individuals in a surface of ~2,000 m² is of 1
individual trackway per 333.33 m² or ~0.018 trackway per m²—a
very low concentration, notwithstanding the very good (but
eroded) exposure. All of the main 5 trackways have directions
concentrated in the II and III quadrants, as in the lower levels
(4). Here, the 4 *Moraesichnium barberenae* trackways lie in the
III quadrant, whereas the *Sousaichnium pricei* large trackway lies
in II, crossing all of the other 4 trackways (fig. 4.68). It is possible
that 3 *Moraesichnium* makers (SOPP 2, 3, and 4) were in a pack,
as they belonged to the same ichnospecies and very probably to
the same Linnaean species. However, 1 was walking (SOPP 2,
the holotype) and 2 were running (SOPP 3 and 4). SOPP 12, a
specimen of the same form, crossed the mudflat in a different
moment, perhaps before the other, when the area was covered
by shallow water or at least was waterlogged. SOPP 5, also of
the same form, crossed the area in a precedent time, that one
represented by the lower adjacent level. The strike of this upper
level is ~N40°W; the dip is 6° to the south. More isolated tracks
were found on this outcrop during the 2008 expedition, along
with some small reptilian bones (S 06 44.074, W 38 15.659).
Level 2, cropping out with a surface of 50 × 2 m, with strike E–W
and dip 6° to the south, is a stratum of siltstone with mud cracks,
showing a short trackway of 3 shallow footprints, attributable to
a small theropod, heading for N44°W (SOPP 13; fig. 4.65). The
concentration of tracks is 0.01 per m². However, there are also
some fragments of unclassifiable footprints. Theodolite landmark
E-2.
Level 1 presents a surface of 17 × 25 m (about 425 m²), with ripple
marks whose crests point to ~N55°E and primary and secondary
mud cracks. The following assemblage was found: one probable
theropod trackway of 2 footprints (SOPP 14; fig. 4.79C); 1

quadrupedal trackway (SOPP 15, underprint; fig. 4.78) of 4 hand-foot sets; and 1 short trackway of 2 incomplete prints, probably attributable to an ornithopod (SOPP 17). There is also a long furrow (SOPP 16; fig. 4.78D) that could perhaps be interpreted as the impression of the tail of a large dinosaur, but it is not associated with a trackway or footprints. Theodolite landmark E-1. These are all very different animals, but the directions of their tracks (and of the furrow) lie on a sector of only 72° (N7°W–N65°E), which also comprises the direction of the ripple mark crests (N55°E). This is probably a preferred corridor, influenced by the landscape. The concentration of individuals here is of 1 track per 10 m², or 0.1 individual per m².

Morphological Description and Discussion

SOPP 1 (FORMERLY SOPP A)

This is a long and fine semibipedal trackway with 52 consecutive excavated hind-foot prints and ~25 excavated forefoot prints (figs. 4.63, 4.66, 4.67, 4.70, 4.72A, 4.72D, 4.78). By January 2008, 48 m of the trackway had been excavated, and besides the 15 hind-foot prints seen and published by Moraes (1924) and by von Huene (1931), some 37 more were excavated between 1975 and 2008. Eventually, more footprints can be excavated. The local popular name of this particular trackway, very evident and so known by common people, is *o rastro do boi* (= the track of the ox) because of the hoof-shaped toes.

The hind prints are tridactyl, almost symmetrical, digitigrade (not plantigrade, according to Moraes [1924] and von Huene [1931]). The main stress was leaned on the sole pad or cushion; among the toes the stress lay on the III or IV digits. The toes are short and bulky, oval shaped, without claws or nails, and the individual digital pads are not seen, as each toe is hooflike. The digit length increases from the II to the IV, but the toe III lay in a more anterior (distal) position, so it seems to be the longest one. Casts of skinfolds are very evident among the toes and between the toes and the sole pad, which is roundish and large with regard to the digit surface. The posterior or proximal margin is heel shaped, and some drag marks can be seen behind it. The total digital divergence is very low (average = 25°).

The manus prints (figs. 4.69A, 4.69C, 4.69D, 4.70D) are very small, oval or elliptical in shape, with the main axis (probably the transverse axis) diagonal to the midline, with less than one tenth of the pes as for the area. Generally, no morphological details are seen, but very probably it represents just the impression of the tips of some fingers.

The pace angulation is high (average = 165°); so is the stride (average = 187 cm), and the ratio SL/FL is 4.24. The internal trackway width has negative values (average =-18.5 cm). The trackway (and so the midline) is characteristically serpentine, generally with the footprints three by three, periodically more at the left and more at the right. The footprint long axis is parallel to the midline or just slightly bent inward or outward.

After the publication of the diagnosis of this new genus and species (Leonardi 1979a), it was discovered (Leonardi 1994) that the track is not bipedal but semibipedal (or, alternatively, semiquadrupedal), with strong heteropody, large pedes, and very small manus. The trackmaker leaned the hand on the soil very slightly; generally, just the right manus was impressed in this trackway (Leonardi 1994, 2015).

Both the forefoot and the hind-foot prints show a strong displacement rim. Those of the pedes are high and wide, sometimes partially collapsed inside the footprints; they are cracked by circular mud cracks that follow concentrically the footprint outline, and by radial cracks (cf. Schanz et al. 2016, fig. 19.4, 376).

Trackway SOPP 1 is the holotype of *Sousaichnium pricei* Leonardi, 1979. This trackway is attributed to the Ornithopoda, most likely a member of the Iguanodontidae. It can be compared to similar tracks attributed to this family but is different from them. A judicious comparison between the SOPP 1 prints and the skeletal taxon *Iguanodon* Mantell, 1825 (fig. 4.80) shows that the feet of *Iguanodon* do not fit exactly in the *Sousaichnium* footprint but come close (cf. Henderson 2017). The toes of the Brazilian prints have roundish distal extremities, as one would infer for *Iguanodon*. However, the sole pad of SOPP 1 is larger than expected for a footprint of *Iguanodon* and equidistant from the three toes. There is a small indentation behind digit II, but this is not as distinct from the impression of digit III as is expected for an *Iguanodon* track; the total and partial digital divergences are lower, digit III is shorter, and all three free digits are relatively shorter. The pace angulation of our trackway is higher than in trackways attributed to *Iguanodontipus* Sarjeant, Delair and Lockley, 1998.

Compared with *Caririchnium* Leonardi, 1984, in *Sousaichnium pricei* the footprints are relatively longer and narrower; the digit III impression is clearly shorter; the hypexes are obtuse, roundish, and located more distally. The sole pad of *Sousaichnium* is longer and relatively larger, the toes are narrower, and there are no nails. The three hoof impressions here are more divergent, and there is a greater distance between them and the sole pad. This trackway was made by a semibipedal dinosaur, not a quadruped like the *Caririchnium* trackway. A trackway from the Albian-Cenomanian(?) of Picún Leufú (Neuquén, Argentina) was attributed to the same ichnogenus, and the species *Sousaichnium monettae* was instituted (Calvo 1991). *Sousaichnium pricei* can also be compared to numerous tracks on the Richardson Ranch, south of Pritchett, in eastern Colorado, visited by Leonardi in 1986.

To sum up, the trackmaker here was a rather large Ornithopoda, probably an Iguanodontidae, walking with an estimated speed of 2.6–4.2 km/h, with the body surely almost parallel to the ground. The animal's gait was undulating and serpentine. It put down its hand partially (always

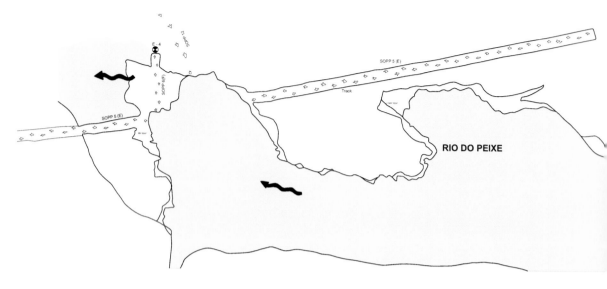

Figure 4.66. *(above and facing)* Map of the main locality of the park Vale dos Dinossauros, the rocky pavement at Passagem das Pedras (SOPP), the "ford of the stones." This is where the first two tracks (the crossing of the beginning of SOPP 1 and SOPP 2) were found before 1924 by Luciano Jacques de Moraes. In December 1975, Leonardi began his research on dinosaur tracks of the Sousa and other basins by excavating the other tracks from this site. Compare with figures 4.62–4.63. Section *A* represents the part of the Passagem das Pedras where the lower and oldest layer SOPP 4 crops out or has been excavated; section *B* is the part of the pavement constituted by the upper and more recent layer SOPP 3. Survey and drawing by Leonardi.

just the right hand in this individual), without placing much weight on it; the dinosaur's weight was carried almost exclusively by the hind-feet. The animal did not drag its tail on the substrate. After the footfall, at the moment of retraction, the foot was extracted from the very plastic but firm mud with a backward movement, sometimes leaving, this way, the impression of the toes not in the surface but inside the mud, like conical tubes. J. O. Farlow (Farlow et al. 2015) found similar prints in which the toe passed through the soil beneath the surface, producing "toe tunnels," in the Main Tracklayer at Paluxy River, Dinosaur Valley State Park, Somervell County, Texas. These cones or "tunnels" at Passagem das Pedras were eventually filled in by sediment of the overlying stratum of red mudstone, so one can distinguish very well the red infilling cast from the dark gray mold. The infilling was in many cases cleared away.

The estimated total length of the trackmaker was probably 7–8 m. It is possible that footprints of this kind were made by a form similar to *Ouranosaurus* from the Niger republic (Taquet 1976; Bertozzo et al. 2017), also in Gondwana. The trackway is in situ, inside the Park *Vale dos dinossauros* at Sousa. However, one plaster cast of hind-foot print 12 is kept in the collections of the Câmara Cascudo Museum at Natal (Rio Grande do Norte, Brazil; fig. 4.70A).

The extremely important monograph by Díaz-Martínez et al. (2015) dealing with ornithopod tracks, which takes a rather "lumping" approach, considers (18) this ichnotaxon to be a nomen dubium. We do not agree with this advice, just in this case, on the following grounds:

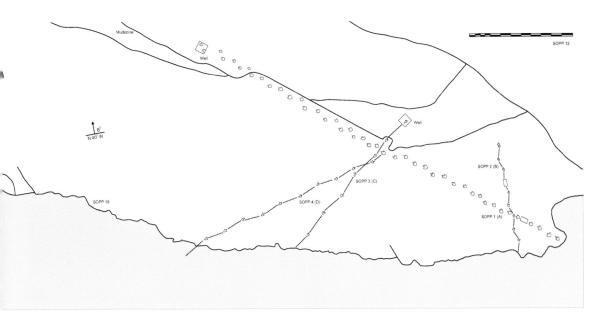

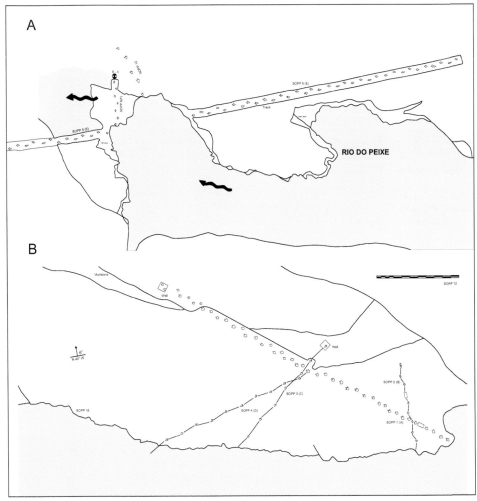

Figure 4.67. The holotype trackway *Sousaichnium pricei*, Leonardi, 1979, in situ at Passagem das Pedras (SOPP), during its excavation in 1975. The trackway was buried beneath 2 m of fluvial sand and gravel and in many places covered by the overlying rock layers. See also figures 2.2A, 2.6A, 2.6B, 3.31–3.33, 3.35, 3.36, 4.68–4.70, 4.72A, 4.72D, 4.78, 4.80, and 4.105A.

1. Many of the footprints are very well preserved; see, for example, figures 4.68A, 4.69B, 4.69C, 4.70A).
2. The footprint is not so elongated. Its mean length is 44.1 cm, and the mean width is 41.9 cm, so the ratio FL/FW corresponds to 1.05.
3. Only a few of the footprints show mud collapsing inside the footprints (only three among the 17 footprints illustrated in fig. 4.78 show this phenomenon).
4. The fact that the print lacks a claw impression is very typical of ornithopod footprints.
5. We do not think the large proximal pad represents metapodial marks but rather a cushion of connective and muscular tissues.
6. The diagnosis by Leonardi (1979a) was not analyzed by Díaz-Martínez et al. in their monograph.

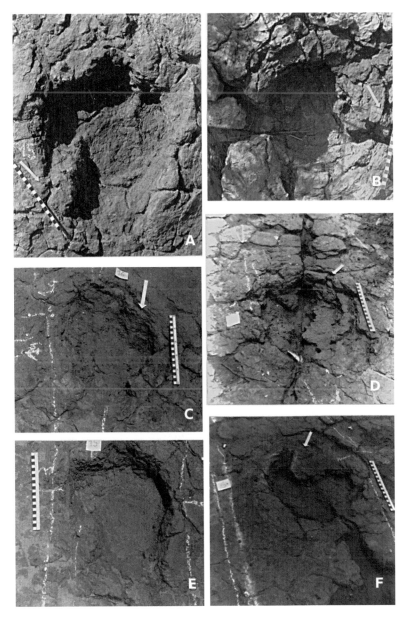

Figure 4.68. Footprints of trackway *Sousaichnium pricei*, SOPP 1, Passagem das Pedras (SOPP). *A*, SOPP 1/17 (right pes); note the very small handprint (a tiny oval shadow) just in front of the hind-foot print and the large incision behind the second toe. *B*, SOPP 1/1 (right pes), the first footprint discovered in this trackway, at the margin of the river bank. *C*, SOPP 1/16 (left pes). *D*, SOPP 1/2 (left pes). *E*, SOPP 1/15 (right pes). *F*, SOPP 1/6 (left pes). In footprints 2 and 6 (*D* and *F*), a broad sheet of mud collapsed into the footprint after the animal extracted its foot from the substrate.

However, it is true enough that there are no more tracks of this kind in the numerous outcrops of the Rio do Peixe basins.

This ichnotaxon is classified, anyway, within the ichnofamily Iguanodontipodidae Vialov, 1988, sensu Lockley et al. 2014b, as it presents diagnostic features that allow it to be classified in that ichnofamily (Díaz-Martínez et al. 2015).

SOPP 2–5 AND SOPP 12

Moraesichnium barberenae Leonardi, 1979 (figs 4.63, 4.66, 4.71–74, 4.78). This is a group of 5 trackways, along with some isolated footprints

Figure 4.69. Footprints and handprints (the latter present only on the right side of the trackway) of *Sousaichnium pricei*, Passagem das Pedras (SOPP). *A*, Right foot-hand set SOPP 1/11: note the very small handprint (outlined in red) at the top right of the photo. *B*, Left hind-foot print SOPP 1/20. *C*, Right foot-hand set SOPP 1/21; note the small, oval handprint (outlined in red) at the upper right corner. *D*, Right foot-hand set SOPP 1/23; once again the small oval handprint is outlined in red at the top.

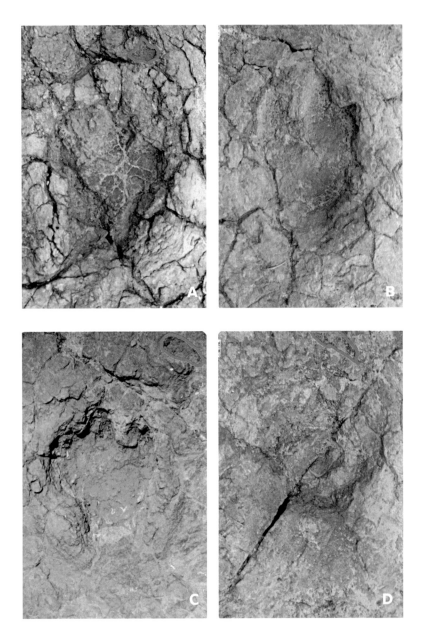

scattered among other localities of this basin. They all share a number of distinctive features.

The footprints are tridactyl, symmetrical, and arrowhead shaped, with a long sole or "plant" constituting more than half the footprint length and area. The deepest part of the print (indicating where the dinosaur carried the most weight on its foot) occurs in a proximal row of digital pads, with the digital pad of digit III being more deeply impressed than those of digits IV and II (and with digit II having the most shallowly impressed pad). The toes are long and narrow, sometimes with distinct digital pads, especially in digit III but sometimes in digit IV. The impression of digit III is the longest, and digit IV is slightly longer than digit II. The total

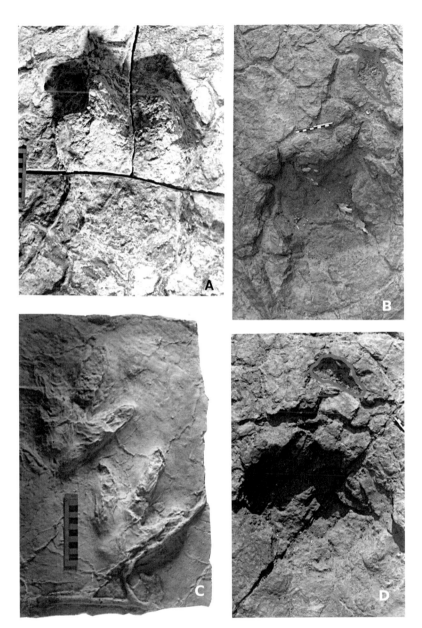

Figure 4.70. Passagem das Pedras (SOPP) and Piau-Caiçara (SOCA). *A*, Plaster cast of footprint SOPP 1/20 (left pes) of *Sousaichnium pricei*; note the three blunt hooves, highlighted by their shadows; Courtesy of the Câmara Cascudo Museum of Natal (Rio Grande do Norte, Brazil). *B*, Unnumbered right foot-hand set of the same trackway; note the very small handprint (outlined in red) in the upper right corner. Graphic scale in centimeters. *C*, Plaster cast of the set of both pedes of a theropod at Piau-Caiçara (SOCA 1364/1–2; compare with figs. 4.45, 4.46). Courtesy of the Câmara Cascudo Museum. Graphic scale in centimeters. *D*, Right foot-hand set SOPP 1/23, with the handprint outlined in red.

divergence between digits II and IV is variable but sometimes very high for a theropod (55° to 88°), and the partial divergences (interdigital angles II^III and III^IV) are subequal. The claw (or, better in this case, nail) impressions are very small and rarely observable and are mostly visible in the imprint of digit III. There are no tail impressions. The footprint long axis is often angled slightly inward with regard to the midline but sometimes is parallel to the midline or even angled slightly outward.

The trackway is narrow, the pace angulation is medium to high (148° to 172°), and the stride is long to very long (214–482 cm), so the SL/FL ratio is highly variable (5.7–13.8), as is, in consequence, the inferred stride/hip height ratio (1.2–2.8). These values are so different in these trackways, four

Figure 4.71. The holotype trackway *Moraesichnium barberenae* Leonardi, 1979 (SOPP 2), attributed to a walking theropod, as it was uncovered at Passagem das Pedras (SOPP) in 1976. *A*, The trackway at its crossing with trackway *Sousaichnium pricei*. *B*, Right footprint SOPP 2/5. *C*, Left footprint SOPP 2/2. This track is nearly completely collapsed. Note the digital pad impression and claw mark of digit III. *D*, Right footprint SOPP 2/3; graphic scales in centimeters. See also figures 4.63, 4.66, and 4.78.

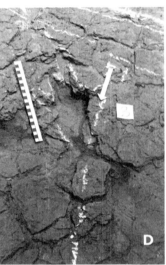

of which were probably produced by the very same Linnaean population, because they are functions of different gaits and speeds. We think this clearly demonstrates that when instituting an ichnotaxon, one must not posit too restricted a range for trackway parameters, or else one risks creating more than one ichnotaxon for trackways made by different individuals, or even the same individual, of a zoological (Linnaean) species.

Frequently at this outcrop, the mud collapsed after the foot was lifted off the substrate, so the toe impressions, especially of digit III, look excessively narrow, like a rima or fissure.

Footprints of this kind, especially SOPP 2, which had long been exposed on the riverbed, are known to local residents as the *rastro da ema*

Figure 4.72. Passagem das Pedras (SOPP). *A*, The crossing of 2 trackways of running (~12.8 km/h and ~22 km/h, respectively) predatory dinosaurs of the same kind, paratypes (SOPP 3 and 4) of *Moraesichnium barberenae*, with the trackway of a walking ornithopod, the holotype of *Sousaichnium pricei* (here footprints 20 to 23 of the trackway). *B*, A section of the paratype (SOPP 5) of *Moraesichnium barberenae* in the course of excavation. Behind and to the right of the human figure, on the other side of the dry riverbed, the trackway continues in a deep trench where the first (exposed) footprint of the trackway was first seen. *C*, Same as in *A*, seen from the other side; *D*, Trackway *Sousaichnium pricei*. Photos in *A* and *C* courtesy of Franco Capone. See also figures 4.63, 4.66, 4.72*A*, 4.72*C*, 4.73, and 4.78.

(= the track of the *ema*). *Ema* is the Brazilian popular name for the South American ostrich *Rhea americana*, called *ñandú* in the Spanish-speaking countries of the continent. The ema was formerly a common animal in this region, but it is unfortunately quite rare today.

Moraesichnium barberenae tracks can be compared with many long-heeled or metatarsal theropod prints; Kuban (1991b; see also Pérez-Lorente 2015 and Xing et al. 2015c) offers a good discussion on this subject. These long-heeled tracks pose a rather difficult problem. Some authors (e.g., Kuban 1991b; see also Citton et al. 2015) think they were produced by semiplantigrade theropods, representing trackmakers whose metapodia could be lowered to the substrate level, like the toes, producing a true

Figure 4.73. Paratype track-way (SOPP 4) *Moraesichnium barberenae*, in situ, Passagem das Pedras (SOPP)—the trackway of a running large theropod. *A*, Right footprint SOPP 4/3. *B*, Left footprint SOPP 4/4. *C*, Left footprint SOPP 4/2. The toe impressions frequently collapsed after their extraction by the trackmaker's foot. The footprints are barely visible due to mud cracking shortly after they were made, and also to modern fluvial erosion of the riverbed; *D*, Right footprint SOPP 4/7 during excavation, with a contact goniometer. *E*, The trackway during the digging in 1976. *F*, Right footprint SOPP 4/5. Graphic scales in centimeters. See also figures 4.63, 4.66, 4.72*A*, 4.72*C*, and 4.78.

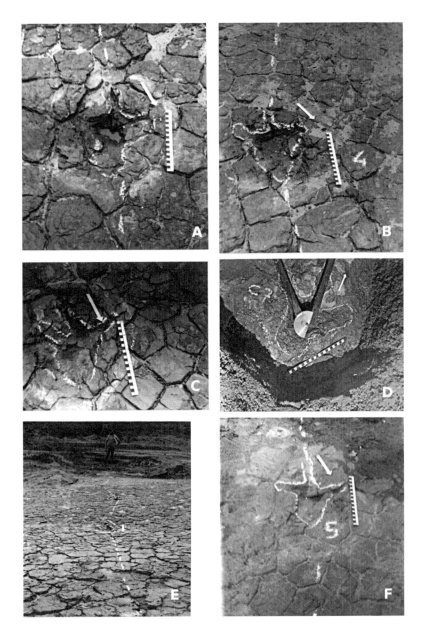

tarsal-metatarsal pad or metatarsal pad, in a true plantigrade or semiplantigrade manner. More evident cases of quite plantigrade footprints are those of some sitting and quiescent *Anomoepus* trackmakers (Lull 1953; Avanzini and Leonardi 1999; Avanzini et al. 2001c). However, here, as for *Moraesichium* trackways, it is a matter of walking and running theropods with elongated pedes, imprinting elongated metatarsal footprints.

Citton et al. (2015) offer a good discussion on this kind of walking, running, and resting. They observe that the term "crouching" was used in literature with two different meanings: 1) for true resting tracks, with the impression of ischial callosity and hands, and 2) for isolated footprints

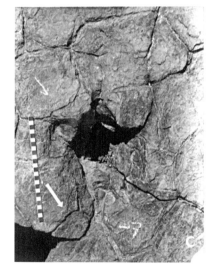

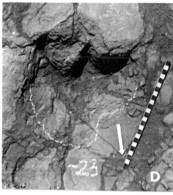

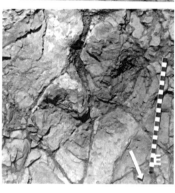

Figure 4.74. Paratype trackway (SOPP 5) *Moraesichnium barberenae*, the portion on the left of the Peixe River, Passagem das Pedras (SOPP); see also figures 3.32*F* and 4.72*B*. *A*, Overall view of the trackway and its footprints. *B*, Right footprint SOPP 5/-22. *C*, Left footprint SOPP 5/-7. *D*, Left footprint SOPP 5/-23. *E*, Right footprint SOPP 5/-24. The numbers assigned to this part of the trackway are negatives because the footprints were uncovered in the opposite direction from the direction in which the dinosaur walked. Toe marks in this trackway are frequently collapsed. See also figures 3.32*F*, 4.63, 4.66, 4.72*B*, and 4.78.

or complete trackways showing elongated footprints, representative of a squatting walking (see Lockley et al. 2003; Milàn et al. 2008).

Alternatively, according to Lockley et al. (2003), the definition of crouching is more restrictive and requires the impression of both the metatarsals and ischial callosity in the resting phase. In this case, the authors used the term "complete crouching." Lockley et al. (2003) then considered true crouching as a different form of behavior compared to a continuous walk in a squatting position, possibly linked to a firmer substrate (at least in the known examples), with the dinosaur stopping to rest at least for a certain period of time. Milàn et al. (2008) considered crouching or "full crouching" the condition in which parallel metatarsal traces, the ischial callosity, and the manus prints are recorded in the substrate. Under this definition, genuine crouching tracks appear to be relatively rare compared to the complete ichnological record of footprints with elongated metatarsal (see Lockley et al. 2003 and Milàn et al. 2008).

In the case of *Moraesichnium* tracks, there are 5 trackways at Passagem das Pedras, all of them with long-heeled footprints, along with a number of isolated footprints of the same aspect. Two of these trackways show a walking gait; 2 of them show a slow-running gait; and one (SOPP 3) presents a running gait (we calculated about 21 km/h). It is possible, and it seems now evident, that some theropods could not only rest but also walk on their long-heeled feet, imprinting their hind-foot prints not only with their digits but also with the metatarsi, with a "squatting walking" or "crouching walking" gait, the use of which is difficult to explain. However, it is more difficult to imagine a theropod slow running (at 7.9 km/h and accelerating to 11 km/h, like the maker of SOPP 12, or at 13 km/h like the maker of SOPP 4) or even truly running (at 21 km/h, like the maker of SOPP 3) while maintaining a "squatting" gait.

These trackways were made by different individuals of a medium-size, probably lightly built bipedal theropod that had the capacity for running at high speeds. They were named *Moraesichnium barberenae* by Leonardi (1979a), who, by an evident *péché de jeunesse* (= youthful indiscretion), attributed them to Hypsilophodontidae. They were later reattributed by him and others to Theropoda (Leonardi 1989a, 1989b, 1994; Kuban 1991b; Leonardi et al. 1987a, 1987 b, 1987c).

SOPP 2 (FORMERLY SOPP B)

A holotype of *Moraesichnium barberenae* (figs. 4.63, 4.66, 4.71, 4.78). This is a trackway of 9 or 10 footprints, over an interval of 10 m, of rather fair quality. It was imprinted in almost dry mud, so it is rather shallow; the last 2 footprints are very shallow, and beyond them the trackway fades away. The pace angulation is very low for this ichnogenus (average = 156°); the stride and pace are short (average = 211 and 106 cm, respectively). The calculated speed is 5.4 km/h. The seventh footprint is presently lacking because it was cut out from the outcrop by Moraes in the 1920s and supposedly sent to the American Museum of Natural History in New York, but Leonardi did not find it there in 1985. The impression of digit III is frequently collapsed. The digital pads and the tiny claw marks are better seen in this than in the other specimens of the ichnotaxon. The ninth and tenth footprints slightly overlap each other, and it is not certain that they were made by the same individual; if not, the second animal was very similar to the primary trackmaker.

SOPP 3 (FORMERLY SOPP C)

A paratype of *Moraesichnium barberenae* (figs. 4.63, 4.66, 4.72A, 4.72D, 4.78). This is a long (16 m) trackway with 7 excavated footprints. The last

Figure 4.75. *(left)* Photomosaic of the holotype (SOPP 6) of the rare bipedal ornithopod trackway *Staurichnium diogenis*, in situ at Passagem das Pedras (SOPP). The metric tape is in centimeters. See also figures 4.63, 4.66, and 4.76–4.78. Photos and mosaics by Leonardi.

Figure 4.76. *(above)* Holotype of the bipedal ornithopod trackway *Staurichnium diogenis*, Passagem das Pedras (SOPP; see also fig. 4.75). *A*, Plaster cast of left footprint SOPP 6/5. Courtesy of the Câmara Cascudo Museum of Natal (Rio Grande do Norte, Brazil). *B*, Claude Luis Aguilar Santos casting the same footprint in the field. *C*, Ripple marks in the marls of the surrounding rock. *D*, Oblique view of the holotype trackway in situ. *E*, The same trackway under different lighting conditions. Compare also with figures 4.63, 4.66, 4.77, and 4.78.

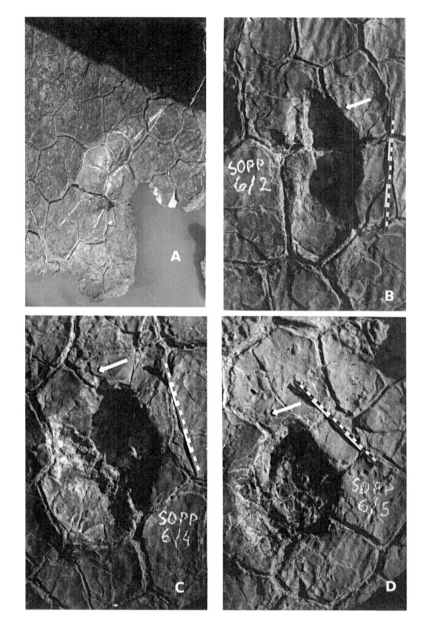

footprint was found by means of digging of a well in the upper strata by Leonardi. The stride and the pace angulation are very high (average = 483 cm and 174°54', respectively), and as a consequence the trackway external width is very narrow, pointing to a very high speed (V = ~21 km/h). The stride length/footprint length SL/FL ratio averages 13.8, and the estimated stride/hip height ratio SL/h averages 2.8. This trackway has one of the quickest estimated speeds of the dinosaurs of the Sousa basin—indeed, one of the fastest estimated dinosaur speeds we know and accept in the world. It is strange, then, that the footprints of this trackway, pointing as they do to a very quick gait, are no deeper than those of the SOPP 2 trackway, which was made by a slowly moving dinosaur. This

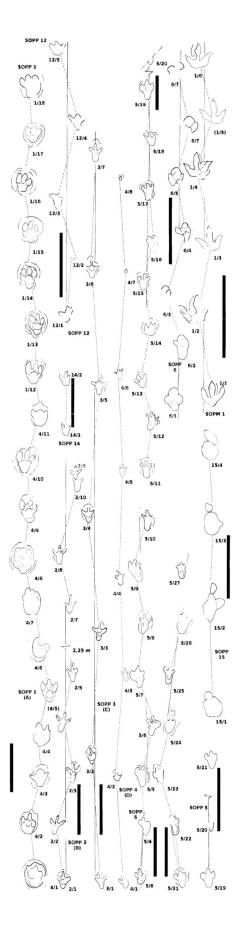

Figure 4.78. The main trackways from the Passagem das Pedras (SOPP) tracksite. Graphic scales = 1 m. Survey and drawing by Leonardi.

probably reflects the substrate conditions at the time the dinosaur crossed the surface.

The total digital divergence is higher than in SOPP 2 (average = 88°40' and 77°23', respectively); this probably is due to the different gait. The long axis of the footprints in this trackway of a rapidly moving animal is more parallel to the midline than those of SOPP 2. This fact probably also depends on the different gait.

SOPP 4 (FORMERLY SOPP D)

This is a 21-m-long trackway with 10 footprints (figs. 4.63, 4.66, 4.72A, 4.72D, 4.73, 4.78). These are shallow, badly imprinted, and quite eroded. In the last 3 footprints one can see only the tip of digit III. The pace angulation is high (average = 171°41'); the stride is long, relative to the small size of this specimen (average = 244.6 cm). The SL/FL ratio consequently is high (average = 12.2), as is the stride/hip height ratio (average = 2.5–3), pointing to a rather high speed. This is calculated as 13 km/h. The footprints are relatively small for this population; the probable figure for the pes length is in fact 20 cm.

SOPP 5 (FORMERLY SOPP E)

This trackway is the paratype of *Moraesichnium barberenae* (figs. 4.63, 4.66, 4.72B, 4.74, 4.78). It is also a long, slightly sinuous trackway. It was discovered as 1 footprint and half of another, cropping out on the lower track-bearing stratum of this locality, on the left margin of the Peixe River. One trench was excavated (1975), and 7 footprints were found. The same level and the same trackway were found while digging on the other (right) side of the river (9.5 m apart; 7 footprints); another trench was excavated (20 m; 1975; 1977; 1983), and 20 more footprints were found. The exposed trackway is now a total of 56 m in length and comprises 34 excavated footprints on the right side of the river and an additional 13 footprints on the left side (2008). The trackway is interrupted by a large gap near the deepest part of the course of the river; eventually, more footprints can be found by excavating at both ends.

The stride is longer than in SOPP 2, and the external width is narrower, pointing to a walking gait (V = 5.6 km/h). The depth of the footprints is variable, from 1 to 10 cm, depending on the plasticity of the fresh substrate at the time the prints were made.

Many footprints have the long axis parallel to the midline, but some of them angle slightly inward or outward. The outline of the footprints is variable; many of them have the normal long-heeled outline, but some of them lack the heel and are short, with the typical theropod short-heeled outline. The footprints do not show claws or nails. The inner and outer toes (digits II and IV) frequently show a small angle between them, but in other cases they are abducted like in other specimens of this ichnotaxon. The consequent variability of the total digital divergence is high (σ for this trackway = 18; for SOPP 2, σ = 11; for SOPP 3, σ = 14). This higher variability, like its low mean value (55°18'), is probably a function of the variable plasticity of the substrate, demonstrated also by the variability of the depth of the footprints and the extent of development of displacement rims. Such sediment bulges are always found in the front of the footprints but otherwise are very different among the different footprints.

SOPP 6

This is a 7-m-long trackway with 7 excavated footprints of a medium-size bipedal dinosaur (figs. 4.63, 4.66, 4.75, 4.76–4.78). The pace angulation is not very high (average = 147°); the average index SL/SF (SL/FL) is = 5.7; the average index SL/h (stride length/estimated hip height) is = 1. The foot axis is slightly and variously angled with regard to the midline. It was classified as *Staurichnium diogenis* Leonardi, 1979.

The footprints are tridactyl, symmetrical, and digitigrade, formed by the round or oval imprints of three hoof-shaped toes separated by sharp edges representing skin folds or wrinkles, and by a variously shaped sole pad imprint. The surface of the central hoofmark (digit III) is larger than the others, and the lateral hoofmark (digit IV) is larger than the medial (digit II). The digit III impression is a large, deltoid hoofmark whereas the II and IV digit imprints are smaller and narrower (the III hoofmark is twice as wide as that of digit II), with rounded distal margins. The total digital divergence is 50°–55°.

Behind the three hooves there is sometimes the impression of a "plantar" pad, which is rounded or oval, with a longer anterior-posterior than transverse axis. This pad is rounded in the first footprint of the trackway but more elliptical in the second. In other places (at least in the fifth print) there is instead a triangular indentation between the proximal margins of the inner and outer hoofmarks (II and IV). The hooves are separated by a deep wrinkle of the skin and/or of horny matter. The hypexes are acute (print 5). Sometimes one can observe some morphological details of the hoof impression: an elongated crest, parallel and in sagittal position along the long axis of the digit III; one can also see an elliptical elevation in the center of the large hoof of digit III.

Taken as a whole, the footprints sometimes show a very characteristic clover shape or cross shape. However, where the quality of the impression is good, especially in the artificially excavated trench, where the tracks had not yet been eroded, the pattern is different from that of a four-leaf clover, so the original drawing of the trackway (Leonardi 1979a) and the diagnosis must be partially changed (see fig. 4.78).

The trackway is attributable to an ornithopod, different from the other known trackmakers. Other similar specimens were found in this basin — for example, at Piau-Caiçara farm (fig. 4.22, SOCA 43, and especially SOCA 46; fig. 4.25A, 4.25B, and 4.25D), at Saguim Farm, and at Floresta dos Borba (SOFB 4).

Díaz-Martínez et al. (2015) consider this ichnotaxon to be a nomen dubium. We agree with this opinion because of the first reason, that is, the poor quality of the impressions. However, we do not agree that this ichnotaxon had been defined on the basis of the metatarsal impressions (Leonardi 1979a). This ichnotaxon is classified, anyway, within the ichnofamily Iguanodontipodidae Vialov, 1988, sensu Lockley et al. 2014b, as it presents diagnostic features that allow it to be classified in that ichnofamily (Díaz-Martínez et al. 2015).

SOPP 7

An isolated theropod left 1 print, medium-size, deep, well impressed but poorly preserved, and longer than it is wide (L/W = 1.4), with medium digital divergence and a well-developed "plantar" portion (fig. 4.79A). The digit III toemark is long, thin, and sinuous, with a stout triangular claw, forming the deepest part of the print. Digits II and IV are very short; both hypexes are wide and rounded. Two and four digital pads, respectively, are seen in the impressions of digits II and IV. The distal extremities of these toemarks are slightly spatulate. The displacement rims are high, especially between the digits, as an interdigital pressure wedge.

SOPP 8

A large, isolated theropod footprint, perhaps a right, of poor quality, wide (L/W = 1), with very high total digital divergence (~105°); a bilobate proximal margin with a median indentation; digit II short and narrow; digit III long and very wide; digit IV long and wide (fig. 4.79B). There may be a spur-shaped digit I impression on the left proximal margin, but the poor print quality prevents certainty about this. There was collapse of mud into the digit III mark. The footprint is most deeply impressed in the center of the posterior part of the print, corresponding to the large digital-metatarsal pad of digit III (see the large track SOPE 2 at the Pedregulho site).

SOPP 9

An isolated, incomplete, poor-quality footprint, perhaps a right; tridactyl, with a long elliptical heel; the right outer toe is probably digit IV (fig. 4.65). It is somewhat, but not entirely, similar to *Moraesichnium barberenae*.

SOPP 10

An isolated, incomplete, poor-quality, tridactyl footprint. The impression of the III toe is not preserved (fig. 4.65). The inner and outer toe marks are thick, blunt, and slightly abducted. The proximal margin is bilobate and probably corresponds to the posterior margin of the digital-metatarsal pads of digits II and IV. It very probably was made by a theropod, but could be also the footprint of an ornithopod with relatively thin toes because of collapse of the mud.

This is a short trackway of 2 footprints, with a long step probably indicating a quick gait
(fig. 4.65). Its footprints are badly preserved and incomplete, medium-size, and tridactyl. The mark of digit III is proximally faint but is distally spatulate, sinuous, with V-shaped clawmarks. The inner and outer toes are shorter than digit III and are stout and sometimes spatulate. There is a long heel: the footprint as a whole can be described as lozenge shaped. It was likely made by a theropod.

SOPP 12

This trackway lies in the main course of the river, and its surface is badly eroded (figs. 4.66, 4.78). It is a trackway of 5 footprints with a variable pace angulation, generally medium (average = 148°10'). The bipedal trackmaker was increasing its speed (from 7.9 to 11 km/h; fig. 5.1), but, strangely enough, the pace angulation was diminishing, even as the stride increased. The footprints are similar in general outline to *Moraesichnium barberenae*, to which this track can be attributed.

SOPP 13

A short trackway of 3 small to medium-size, fairly well preserved footprints (fig. 4.65). The high index SL/FL (~7), the very high pace angulation (almost 180°), and the very low trackway width all point to a rather quick gait; the speed is calculated in 6–7 km/h. Of particular interest is the shape variability of the 3 footprints in the same trackway: one would possibly attribute them to three different forms, if they had been found as isolated prints, and an ichno-splitter would perhaps institute three new taxa for them.

The best preserved of the 3 prints is the second: it is a tridactyl footprint, shallow but well imprinted, and longer than it is wide (L/W = 1.16), with a very low total digital divergence (30°). The hypexes are narrow but rounded, and the impressions of digits II and IV are spatulate; the tips of all the toemarks are blunt, without claws. Digit II is narrower and more pointed than the others and only slightly abducted from the impression of digit III. The impression of the third toe is wider and stouter than that of digit II. Perhaps this greater width of the digit III impression is due to a lateral movement of the toe as it was impressed into the sediment, causing it to slide sideways in the plastic substrate, and so widening its impression; the digit IV impression is also stout. The proximal margin of the print is trilobate; this outline perhaps corresponds to the posterior margins of the digital-metatarsal pads of the three digits.

Notwithstanding the width of the digit III and IV toe impressions, and the lack of apparent claw marks, this seems not to be an ornithopod track because of the very high pace angulation and the high SL/FL ratio. The trackway is attributed to a theropod.

This is a short trackway of 2 poorly preserved footprints (fig. 4.79C). It has a relatively short stride (index SL/FL = 3.4), but the extrapolated pace angulation would be high (~170°). The footprints are long, with a low digital divergence; the toe marks lack evident claws, but the impression of digit III is long, sinuous, and sharp. It is attributable to a medium- to large-size theropod, despite the relatively short stride.

SOPP 15

This trackway is composed of undertracks of a quadruped (figs. 4.78, 4.79F–G). Trackways of quadrupedal dinosaurs other than sauropods are very rare in the Sousa basin. Unfortunately, this one is very badly preserved and difficult to attribute to a maker. The pace angulation is very high for a quadruped (average = 174°). There are 4 hand-foot sets. The hand and the foot impression in each set all show marginal overlap; the long axis of the hand-foot set is parallel to the trackway midline. There is evident heteropody; the pes impression is twice as long as that of the manus and has a mean surface four times larger than that of the hand. Because these are subtracks, prints of the hands and feet are merely concavities without morphological details; however, the pes is elliptical and almost always longer than wide (mean index L/W = 1.36), and the hand is rounded. As for attribution, two hypotheses are viable: this could be the subtrackway of a small quadrupedal ornithopod or that of an ankylosaur, especially a nodosaurid.

SOPP 16

This is a long, semicylindrical depression in the form of a furrow, 2 m long and with gradually increasing width, from 20 to 30 cm, from W to E (fig. 4.79D). The trace is semielliptical in cross section, with a depth of less than half the width (3.5–5.3 cm). The surface is smooth, without furrows or irregularities; the intersection between this depression and the flat surrounding surface is abrupt. In another environment one could not exclude the hypothesis that this is a rill furrow, but in the Sousa basin such features do not exist. Conceivably, this could be the impression of the tail of a medium-size or large dinosaur, but this is far from certain. Tail impressions of dinosaurs are very rare the world over; however, for Lower Cretaceous, see, for example, a lot of them in La Rioja, Spain (Pérez-Lorente 2015) and some in Gansu Province, China (Fujita et al. 2012).

One observation in favor of this hypothesis is that mud cracks here follow the margin of the impression, as they often do with the margins of footprints (Schanz et al. 2016, fig. 19.4, 376). Two observations against this hypothesis are that there is no displacement rim associated with it and that there are no footprints of a large dinosaur associated to it.

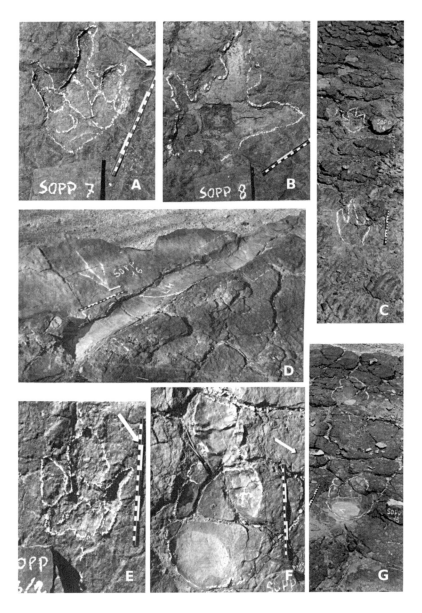

Figure 4.79. Passagem das Pedras (SOPP). Upstream of the main rocky pavement of SOPP, where the park is situated, additional tracks were found in older siltstone and shale of the same Sousa Formation. *A*, Theropod track SOPP 7. *B*, Theropod track SOPP 8. *C*, Theropod trackway SOPP 14. *D*, Rill mark or dinosaur tail impression SOPP 16. *E*, Theropod track SOPP 13/2. *F*, Hand-foot set of trackway SOPP 15, likely undertracks. *G*, A short trackway (SOPP 15) of 4 hand-foot sets (probably undertracks), possibly attributable to a young, quadrupedal ornithischian. Graphic scales in cm. See figures 4.62 and 4.63 (top).

SOPP 17

This is a short trackway of 2 footprints, very badly preserved, produced by a slowly moving, bipedal, and probably heavy trackmaker (fig. 4.65). The index SL/FL averages 4. If these are two successive prints of the opposite feet, the pace angulation was probably near 120°, with a high trackway width. The feet of the trackmaker were probably tridactyl, but as preserved the footprints have only two toemarks. We cannot hypothesize a case of functional didactylism because the first footprint lacks the digit II impression, and the second lacks that of digit IV. Thus, the presence of only two toes in these footprints is due to the bad quality of the prints at the time they were emplaced or their degradation during subsequent preservation.

Combining the few morphological details of both prints, one can reconstruct a tridactyl footprint, slightly asymmetrical, with the main weight carried on the tips of the toes. These are wide and blunt, hooflike. The total interdigital divergence is medium (90°). Thus, these seem to be rather typical ornithopod footprints, very similar to *Iguanodontichnus* Sarjeant et al., 1998. Alternatively, these could both be left footprints, with the intervening right footprint lacking. If so, the pace angulation would be lower (~90°–100°) and the trackway width even greater.

SOPP 18

This is a very small, isolated footprint, one of the smallest of the Sousa basin (length 5.6 cm), located at the SE margin of the pavement of Passagem das Pedras. The only footprints in the basin smaller than this one are the lacertoid footprint SOES 15 from Serrote do Pimenta and the tiny footprints of the Caiçara-Piau farm, attributable to half-swimming turtles. SOPP 18 is impressed into a very smooth, gray-greenish siltstone surface showing primary mud cracks. The footprint is as long as it is wide (L/W = 1) and is probably a right and tridactyl, with the deepest part of the print impressed slightly on the lateral (right) margin. The print is mesaxonic, with a medium digital divergence (64°). The impressions of digits II and III are relatively short, respectively bearing two and three digital pads. The impression of digit III is sinuous and sharply pointed at its terminal end. The impression of digit IV is the longest and stoutest of the three, with a claw mark and a single, long digital pad.

The footprint can be attributed to a very small, bipedal, and tridactyl dinosaur, probably a juvenile or a hatchling, tracks of which are rare. A more exact classification seems to be impossible. The specimen was unfortunately lost, like footprint SOCA 132 bis and other slabs.

Diminutive tridactyl dinosaur tracks are rare in Jurassic and Cretaceous terrains all over the world. There are, however, recent reports of very small dinosaur tracks, *Minisauripus* (theropod, ~2.5 cm long), and other forms from the Early Cretaceous (Barremian-Albian) of Shandong Province, China, as well as from two localities in South Korea's Haman Formation, also considered to be Early Cretaceous (Aptian-Albian) in age. Other small tracks from the Sichuan fauna include *Aquatilavipes sinensis* (2.5 cm long, a possible junior synonym of *Koreanornis hamanensis*) and *Grallator emeiensis* (2 cm long), all of them found associated with *Minisauripus* at both the Korean and Chinese localities (Lockley et al. 2008; Li et al. 2015). Jim Farlow notes that he wouldn't be surprised if many of the Mesozoic footprints attributed to birds turn out to have been made by baby nonavian dinosaurs (J. O. Farlow, personal communication, 2015). Some diminutive theropod footprints were found also at the Lower Jurassic (Hettangian-Sinemurian) Lavini di Marco site (Trento, Italy), after the book by Leonardi and Mietto (2000) was issued.

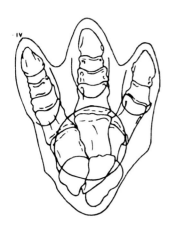

Disegni che ho preparato per il lavoro del 1979, e che poi per qualche motivo non ho pubblicato (incertezza)

Orma tipo di Sousaichnium pricei, con sovraposte ossa di ? Iguanodon, immagin.

Idem, con sovraposte ossa simili a quelle di Iguanodon, ma adattate all'orma tipo di Sousaichnium

vedi didascalie Leonardi 1979 vedi anche dietro

SOPP 19

Natural erosion by sun and rain in the years after our excavations has exposed a new trackway in the main rocky surface, about 16 m E from the crossing of SOPP 4 and 5 with SOPP 1, heading ~S60°W. It has 8 footprints of poor quality, bearing three rounded hoof impressions. The stride is about 115 cm, and the pace angle is ~130°. The trackway is attributable to an ornithopod.

SOPP 20

Another recently (2018) exposed trackway of 12 footprints crosses the main trackway SOPP 1 near footprint 43 of the latter and moves toward S45°W. The footprint length is 30 cm; the pace is 165 cm; the stride 318 cm; and the pace angle is about 175°. The trackmaker was a theropod.

Figure 4.80. *A, B,* Preliminary and vintage interpretive sketches of the first trackways of Passagem das Pedras by Leonardi (1978) for comparison to the skeletons of iguanodontids. *A,* Compares a "mean" *Sousaichnium pricei* footprint with the skeleton of a corresponding left foot of Iguanodon Mantell, 1825. *B,* Shows a hypothetical reconstruction of the foot skeleton to fit the mean footprint of *Sousaichnium pricei.*

Still another recently (2018) uncovered trackway of 14 footprints crosses the main trackway SOPP 1 near its footprint 35 and heads either S48°W or in the opposite direction. The new trackway has the following mean parameters: pace 124 cm; stride 232 cm; pace angle ~135°; footprint length ~30 cm. The prints are of poor quality and unclassifiable; perhaps the trackway can be attributed to an ornithopod.

Summing up, exposed at eight levels, this ichnofauna, with its seven dinosaurian associations, includes about: 13 large theropods, 2 small theropods, 1 very small footprint of a bipedal dinosaur, 4 graviportal ornithopods, 1 quadrupedal Ornithischia *incertae sedis*, a score of uncertain dinosaur tracks, and a possible tail impression. Altogether, 42 individual dinosaurs are represented. Over time, more bipedal (theropod) dinosaur trackways continue to appear on the main layer because of progressive natural erosion.

The Behavior of the SOPM-SOPP Ichnofauna as a Whole

The directions of travel recorded by all of the dinosaur tracks of the bed of the Peixe River between Passagem das Pedras and Poço de Motor sites in the deeper (older) levels lie mainly in the second and third quadrants, with a tendency to shift from E to W; however, upstream of the large pavement of Passagem das Pedras, the trend passes to the fourth and first quadrants. At level 5 several tracks have directions in the fourth and first quadrants.

The overall distribution of the SOPM-SOPP site is random. However, some concentrations do exist (10 tracks, i.e., 29.4% of the sample) in the sector 135°–199° (64°, or 17.7% out of 360°). This is a very different situation from that of the Piau-Caiçara locality.

Piau II (SOPU)

This locality is situated on the former Piau farm, in the municipality of Sousa, on the unimproved road Lagoa do Canto—Piau-Caiçara, near the Cupim creek, a tributary of the Peixe River, on the farm of Manuel Boaventura (Manu), approximately 1 km N of the principal Piau-Caiçara locality (above, SOCA) and 10.5 kms WNW of Sousa (S 06°43'52", W 38°19'37"). The locality was discovered by Leonardi with Anna Alessandrello of the Museo Civico di Storia Naturale of Milano (Italy) and Carvalho in December 1988.

SOPU 1 and SOPU 2

Two footprints, probably attributable to half-swimming theropods, associated with invertebrate bioturbation features, were found on a low flagstone fence—a kind of fence that's common in this region (fig. 4.81A). The pieces were not collected.

References: Leonardi 1994.

This site is an abandoned meander of the Peixe River, approximately 700 m long. The river presently follows a more recent alternative course (Rio Novo, or New River). In the old meander cutoff there is a pool during the drought season. It is situated in the municipality of São João do Rio do Peixe, 4.3 km SE of the homonymous town (S 06°45'16", W 38°24'50"). Here a large detached slab of dark gray siltstone was found (approximately 5 m² and 30 cm thick) with a short trackway of a medium-size to large theropod (ANPV 1). The Poço da Volta site is a very good locality in terms of outcropping rocky pavements, with many siltstone levels bearing mud cracks and ripple marks—but with no tracks, apart from the previously mentioned detached slab.

References: Leonardi 1985b, 1989a, 1989b, 1994; Leonardi et al. 1987a, 1987b, 1987c.

ANPV 1

This is a short trackway of 3 footprints, attributable to a theropod of medium size (fig. 4.81C). The first footprint is represented only by the distal end of digit III; the second and third prints are complete and are partially infilled with a natural cast. The digit impressions are stout and show claw impressions. Some of the toe marks are crooked. The pace angulation is very high (179°). The trackway was left in situ and was still there in 2013.

ANPV 2

Some 200 m downstream of the location of ANPV 1, near a path and a ford, there is an incomplete and unidentifiable roundish footprint with a diameter of about 36 cm and a displacement rim, but without morphological details. In 2003, Luiz Carlos da Silva Gomez of Sousa discovered 6 more theropod tracks at this locality (communication by letter; we have not seen these prints).

This locality is situated on the small Saguim farm, in the municipality of Sousa (figs. 4.82, 4.83) on the Lagoa do Canto-Piau unimproved road, about 2.5 km SW of Lagoa do Canto on the Sousa-Uiraúna road. The track-bearing layer crops out along a stretch 80 m long on the northern side of the road, beside the main farmhouse (S 06°43'24", W 38°19'16"). The footprints were discovered on December 6, 1988, by Leonardi, Anna Alessandrello of the Museo Civico di Storia Naturale di Milano, Italy, and Carvalho.

The outcrop consists of a sequence of siltstone beds of the Sousa Formation, of deep purple color (strike N60°E; dip 28° to the south). The footprints are situated on only one of these layers. Some of the beds show invertebrate bioturbation, among them the underside of the track-bearing layer. In a dry well behind the main house, where oil was found later, is a 4-m exposure of the Sousa Formation: chocolate-colored to deep purple siltstones with a 20-cm intercalation of greenish siltstone, with some

Figure 4.81. Localities Piau II (SOPU), Poço da Volta (SOPV), and Zoador (ANZO, Sousa Formation). *A*, Piau II is a locality along the dirt road between Piau and Lagoa dos Estrelas. Ripple-marked slabs have been turned on edge to make low stone fences; the rock surfaces have rare tracks made by half-swimming theropods. *B*, ANZO 1, a rare tetradactyl theropod right pedal footprint; erosion has revealed the lamination of the infilling upper layer. Courtesy of Luiz Carlos da Silva Gomes of Sousa. *C*, ANPV 1/1, the second footprint of a short theropod trackway found on a large detached slab at Poço da Volta (São João do Rio do Peixe). The caliper is open to 5 cm. *D*, The same footprint as in *B*, but with different lighting.

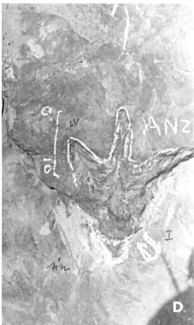

bioturbated levels. Neither ripple marks nor mud cracks were seen. Five isolated footprints on one narrowly outcropping layer were discovered here. The footprints will be described from S to N.

Reference: Leonardi 1994.

SOSA 1

This is a fine footprint of a large theropod, deeply imprinted, with stout digits and long, acuminate claws (figs. 4.82, 4.83A). The deepest part of the print is in digit III. The footprint is similar to ANJU 1 and 2, from the Juazeirinho locality (ANJU). Like them the track is quite symmetrical,

Figure 4.82. The Sítio Saguim (SOSA: Sousa Formation, Sousa), with one of the footprints discovered there. Leonardi and Anna Alessandrello of the Museo di Storia Naturale of Milan, Italy, are explaining the tracks to the farmer and his family, 1988. See also figure 4.83. Photo courtesy of Franco Capone.

with a rather large metatarsophalangeal region, as is usually observed in large theropod footprints from post–Lower Jurassic strata (Xing et al. 2014b). The print can be classified as *Eubrontes* (E. Hitchcock, 1865). The trackmaker was heading S28°W.

SOSA 2

An incomplete, isolated footprint, probably an undertrack, and attributable to a theropod (fig. 4.83E). Retroscratches produced by the claws during retraction of the foot are visible. Because of its small size, the trackmaker either belonged to a small-bodied species or was a young individual of a larger form.

SOSA 3

An incomplete footprint (probably a right) attributable to an advanced ornithopod and very similar to *Staurichnium diogenis* (probably a nomen dubium; fig. 4.83B). It is also reminiscent of some of the footprints of the Rio Piranhas Formation from the same basin (see below) as well as some prints from the Picún Leufú locality in Neuquén Province, Argentina (Calvo 1991). The footprint is tridactyl, with rounded, hooflike toe impressions, with digit III leaving the largest impression of the three. Both the "heel" and the hypexes are rounded. The print is deepest in the impressions of digits II and III. The trackmaker was headed approximately N32°W.

Figure 4.83. Sítio Saguim (SOSA). Alongside the unimproved road, in a very narrow outcrop, a number of tracks were found. *A*, A fine isolated theropod footprint (SOSA 1, probably left pes) with long claws. *B*, SOSA 3, an isolated ornithopod footprint, reminiscent of *Staurichnium diogenis*. *C*, SOSA 4, a left footprint attributable to a theropod with broad toes and distinct claws. The great width of the toes is probably extramorphological, perhaps due to mud collapse. Scratches produced by the toe claw during retraction of the foot can be seen in the impression of digit II. *D*, Shallow theropod track SOSA 5. *E*, SOSA 2, an incomplete, isolated left footprint attributable to a theropod, and probably an undertrack. See also figure 4.82. Graphic scales in centimeters.

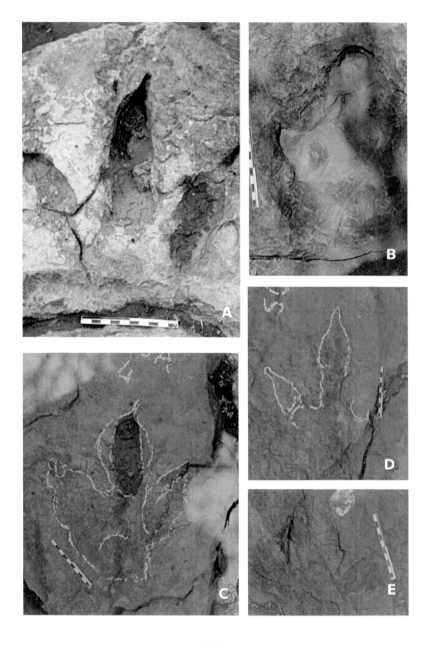

SOSA 4

A left footprint attributable to a large theropod. The print has broad toe marks, but their great width is probably extramorphological (fig. 4.83C). Digits II and III have just one large digital pad, but digit IV seems to have four pads. Digits II and III are closer together than digit III is to digit IV. Claw marks are evident on all three digits. Inside the digit II impression is a scratch, probably produced by the claw during retraction of the toe. The footprint is deepest in digit III. The trackmaker moved slowly northward.

An isolated, shallow, and incomplete footprint attributable to a large theropod (fig. 4.83D). The digit impressions are spatulate with well-developed claw marks. The animal headed approximately N55°E.

All of the above-described specimens, found in situ, were collected and placed in the small museum of the Dinosaur Valley park in Sousa, Paraíba.

To sum up, this small footprint assemblage records passage of 4 large theropods (3 adults and perhaps 1 youngster) of at least two different kinds, and a single ornithopod. The numerical dominance of carnivorous dinosaur footprints is a rather common situation in the Rio do Peixe ichnofaunas, as will be discussed in a later chapter.

The distribution of the trackmakers' directions of travel seems random, but 3 of the 5 footprints show directions in the first and third quadrants of the compass, as is true of many other trackways in this basin. Of course, the travel directions at the present site are only approximate, based as they are on isolated footprints.

Tapera (APTA)

An important ichnofauna including scores of individual tracks was recorded from a large outcropping rocky surface by José Ildo Ferreira of Aparecida and communicated to Leonardi by Robson de Araujo Marques of Sousa. The locality, in the Fazenda Tapera of Antônio Nóbrega, was briefly visited twice by Leonardi in August 2005. He did not know, however, that the site had already been discovered and published by Hebert B. N. Campos in two abstracts in two local journals (Campos 2004, 2005). The site (S 06 46.188, W 038 06.695) is in the municipality of Aparecida, E of Sousa, in the bed of the Peixe River, about 250 m N of the bridge over the national highway BR 230 and about 250 m downstream of locality APVR II (fig. 4.84A). The rocks belong to the Sousa Formation. They show ripple marks and mud cracks and have a black weathered surface.

On the basis of Leonardi's brief examination of the tracks, the ichnofauna included three or four forms attributable to theropods (figs. 4.84B, 4.84D, 4.84E), one of them very small (fig. 4.84E); one or two forms of Sauropoda (fig. 4.84C); and perhaps one isolated ornithopod track.

More recently, crocodilian traces (tracks and body imprints on the mud; fig. 4.84B, 4.84F) were recorded from this site: Campos, Silva, and Milàn (2010) describe the track assemblage of the Tapera ichnosite as including isolated theropod tracks and traces produced by large crocodilian trackmakers; they interpret these last as half-swimming tracks and resting traces, possibly produced by Mesoeucrocodylia. These tracks are not the first record of crocodylian tracks in Brazil, if the trackmaker of the hand-foot pair of Batrachopodidae Lull, 1904, from the Piau-Caiçara (SOCA) site, level 13/3 (SOCA 13334–13335; Leonardi 1984c, 60; 1994, 58), is correctly attributed to the order Crocodylia Gmelin, 1788. These crocodiloid traces should be usefully compared with images, data, and commentary of the papers by Abbassi et al. 2015; Mateus et al. 2017; Kim et al. 2020; and especially Farlow et al. 2017.

References: Campos 2004, 2005; Leonardi and Santos 2006; Campos et al. 2010.

Várzea dos Ramos (APVR I; formerly SOVR)

Várzea dos Ramos is situated in the Aparecida (formerly a district of Sousa) municipality, in the Peixe River rocky bed, about 12 km E of Sousa. The locality is reached by going 13 km E from Sousa on the Sousa-Aparecida federal highway (BR 230) and then turning left (N) on a dirt road, going about 2 km, then turning right (E) on a path in a Carnaúba palm grove until reaching the Peixe River. The coordinates are S 06 46.158; W 038 06.673 (fig. 4.85).

The locality was discovered in September 1985 by Wilton Viana Barbosa Jr., during a geology field trip led by J. M. Mabesoone of the School of Geology of the Federal University of Pernambuco at Recife (UFPE). It was visited and surveyed by Leonardi, Maria de Fátima C. F. dos Santos, and Claude Luis Aguilar Santos in December 1985.

The outcrop is a large rocky pavement of the Sousa Formation, with a surface of 160 × 20 m (approximately 3,200 m²; strike N30°E; dip 8° to the north). The rock here is a yellowish-brownish siltstone with a dark gray or black varnish on the upper surface. Ripple marks are present, crests of which are variably oriented (main directions: N10°E and N65°W), along with small mud cracks. The strongly laminated upper layer contributed to heavy erosion of the tracks, which in consequence are generally of poor quality.

To the south, downstream from this outcrop, there is a different layer of mudstone with footprints; this unit is brown in color, with primary and secondary mud cracks. This layer is separated from the first rocky pavement by a number of small faults. Much of the information about this ichnofauna is taken from Santos and Santos (1989).

References: Santos and Santos 1989; Leonardi 1994; Leonardi and Santos 2006.

More than 60 footprints occur at the main site (fig. 4.86A–C), a few of which are in short trackways; the remainder are isolated footprints as well as many additional incomplete and unidentifiable prints, altogether representing almost 60 bipedal dinosaurs (theropods):

SOVR 29 and SOVR 36

A trackway of 2 footprints (pace 80 cm).

SOVR 30, SOVR 33, SOVR 34

A trackway of 3 footprints (pace: respectively 97 and 94 cm; stride: 188 cm; pace angulation: 177°).

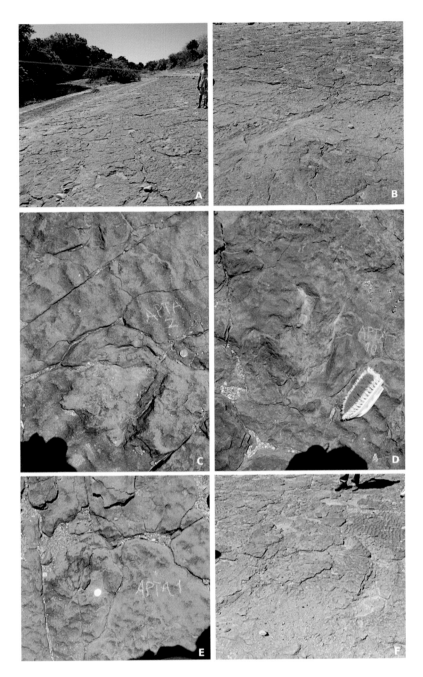

Figure 4.84. Tapera site, Sousa Formation, municipality of Aparecida (Paraíba), formerly a district of Sousa. *A,* The track-bearing rocky pavement in the bed of the Rio do Peixe, near its mouth with the Piranhas river, at Tapera farm. *B,* A section of the pavement with theropod tracks. *C,* An incomplete hand-foot set of a small sauropod. The coin has a diameter of 2.3 cm. *D,* A large theropod footprint. *E,* A very small theropod footprint. The coin has a diameter of 2.3 cm. *F,* Another view of the rock surface seen in *B* with a possible body impression of a crocodile.

SOVR 60–61

A trackway of 2 footprints (pace 93 cm).

The others are isolated footprints. Almost all of the footprints are attributable to large theropods except for a large footprint (65 × 57 cm) lacking morphological details but with a strong displacement rim, which could be attributed to the ornithopods or, more probably, to sauropods.

Figure 4.85. The black siltstone surface of the Sousa Formation in the bed of Peixe River at Várzea dos Ramos Farm, municipality of Aparecida (Paraíba; code APVR I, formerly SOVR I). The ichnofauna contains some 60 theropod tracks and 1 large sauropod footprint. Reproduced from Leonardi and Santos (2006). See also figures 4.86C–E.

On the second smaller pavement, along with a large print made by a half-swimming theropod (SOVR 62), a short trackway of 2 footprints was found (SOVR 63; pace 104 cm). These last footprints have very widely separated and sometimes plump toes, because of which they could be attributed to ornithopods; but the long pace would be more typical of theropods.

The theropod footprints of the Várzea dos Ramos might be divided into at least four groups:

1. Some rare, well-preserved footprints with long, pointed toes with narrow interdigital angles; digit III toe has a long claw (for example, SOVR 51). From a morphological point of view, they could be classified as *Grallator* because of the length of digit III or as a slim form of the ichnogenus *Eubrontes*, such as *Eubrontes tuberatus* (E. Hitchcock, 1858). On the other hand, because of the prints' relatively large size and their geographic origin and age, their trackmakers could hardly be the same as the makers of the classic Connecticut Valley Early Jurassic ichnogenera, in spite of the similarity to *Grallator* or *Eubrontes*. This group of prints is here attributed to large theropods.

2. A few footprints with widely divergent, frequently slim toes (e.g., SOVR 63). Sometimes there is only the impression of the digits (SOVR 63); in other specimens, there is the impression of a large, wedge-shaped "plantar" cushion. These prints are also attributed to large theropods.

3. A large number of poorly preserved footprints (many of them probably undertracks) with a smaller digital divergence than in footprint

group B and a large, wedge-shaped plantar cushion (e.g., SOVR 10 and 14). Classification of the footprints of this group is difficult, but they are large theropod footprints of bad quality.

4. A few footprints with stout, well-defined, padded toes with tips that are rounded or, rarely, with claw marks. They can be interpreted as incomplete footprints classifiable from a morphological point of view as *Eubrontes sp*. They are also attributable to large theropods (e.g., SOVR 23).

Apart from the large footprint questionably attributed to an ornithopod, or better to a sauropod, the about 60 tracks at this site represent an assemblage of theropods of medium to large size by the Peixe River standard, with a fairly uniform size distribution. No prints of very small/young individuals are recorded. Footprint lengths range from 17 to 35 cm, but the majority fall between 23 and 32 cm; print width ranges from 12 to 27 cm, with most of the prints between 17 and 25 cm in width (see diagram in Santos and Santos 1989).

Várzea dos Ramos II (APVR II)

This locality is situated in the bed of the Peixe River on the Várzea dos Ramos farm, in the municipality of Aparecida (Paraíba), formerly a district of the Sousa municipality, downstream from the previously published Várzea dos Ramos locality (APVR, formerly SOVR; figs. 4.86A, 4.87). It is 15 km ESE as the crow flies from the center of Sousa and 1.5 km WNW of Aparecida; its coordinates are S 6°46'38", W 38°05'42". The locality was discovered in 2003 by Leonardi.

Reference: Leonardi and Santos 2006.

At this site, on a dark-gray rocky pavement, there is a theropod trackway (figs. 4.86A, 4.86E, 4.87). The trackway contains 8 footprints and has the following approximate mean parameters: pace 90 cm, stride 180 cm, pace angulation 175°, footprint length 30 cm, footprint width 27 cm, and interdigital divarication (II^IV) 90°.

Várzea dos Ramos III (APVR III)

This locality is situated in the bed of the Peixe River inside the Várzea dos Ramos farm, in the municipality of Aparecida (Paraíba), downstream from the already published Várzea dos Ramos locality (APVR, formerly SOVR; fig. 4.86B). It is 14 km ESE as the crow flies from the center of Sousa and 2.5 km WNW of Aparecida; its coordinates are S 6°46'38", W 38°06'15". The locality was discovered in 2003 by Leonardi.

Reference: Leonardi and Santos, 2006.

At this place the river has high banks, exposing a large number of layers—a very unusual situation for this river (fig. 4.86E). At the foot of the hydrographical left bank, on a rippled red surface, 6 isolated theropod footprints were found, all of them with very narrow digits because of collapse of the mud. Accidentally, this rocky pavement with its collapsed tracks is rather similar to that illustrated by Xing et al. (2016b; in fig. 3).

Figure 4.86. *A*, Várzea dos Ramos II (APVR II), Sousa Formation, Aparecida. A theropod trackway is exposed in the riverbed, perpendicular to the river channel. *B*, Várzea dos Ramos III, Sousa Formation, Aparecida (APVR III). At this site, the river has high banks—a very rare situation in the Peixe River. The small ichnofauna exposed here contains a number of large and small theropod footprints. *C*, Robson Araújo Marques points to a fine pair of theropod tracks at Varzea dos Ramos I (APVR I). *D*, One of the numerous theropod tracks from Várzea dos Ramos I. *E*, The sixth footprint of the theropod trackway from Várzea dos Ramos II, on the dry bed of the Peixe River.

Four footprints of half-swimming theropods also occur in the same surface at the APVR III site.

Zoador (ANZO)

The outcrop (strike N89°W; dip 11° to the south) is situated on small rapids in the bed of the Peixe River, 3.4 km ESE of the center of the town of São João do Rio do Peixe, and in the territory of that municipality (S 06 45.301, W 038 24.595). The place was discovered by Leonardi in 1984.

References: Leonardi 1985b, 1989a, 1994; Leonardi et al. 1987a, 1987b, 1987c.

At this site is an isolated left footprint of poor quality (ANZO 1; fig. 4.81*B*, 4.81*D*), filled in by sediment of the upper contiguous layer. The

Figure 4.87. At Várzea dos Ramos II, on a dark-gray rocky pavement crossing the Peixe River, a theropod trackway was discovered. Sousa Formation. From Leonardi and Santos (2006).

infilling is laminated, and the lamination is evident because of differential erosion. The footprint is mesaxonic and tetradactyl, the latter condition being a rather rare phenomenon in our Paraíban ichnofauna. The footprint is longer than it is wide (index L/W = 1.12). A shallow heel-shaped concavity at the proximal margin is probably not a heel at all but rather an extramorphological furrow caused by the foot dragging. The general outline of the footprint is wedge shaped and strongly symmetrical. Digit I is impressed as a small spur, thin, blunt, and relatively short (38 mm). The other three toes are short compared to the plantar region. The impression of digit IV is larger than the others; all three main digit marks are blunt tipped. The interdigital divergence II^IV is wide (approximately 80°). Because of the infilling, digital pads cannot be observed. The digit III axis points S45°W. The footprint was left in situ.

In addition, there are at this site more possible but uncertain dinosaur footprints.

Summary of the Sousa Formation Ichnofauna

The 21 sites from the Sousa Formation (figs. 3.10, 3.32, 4.1–4.87) altogether represent at least 66 stratigraphic and ichnofossiliferous levels, which preserve an overall ichnofauna of 296 theropods, among them 249 large theropods, 31 small- to medium-size theropods with a particularly long III digit, and 16 other theropods. In addition, there are 16 sauropod trackways, 19 trackways attributed to graviportal ornithopods, and 1 trackway of a small quadrupedal ornithischian, for a total of 20 individual tracks attributed to ornithischians. Beyond that, there are about 38 individual tracks of dinosaurs that are of uncertain identity or completely unidentifiable.

Altogether, the recorded dinosaurian individuals number more than 370. Finally, smaller vertebrates are represented by 1 batrachopodid set, crocodilian traces (tracks and body imprints in the mud), and a large number of small chelonian tracks.

Antenor Navarro Formation

Arapuã (UIAR)

On the Arapuã farm (not the farm of the same name marked on the SUDENE [*Superintendência do Desenvolvimento do Nordeste*] scale 1:200,000), in the municipality of Uiraúna, in the bed of an unimproved road between Ipueiras and Pocinhos, 1 km S of Ipueiras, in the Uiraúna-Brejo das Freiras basin, are some large, rounded footprints of poor quality with displacement rims, on a surface of a layer of the Antenor Navarro Formation, along with ripple marks. The footprints are unidentifiable. This site was discovered by Leonardi and Carvalho in August 1987. Prior to this study, the site has been unpublished.

Aroeira (ANAR)

This site is situated in the municipality of São João do Rio do Peixe, on the São João do Rio do Peixe-Morada Nova dirt road, 11 km ENE from São João do Rio do Peixe, near the northern boundary of the basin (S 06°41'44", W 38°22'06"), on the Aroeira farm. The locality is several hundreds of meters N of the boundary between the Sousa Formation (to the south) and the Antenor Navarro Formation (to the north). This last formation presents itself in this region, as usually, as coarse sandstones of grayish or yellowish color; but locally at Aroeira the rock is greenish-gray. The outcrop is a few square meters of sandstone covered by a thin layer of siltstone on the upper surface. It crops along the side of the previously mentioned dirt road. The outcrop was discovered by Leonardi in December 1984 but was not excavated, studied, or otherwise sufficiently recorded.

References: Leonardi 1985b, 1989a, 1989b, 1994.

A short trackway of 3 or 4 large footprints is associated with some small, unidentifiable dinosaur footprints. The main trackway (ANAR 1; fig. 4.89E) shows large (40–50 cm long and wide), rounded footprints with high displacement rims but lacks morphological details. The footprints are very similar to those attributed to sauropods at Serrote do Letreiro (see below) and at Engenho Novo (above). At first glance, the trackway seems to have been made by a biped, but it was likely made by a quadruped, with complete overlapping of manus by pes prints; if so, it is probably attributable to a sauropod. Alternatively, if the trackmaker was in fact a biped, it would likely have been a very large ornithopod.

At Baleia farm (owner José Bastos, 1987), in the municipality of Uiraúna, about 1 km SSW of Uiraúna, two doubtful, very shallow theropod footprints were found. Antenor Navarro Formation. The layer has a strike parallel to the marginal fault and dip ~20° toward NNW.

Baleia (UIBA)

    This site was discovered by Leonardi and Carvalho in August 1987. Unpublished.

In the small Pombal basin E of the Sousa basin (fig. 1.10), a probable but poor-quality, unidentifiable dinosaur footprint was discovered by Leonardi and Carvalho in December 1988 at the locality of Grotão, municipality of Pombal, in the main sub-basin of Boa Vista. The rocks here consist of coarse Antenor Navarro sandstone. Unpublished.

Grotão (POGR)

This is the only relatively good site thus far discovered in the Brejo das Freiras-Uiraúna basin (Carvalho 1996a), with 4 badly eroded footprints or short trackways attributable to theropods. The locality is situated on the Pocinho farm, Uiraúna Municipality, on the unimproved Uiraúna-Baleia-Ipueiras-Pocinho-Varginha road, 7 km S of Uiraúna, about 1 km W of the Uiraúna-São João do Rio do Peixe road (S 6°35'13", W 38°25'15"). The outcrop is in the Uiraúna-Brejo das Freiras (Triunfo) basin and consists of fine sandstones, immature coarse sandstones, and conglomerates of the Antenor Navarro Formation, with upward fining of grain size (fig. 4.88A). Cross bedding of both tabular-planar and trough types occurs. Some mud cracks are seen at the top of the sequence in a layer of argillaceous siltstone.

Pocinho (UIPO)

    The footprints were discovered by Leonardi and Carvalho on August 6, 1987. They are the very first tetrapod fossils discovered in the Uiraúna-Brejo das Freiras basin. The prints were left in situ because of their poor quality and the thickness of the beds.

    References: Carvalho 1989; Leonardi 1994.

UIPO 1

An incomplete, poor-quality, tridactyl footprint, possibly a right, attributable to a large theropod. The toe marks are broad with broad claws. The digit III impression is shorter than usual.

UIPO 2

A badly eroded, isolated, tridactyl footprint with short, broad, outward bending digits and a long and wide "plantar" region (fig. 4.88B). The print is attributable to a large theropod.

UIPO 3

A large, badly eroded, isolated, tridactyl footprint. Unclassifiable. A nail or claw mark on digit IV would perhaps point to a large theropod as the maker, but the print is otherwise unidentifiable.

Figure 4.88. Sandstones of the Antenor Navarro Formation, Poçinhos farm (UIPO), near Uiraúna in the Uiraúna-Brejo das Freiras basin. Tracks are very rare here. *A*, Carvalho searching for tracks in the very coarse and rather unproductive sandstones of the unit. The only prints found here were theropod tracks of poor quality. *B*, Theropod print UIPO 2.

UIPO 4

A short trackway of 2 footprints and an isolated footprint from a small flagstone quarry at the same locality, attributed to a theropod.

Riacho do Cazé (SORC)

This site is situated at the foot of the northern *serrotes*, in the municipality of Sousa, 4.2 km NNE of Sousa, S 06 43.153, W 38 14.548, near the northern border of the basin. It shows terrains associated with the very coarse-grained sedimentary rocks of the Antenor Navarro Formation (figs. 4.89A–D, 4.90). It was discovered by Leonardi in 1979.

Reference: Leonardi and Santos 2006.

Two uncertain large footprints, possibly in a trackway but not contiguous, were discovered in this site. They present wide and high displacement rims, similar to the tracks attributed to sauropods at Serrote do Letreiro. A line joining both footprints is oriented parallel to about N20°E. Some new, poor-quality sauropod and theropod footprints were subsequently found in December 2003 (Leonardi and Santos 2006).

Riachão do Oliveira (SORO)

This site (fig. 3.15A; S 06 43.347, W 38 14.079), located E of Benção de Deus hill (S 06 42.829, W 038 14.636), was reported by Luiz Carlos da Silva Gomes of Sousa to Leonardi (who then visited it on August 8, 2005) because of large fossil plants (figs. 3.15B, 3.15C) in a stone quarry. The coarse sandstones belong to the Antenor Navarro Formation. One surface of the same formation, along the Riacho de Santa Rosa upstream of the quarry, probably shows a shallow and poor-quality ornithopod footprint (fig. 3.15D).

Serrote do Letreiro (SOSL)

This locality is situated almost at the top of the Serrote do Letreiro, so called because of the presence of native engravings, associated with the dinosaur tracks, on the rock surfaces. It is at the northwestern margin of the Sousa basin, at the eastern side of the Lagoa dos Estrela-Pereiros country road, which is a byroad of the main Sousa-Uiraúna road (starting at the km 12 marker of this road); about 0.8 km from the main house of the Lagoa farm and 10.5 km NW of Sousa (S 06 41.602, W 038 18.498; figs. 1.11, 3.5, 3.8, 3.9, 4.91–4.94). The conglomerates and sandstones of this outcrop belong to the Antenor Navarro Formation and correspond to an alluvial fan paleoenvironment. The locality was discovered by Leonardi in 1977 and was studied by him, and later by Carvalho for his thesis (1989).

The locality is a trapezoid-shaped rocky pavement of about 7,200 m² (a strip of 240 × 30 m, with its lower part in the bed of a gully), with three different rock layers gently sloping nearly westward (strike N18°W to N24°W; dip 5° to 10° toward the south). Each of the three layers preserves a track assemblage.

References: Leonardi 1979b, 1979c, 1980b, 1980c, 1980d, 1981a, 1981c, 1989a, 1989b, 1994; Carvalho 1989, 2000b.

The track-bearing levels cropping out at this site (fig. 1.11) record 3 very different track assemblages, despite their having accumulated in the same environment over a geologically short interval. The first animals to cross the site, leaving their tracks in the lowermost rock level, were a herd of sauropod with some accompanying theropods, almost all of them progressing, more or less at the same time, in a southeasterly direction. In the next higher exposure, a single large ornithopod slowly walked to about S40°W. In the uppermost level, a number of theropods crossed the ancient alluvial fan, moving in almost random directions (Leonardi 1979b).

These ichnoassociations accumulated in a small area over a short span of time, during which the landscape is likely to have changed very little or not at all. The differences in track assemblages among the three levels in this constant environment are probably due to the nature of the dinosaurs themselves. Large animals require considerable space to meet their

Figure 4.89. Riacho do Cazé (SORC) and Aroeiras (ANAR). *A–D*, Landscape and rock outcrop features at Riacho do Cazé, Sousa. The flora is dominated by the cactus *xique-xique* (*Pilocereus gounellei*) and the bromeliacean *macambira* (*Bromelia laciniosa*). The rocks are very coarse sandstones and gravels of the Antenor Navarro Formation, deposited in marginal fans near their source. It is not easy finding tracks in these coarse, high-energy facies, but some poor-quality sauropod and theropod prints occur in finer sandstone or siltstone surfaces. *E*, Unexcavated and unstudied trackway of a large animal, probably a sauropod, from the Antenor Navarro sandstones, Aroeiras (ANAR 1; São João do Rio do Peixe).

needs. A herd of some scores (at least) of sauropods, for example, would need to move around a very large home range to find enough fodder, especially in a semiarid environment, as here. The chance that the same kinds of large animals would leave their footprints in three contiguous rock layers that accumulated in the same small patch of ground seems practically nil.

Level A: Assemblage SOSL 4–5

Layer A (upper level) presents a cluster of footprints and trackways, representing about 37 individuals of the same theropod footprint morphotype and possibly the same Linnaean species (group SOSL 4–49996 and

Large Sauropoda

Large Theropoda

Figure 4.90. Very coarse sandstone surface of the Antenor Navarro Formation at Riacho do Cazé (SORC), where some rare tracks of large sauropods and large theropods were found. From Leonardi and Santos (2006).

SOSL 5; figs. 4.91, 4.92*F*, 4.93) and, downhill, the tracks SOSL 9 and SOSL 10 (fig. 4.92*D*). There are 8 short trackways and 29 isolated footprints. The sample includes an isolated trackway of 3 footprints (SOSL 5) as well as a dense grouping of tracks (SOSL 4–49996), which are concentrated in a surface of 6 × 4 m; so many footprints are present that there is overtrampling (number of individual footprints/m² = 1.5).

SOSL 40, 43, 4999 are short trackways of at least 2 or 3 footprints. (Please see the drawing of these and other footprints in fig. 4.93.)

SOSL 41 is an incomplete trackway of 6 footprints (fig. 3.37*C*).

SOSL 42 is a trackway of 5 footprints (figs. 3.38*A*, 3.38*B*).

SOSL 44 is a trackway of 10 footprints (figs. 3.37*B*, 3.37*D*, 3.38*C*).

SOSL 5, larger than the others, is an isolated footprint with large digital divergence (fig. 4.92*F*).

The other tracks are isolated and often incomplete footprints. All of them are very shallow. Some of them are eroded; others are filled in with the coarser sandstone of the upper contiguous layer. This sometimes gives an effect of reverse relief.

SOSL 3 is an isolated, strange, and doubtful footprint (fig. 4.93) consisting of four polygonal pads that can be compared tentatively to the manus of some forms of *Batrachopus* E. Hitchcock, 1845 or with some incomplete dinosaurian footprint, such as those illustrated by Bronson and Demathieu (1977). However, it may not be a footprint at all but merely an artifact of erosion of a small cluster of pelitic intraclasts.

SOSL 7, SOSL 9—These are 2 isolated and unidentifiable footprints.

SOSL 10 (fig. 4.92C)—These are 2 superimposed theropod tracks of the same kind as each other.

Associated with the tracks, there is also a possible impression of a tail (SOSL 49997), 80 cm long, regularly tapering, nicely curved. However, it is not associated with a trackway.

All the footprints are very similar in morphology, showing about as much variability as one would observe among footprints of a single trackway (e.g., SOSL 44, fig. 4.93). What variability there is among footprints is due mostly to substrate conditions and trackmaker gait. The trackways are narrow, indicate slow walking (average SL/FL = 4.77), and are straight. The footprints are tridactyl and mesaxonic and generally longer than they are wide (average L/W ratio = 1.24). The digits are stout and broad, with strong claws. Digit III is the longest, and digits II and IV are subequal in free length. However, digit II is sometimes separated from the other two by a gap and/or by a characteristic indentation along the proximal-medial outline of the footprint. In the best specimens, digital pads can be seen.

Despite the uniformity of shape, the size range of footprints is relatively high (footprint length range is 16.6–30.1 cm, with a high standard deviation). This could indicate the presence of young (e.g., SOSL 492, 499, and 4993) as well as older adult theropods of the same kind occupying the site; the uniformly shallow depth of all the impressions suggests animals crossing the fan at more or less the same time. This is an unusual situation for the Rio do Peixe basins.

The direction of travel of 21 of the theropods was measured. Even if the dinosaurs were indeed at the site at about the same time, their travel directions do not indicate organized gregarious behavior, although almost all of the individuals (86.4%) were heading in a northerly direction (compass sector 220°-40° [clockwise]), with concentrations to the north and northwest. Thus, one can imagine 40 or so bipedal dinosaurs crossing the alluvial fan at the border of the basin, exploring it and searching for small game or invertebrates.

These tracks can be attributed to small to medium-size theropods of a South American family; from a morphological point of view, the tracks can be classified as a small form of *Eubrontes* isp. or, less probably, because of the relatively short toe III, as a large form of *Grallator* isp. Footprints of this kind have a relatively large metatarsophalangeal area, as is usually observed in the large theropod footprints from post–Lower Jurassic strata, and in them the metatarso-phalangeal pads are aligned with the axis of digit III, a feature very common in the Lower Cretaceous theropod (*Eubrontes*-type) tracks (Xing et al. 2014a).

Level B

Layer B (middle level) shows just 1 short trackway consisting of 2 noncontiguous footprints (SOSL 8; fig. 4.92E). The first footprint is well preserved, well impressed, and deep (13 cm). It is wider than it is long (length 35.5 cm × width 40 cm), tridactyl, and mesaxonic; it has broad,

Figure 4.91. Photomosaic of the Serrote do Letreiro (SOSL) tracksite. This rocky surface at subsite 4 shows a lot of theropod tracks, often infilled, along with Indian petroglyphs. Graphic scales in centimeters. Photos and mosaic by Leonardi. As for this cluster of theropod tracks, see also figures 3.9C, 3.9D, 3.27B–D, and 4.93.

Figure 4.92. Additional tracks from Serrote do Letreiro. *A*, One of the hand-foot sets from sauropod group SOSL 12–39, 50, 400–489, at the Riacho do Pique. *B*, A short and narrow sauropod trackway of the same group as above. The wooden caliper is 50 cm long. *C*, SOSL 1, a sauropod set, the very first footprints found at Serrote of Letreiro. *D*, SOSL 10, probably 2 overlapping theropod footprints. *E*, A short trackway (SOSL 8) attributed to the ichnogenus *Iguanodontipus*. *F*, Theropod footprint SOSL 5, with a wide divarication between digits II^IV.

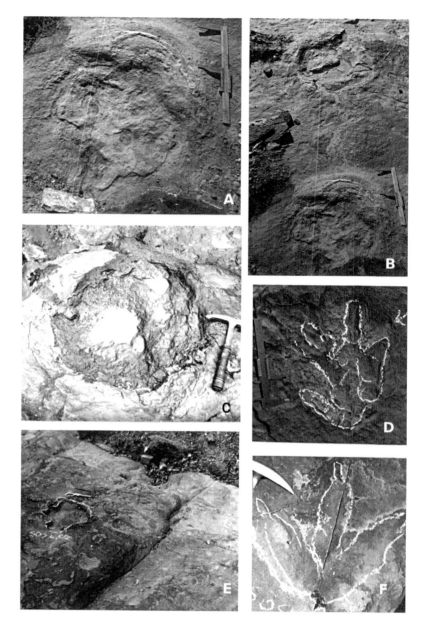

hooflike toes and a wide "plantar" region. The trackway can be assigned to the ichnogenus *Iguanodontipus*, first described from the Berriasian of Dorset (southern England) by Sarjeant et al. (1998; see also the Lower Cretaceous *Iguanodontichnus* trackways of Algarve basin in Portugal, in Santos, V. F., et al. 2013). The trackmaker was some form of South American graviportal ornithopod.

Level C

Layer C (the lowermost level) presents a cluster of about 40 footprints, including short trackways and isolated footprints of a herd of 16 sauropods and some tracks attributable to large theropods, in a long outcrop exposed

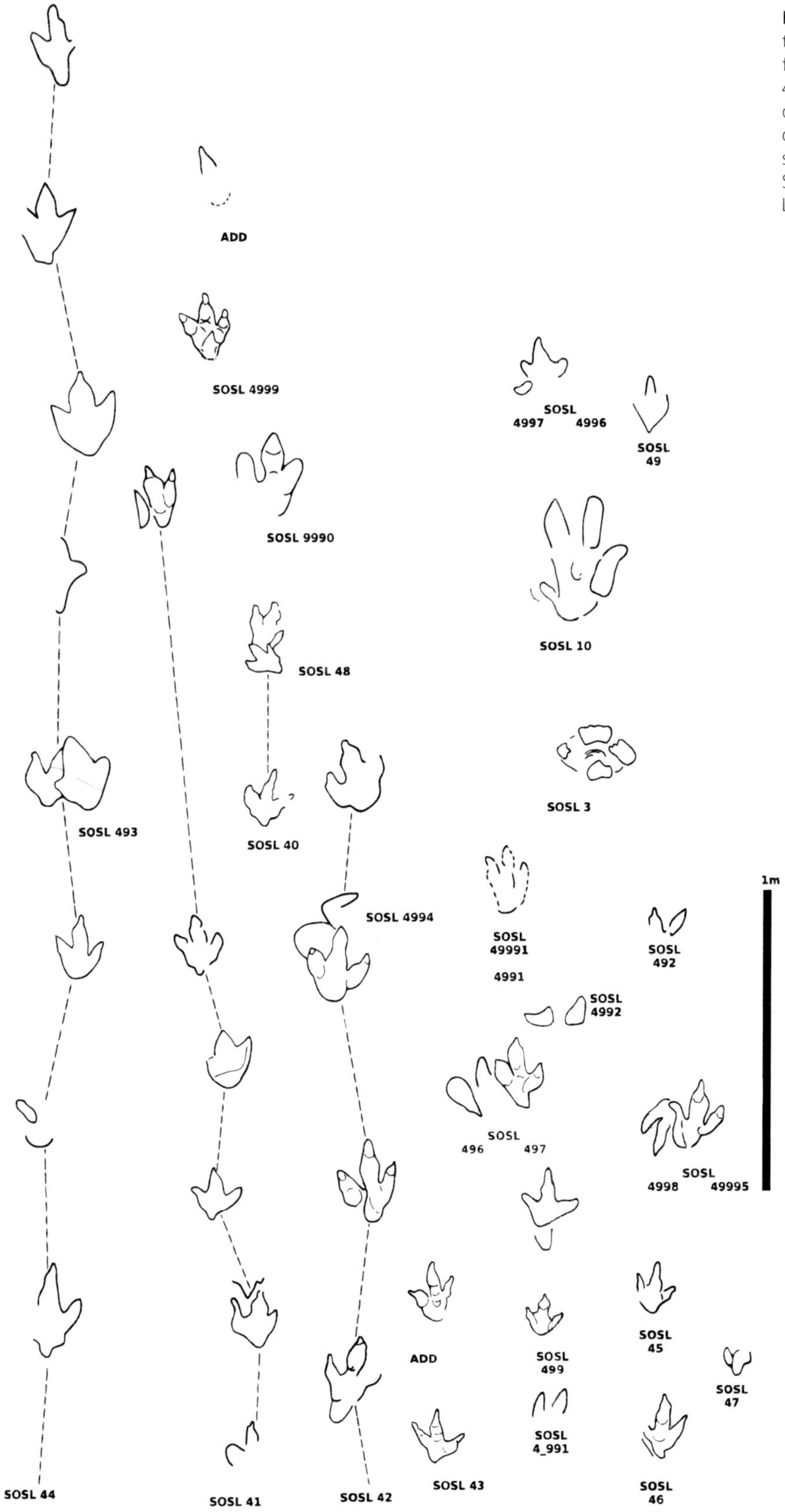

Figure 4.93. The main trackways and isolated footprints from subsite 4 at Serrote do Letreiro, drawn parallel to each other and at the same scale (bar = 10 cm). Sousa. Drawing by Leonardi.

ADD

SOSL 4999

SOSL 9990

SOSL 4997   SOSL 4996

SOSL 49

SOSL 48

SOSL 10

SOSL 493

SOSL 40

SOSL 3

SOSL 4994

SOSL 49991

4991

SOSL 492

SOSL 4992

1m

SOSL 496   497

SOSL 4998   49995

ADD

SOSL 499

SOSL 45

SOSL 47

SOSL 44

SOSL 41

SOSL 42

SOSL 43

SOSL 4_991

SOSL 46

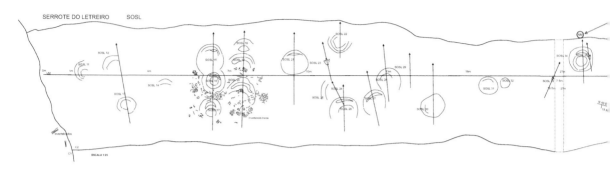

Invertebrate
icnofossil

Sauropod

Theropod
/ Ornithopod

Direction

Figure 4.94. Map of the Serrote do Letreiro (SOSL) tracksite, bed of the Riacho do Pique. This locality contains a score of sauropod prints as well as rarer theropod tracks. Map by Carvalho (1989).

in the bed of a temporary rivulet, the Riacho do Pique (footprints SOSL 2, 11–39, 50, 400–489; figs. 4.92A–C, 4.94; Carvalho 1989, 2000b).

SOSL 2

A short trackway of 4 footprints, probably attributable to the same kind of theropod form responsible for the group SOSL 4 and 5. The first, third, and fourth footprints are poorly preserved and by themselves would be unclassifiable; the second one is also difficult to interpret. It very probably represents the overlap of 2 theropod footprints headed in different directions. Quite unfortunately, it was first interpreted and published by Leonardi (1979b) as if it were just 1 footprint and ascribed to *Isochirotherium* sp. Then it was attributed to a large bipedal "thecodont" because of the apparent presence of a fifth digit in abduction and other features. In the same publication Leonardi used the supposed presence of *Isochirotherium* sp. and the lack at that time of any other means of dating the Antenor Navarro Formation to interpret this formation as Triassic. As a consequence, the other formations of the Rio do Peixe Group were considered to be Jurassic instead of Cretaceous. It was an unfortunate blunder, but it must be remembered that previous authors had often classified this formation as Paleozoic (Devonian to Permian) due to a lack of body fossils and its general aspect, and also because they had not yet discovered the dinosaur tracks. Its reassignment to the Mesozoic happened only after the discovery of its dinosaur footprints.

ASSOCIATION SOSL 1

See figure 4.92C and pages 2, 11–39, 50, 400–89, referred to as "group SOSL 1" in Leonardi 1979b (figs. 2.2C, 3.37A, 4.92A, 4.92B, 4.94)

    This occurrence has at least 38 footprints or sets of footprints in partial overlap. Of these, 6 are theropod, 4 are unidentifiable (a poorly preserved

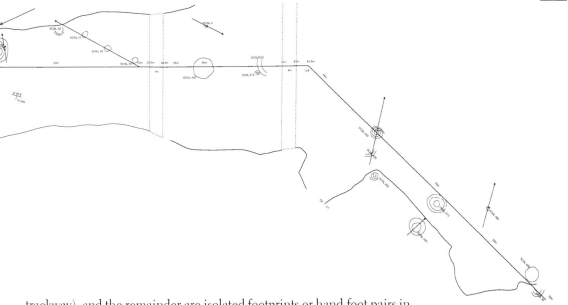

trackway), and the remainder are isolated footprints or hand-foot pairs in the bed of a small temporary creek named Riacho do Pique and attributed to sauropods (Carvalho 1989; Leonardi 1994). These last show the following features (1) manus prints are concave crescents with the convex margin directed forward, that is, in the direction in which the trackmaker was going; (2) pes prints are rounded features with a diameter of 40–100 cm; and (3) there are wide displacement rims around the pair and between the handprint and the footprint. The manus print was probably crescent shaped by nature but was often compressed in an antero-posterior direction by the displacement rim of the pes, and so it became very narrow; sometimes it is little more than a curved groove between both displacement rims. The hind-foot print often presents a reverse relief, which might be interpreted as a suction structure, produced by a flat or concave hind-foot sole on the cohesive mud-sand substrate. The footprints are very shallow, and the wide displacement rims are low, so one cannot avoid the impression that the waterlogged, muddy sand structure collapsed after foot retraction. This considerably reduced the areal extent of the primary footprint, which was partially occupied by the collapsed rim. Morphological details like claw impressions are not preserved.

These footprints are identified as those of sauropods because of their large size, the presence of both manus and pes prints and the overlap thereof, the crescent-shaped handprints, and the characteristically huge displacement rims associated with the pes prints. Because of the poor impression and preservation and lack of morphological details, the material cannot be attributed to an ichnogenus, and the trackmaker cannot be attributed to a special family either.

All the footprints of this kind that allow distinction between fore and hind-foot prints (thirteen in all) show that the trackmaker was headed toward an azimuth very close to SE; all of them occur in a narrow sector of 37°, from S41°E to S78°E. Because all the prints, complete or incomplete, well or badly preserved, present the same general aspect and the same phenomena of suction and collapse, and many of them have the antero-posterior compression of the hand and the same general direction, it is likely that all of these tracks were imprinted at the same time and do represent a herd of sauropods proceeding together in a southeasterly direction.

The number of individuals is not easy to determine. Footprints SOSL 15–20 are associated in two rows headed approximately S65°E and may constitute a single trackway or, probably, 2 overlapping trackways; some alignments and/or associations of prints can also be observed in the groups SOSL 12–13; 22 and 26; and 27 and 28. The herd would thus preserve a record of about 16 individuals. Across the front end of the exposure (in terms of the direction that the dinosaurs were going), the dinosaurs were spread out at 1 animal per 3.3 m, over a distance of about 50 m. On the northern (left) wing of the herd the trackmakers were closer together, about 10 individuals over 28 m, or 1 animal per 2.8 m.

In addition to the footprints of the herd, a number of isolated, deep sauropod footprints with large and high displacement rims are scattered downhill in the bush; these were the very first tracks discovered by Leonardi at the site in 1977 (fig. 4.92C).

Another trackway, consisting of 4 prints, was made by an unidentifiable biped (SOSL 36–39). This animal traveled in a different direction (N47°E), and its prints preserve no morphological details. The trackway could have been made by a quickly moving bipedal dinosaur that crossed the (waterlogged) fan at a different time and/or on a different sediment layer, in which case this trackway would be an undertrack.

This footprint assemblage also includes 5 theropod tracks: 4 small to medium-size isolated footprints (SOSL 29, 33, 35, 37, 39, 48o; fig. 4.94) and 1 trackway of 2 footprints (SOSL 23–24). To these the previously mentioned SOSL 2 trackway must be added. These trackways are all of poor quality, but they are all attributable to medium-size theropods. The preservation quality of the footprints and the presence of mud-collapse features are similar to those described for the sauropod footprints, and the displacement rims are particularly wide. One of the footprints was impressed on the displacement rim of a sauropod footprint, so at least one of the large theropods crossed the fan after the passage of the sauropods. Other than SOSL 2, which headed S41°W, the other theropod tracks show in this level the same general direction of travel as the sauropod herd, with azimuths in approximately the same sector of 100°–140° (more exactly, S50°E–S75°E).

The sauropod group was probably a true herd with gregarious habits, as previously described. It was directed to the southeast, coming from the highlands outside the basin and entering it diagonally, as reconstructed in figure 4.100. Because of this direction of travel, and because sauropod tracks were found at Serrote do Pimenta (SOES) (with travel directions parallel to the north–south axis but without evidence of which of the two directions the animals were heading, whether entering or leaving the basin) but also inside the basin, in terrains of the Sousa Formation at Caiçara-Piau and at Engenho Novo, it is unlikely that the herd was leaving the basin for the highlands, where the huge plant eaters could graze on Araucaria trees, which were more common in the hills. These large animals probably needed a wide territory and lived in a variety of environments, as much in the valley as on the hills.

The theropods of this level C were perhaps operating in a pack; they surely crossed the fan in the same direction as the sauropod herd, over more or less the same interval, and at least one of them certainly crossed the site after one of the sauropods. One cannot say if they really were actively following the sauropod herd. However, this association of sauropod and theropod tracks is common, and in this case it also could point to an interaction between a sauropod herd and a small pack of theropods in a hunter/hunted relationship. This is the only known example of possible interaction between predators and plant eaters in the Rio do Peixe basins.

The ratio of herbivore-to-carnivore trackways at this level and ichno-association is 3:1—different from other cases, such as the Toro-Toro site (Upper Cretaceous of Bolivia; Leonardi 1984b), where 8 sauropods were "followed" by 32 large theropods, with an herbivore:carnivore ratio of 4:1. The number of theropods seems to be too low at Serrote do Letreiro to have enabled them to handle a herd of about 16 large sauropods. Most probably, the theropods would have been going after only 1 of the sauropods.

Serrote do Pimenta (or Serrote Verde or Estreito Farm) (SOES)

This locality is situated at the foot of the Serrote do Pimenta (figs. 4.95–4.108), on the Estreito farm, along the unimproved and very bad road to Serrote Verde, between 240 and 250 m above sea level, at the northern border of the Sousa basin. It is situated in the Sousa municipality, 6.5 km as the crow flies NE of the town of Sousa. It is not easy to reach this locality because landowners do not always permit entry to their farms (sometimes using guns to reinforce this restriction), necessitating changes in itinerary. It is particularly difficult to get there in the rainy season due to flooding. Generally, one reaches the place by leaving Sousa by the outskirt of Prazeres and driving for 2 km along the old road to São Francisco (Paraíba) and Alexandria (Rio Grande do Norte), then entering a farm road to Estreito and taking that to Serrote Verde. The locality has the following coordinates: S 06 43.309, W 038 11.736. In this place, the road and the contiguous *caatinga* (desert scrub vegetation) to the right (E) are covered by a rocky pavement. The layers are alternating yellow, fine-grain sandstones, siltstones and mudstones that were deposited at the heterotopic boundary between the Antenor Navarro Formation (alluvial fans) and the Sousa Formation (flood plain) (fig. 4.100).

This locality was discovered by Leonardi in 1979. Excavations were carried out by him and his collaborators: Giancarlo Ligabue, Philippe Taquet, Diogenes de Almeida Campos, and Luis Carlos Godoy (Ligabue expedition in 1983; Bonaparte et al. 1993), as well as by Carvalho and

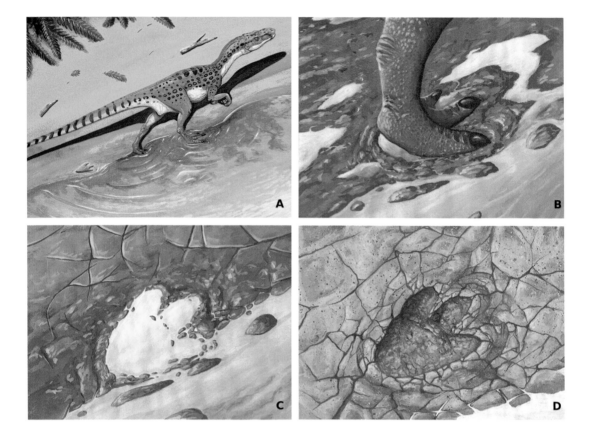

Figure 4.95. Mechanism of origin of a theropod track. The imprint is produced in soft mud covered by thin microbial mats. *A,* Taphonomic scenario for the footprint preservation. *B,* Theropod print on a wet surface colonized by biofilms. *C,* Biofilms continue to develop on the surface, biostabilizing and promoting imprinting of the footprint. *D,* During a draught phase, calcification of the mat occurs, followed by mud cracking. Art by Ariel Milani Martine.

Anna Alessandrello (*Natura Oggi* magazine [Milano, Italy] expedition in 1988). All the rocky pavements surrounding this site were surveyed with no results, despite the fact that there are extended, good outcrop surfaces of the same lithological material.

References: Leonardi 1984b, 1985a, 1989a, 1989b, 1994; Leonardi et al. 1987a, 1987b, 1987c; Lockley 1987; Carvalho 1989; Carvalho and Leonardi 1992.

An interesting set of dinosaur tracks is distributed over a wide area of about 200 × 150 m, at three main levels (fig. 2.10). The separation between the three levels is not always evident because of erosion, soil cover, and the sparse cactus and other thorny plants of the semiarid caatinga. Many of the tracks are concentrated on the surface of a stratum over an area of 26 × 14 m (including the theodolite landmarks E-6 and E-7) on the right side (SE) of the road. This main pavement is partially covered by regolith and soil and partially uncovered by rivulets; it was also to some extent excavated by digging a trench and then following the main trackway SOES 9 (theodolite landmark E-6).

SOES 1

A long trackway of 3 large theropod footprints (figs. 4.101A, 4.104A, 4.104C). The first footprint is represented just by the impression of digit

III. The second print, in contrast, is particularly good, deep and well preserved, with stout, well-padded toe impressions and claw marks that are triangular and broad. The pace angulation is very high (180°), and the stride is very long (475 cm); the SL/FL ratio is 13.6 and the SL/h ratio is 2.8. The estimated speed of the trackmaker is ~22 km/h. This trackway has the fastest estimated speeds of dinosaurs in the Sousa basin.

SOES 6

A short trackway of 3 footprints, with low pace angulation (147°30'; fig. 4.108E). The tridactyl footprints angle inward (negative rotation)—a rare situation in theropod footprints of this basin; they show long, strong claw marks. This trackway is impressed in the rocky pavement of the dirt road, about 50 m uphill from the sauropod herd trackways. It is attributed to a large theropod.

SOES 7

A hand-foot set of a quadrupedal, nonsauropod and nonornithopod dinosaur of medium-large size (fig. 2.10, theodolite landmark E8; figs. 4.101D, 4.104B, 4.104D). Tracks of quadrupeds are relatively rare in the Sousa basin, except for those of sauropods (with a total across the basin of about 59 individuals, including isolated footprints) and 5 trackways of quadrupedal or subquadrupedal iguanodontids. Apart from these, there are at present just this one set and the short trackway (undertrack) SOPP 15.

Unfortunately, SOES 7 is just one set, not a complete trackway. The pes print almost completely overlaps the handprint, in which only two curved and pointed fingers are present. The pes print is large, slightly wider than it is long (FL/FW = 0.9), and tetradactyl, with a wide and deep plantar pad and four short, pointed toes. The leftmost toe mark, possibly digit IV, is stout and wide; the second one from the left, perhaps digit III, is curved; the other two toe marks are spatulate. The total divergence between the innermost and outermost toe marks is low (about 35°). The original manus-pes set was left in situ; a plaster cast of it is kept in the collections of the Museo Câmara Cascudo of Natal (Rio Grande do Norte, Brazil; fig. 4.104B, 4.104D). This set is attributable to an ankylosaur and was the first ankylosaur track reported from South America (Leonardi 1984b), discovered some 14 years before those of Cal Ork'o, Sucre, Bolivia (Meyer et al. 1999; McCrea et al. 2001; Meyer et al. 2018).

SOES 8

This is a very badly preserved footprint similar to those of *Caririchnium magnificum*, located immediately north of SOES 9.

SOES 9

Caririchnium magnificum, Leonardi, 1984, holotype (figs. 4.96, 4.101B, 4.101G, 4.102, 4.103, 4.105A, 4.105C, 4.106G, 4.107). The trackway is of a quadrupedal dinosaur with high heteropody. The handprint is very

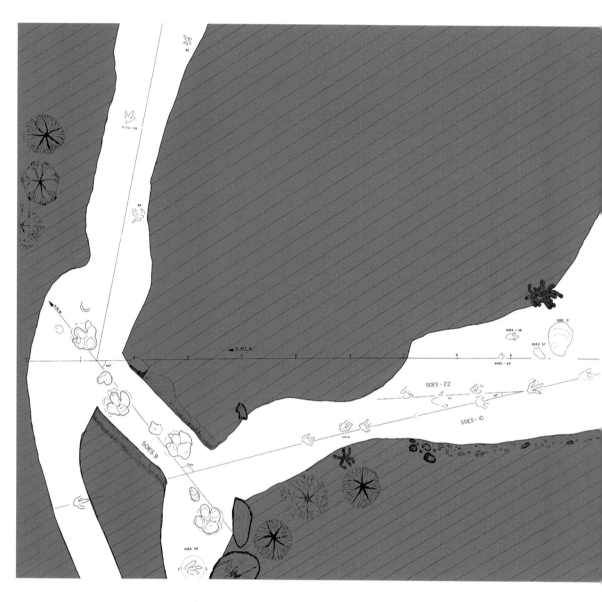

Figure 4.96. Serrote do Pimenta tracksite, Fazenda Estreito (SOES; figs. 1.11, 2.1*B*, 2.2*F*, 4.98–4.108), Antenor Navarro Formation, Sousa. This map covers an area of about 26 × 14 m (including the topographical points E-6 and E-7) on the right side (southeast) of the unimproved road that runs from Serrote Verde to Alexandria (Rio Grande do Norte). This main trackway surface is partly covered by regolith and soil but partly exposed by rivulets. The site was also partially excavated along the main trackways, especially trackway SOES 9 (theodolite landmark E-6). The site contains trackways and isolated footprints of about 30 large theropods, a fully quadrupedal ornithopod trackway, and a rare lacertoid footprint. Survey and drawing by G. Leonardi.

small, elliptical or rounded, with a highly variable outline, and probably represents contact of the manus with the substrate in a variety of manners; sometimes the impression of the tips of two or three fingers can be seen.

The foot is massive, symmetrical, tridactyl, and slightly longer than it is wide (average ratio FL/FW = 1.17). The impressions of digits II and IV

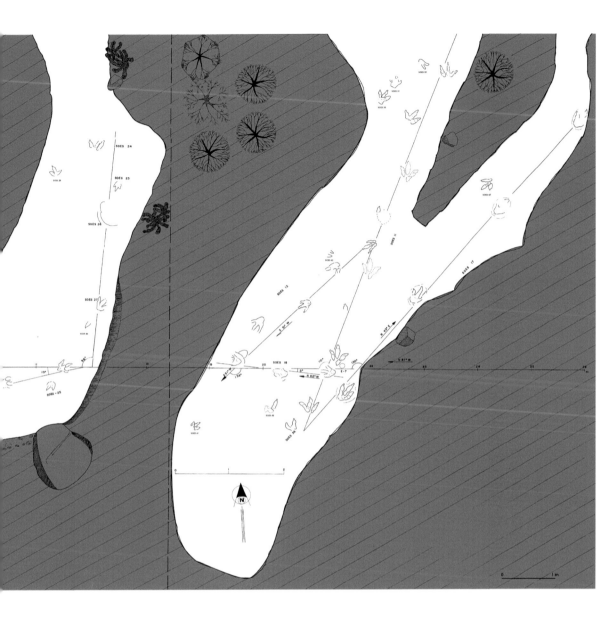

are subequal in length and show large and blunt triangular hooves; digit III is the longest of the three and shows an elliptical to pentagonal blunt hoof. Digital pads are not distinct, but there are some small, rounded concavities, perhaps corresponding to corns or callous plates. The total digital divergence is low (II^IV, average = 35°50°). The small nail impressions are probably wider than they are long. The "plantar" pad is rounded to a pentagonal shape. Wrinkles of the skin along the toes and between them and the sole are represented by distinct ridges in the footprints. The pes long axis is nearly parallel to the midline or slightly inwardly rotated.

The pace angulation is high for a quadruped (hand average pace angulation = 169°52'; hind-foot average pace angulation = 147°30') and the trackway is narrow; the inner trackway width takes negative values (average = -11.5 cm), indicating that the inner edges of the footprints intersect

Figure 4.97. Paleoenvironmental reconstruction of the depositional environment of the Sousa Formation, for a site in the Sousa basin such as Passagem das Pedras. Two large theropods follow and prepare to attack a small herd of ornithopods along the bottom of the drying lakebed, imprinting their deep trackways in the mud. Redrawn in India ink by G. Leonardi; modified from an original color picture created in 1983 by Renzo Zanetti and published by Bonaparte et al. (1993).

the trackway midline. The ratio SL/FL is = 4.8; the calculated speed is 3.6–5.6 km/h. The trackmaker was moving across firm ground, and so the footprints are shallow, with a low though ubiquitous displacement rim.

This trackway was the basis for the new ichnogenus and new ichnospecies *Caririchnium magnificum* Leonardi, 1984, first attributed by Leonardi (questionably but hopefully!) to a stegosaur (Leonardi 1984b; Bonaparte et al. 1993) but later more correctly assigned to a quadrupedal iguanodontid (Leonardi et al. 1987a, 1987b, 1987c; Leonardi 1994; fig. 4.103), as Philippe Taquet had already suggested to Leonardi while performing fieldwork during the 1983 Ligabue expedition (Bonaparte et al. 1993).

The trackway remains in situ and ought to be urgently protected. A plaster cast of the first hand-foot set is kept as a plastotype in the collections of the Câmara Cascudo Museum of Natal (Rio Grande do Norte, Brazil; fig. 4.105A). Additional specimens of this ichnotaxon occur at other localities in the Sousa basin, such as Floresta dos Borba and Baixio do Padre.

Recently (Menezes et al. 2019), 1 *Caririchnium* isp. hind-foot print was found in the Aptian sediments (alluvial plains, Itapecuru Formation) of the Parnaíba Basin, in an area located in the Itapecuru River valley, in the Guanaré oucrop, outskirts of Itapecuru-Mirim, State of Maranhão, Brazil. Sauropod footprints would be associated in that area.

This ichnogenus *Caririchnium* was recently considered to be a valid ichnotaxon by Díaz-Martínez et al. (2015, 12). It had great success; in fact

Figure 4.98. Maria de Fátima C. F. dos Santos observes the then (1980) recently excavated trackways 101–104 from a herd of 8 sauropods on the unimproved road going to the Serrote do Pimenta site (SOES). Compare with points E-1 and E-2 of the map (fig. 2.10).

it presently consists of five ichnospecies. Four of them were considered valid ichnotaxa by the same authors: *C. magnificum*, Leonardi, 1984; *C. leonardii*, Lockley, 1987, from the "mid" Cretaceous of Colorado; *C. lotus* Xing, Wang, Pan and Chen, 2007 (see also the splendid monograph by Xing et al. 2015b), from the "mid" Cretaceous of China; *C. kyoungsokimi* Lim, Lockley and Kong, 2012, from the "mid" Cretaceous of Korea. One of them was considered to be nomen dubium: *C. protohadrosaurichnos* Lee, 1997, from the Cenomanian of Denton County, Texas. The genus *Caririchnium* had 29 citations, according to Díaz-Martínez et al. (2015); it has at least 304 citations today (2021).

This ichnotaxon is classified by Díaz-Martínez et al. (2015) along with *Iguanodontipus* Sarjeant, Delair and Lockley, 1998 and *Hadrosauropodus* Lockley, Nadon and Currie, 2003, within the ichnofamily Iguanodontipodidae Vialov, 1988, sensu Lockley et al. 2014b. Nonetheless, the tracks of

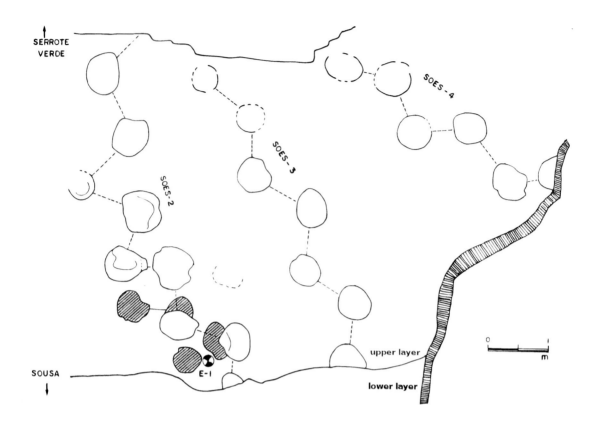

Figure 4.99. Locality Serrote do Pimenta (SOES). Four of the 8 sauropod trackways found on the dirt road on the northwestern side of the site. Four other trackways, parallel to these, were excavated later (1988), to the right (southeast) of the trackways illustrated here. Altogether, the 8 trails record the passing of a sauropod herd. E-1 is a topographical reference point. The present-day codes for the trackways herein represented are SOES 100 (formerly SOES 2); SOES 101 (formerly SOES 1); SOES 102 (formerly SOES 3); and SOES 103 (formerly SOES 4).

some of the ichnotaxa that are there considered nonvalid (among them *Sousaichnium* and *Staurichnium* from Passagem das Pedras, Sousa) according to the above authors present diagnostic features that allow them to be classified within Iguanodontipodidae. A complete emended diagnosis for the ichnofamily Iguanodontipodidae and for the ichnogenus *Caririchnium* (and its type ichnospecies *C. magnificum*) can be seen in Díaz-Martínez et al. (2015, 20–24 and 26–32, respectively).

Lockley (1987) instituted a second ichnospecies for this ichnogenus — *Caririchnium leonardii* — for some fine trackways in situ at the Alameda Avenue (Dinosaur Ridge) site, about 30 km SW of Denver, Colorado, from rocks of the Dakota Formation (Late Cretaceous). It is now classified as *Hadrosauropodus leonardii* Lockley, 1987 (Díaz-Martínez et al. 2015, 34–37).

The features of the trackway and the general outline of the footprints are really very similar to *C. magnificum*; even the wrinkles that divide the pads are the same. The nails or small hooves at the distal ends of the digital pads are the very same, but sometimes they penetrate down into the sediment. However, the general outline of the toes is more pentagonal or spatulate in *C. leonardii*. As for the handprints, those of *C. magnificum* are similar to those of a quadrupedal trackway of 8 hand-foot sets, attributable to iguanodontids, displayed in the Denver Museum of Natural History under the feet of a *Trachodon* skeleton. This last trackway comes

G. Leo dis.

from the Dakota group near Lamar, Colorado. Other interesting hind-foot prints, useful for comparison with those of *C. magnificum*, are some of the innumerable tracks at the Richardson Ranch, south of Pritchett, eastern Colorado, visited by Leonardi in 1986.

The *Caririchnium* morphotype has also recently been recognized in Asia: from the Feitianshan Formation, Lower Cretaceous of Sichuan Province, southwest China (Xing et al. 2014a); from the Lower Creta-ceous Zhaojue dinosaur tracksite, Sichuan Province, China (Xing et al. 2015a); from the Longjing tracksite, in the upper member of the Jiaguan Formation, Lower Cretaceous (Aptian–Santonian or Berriasian–Santo-nian) in Sichuan basin, southwest China (Xing et al. 2015e, 2016c); from the Lower Cretaceous Kitadani Formation, Fukui, Japan (Tsukiji et al. 2017); and from Upper Cretaceous (Campanian to Maastrichtian) lacus-trine deposits on islands in the vicinity of Yeosu, southernmost Korea, at about 30 track-bearing levels, representing the youngest dinosaur track re-cords in Asia (Paik et al. 2006, 461). However, in this last occurrence, the ornithopod tracks ought to be better classified as *Hadrosauropodus* (Díaz-Martínez et al. 2015). A dwarf form of *Caririchnium* isp. would come from the Tirgan Formation, Kopet-Dagh region, northeastern Iran. This case of dwarfism, along with that of the accompanying dinosaur ichnofauna, would be explainable by the island rule; an island in the northern Tethys (Abbassi et al. 2018). *Caririchnium* was also (unsuccessfully) compared to

Figure 4.100. Paleoenvi-ronmental reconstruction of Serrote do Pimenta in the Early Cretaceous. A large pterosaur watches from above as a herd of a dozen sauropods, possibly titanosaurids, descends from an Araucaria grove covering a plateau into the Sousa basin. The dinosaurs cross an alluvial fan to reach a small temporary lake at the base of the fan. This locality is on the northern margin of the basin, very near the crystalline, gneissic-mig-matitic Precambrian highland source of the fan sediments that will one day become part of the Antenor Navarro Forma-tion. Drawing by Leonardi.

Figure 4.101. Footprints from Serrote do Pimenta (SOES). *A*, SOES 1/2, a fine track of a running, large predator, *B*, A hand-foot set of *Caririchnium magnificum*, attributed to a large quadruped ornithopod. *C*, SOES 15, a rare (at Rio do Peixe basins) lacertoid footprint. *D*, A rare handset of an ankylosaur, probably a nodosaurid (SOES 7); it was, when discovered, the first ankylosaur track reported from South America. The adjacent symbol is not a petroglyph but a reference point (E-8) for the theodolite. *E*, SOES 17/3 (right pes), from the trackway of a large running theropod: the lozenge shape is common in eroded theropod footprints. *F*, A large footprint (SOES 21), attributable to either a sauropod or, more probably, to a large ornithopod. Nearby are the theropod trackway SOES 10 and other scattered theropod footprints (SOES 20 at left, SOES 22 at right, etc.) discovered during the excavations and survey in 1980 and 1983. *G*, Another view of *Caririchnium magnificum*, as seen during the excavations in 1983.

some of the ornithopod tracks from Obernkirchen (northern Germany), mentioned as ?*Caririchnium* isp. (Böhme et al. 2009; cf. Richter and Böhme 2016 and Hornung et al. 2016).

So *Caririchnium* seems to have been a long-lasting morphotype being recorded at least all along the Lower Cretaceous terrains of the two Americas, Europa, and eastern Asia. There is still no record of *Caririchnium* or of other large ornithopod tracks in the other continents of "Gondwanaland," for example, Africa and Australia.

This phenomenon is strange because this way, the same ichnogenus *Caririchnium* would be applied to very different trackmakers, especially from the point of view of the toe structure, including the Iguanodontidae (and related forms) and the Hadrosauridae with the hind-toe phalanges in general more dorsoventrally compressed and wider than they are long, compared to the iguanodontids, and with all three ungual phalanges spade shaped, differently from the iguanodontids. Iguanodontidae and related ornithopods are found in Lower Cretaceous terrains and Hadrosauridae in the Upper Cretaceous terrains. However, in ichnology, one deals only with the morphology of the tracks.

Note that the Iguanodontidae and related ornithopods, both from Laurasian and Gondwanan continents, are known from the Late Jurassic through the Early Cretaceous, and the Hadrosauridae are found perhaps from the Cenomanian and surely from the Santonian and during the last stages of the Late Cretaceous till the Late Maastrichtian (Weishampel et al. 2004, 432–438).

Díaz-Martínez et al. (2015), however, attribute to the *Caririchnium* ichnogenus only specimens from the Early Cretaceous (Berriasian-Albian), not because of a temporal choice but after the revision of the material. The late Early Cretaceous and Late Cretaceous large ornithopod tracks (Albian-Maastrichtian) are attributed by them to *Hadrosauropodus*. So, *Caririchnium*, and particularly the type ichnospecies *C. magnificum*, would span only all the Early Cretaceous (Díaz-Martínez et al. 2015, 40).

SOES 10

A long trackway (12 m) with 8 discontinuous footprints (figs. 4.96, 4.101F, 4.106B, 4.108F, 4.108G). The digital divergence is relatively high. A relatively long step length points to a rather quick gait (V = 7 km/h). The trackway is attributed to a large theropod.

SOES 11

A long trackway (9 m) of 9 footprints (figs. 4.96, 4.108A, 4.108B). The toe marks are stoutly constructed, and their distal ends bend away from the footprint midline. Digits II and III of both left and right footprints are nearly parallel, with very low digital divergence; digit IV shows a greater separation from digit III (mean digital divergence: II^III=19°30'; III^IV=52°45'). The total divergence between digits II and IV is high (average = 72°). The claw marks are long and triangular. The stride is long (average = 197 cm), indicating rapid movement of the trackmaker, which was probably a large theropod.

SOES 12

A trackway of 5 footprints (figs. 4.96, 4.108C, 4.108D). The individual prints are quite variable in shape, but they are generally long, with a high digital divergence. The trackway pattern is irregular, with a highly variable pace angulation (σ = 12). The trackmaker was a theropod.

SOES 15

An isolated small footprint of poor quality, probably because of its diminutive size and occurrence in coarse sedimentary rock (figs. 4.96, 4.101C). It

Figure 4.102. Photographic reconstruction of a portion of the holotype trackway *Caririchnium magnificum* (attributed to a quadrupedal ornithopod) in situ at Serrote do Pimenta (SOES).

Figure 4.103. *(facing)* Restoration of the probable maker (a Gondwanan ornithopod) of the fine trackway *Caririchnium magnificum* at Serrote do Pimenta (SOES), Baixio do Padre, Floresta dos Borba, and other sites in the Sousa basin. Painting by Ariel Milani Martine.

Figure 4.104. *(left)* Casts of footprints from Serrote do Pimenta (SOES) in collections of the Câmara Cascudo Museum of Natal, under different lighting conditions. *A, C,* Footprint SOES 1/2, made by a large, running theropod. *B, D,* SOES 7, the ankylosaur hand-foot set. Courtesy of the Câmara Cascudo Museum. Graphic scales in centimeters.

is interpreted as a left hind-foot print of a lizardlike reptile. As preserved, the print is 3.8 cm long and 2 cm wide. There is a plantar pad (1.5 cm long and 1.2 cm wide) and three or perhaps four toes. What are interpreted as the impressions of digits III and IV are long and curved inward. The mark of digit II is short and bent inward; the impression of digit I is not distinctly seen, and digit V is not impressed, probably because the lacertoid footprint was disrupted on its lateral side by the impression of another small, unidentifiable footprint. The small slab is kept in the museum of the Dinosaur Valley Natural Monument in Sousa.

Figure 4.105. Serrote do Pimenta (SOES). *A*, Plaster cast of a hand-foot set of *Caririchnium magnificum* (SOES 9/1, left pes and manus). Courtesy of the Câmara Cascudo Museum. *B*, Sauropod trackways SOES 100 and 101 on the rocky roadbed to Serrote Verde, at Serrote do Pimenta, during the Ligabue expedition in 1983. *C*, Another view of the same footprint set as in *A* in situ, greased prior to preparation of the plaster cast of *A*. *D*, Fiberglass replicas and tracks at Serrote do Pimenta in 1988.

SOES 16

An incomplete footprint of an unidentifiable bipedal dinosaur, near trackway SOES 6, on the road pavement (theodolite landmark E-9).

SOES 17

A long trackway of 4 footprints, attributable to theropods, similar to those of SOES 1 (figs. 4.96, 4.101*E*). The footprints are large, with long, stout toe marks bearing long claws. There is low digital divergence (42°). The trackway has very long strides (average = 445 cm; SL/FL ratio 11.6, SL/h ratio 2.4) and a very high pace angulation (average = 175°), pointing to

Figure 4.106. Serrote do Pimenta (SOES). *A,* Theropod footprint SOES 29. *B,* Large theropod footprint SOES 10, part of a fine trackway. *C,* The landscape at Serrote do Pimenta (Fazenda Estreito) near the northern margin of the Sousa basin, with two fiberglass models from our lab. In the foreground, part of the main track-bearing surface, with *xique-xique* cactus and thorny scrubs of *jurema preta (Pithecolobium* sp.); in the background, the green floodplain of the Peixe River can be seen. The hills in the far distance are along the southern margin of the basin. *D,* Partial view of the parallel trackways of a herd of at least 8 sauropods (SOES 100–107). Here the track SOES 100 tramples on the smaller SOES 101. *E,* The same as in *D,* from another point of view. *F,* A portion of the main track-bearing surface, with many tracks, among them large isolated ornithopod footprints, and many theropod tracks. *G,* Holotype of *Caririchnium magnificum,* in situ. Note the strong heteropody.

a running gait (V = 16.3 km/h). The trackmaker was very probably a theropod.

SOES 18

A short trackway of 3 tridactyl, badly preserved footprints, probably made by large theropods (fig. 4.96).

SOES 21

A poorly preserved footprint, probably similar to prints of *Caririchnium magnificum* (figs. 4.96, 4.101F).

Figure 4.107. Trackway *Caririchnium magnificum* during the second phase of excavation in 1988. This is a long trackway (25 excavated meters) of a quadruped ornithopod, with 7 hand-foot sets and, after a long gap, several additional sets of poor quality. Here, as in many other exposures, the surface of the coarse sandstone of the Antenor Navarro Formation, deposited very near the source of its sediments, is covered by a "varnish" of fine siltstone, which very probably was impregnated and covered by a superficial film of microbial mats before being buried by the next, and younger, layer of sediment. Courtesy of Franco Capone.

SOES 22

A short trackway of 3 badly preserved tridactyl footprints (fig. 4.96). The individual prints are noticeably variable in size and in the divergence of the footprint long axis from the trackway midline. The pace angulation is particularly low (144°). The trackway is probably attributable to a large theropod.

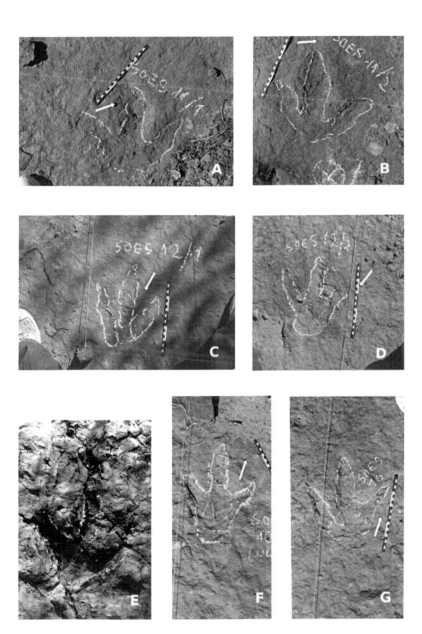

Figure 4.108. Theropod footprints from Serrote do Pimenta (SOES), mainly from the main track-bearing surface. *A*, SOES 11/1. *B*, SOES 11/2. *C*, SOES 12/1. *D*, SOES 12/2. *E*, SOES 6, a short trackway of 3 footprints impressed in the rocky pavement of the dirt road, about 50 m uphill from the sauropod herd trackways. *F*, Unnumbered footprint from trackway SOES 10. *G*, Final footprint from trackway SOES 10. Graphic scales in centimeters.

SOES 26

An incomplete footprint, probably attributable to *Caririchnium magnificum*.

SOES 19, 20, 23, 24, 25, 27, 28, 29, 30, 31, 32, 33, 34, 35, 37, 38, 39

These numbers designate a series of isolated tridactyl footprints, some of which (SOES 20, 23, 24, 25, 28, 31, 37, 38, 39) are badly preserved (fig. 4.96). They are probably attributable to large theropods.

Some of the trackways in this series were formerly labeled SOES 1–4, but others are new trackways (figs. 4.98, 4.99, 4.100, 4.105B, 4.106D, 4.106E). This group of 8 trackways records the crossing of a herd of sauropods. Four of these subparallel trackways are impressed into the rocky pavement of the dirt road of Serrote do Pimenta (figs. 4.98, 4.99) and were previously published (Leonardi 1994; Leonardi et al. 1987a, 1987b, 1987c). Excavation on the right (SE) side of the road during the *Natura Oggi* magazine expedition (1988) with Leonardi, Carvalho, and Anna Alessandrello found 4 new trackways, all of them subparallel to the former. Excavation on the left (NW) side of the road would probably uncover more trackways of the same group.

The 4 trackways on the road respectively consist of 9 (SOES 100), 4 (SOES 101), 7 (SOES 102), and 9 (SOES 103) oval or rounded footprints (figs. 4.98, 4.99). Trackway 100 overlaps trackway 101, whose trackmaker therefore crossed the sand flat before the maker of trackway 100. The trackways generally have a relatively high pace angulation (~120°), and in most of them the individual footprints are more or less regularly spaced. However, SOES 100, which has the largest footprints, shows a very irregular trackway. The standard deviation of its stride is $\sigma = 24.9$, while those of the other parallel trackways have $\sigma = 6.5$ and 12, respectively. The pace angulation of SOES 100 is lower (average = 93°08') than the angulations of the other 2 trackways (118°48' and 121°15'), and the standard deviation of its pace angulation is higher (17.1) than the angulations of the other parallel trackways ($\sigma = 8.2$ and 3.6).

The 4 trackways excavated in 1988 to the right (SE) of the former 4 exposed on the road consist of 4 (SOES 104), 6 (SOES 105), 2 (SOES 106), and 5 (SOES 107) excavated footprints; these trackways and footprints are very similar to those of the road and parallel to them. The footprints lack morphological details and are quite probably subtracks. Because the trackways do not show manus-pes sets, they give the appearance of having been made by bipeds. However, their shape and large size, as well as the evidence of gregarious behavior the trackways provide, suggest that these are in fact sauropod footprints. If so, they would seem to be pes prints, with manus prints having been completely overprinted by the hind-feet.

To sum up, this is one of the best and most abundant ichnofaunas of the basins. The site has three main levels. The main pavement contains the tracks of about 30 large theropods, a fully quadrupedal iguanodontid trackway, 2 isolated footprints of the same kind, and a rare lacertoid footprint. On the rocky pavement of the dirt road on two contiguous layers are the trackways of 8 sauropods in a herd, 1 theropod trackway, and 1 hand-foot set attributable to an ankylosaur. There is also 1 isolated fine trackway of a running large theropod. Altogether, the site shows the tracks of 41 individuals: 28 large theropods, 3 ornithopods, 8 sauropods, 1 ankylosaur, and 1 lacertoid hind-track.

The site is located at the northern margin of the Sousa basin, 14.7 km WNW of Sousa as the crow flies (S 06 41.055, W 38 20.733; figs. 4.109, 4.110A). The conglomerates and yellow sandstones of this outcrop around the small hamlet of Floresta dos Borba show a largely extended rusty rocky surface (strike N40°W, dip 6°–10° to the south) and belong to the Antenor Navarro Formation. Leonardi performed a short on-the-spot investigation here, on the very last day of the expedition in December 2003. He conducted another visit on August 13, 2005. However, there is scope for considerably more survey and fieldwork at this site.

The occurrence of tracks in the hamlet of Floresta dos Borba and environs was reported to Leonardi by Robson Araújo Marques and Luiz Carlos da Silva Gomes.

Reference: Leonardi and Santos 2006.

In the exposure, the following dinosaur tracks were discovered from W to E:

Three Large Theropod Isolated Tracks

SOFB 1

A large, very deep, tridactyl, wide right footprint, 45 cm long, 52 cm wide, with a high digit divarication (II^IV), up to 84° (figs. 4.110D, 4.111A). The trackmaker was heading ~S75°W.

SOFB 2

An incomplete theropod footprint in which only the free part of the digits is impressed.

Floresta dos Borba (SOFB)

Figure 4.109. Tracksite Floresta dos Borba (SOFB; see also figs. 4.110–4.112), Antenor Navarro Formation, Sousa, located almost at the foot of the Serrote do Jerimum, at the northern margin of the Sousa basin. A preliminary assessment of the ichnofauna indicates that it consists of different forms of large theropods, sauropods, and two forms of large ornithopods. From Leonardi and Santos (2006).

Large Ornithopoda

Large Ornithopoda

Large Sauropoda

Large Theropoda

Figure 4.110. Floresta dos Borba. *A*, The savannah plain surrounding the tracksite. *B*, A fine theropod trackway (SOFB 6) with a long stride, high step angle, and long, thin (perhaps collapsed) digits. *C*, Single footprint from trackway SOFB 6. *D*, SOFB 1, an isolated large theropod footprint. *E*, Isolated ornithopod footprint SOFB 4, very similar and attributable to *Staurichnium diogenis*. *F*, A large, isolated ornithopod footprint SOFB 5, similar to *Caririchnium magnificum*. Graphic scales in centimeters.

SOFB 3

A large, very shallow tridactyl left footprint, recognizable only because the material inside the footprint differs in color from the surrounding rock (fig. 4.111B). The print is 39 cm long and 46 cm wide, with high digit divarication (II^IV), up to 76°, and a very wide displacement rim diameter (up to 62 cm). The animal was heading ~S85°W.

SOFB 1 and SOFB 3 are among the largest theropod footprints of the Rio do Peixe basins. A number of other theropod tracks were discovered on the same surface in 2005, among them a fine trackway with collapsed footprints (SOFB 6; fig. 4.112B, 4.112C); the general outlines of the footprints and the very long claw marks are very similar to SOCA 1351 of Piau-Caiçara.

Sauropod Tracks

A large number of sauropod tracks, many of them big, deep, and with high displacement rims. Among these are some hand-foot sets (fig. 4.112A–F). The hind-foot prints reach lengths of 90 cm and widths of 60 cm. The small, horseshoe-shaped handprints are barely discernible, being almost completely or fully squashed by the anterior part of the displacement rim

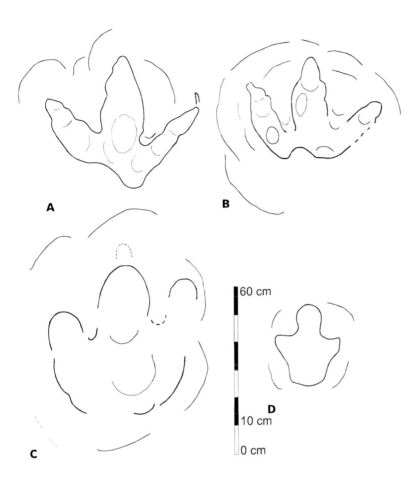

Figure 4.111. Theropod and ornithopod footprints from Floresta dos Borba. *A,* SOFB 1, an isolated large theropod footprint. *B,* SOFB 3, another isolated large theropod footprint. *C,* A very large isolated ornithopod footprint (SOFB 5), attributable to *Caririchnium magnificum. D,* Isolated ornithopod footprint SOFB 4 *Staurichnium diogenis.* From Leonardi and Santos (2006).

60 cm

10 cm

0 cm

A

B

C

D

of the foot. Some hind-foot tracks show claw impressions. However, a few sauropod footprints at this site are relatively shallow. At least one shows positive topographic relief with its surroundings, due to the suction effect of the foot as it was removed from the sediment, similar to what was seen in some footprints from the sauropod herd at Serrote do Letreiro in the rocky bed of the Riacho do Pique (Carvalho 2000b; see above). These sauropods were probably gregarious and traveling in a herd, but no clear direction of travel is discernible in many of them. The main group of sauropod footprints is concentrated inside the hamlet.

Ornithopod Tracks

At this site, there are some rare and fairly well-preserved ornithopod footprints. Unfortunately, they are isolated. Among them are:

SOFB 4

A well-imprinted tridactyl footprint with a wide digit III hoofmark and small hoofmarks at the tips of digits II and IV. This print is 29 cm long and 24 cm wide. It is most likely attributable to *Staurichnium diogenis* (probably a nomen dubium). The dinosaur was heading more or less to the west (figs. 4.110E, 4.111D).

Figure 4.112. Heavily trampled surface of footprints of sauropods with rounded feet at Floresta dos Borba. *A*, The very small hamlet of Floresta dos Borba can be seen in the background. In the foreground is part of the pavement with the sauropod tracks. *B–C*, The tracks of the sauropod herd. *D–F*, Three different specimens of sauropod hindfoot prints showing anatomic details, particularly marks of the claws.

A large, very shallow, tridactyl footprint, 56 cm long and 53 cm wide, with a wide but very shallow displacement rim (figs. 4.110*F*, 4.111*C*). The print is attributable to *Caririchnium magnificum*, and its maker was heading in a northerly direction.

Altogether, this is a varied ichnoassociation that would be worth a complete survey and a detailed study, including that of the relationships among the rocky levels and the dinosaur assemblages. Preliminary observations point to a number of fine theropod trackways and footprints (at least 4, of at least three different forms), 3 of them very large; a great number of large and roundish sauropod footprints and hand-foot sets, many of them in a herd; and, finally, 2 fine ornithopod tracks of different kinds.

Summary of the Antenor Navarro Formation Ichnofauna

Altogether, the 10 sites from the Antenor Navarro Formation sample at least 18 stratigraphic levels that preserve the following ichnofauna (figs. 4.88–4.112): 89 footprints and trackways identified as made by saurischians: 53 large theropods and 36 sauropods. As for the 10 sets of tracks assigned to ornithischians, there are 9 graviportal ornithopods (3 of them quadrupedal) and 1 quadrupedal ornithischian, very probably an ankylosaur. There are about 10 dinosaur tracks that are unidentifiable or of uncertain affinities. Altogether, the recorded dinosaurian individuals number more than 109. In addition, there are 2 nondinosaurian tracks: a batrachopodid footprint and a lacertoid footprint.

Rio Piranhas Formation

Altogether, the six sites from the Rio Piranhas Formation include a minimum of about 12 levels; however, it is very difficult to distinguish the levels or layers in this formation (figs. 4.113–4.124).

The locality of Cabra Assada is a small hamlet situated at S 06 49.892, W 038 23.999, about 12 km SSE of São João do Rio do Peixe, in the territory of this municipality, along the road between São João do Rio do Peixe and Marizópolis (fig. 4.113).

The footprints were found in stone fences in the village. The flagstones of these fences came from an outcrop near the village, in a tributary creek of the Peixe River, where a small quarry had been dug. This site is accessible only in seasons of protracted drought. The slabs were discovered by Leonardi in 1984; he collected them, along with Maria de Fátima C. F. dos Santos and Claude Luis Aguilar Santos, in 1985. They are kept in the ichnological collections of the Câmara Cascudo Museum of Natal (Rio Grande do Norte, Brazil).

The slabs are siltstones and sandstones containing beds of fine pebble conglomerates; the fine-grain sandstones are a yellowish color, and the coarse sandstones are gray. The grains are quartz and potassium feldspar, and the cement is more silica than carbonate. Ripple marks, primary and

Cabra Assada (ANCA)

Figure 4.113. Cabra Assada farm, Cabra Assada ichnosite (ANCA; Rio Piranhas Formation), along the road between São João do Rio do Peixe and Marizópolis. Theropod and ornithopod footprints, as well as fine ripple marks, occur in the stone slabs erected as a fence.

secondary mud cracks, and trails and burrows of invertebrates are also present (Santos and Santos 1987b).

Because of the coarse-grain sediment and the position of this unit at the southwestern margin of the basin (just 1 km N of the boundary), this outcrop can be attributed to the heteropic interfingering of the Sousa Formation (argillites and siltstones) and the Rio Piranhas Formation (sandstones and conglomerates). The environment was that of a transition between an alluvial fan and the flood plain.

References: Leonardi 1985b, 1989a, 1994; Leonardi et al., 1987a, 1987b, 1987c; Santos and Santos 1987b.

ANCA 1

This trackway features a reverse print (negative copy, convex hyporelief) of an incomplete footprint of poor quality, probably a left, from a large, stout-toed theropod (see Lockley and Xing 2015). The digit III impression is rather short. The impressions of II and III digits are separated, but those of III and IV join at the bases of the toe marks. There is a relatively wide total digital divergence (II^IV). Footprints of this shape were also found at Serrote do Pimenta, in the Antenor Navarro Formation, and at Caboge farm (SOCV 7) in the Rio Piranhas Formation, in this same basin.

ANCA 2

A reverse print of a left footprint of a relatively small theropod. The distal end of digit II curves medially. There are three digital pads in the impression of digit IV, two in III, and one in II.

ANCA 3

A reverse print of a probable left theropod footprint. The print is tridactyl or perhaps tetradactyl, with the possible presence of the first toe in abduction as a spur. The print can perhaps be classified in the morphofamily Gigandipodidae Lull, 1904, because the footprint pattern looks very similar to *Gigandipus* E. Hitchcock, 1855 and to *Hyphepus* E. Hitchcock, 1858, from the Lower Jurassic of the Newark Group in Connecticut and Massachusetts. In the Sousa basin, a single rare trackway of this kind was found at Piau farm (SOCA 1350).

ANCA 4

A reverse print of a shallow left footprint of good quality. Digits II and III have well-developed claw marks, with that of digit II directed medially. This footprint is very similar to tracks SOPI 1 and 2 from the Piedade locality of the Sousa Formation. The rock in which the print is preserved is a fine-grain sandstone, albeit with some coarse clasts. Associated with this footprint are two isolated digit impressions and some invertebrate trails. The footprint can be attributed to a large theropod, morphofamily Eubrontidae Lull, 1904; ichnogenus *Eubrontes* E. Hitchcock, 1845.

ANCA 5

An incomplete footprint of poor quality (mold—that is, concave epirelief), attributable to a large ornithopod. There are hooflike toe marks with longitudinal rises; the third toe has a groove between two rises. These features can be interpreted as morphological details of the animal's hooves. This footprint has a general outline comparable to that of *Sinoichnites youngi* Kuhn, 1958, discovered and described by Teilhard de Chardin and Young (1929) from the Upper Jurassic of north Shensi, China, also attributable to an iguanodontid ornithopod; however, this ichnoform is correctly considered to be a nomen dubium by Díaz-Martínez et al. (2015).

ANCA 6

A tridactyl footprint (concave epirelief) of a large bipedal dinosaur. The distal extremities of two of the toe marks (digit III and one of the peripheral toes) are not distinct, but there is a well-developed "plantar" pad. Digital pads are present on digit III. The footprint outline is reminiscent of some *Iguanodontipus* footprints, but there is a small nail or claw mark. Distal ends like this do occur in *Iguanodontipus* but are also seen in some of the stout-toed footprints from the Piau farm that are attributed to large theropods. Consequently, the affinities of this footprint are uncertain.

ANCA 7

An incomplete footprint, preserved as mold (concave epirelief) and cast (convex hyporelief) in two different flagstones. The digit impressions are very stout, with hooflike terminal ends. Digital pads are not present; each toe mark consists of a wide callous pad, with some longitudinal structures or furrows parallel to the longitudinal axis of the digit. These furrows do not seem to be scratches. Wide displacement rims surround the print. A similar footprint was found at Caboge on the Curral Velho farm, in the same basin, and also from the Rio Piranhas Formation. The print can be attributed to an ornithopod.

ANCA 8

A reverse print of a large, stout-toed footprint with a wide digital divergence and a long, curved claw mark on one toe. The maker was a large theropod.

ANCA 9

A reverse print of a small, incomplete footprint of poor quality. The peripheral toes are rather divergent distally and curve outward. The print may have been made by a small theropod.

ANCA 10

A reverse print of small size, possibly a right, with a well-impressed digit III mark. The print was probably made by a theropod and may be assigned to *Eubrontes* E. Hitchcock, 1845.

This is a slab with 3 footprints, preserved as negative natural copies (casts). Two of them (ANCA 11 and 12) are short, plump, rounded footprints with general outlines very similar to those of prints attributed to *Iguanodontipus*; however, they show some true claw marks. So, they could be theropod subtracks or, alternatively, ornithopods provided with long ungual phalanges. This is a typical case of the difficulty of distinguishing footprints of large theropods from those of ornithopods, which happens in a surprising number of cases. Footprint ANCA 13 is incomplete, but it seems to belong to Eubrontidae Lull, 1904.

To sum up, the 13 individual bipedal dinosaurs represented by this footprint assemblage surely lived in this same place over a brief period, impressing their tracks into different but closely spaced sediment layers. Of their trackmakers, 7 were probably large theropods, 2 were ornithopods, and 4 were of uncertain affinities. The theropod footprints are sometimes rather typical in appearance and easy to identify, but in other cases they are subtracks and/or badly impressed and preserved, such that their shape is deformed, giving their toes a misleadingly broad appearance. The attribution of such prints to large theropods is based principally on the presence of claw marks because we have no trackway parameters on which to rely for assistance in identification. If assignment of most of the footprints to large theropods is correct, they represent 54% of the sample; ornithopod prints then constitute 15% and completely unidentifiable footprints 31%. Among the large theropods, we have at least four different forms; there are two forms among the ornithopods. No footprints attributable to sauropods are present in this sample.

Nothing can be said about the trackmakers' associations or directions of travel because the tracks were collected in isolated flagstones.

Curral Velho (SOCV)

The ichnolocality Curral Velho is located on the homonymous farm, about 8 km as the crow flies SSE from the town of Sousa, in this municipality, in the territory of the former large Caboge farm, near hill 233 (S 06 49.013, W 038 12.357; figs. 4.114–4.116). At this locality, the Rio Piranhas Formation rocks typically form low cuestas of white to grayish cross-laminated beds of fine conglomerates, with rare layers of siltstone and mudstone. On the surface of finer-grain beds are primary and secondary mud cracks and, rarely, ripple marks. The white color indicates a stratigraphic position in the upper portion of the Rio Piranhas Formation. The site was discovered by Leonardi in 1979.

References: Leonardi 1981a, 1981e, 1987b, 1994; Leonardi et al. 1987a, 1987b, 1987c.

Distributed across a relatively small surface are 12 isolated footprints of bipedal, tridactyl dinosaurs of medium to large size. All of the footprints, however, are of poor quality. They are filled in by sediment of

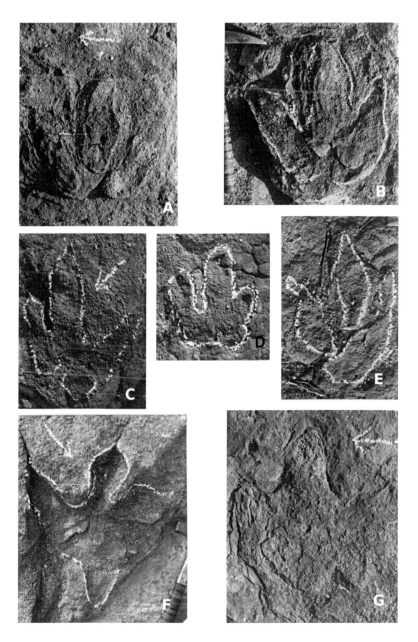

Figure 4.114. Tridactyl dinosaur footprints (mainly theropods) from Curral Velho (SOCV; see also figs. 4.115B–F, 4.116), Rio Piranhas Formation, Sousa. A, SOCV 1, an infilled print with apparent reverse relief. B, SOCV 2, another infilled print in apparent reverse relief. C, Infilled footprint SOCV 7. D, Infilled footprint SOCV 5. E, Infilled print SOCV 6. F, Eroded footprint SOCV 4. G, Infilled footprint SOCV 11, possibly made by an ornithopod. The blunt toes of these footprints would seem to make them attributable to small ornithopods, but the general structure of the footprints, with a very low II^IV digital divarication, makes attribution to theropods more likely in most cases. See also figure 4.116.

the overlying layer, so all that can be observed of the footprints is their outline. In some cases, the sediment that filled in the prints is coarser and more resistant than the tracklayer itself, and so there is an apparent inversion of relief, with the material inside the footprint being topographically higher than the surrounding rock.

Large Theropods

Group SOCV 1, 2, 6, 7. Four isolated footprints of similar form, surely attributable to theropods of medium to large size.

Figure 4.115. Ornithopod footprints from Curral Velho (SOCV; see also figs. 4.114, 4.116) and Mãe-d'Água (SOMD; see also fig. 4.116), Rio Piranhas Formation, Sousa. *A*, SOMD 3, a print with a very long digit III impression. *B*, SOCV 8 (left pes?). *C*, Infilled footprint SOCV 9, in pseudo-reverse relief. *D*, Infilled print SOCV 10, in apparent reverse relief. *E*, A group of ornithopod footprints at Curral Velho, among them SOCV 8, 9, and 10. *F*, The very long-heeled SOMD 4, attributed to an ornithopod because of the blunt and rounded hoof impressions; it also has a very long III digit. Graphic scales in centimeters.

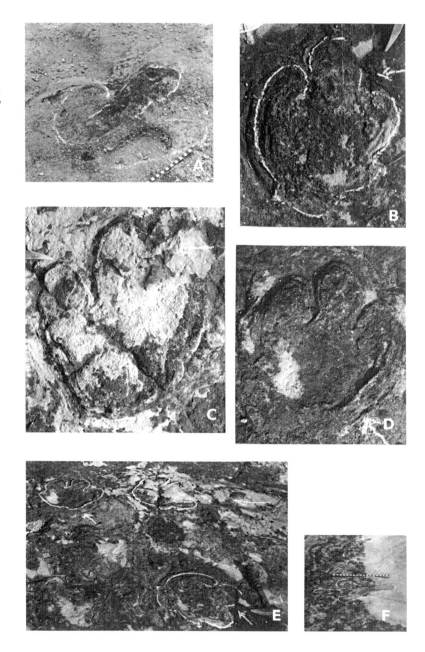

SOCV 1 (FIGS. 4.114A, 4.116)

Tridactyl, probably a left, with short but stout toes and pointed but indistinctly impressed claw marks. The digital impressions are quite divergent (interdigital angle II^IV = 60°) and separated from each other at their bases. If this separation is not an artifact of preservation of the footprint as an undertrack, the dinosaur was walking with the proximal end of the digital part of the foot clear of the ground. The footprint is otherwise a typical theropod print of medium size.

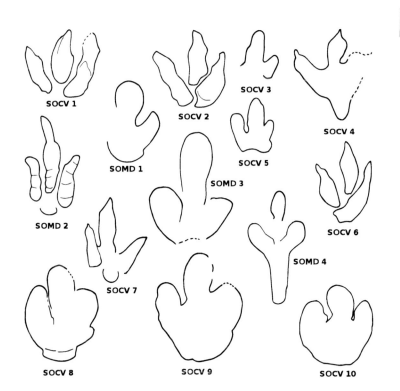

Figure 4.116. Isolated footprints from the two localities Curral Velho (SOCV) and Mãe d'Água (SOMD), found in 1979 in the southern part of the Sousa basin, municipality of Sousa. Prints SOCV 1, 2, 6, and 7 are attributed to medium to large theropods. SOCV 3 and 5 are probably attributable to small bipedal ornithischians. SOCV 4 is interpreted with uncertainty as a theropod track. SOCV 8, 9, and 10 are attributed to ornithopods. SOMD 1 is an incomplete ornithopod track. SOMD 2 is a large theropod track with very long claw marks, especially of digit III. As for SOMD 3 and 4, their roundish digital pads point to ornithopods as trackmakers; the long-heeled outline of SOMD 4 is, however, quite uncommon. Graphic scale = 10 cm. Drawing by Leonardi.

SOCV 2 (FIGS. 4.114B, 4.116)

Like SOCV 1, another tridactyl footprint. The proximal end of the impression of digit III is distal to the proximal confluence of the two peripheral digits, suggesting the combined bundle of metatarsals II–IV was somewhat arched, with its dorsal side convexly arched. Consequently, the impression of digit III toe is shorter than the other two. The peripheral toe impressions are about the same length. The footprint shows apparent reverse relief.

SOCV 6 (FIGS. 4.114E, 4.116)

A right footprint morphologically similar to SOCV 1 and SOCV 2, except with longer and thinner toe marks and a slightly less distal divergence of the peripheral toe marks (total interdigital angle II$^\wedge$IV = 57°). The fourth digit is noticeably longer than digit II.

SOCV 7 (FIGS. 4.114C, 4.116)

A right footprint showing both similarities with and differences from the preceding three prints. The general shape and size of the toe marks are similar, but the digital divergence is less (II$^\wedge$III = 15°; II$^\wedge$IV = 54°). Digits II and III are not separated but join at a hypex distal to the junction of digits III and IV. Digit IV shows a rounded proximal pad.

SOCV 11 (FIG. 4.114G)

An isolated, poorly preserved, tridactyl footprint, probably a right, with short and blunt toe impressions. Digit III is noticeably longer than the other two. The right toe mark (most probably digit IV) is separated from digit III not only in the free portion of the print but also more proximally. Digit II shows a characteristic bend along its proximal-medial outline. Digital divergence is low, both between adjacent toe marks and overall. The blunt toe marks notwithstanding, this print is probably attributable to a large theropod, but we cannot exclude the possibility that it was made by an ornithopod.

SOCV 15.

A footprint similar to SOCV 11. There is a long claw on the rightmost toe mark (probably digit IV, in which case this is a right footprint).

Uncertain Theropod

SOCV 4 (FIGS. 4.114F, 4.116)

A large, poorly preserved, not infilled, tridactyl right footprint, probably an undertrack. Digit impressions are short and broad; those of III and IV are spatulate. Digit II is quite a bit shorter than IV, whose proximal extremity forms a long heel. Digit III is very short. All of the toe marks appear to be terminally pointed, but it is uncertain if these represent true claws or extramorphological erosion of small fractures in the rock. The print was probably made by a large theropod or, less likely, an ornithopod.

Ornithopods

Group SOCV 8, SOC 9, and SOCV 10 (respectively, figs. 4.115B, 4.115E, 4.116; figs. 4.115C, 4.115E, 4.116; figs. 4.115D, 4.115E, 4.116) include three medium-size to large isolated prints that are quite similar in morphology, all of them probably tridactyl, with a rounded general outline. Their toe marks are short, broad, rounded, and hooflike, with a spatulate outline. The digital divergence is not easily measured but appears to be low. The hypexes are often wide and/or rounded. There is a large circular or semi-circular "plantar" pad. These footprints come from the same level and from the same restricted area (a few m²) and probably belong to three individuals of the same Linnaean species.

These footprints are quite similar to *Sousaichnium monettae* Calvo, from the Limay Formation, Neuquén Group, Upper Cretaceous (Cenomanian), from Picún Leufú (Neuquén, Argentina), now considered to be a nomen nudum by Díaz-Martínez et al. (2015). They are also something like *Sousaichnium pricei* except that the spatulate toes of these tracks from the Rio Piranhas Formation discriminate them from *S. pricei*. Yet another similar morphotype is known from the Upper Jurassic or Lower Cretaceous of the State of Michoacán, Mexico (Ferrusquia-Villafranca et al. 1980). SOCV 8–10 also show general similarity to tracks from the US, such as "*Dinosauropodes bransfordii*" Charles Strevell, 1932 and "*Dinosauropodes magrawii*" Strevell, 1932 of the Upper Cretaceous of Utah, although the toe marks of the North American forms are more pointed, as is also the case with some footprints from the Aptian of Logroño, Spain (Cladellas and Llopis 1971).

SOCV 8

The toe marks are particularly short and rounded, without subdivisions (figs. 4.115B, 4.115E, 4.116). A small salient along the right margin could be interpreted as a first digit and then compared to that of *Iguanodon* sp. in Haubold (1971, fig. 54/5). However, a long first digit is improbable in a so derived form, and the skeletal taxon *Iguanodon* had lost digit I. The proximal margin of the print shows a very peculiar, annular "heel" mark, which. differentiates this footprint from the other two in this group. There is also a topographic inversion of relief due to infilling and differential erosion.

SOCV 9

A large footprint with apparent inversion of the relief (figs. 4.115C, 4.115E, 4.116).

GROUP SOCV 3 AND SOCV 5

These are small and incomplete but probably tridactyl footprints (respectively, figs. 4.115A, 4.116; 4.115D, 4.116). The toes are not very divergent and are blunt and clawless. The digit III impression is rather longer than the peripheral toe marks. The poor quality of the prints makes identification difficult, but their form is similar to what one would expect for small-size ornithopods, contrary to the opinion of Leonardi (1987b). This form is similar to *Wintonopus latomorum* Thulborn and Wade 1984, from the "Mid" Cretaceous of Queensland, Australia, attributed to small ornithopods and personally seen by Leonardi in 1997 (see also Romilio and Salisbury 2012; it is presently considered to be swim tracks: Milner and Lockley 2016).

SOCV 5 is a left pes print; SOCV 3 is incomplete and probably also a left print. The azimuth of this last footprint is parallel to the crests of ripple marks in an adjacent layer.

Despite the small sample size (12 individuals), the ichnological material here is rather diverse, suggesting the presence of both large theropods and ornithopods. There were 7 large theropods with at least three different forms; 3 large ornithopods, probably of the same form; and 2 probable small ornithopods. All of the trackmakers were bipedal dinosaurs—common in the Sousa basin—of a wide range of sizes.

**Lagoa do Forno I
(SOLF I)**

On this farm, Leonardi found two new localities in December 2003. The first (SOLF I) is situated at S 06 48.066, W 38 10.039, 7.1 km SE from the center of Sousa as the crow flies and 1,500 m along the road from the octagonal chapel of the hamlet Lagoa do Forno, in the municipality of Sousa. This locale is very probably at the boundary between the Sousa and Rio Piranhas Formations (figs. 4.117–4.119).

References: Leonardi and Santos 2006.

In the roadbed, almost completely erased by the bulldozers that cut the road, are at least 15 footprints (figs. 4.117A, 4.118). These footprints are almost always seen in a horizontal section because they were cut across by the machines. In addition to the dinosaur tracks, there are also grooves incised into the soft sandstone by ancient ox-pulled chariots traversing along an old trail (fig. 4.117A) as well as a long, deep cattle path carved into the sandstone by hooves.

SOLF I 1

A sauropod footprint with a displacement rim (fig. 4.117D).

SOLF I 2

A sauropod track with a displacement rim. Probably a handprint (fig. 4.119).

SOLF I 3

A sauropod handprint with a relatively high displacement rim (fig. 4.119).

SOLF I 4

A large sauropod footprint with a displacement rim (fig. 4.119).

SOLF I 5–7 and SOLF I 9

Some small to medium-size isolated theropod footprints of poor quality (fig. 4.119).

SOLF I 8

An incomplete theropod footprint that appears to have been left by a swimming animal.

This kind of exclusive sauropod and large theropod assemblage is not so common in the localities of the Rio do Peixe basins, where there are frequently different alternatives; we found this ichnofauna in Lagoa do Forno as well as at Engenho Novo, Várzea dos Ramos I, Baixio do Padre, Fazenda Paraíso, and Riacho do Cazé. However, this kind of saurischian-dominated ichnofaunas is common in many ichnosites the whole world over. Two newly found and described ichnosites or clusters of sites are the Agua del Choique section from the Upper Cretaceous of Mendoza (Argentina; González Riga et al. 2015) and the Bajiu tracksite at the top of the Feitianshan Formation (Lower Cretaceous) of southwestern Sichuan, China (Xing et al. 2016d).

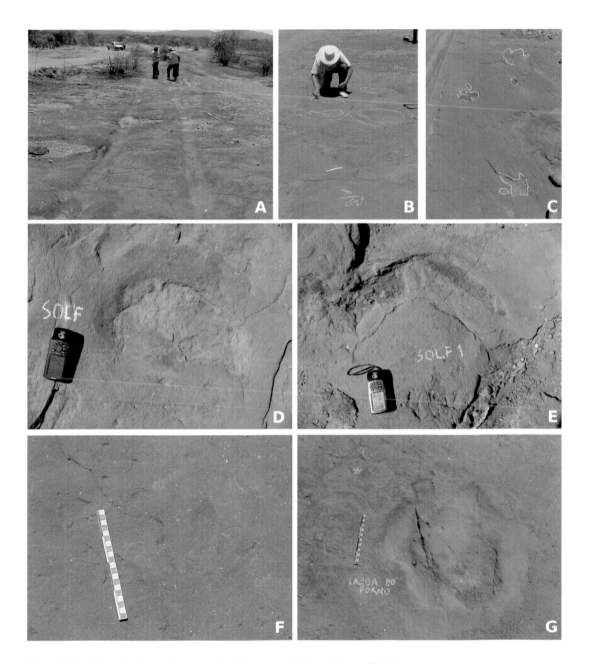

Figure 4.117. Isolated footprints of sauropods and theropods, in the sandstone of the Rio Piranhas formation at Lagoa do Forno (SOLF), Sousa. *A*, Alongside the present-day dirt route between Sousa and Acauã, at the locality Lagoa do Forno I (SOLF I), are the remnants of an older road with the ruts of ancient carts or wagons cut into the pink sandstone, in which fossil footprints occur. *B* and *C*, Sauropod and theropod tracks. *D*, A large horseshoe-shaped handprint of a sauropod. The length (from the GPS device) is 16 cm. *E*, A hand-foot set of a sauropod, the hindfoot filled in with sediment. *F*, A theropod footprint at SOLF I, graphic scale in cm. *G*, A large and isolated, poor-quality, probable theropod track at Lagoa do Forno II (SOLF II)—a different locality, as one can see by the different color of the sandstone. The print is situated in the road, near the chapel of the farm. graphic scale in centimeters.

Lagoa do Forno II (SOLF II)

The second site is 8.4 km SE of the center of Sousa (S 06 48.563, W 38 10.492), near the aforementioned octagonal chapel of the hamlet Lagoa do Forno, in the municipality of Sousa, very probably at the boundary between the Sousa and Rio Piranhas Formations.

References: Leonardi and Santos 2006.

SOLF II 1

One isolated, very large theropod footprint with a deep and wide displacement rim all around it. (figure 4.117 G).

Summing up, the Lagoa do Forno (I and II) as a whole is a low-diversity ichnoassociation composed of 9 sauropod and 6 theropod footprints, all of them isolated, without any indication of gregariousness or other interaction among the trackmakers. This is one of the few ichnofaunas in the Rio do Peixe basins where the number of footprints of plant-eating dinosaurs exceed those of meat eaters.

Mãe d'Água (SOMD)

This locality (figs. 4.120, 4.121A) is situated 7 km SSE of Sousa as the crow flies, in the territory of this municipality, on the Mãe d'Água farm, part of the former Caboge farm (S 06 49.320, W 038 12.045). It lies about 1 km NNW from the neighboring locality Curral Velho (SOCV). At Mãe d'Água, the sandstone of the Rio Piranhas Formation are pink in color, indicating that this outcrop very probably belongs to the lower section of the formation. The landscape here shows low cuestas (very typical of this formation's topography; figs. 4.120–4.121A) of medium-grain to coarse sandstones with alternating beds of siltstones, mudstone, and fine conglomerates.

Figure 4.118. At least 6 large sauropod and 4 large theropod footprints occur in the roadbed at Lagoa do Forno I, but they were almost completely obliterated by the bulldozer that cut the road. Reproduced from Leonardi and Santos (2006).

Large Sauropoda

Large Theropoda

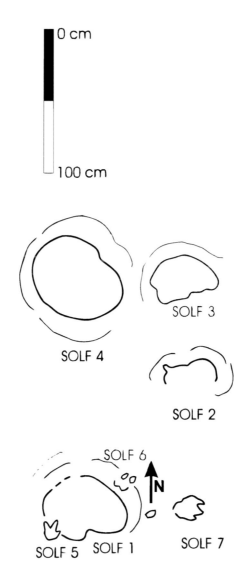

SOLF 9

Figure 4.119. Map of the main road surface at Lagoa do Forno I, with theropod and sauropod tracks in the roadbed. From Leonardi and Santos (2006).

0 cm

100 cm

SOLF 3

SOLF 4

SOLF 2

SOLF 6

N

SOLF 5   SOLF 1

SOLF 7

At this site, the coarse texture of the sandstone does not permit preservation of the finest morphological details. The footprints are filled in with material of the overlying contiguous layer. Erosion has cut down through the rock such that the footprints are observable only as outlines (figs. 4.115, 4.116, 4.120, 4.121).

References: Leonardi 1981b, 1981e, 1987b, 1989a, 1989b, 1994; Leonardi et al. 1987a, 1987b, 1987c; Leonardi and Santos 2006.

Large Sauropoda

Large Theropoda

Small Ornithopoda

Figure 4.120. The low and arid upper structural surface of a cuesta in the Rio Piranhas Formation at Mãe d'Água (SSE of Sousa), where sauropod, theropod, and long-heeled ornithopod tracks were found. From Leonardi and Santos (2006).

The tracks, discovered by Leonardi (1979), comprise a short trackway and 3 isolated footprints (SOMD 1–4, figs. 4.115A, 4.115F, 4.116). Aside from these known tracks, some more poor-quality tracks of sauropods and theropods were discovered at this site in December 2003, among them SOMD 5–8 (figs. 4.121B, 4.121C), all of them isolated.

SOMD 1

An incomplete, isolated footprint with a rounded outline (fig. 4.116). There is a round "plantar" pad and two preserved digit impressions; that of digit III being spatulate, with a long, rounded hoof. The footprint can be attributed to an ornithopod.

SOMD 2

A tridactyl right footprint, relatively well preserved and not filled in to show reverse topography (fig. 4.116). The long, stout toe marks are nearly parallel proximally, but the two peripheral toes distally curve away from the footprint midline (medially and laterally). All three toes show claw marks; those of digits II and III are very long and broad. Digital pads are very evident: two or three in digit II; four in III, with a quite large distal pad; and four in IV. There is also a small proximal ("plantar") pad, perhaps associated with digit III. The relatively long digit III mark is similar to what would be expected in a footprint made by a small theropod, but the large size of the footprint (length: 34 cm) indicates a larger form.

SOMD 3

A large tridactyl footprint lacking digital pads, with rounded, hoof-shaped toe impressions (figs. 4.115A, 4.116). The digit III mark is by far the longest

Figure 4.121. Mãe d'Água Farm (SOMD), Rio Piranhas Formation, Sousa. *A,* The layers of the coarse sandstone on the side of the cuesta; photo Luiz Carlos Gomes da Silva. *B,* A rare, poor-quality theropod track at this locality. *C,* An infilled sauropod hindfoot print at Mãe d'Água. Graphic scales in centimeters.

and is slightly spatulate. Either this print was made by a different form of ornithopod with quite a long digit III or, more probably, the animal dragged the middle toe as the foot was brought forward, leaving an extra-morphological furrow, in which case its foot was likely similar to that of the makers of SOMD 1 and SOCV 8, 9, and 10.

SOMD 4

This is a short trackway of 2 footprints (figs. 4.115F, 4.116). The footprints are large and tridactyl, with long plantar pads (likewise in the case of long-heeled theropod footprints—for example *Moraesichnium barberenae*); three toe marks are shaped like long, round hooves. This is a completely different form. Some similarity with SOMD 3 can be seen in the toe marks. The most similar specimen is an isolated long-heeled footprint appearing in material from the Cretaceous (Albian to Cenomanian) Limay Formation, Neuquén Group, from Picún Leufú, Neuquén Province, Argentina (Calvo 1991). The distal roundish outline of the clawless toes points to the Ornithopoda.

SOMD 5

An isolated large sauropod footprint of poor quality.

SOMD 6

Another isolated large sauropod footprint of bad quality (fig. 4.121C).

SOMD 7

An isolated theropod footprint of poor quality (figure 4.121B).

SOMD 8

A long-heeled footprint attributable to an ornithopod.

All in all, this is a small but varied ichnofauna with more than 8 individuals and 5 different forms: 1 large and 1 medium-size theropod (at least 2 individuals), 1 typical ornithopod (3 individuals), 1 sauropod (2 individuals), and the long-heeled form of uncertain attribution, which can probably also be attributed to the ornithopods (four individuals of this last group). If these identifications are correct, this is a rare case in the Rio do Peixe ichnofauna, in which ornithopods were numerically dominant (75% of trackmakers) over theropods (25%). In addition to the aforesaid known tracks, some more poor-quality isolated tracks of sauropods and theropods were discovered at this site in December 2003, among them SOMD 5–8 (figs. 4.121B, 4.12C).

Paraíso Farm (SOFP)   This site is located in the Fazenda Paraíso, on the former Lagoa da Estrada farm, 50 m from the farm's main gate (fig. 4.122A). This is in the municipality of Sousa, 9.8 km ESE of the city center as the crow flies; 1,100 m ESE from the octagonal chapel of Lagoa do Forno; and 300 m ESE of the iron cross on the dam of the Forno reservoir (S 06 48.793; W

Figure 4.122. *A*, The main rocky surface with dinosaur tracks at Paraíso Farm (SOFP, Rio Piranhas Formation, Sousa). The locality is situated adjacent to tracks of the narrow-gauge Fortaleza-João Pessoa railroad. *B* and *C*, The main theropod trackway, made by a much larger individual than average for theropod trackmakers of the Rio do Peixe basins. *D–F*. Individual footprints in the main pavement at SOFP. Some of the tracks (for example, the one in *E*) might appear to be ornithopod tracks because of what look like rounded hoof marks, but this appearance is deceiving; the rounded shape is due to infilling and does not reflect the anatomy of the trackmaker. Consequently, these footprints can also be attributed to theropods. Graphic scales in centimeters. The length of the GPS device is 16 cm.

38 09.857). The locality was discovered and published, with a preliminary map of the pavement, by Sérgio A. K. Azevedo (1993). A new map of the pavement is presented herein (fig. 4.123).

The outcrop is a medium- to fine-grain yellow sandstone of the Rio Piranhas Formation. The main rocky pavement with tracks (fig. 4.123; 10 × 8 m; strike S64°W; dip 14° to the south) is situated alongside the tracks of the narrow-gauge Fortaleza-João Pessoa railway, now out of use (fig. 4.122A).

References: Azevedo 1993; Leonardi and Santos 2006.

The main pavement of the site contains a fine trackway of at least 8 footprints of a large theropod (SOFP 1, fig. 4.123, 4.124), imprinted in a terrain that was more waterlogged in some places than in others. The animal was heading N45°E. The SL/FL ratio of the trackway is low, about 5.3. Two other footprints form a second trackway (SOFP 2). A third trackway

Figure 4.123. Map of the main rocky pavement of the Fazenda Paraíso tracksite with theropod tracks. From Leonardi and Santos (2006).

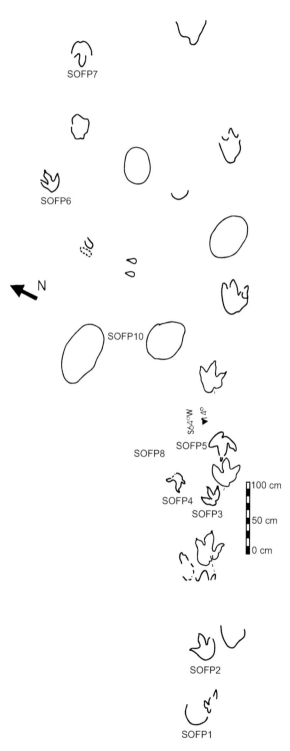

SOFP9

SOFP8

SOFP7

SOFP6

N

SOFP10

S64°W 4°

SOFP5

SOFP8

SOFP4

SOFP3

100 cm

50 cm

0 cm

SOFP2

SOFP1

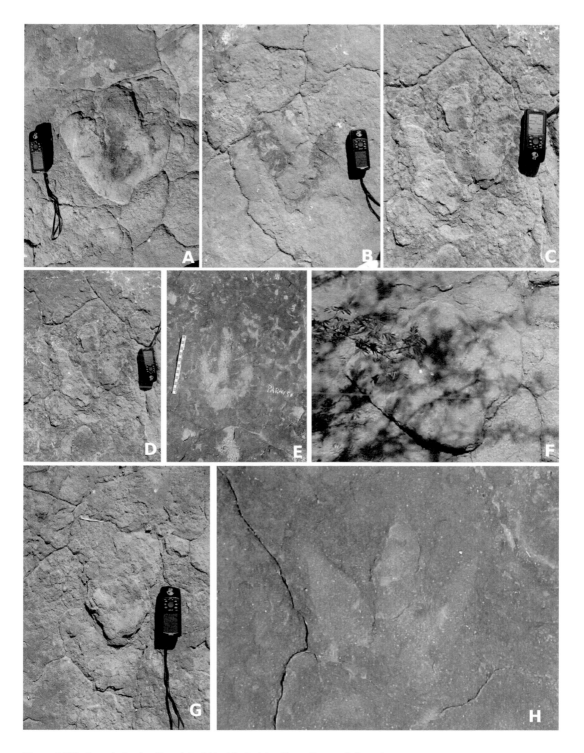

Figure 4.124. Fazenda Paraíso (Sousa). Individual footprints of large theropods from the main exposure and from the surrounding area. The length of the GPS device is 16 cm.

with 3 footprints (SOFP 10) seems to be made of underprints (fig. 4.123). This last dinosaur was heading EESE.

There are also 8 isolated footprints (SOFP 2–9). Almost all of these tracks are filled in with the coarser material of the overlying contiguous layer. Some of them (such as SOFP 9; figs. 4.123, 4.124A) appear similar to ornithopod tracks because of the rounded distal ends of the toe marks, but this is most likely a result of the infilling and not the anatomy of the trackmaker; such footprints are therefore also attributed to theropods (e.g., figs. 4.122E, 4.124A, 4.124F). Another large theropod footprint is located about 40 m along the rails in the direction of João Pessoa. Some other isolated, poor-quality footprints are scattered in the thorny bush of the farm around a small reservoir, to the north of the main pavement. They are attributed to a sauropod (SOFP 11) and to theropods (SOFP 12, 13; e.g., fig. 4.124E).

Altogether, approximately thirty tridactyl, digitigrade footprints were found at this site. Nearly all of the tracks can probably be attributed to large theropods (N = 13) of three or four different forms, except for a single sauropod hind-foot print. Some of the theropods (SOFP 1, 2, 5, 9, 10; figs. 4.122F, 4.123) were very large, and their footprints are among the largest in this basin. Despite the low diversity of the assemblage, this is one of the most fascinating sites in the Sousa basin.

Summary of the Rio Piranhas Formation Ichnofauna

Altogether the 6 sites from the Rio Piranhas Formation sample about 12 stratigraphic levels that preserve the following ichnofauna: 27 large theropods and 5 small theropods (32 theropods), at least 7 sauropods, 2 small ornithopods, and 9 graviportal ornithopods (11 ornithopods) as well as at least 5 unclassifiable or uncertain dinosaurian tracks. Altogether, the recorded dinosaurian individuals in the Rio Piranhas Formation number in excess of 55 (figs. 4.109–4.123).

Data Tables and Statistics

5

This section is a detailed presentation and statistical analysis of dinosaur trackway data (see tables 5.1–5.52 at the end of the chapter). These data originally appeared in three publications of Leonardi et al. (1987a, 1987b, 1987c) and are partly updated here. All of the data are for trackways found and studied before 1985. Other trackways were discovered through January 2015 (Santos et al. 2015), but very few of them were systematically measured, and only a few received identifying trackway codes. All measurements of footprint and trackway parameters were made following the protocols proposed by Leonardi and collaborators (1979d, 1987a), as illustrated in figures 2.3 and 2.4.

The Data Tables

Aside from hundreds of isolated dinosaur footprints, the Sousa basin presents a sample of 82 complete dinosaur trackways (i.e., footprint sequences) that were found and studied through 1987, the data for which are presented here. Their makers are identified as 49 large theropods, 10 medium-size theropods, 3 small theropods with long and sinuous III digits, 7 ornithopods, 8 sauropods, and 1 ankylosaur; there is also 1 trackway of an indeterminate (incertae sedis) quadruped and 3 trackways of unidentified bipedal dinosaurs (either large theropods or ornithopods). In this sample, 11 trackways are quadrupedal (SOES 9, SOES 100–107, SOPP 1, SOPP 15) and represent 13.6% of this sample. Two of the trackways made by quadrupeds are attributed to ornithopods; another is classified as a probable quadrupedal ornithischian. The other 8 trackways made by quadrupeds are attributed to sauropods and, though bipedal in appearance due to lack of manus prints, reflect quadrupedal locomotion with total overlap of pedal and manual footprints.

The Ichnofauna

The data obtained from extrapolation are for short, incomplete theropod trackways of 2 footprints in which the position of a third print was estimated. In this case, the bisectrix (or bisector) of the angle evidenced by the axis of the III toes of the preserved footprints was assumed to be the midline of the trackway. Then, the hypothetical third footprint was located to the same distance: in the X coordinate, the distance is the same as between the reference point of the first footprint of the trackway and the midline (ML); in the Y coordinate, the distance corresponds to 2 times the length of pace between the first and second footprints.

Parameters of the Trackways

As mentioned previously, this locality presents at least 25 levels bearing dinosaur tracks. Most of these trackways (26) are very short and feature 2 or 3 footprints due to the structural and morphological situation of the outcrop, where the surface of the layers is long from one bank to the other but narrow parallel to the course of the river. There are also dozens of isolated footprints belonging to at least 230 individuals. Most of them are theropods and mainly large ones. There are a few ornithopod footprints and a group of footprints, maybe originally distributed in the 7 trackways, belonging to sauropods.

Most of the trackways with 3 or more footprints are relatively regular, as demonstrated by the standard deviation of the stride length, <10. Even so, SOCA 97 indicates an irregular gait, showing fast deceleration (SD of the strides = 39.7); in contrast, SOCA 1331 shows acceleration. SOCA 43 presents regular strides (SD = 9.9), but the step angle is irregular (SD = 47.3) because it corresponds to a short trackway bent to the left. In general, comparing the standard deviation of the stride and that of the step angle, as well as the other trackway parameters, we observe a notable independence between them, which depends only partially on the trackway characteristics. This phenomenon must be attributed also, in some cases, to the different number of measurements that are available for these parameters as well as to the relative values of the parameters. Another observation is that the majority of theropods of the SOCA site paced slowly, with low step angles and short strides compared to those in other ichnosites. However, the breadth of almost all the trackways presents negative values. We can see 53.8% of these trackways show constant foot divergence from the midline (DML; SD = 0); 23% are rather constant (DML; SD ≈ 10), and the other trackways present very high variability.

This locality presents seventeen trackways; among these are 10 pertaining to quadrupeds (8 sauropods, 1 ornithopod, 1 ankylosaur) that comprise 90% of all complete and measured quadrupedal trackways in the basin. In the Sauropoda group (SOES 100–107), we observe a higher irregularity in trackway SOES 100 than in the other seven. This will be argued in the last part of this statistical study. The great difference in the gait characteristics of the represented groups is evident; for example, the step angle in Sauropoda is very low (average = 111°); it is relatively very high in the only quadrupedal ornithopod (average = 147°30' in the foot and 169°52' in the hand; table 5.50); it is almost always very high in the large theropods (average = 171°20'), among which there were some fast runners. Most of the trackways with 3 or more footprints are regular in all parameters, with a standard deviation almost always ≤10. An exception is SOES 100, the previously mentioned sauropod.

In this classical locality we find seven levels with isolated footprints or short, incomplete trackways of 2 footprints each. The longest trackways belong to 6 large theropods, 2 bipedal ornithopods, and a probable quadrupedal Ornithischia.

The manner of gait of the dinosaurs of the main level at Passagem das Pedras sensu stricto will be widely discussed in the last part of this statistic section. One of the trackways (SOPP 15, table 5.37) is quadrupedal, but unfortunately it is an undertrack. Although it lacks morphological details, the trackway is probably too narrow to be classified as ankylosaurian and so, by the general structure and parameters, can be attributed with certain probability to a quadrupedal ornithischian. Indeed, the step angle is very high (average = foot 177°30'; hand 174°), indicating a relatively high speed that was slowing down (therefore the stride length is diminishing) without reducing the step angle.

With the exception of SOPP 12, the trackways in this locality are very regular according to the values of the standard deviation. Almost all trackmakers had a fast walking gait or a running gait, in straight trackways, with stride and step angle high to very high.

The footprint numbering of SOPP 5 (tables 5.35, 5.36) starts at -50— with negative numbers because, in contrast to all the other trackways, this one was excavated (and is in the process of being excavated) in the direction opposite to the animal's procession.

SOSL—Serrote do Letreiro (tables 5.38–5.40)

The main subsite is a small rocky outcrop (~6 × 4 m) that shows 7 trackways and 31 isolated footprints, with a total of 38 theropod individual tracks, attributed to the same kind (and probably to trackmakers of the same Linnaean species). The surfaces of some near and contiguous layers present trackways and isolated footprints of the same kind, a short trackway of 1 ornithopod and several large footprints and trackways of sauropods.

In spite of belonging to the same kind and probably to the same population sensu stricto, the SOSL group 4 and SOSL 5 trackways (from left to right of the table) display greatly varied values: they are regular-to-irregular and quick-to-slow trackways. The inner width is almost always negative; the standard deviation in all the trackways is very constant, resulting in low values of standard deviation, although these are somewhat varied at the population level. This probably included adult and young individuals. As for the group of sauropod tracks at Riacho do Pique, see Carvalho (1989).

SOPM—Poço do Motor (table 5.24)

The locality, although bearing this different name in local use, is the geographical and stratigraphic continuation of the Passagem das Pedras locality. The layers here are lower and older than those of Poço do Motor.

Five short trackways and 1 isolated footprint were found in three levels along 150 m of the riverbed of the Rio do Peixe, with a total of 6 individuals, all of them belonging to large theropods.

SOMA—Matadouro (table 5.12)

The outcrop (6 × 2.5 m) is located in the bed of the Rio do Peixe inside the city of Sousa. All 4 trackmakers were large theropods, but only 2 short trackways are fairly well preserved. Morphologically, the trackways are very different; however, the values of their parameters are very similar.

SOPI—Fazenda Piedade (table 5.22)

This outcrop, with three contiguous levels, shows in a small surface (14 × 3 m) 1 fine trackway with 3 footprints and 7 isolated footprints attributed to 5 large theropods, 1 ornithopod, and 1 incertae sedis (half-swimming track).

Diverse Localities—São João do Rio do Peixe (table 5.5)

The table gathers the data of 7 short trackways and 1 isolated footprint, all of them attributed to large theropods from three localities of the São João do Rio do Peixe municipality. Almost all of the trackways are shallow and badly preserved, with the exception of ANJU 2.

ANEN—Engenho Novo (tables 5.9, 5.10)

In this locality, a main rocky surface with small dimensions (~250 m²) and two contiguous levels, the surfaces of which are only a few centimeters thick, presents 7 trackways and 18 isolated footprints, resulting in a total of 25 individuals. Among them are one hand-foot set of a large Sauropoda (ANEN 10), 1 trackway made by a large, apparently bipedal (or maybe semibipedal) animal (ANEN 8), 1 curved trackway of a small theropod (ANEN 1), 21 trackways and isolated footprints of large theropods, and 1 footprint incertae sedis. The theropod trackways are very regular. ANEN 1 points to constant acceleration, and ANEN 8 is the most irregular trackway. At first, ANEN 8 was tentatively attributed to an ornithopod, although it doesn't show morphological details, due to some similarities with other trackways attributed, with reason, to that group.

Parameters of the Footprints

SOES Estreito; Serrote do Pimenta (tables 5.46–5.49, 5.51)

Table 5.51 shows the data related to the footprints of trackway SOES 9, which is particularly interesting because it is one of the few good-quality quadrupedal trackways with morphological details of the Sousa ichnofauna. The comparison between the data related to the hand and to the foot is interesting because it shows the enormous dimensional difference between them (heteropody). This points to the fundamentally bipedal body structure of the trackmaker, whose group must have recently come back to the quadrupedal gait due to the increasing dimensions and the notable weight. The index of the foot (FL/FW) is smaller (average = 1.09) than that of the hand (average = 1.25) because these last tracks are relatively wider and ellipsoid. Leonardi (1981d) had already observed that in quadrupedal animals—at least those with strong heteropody—the forefeet impressions are more variable in shape and position indicating their being used to probe or explore rather than supporting the animal; the hind-feet's behavior can be defined as "gregarious" in regard to the forefeet. This concept found support in the analysis of the standard deviation of the footprints of the trackway SOES 9. The trackway is very regular, with a standard deviation always smaller than 10.

The 8 sauropod trackways (SOES 100–107) show low variability (SD <10) in footprint length and in footprint width (FL and FW); the other parameters of the footprints—probably subtracks lacking details—could not be measured. The mean values are not very distant; however, there are growth of values from SOES 101 to SOES 102 and to SOES 100. The trackmaker of this last trackway was the biggest in this small herd of at least 8 sauropods (SOES 100–107); this is also true regarding some isolated footprints that represent other trackways (still not completely excavated) of the same herd. The index of the footprints (FL/FW) is substantially constant but with a smooth tendency to increase with individual growth: SOES 101, 102, and 100 have respective indexes of 1.05, 1.12, and 1.17, which means that as the animal size increased, the length of the hind-foot increased more than its width. This is an example of differential growth (fig. 4.99).

In each large theropod trackway of this locality, the footprints have extremely constant dimensions, with low standard deviation (almost always <5). The most variable parameter in the tables, with standard deviation <10 in three cases, is the cross-axis angle, which probably depends on wrong measurements due to the difficulty of reading this parameter correctly, especially in tridactyl footprints with poor impression or preservation.

The measurement of the toes was carried out, when possible, according to both the free length of digit and the length of the phalangeal portion of the digit. One notices that in some large theropods as well as in ornithopods, digit IV is shorter than digit II with both measurement methods. The digital divergence is surprisingly constant except in the case of SOES 10. The partial divergence II^III and III^IV are almost always similar in the same individual track, but SOES 11 is different, having toes II and III characteristically very close while toe IV is always distant from III.

SOSL—Serrote do Letreiro (tables 5.41–5.43)

Most of the tracks correspond, with all probability, to a theropod population sensu stricto. Also, there are an ornithopod trackway (SOSL 8) classified as *Iguanodontipus* Sarjeant et al., 1998; the trackways of sauropods in a herd; and other isolated sauropod footprints.

The theropod footprints of each trackway do not vary much, with the exception of the digital divergences (partial and total) that, in a good number of trackways, are variable enough, as evidenced by the high standard deviation (>10 in 8 cases, >20 in 3 cases). This points to animals with agile and nervous toes. In this population, the free length of digit IV is often shorter than that of digit II, and the indexes II/IV (mean values) are indeed often higher than 1, in the same order as in table 5.42, from left to right: 1.04, 1.16, 1.10, 0.99. The indexes II/IV of the mean length of the phalangeal portion are, however, all lower than 1 (0.74, 0.61, 0.92, 0.63). This fact indicates that from an osteological point of view, the IV digit was longer than II, as one would expect; however, it was probably more connected by ligaments and by the skin to digit III than digit II was, with the result that the free portion of it is proportionally shorter. So, the hypex III-IV is situated in a more distal position than the hypex II-III, and therefore digit IV seems to be shorter in the footprints. Digit III, as in other populations, is at least one-third longer than the other two.

In SOSL 8 (*Iguanodontipus*) the relationship between digits II and IV is the same as seen above for the population of theropods, but digit III is proportionally much shorter.

SOPP—Passagem das Pedras (tables 5.31–5.35)

The data reported in the tables of this locality, as well as the considerations about them, are relative to the diggings of 1987; later on, the excavations of trackways SOPP 1 and SOPP 5 were continued.

The SOPP 1 trackway, the longest of the locality, apparently regular, shows standard deviation of 6 for FL and 4.2 for FW. These relatively high values depend on the fact that the mud, after the trackmaker lifted its foot out, sometimes fell inside the footprint, which, because of this, sometimes looked shorter.

The values of the standard deviations in the other trackways are, in general, much lower (with an exception for the angle values), indicating regular gaits and pointing to footprints impressed in good plastic mud, as well as walking gaits. Unfortunately, we do not have all the data about the footprints of trackways SOPP 3 and SOPP 4, which show a running gait, because they were badly eroded by floods.

The values of the standard deviation for the cross-axis angle are high, probably due to problems with measurements and interpretation. There are high standard deviation values for the digital divergence, pointing to agile, fine, and nervous toes. In these tables, digit IV is much longer than digit II, both in the free length of digit and in its phalangeal portion. In the free length, digit III is longer than the others; however, the difference is less evident when calculating the length of the phalangeal portion. The digital divergence, low in the SOPP 1 (*Sousaichnium pricei*) hind-feet, is characteristically very high in SOPP 2, 3, 4, and 5 (*Moraesichnium barberenae*).

SOCA—Fazenda Caiçara-Piau (tables 5.18–5.21)

These tables contain data relative to the footprints of 26 trackways that, with exception of SOCA 43, 46, and 153, all belong to theropods. In these trackways, the standard deviation values are always low in the length as well as the width of their footprints, indicating straight trackways with a walking gait, all of them very regular. SOCA 97 is an exception regarding the length of the foot. All the parameters of theropod footprints show great variability from one trackway to another, as we shall see better later. The SOCA 46 trackway, attributed to an ornithopod, is surprisingly constant in its values although it is curved. The values of the digits (calculated both as free-digit and as the phalangeal portion of the digit) are remarkably constant, always with very low standard deviation, with both measurement methods.

Digit IV is often longer than digit II both in the free length of the digit and in the length of the phalangeal portion. Digit III, in its free length, is almost always much longer than the others digits. In the SOCA sample, the measurement of the length of the phalangeal portion of the digit is almost always impossible. The digital divergence variability (partial and total) is fairly varied in the theropod association of this locality. Also, the ratio between the digital divergences II^III and III^IV is very variable, which points to a rather heterogeneous association, with the presence of different kinds of theropods.

SOPM—Poço do Motor (table 5.25)

There are 5 large theropod trackways here. The footprint parameters are very constant, with some exceptions, as always, as for the cross-axis angle and the digital divergences. This fact causes rather different standard deviation values (from 1.4 to 26.5). The L/W indexes of the footprints are varied (0.94, 1.27, 1.42, 1.22, 1.43), which probably points to the presence of different kinds of theropods.

Diverse Localities of São João do Rio do
Peixe Municipality (tables 5.6–5.8)

The tables show data on the footprints from 7 trackways in three localities in this municipality: Barragem do Domício (ANBD), Juazeirinho (ANJU), and Poço da Volta (ANPV). All of them are attributable to large theropods. The values of these footprints are extremely regular, with exception of the angular values, whose standard deviation is a bit high (mainly in ANPV 1). The footprints indexes (L/W) are less varied than the preceding samples. The relative length of digit II and IV is variable if we use the method of calculating the free length of the digit. Digit IV is, however, always longer than II in the few cases where the method of calculating the length of the phalangeal portion was used. Digit III is always much longer than the others with the free length of digit method.

ANEN—Engenho Novo (table 5.11)

This locality presents 1 trackway with large dimensions, apparently bipedal (ANEN 8), attributed to an ornithopod or rather a sauropod; a bent trackway of a small theropod (ANEN 1); and 5 theropod trackways of medium-large size. The values of the parameters of the footprints are almost always extremely regular; however, the angular values show the highest standard deviation, as always. Among the footprint indexes (L/W), ANEN 8, a trackway with very large footprints and doubtful classification, has the lowest standard deviation ($\sigma = 1.06$); ANEN 1, attributed to a small theropod, has the highest, as expected, because it is a curved trackway. The index values of the footprints of the medium or large theropods are between 1.17 and 1.47.

The relative length of digits II and IV is very varied in the free length of the digit, but digit IV is always longer than digit II if it is measured calculating the length of the phalangeal portion. The explanation given above is valid here too. Digit III is always longer than the others as regards the free length of the digit. The cross-axis angle is always close to 90°, which is rather typical in theropods, which are almost always strictly mesaxonic. The footprints are substantially symmetrical except for the partial digital divergences. In fact, when calculating the following index—digital divergence II^III/digital divergence III^IV—and extracting the standard deviation, the value obtained is very low (average SD = 0.32). In these five cases, four indexes are inferior to 1, indicating that the most common situation in this sample is that the digital divergence II^III is slightly smaller than the digital divergence III^IV.

SOPI—Fazenda Piedade (table 5.23)

Here there is only one trackway (SOPI 1) with 3 footprints, attributed to a large theropod, and other isolated theropod footprints. In the footprints of SOPI 1, the data are extremely regular and very constant in the digital divergence. Here, too, the index—digital divergence II^III/digital divergence III^IV—is <1 and particularly low (0.47).

SOMA—Matadouro Municipal (table 5.13)

In this locality, there are 2 short trackways of large theropods (along with 2 others of poor quality). The parameters of these footprints are very constant.

We want to emphasize, in summary, some of the most important results of the footprint parameters:

· At least in one case (SOES 9), the probing or exploring function of the forefeet can be confirmed, as well as the "gregarious" function of the hind-feet (Leonardi 1981d). It is just one case, but this is practically the only well-preserved quadrupedal trackway in the Rio do Peixe basins, with the exception of 1 sauropod trackway at Piau.

· The footprints frequently show very constant values in their main parameters—that is, length and width, and length of the digits, with very low standard deviations. The values of the angular measurements are more variable. For the cross-axis angle, this depends on problems of measurements, and for the digital divergence it depends on the digit mobility, especially in theropods. However, there are specimens that show a constant pattern of partial digital divergence, also among the theropods.

· The relationship between the length of digits II and IV is highly variable; however, generally speaking, the following case is more common: digit II is longer than IV if measured with the method of free length of the digit but shorter if measured with the method of the length of the phalangeal portion. Digit III is always longer than the others, at least in the free length of the digit, where more data are available. This fact is more evident in theropods, especially small ones. No observations of this kind can be made for the sauropod footprints.

· As regards the partial digital divergences, the II^III angle of divergence is often smaller than the III^IV one.

· The assemblages of theropods, both small and large, are frequently very heterogeneous in the Rio do Peixe basins, with probably a high index of diversity even within the theropod assemblage.

Statistical Study

The first two sections of this analysis present tables of data for parameters of (1) the trackways and (2) the footprints (tables 5.5–5.52). These two sets of data were kept separate and were separately examined. In fact, as Farlow observes (2018, 320), the parameters of the footprints are different and more important than those of the trackways to distinguish the morphotypes, because the former are beyond the control of the trackmakers and depend on their physical constitution. The way animals place their feet on the ground, that is, their gait, is, on the contrary, largely under their control. They decide whether to walk, trot, or run, if physically they are able to. The animals decide whether to go in a straight line, as they normally do, or for some reason proceed to the right or to the left.

In addition to reporting the raw data, means and standard deviations are tabulated at the end of every series of measurements, together with the maximum and minimum values of the range.

The first part of the present section examines statistical functions related to the main trackway and footprint parameters of the more notable dinosaur ichnoassociations, including isolated prints. The second part of this section discusses graphs, histograms, and bivariate relationships using data from both trackways and isolated footprints; this analysis deals with nearly all the good-quality individual tracks (trackways and isolated footprints) of the Rio do Peixe basins.

Table 5.1 presents and explains the codes of the ichnosites; tables 5.2–5.4 provide a list and trackmaker identifications of all the good-quality trackways and isolated footprints found in the Rio do Peixe basins. The analyzed data were obtained between 1975 and 1987, at which time a 12-year period of intensive field research was finished. Information for trackways and footprints discovered after 1987 are not included in this statistical study. Unlike the other two parts of this section, the analysis reported here takes into consideration not only the 79 trackways but also all of the identifiable isolated footprints, a complete list of which can be found in tables 5.2–5.4 as well as in the diagrams and histograms.

The entire Rio do Peixe sample of tetrapod footprints and trackways represents 537 individual animals, nearly all of them (535) dinosaurs. This count does not include the hundreds of small "half-swimming" footprints referred to as chelonians (see also Hornung et al. 2012b). The tables of data (tables 5.5–5.52) and the present statistical study are based on 365 measured individual tracks.

To compare data across individual trackmakers, we calculated the mean value of each parameter (e.g., the stride, tables 5.5–5.51) for each individual trackway. The mean of the means was then calculated as the average of the individual trackway means.

Passagem das Pedras — SOPP (table 5.53)

Mean values of the stride and of the oblique pace of the trackways from this site are higher than the means and modes of trackways for the basin as a whole. This is because 2 of the 7 theropods at this site seem to have been quickly running and thus taking longer steps than usual. In contrast, the mean value of the trackway external width across trackways at this site is very close to that for the entire basinal sample of trackways. Similarly, the mean of the step angle across trackways is almost identical to the mean for the whole basin. These last two results reflect the fact that these two parameters have limits (footprint width in the first case and 180° in the second), which will be discussed in greater detail below, while the stride, the pace, and the oblique pace can show a higher range of variability on both sides of the mean. The means and modes of parameters of footprint size for Passagem das Pedras are higher than the corresponding means and modes for the trackway sample of the basin as a whole.

The standard deviation of the stride is very high (107.7), possibly due in part to the fact that the 7 individual animals may represent three different kinds of trackmakers. More importantly, however, the trackway sample includes both walking individuals and quick runners.

For mathematical reasons, the standard deviation of the oblique pace is obviously very near half the standard deviation of the stride, given that values of the oblique pace are a little more than half the stride values in trackways of fast or very fast bipedal runners. Consequently, the dispersion of values of the oblique pace is a little more than half that of the stride. The standard deviation is low for both the step angle and the external width due to the above-described maximal values of those parameters. The values of standard deviation are relatively low for both the length and the width of the footprints due to the fact that these dimensions are not, in general, much affected by trackmaker speed (Leonardi 1981d); the feet (and, as a consequence, often also the footprints) have almost the same dimensions no matter what an animal's speed is. There can be another

Table 5.53 Summary Statistics from Passagem das Pedras (SOPP 1-6 and 12)

	N	M	σ_{n-1}	CV %	CI	
					Lower	Upper
Stride	7	244.6	107.7	44	164.8	324.4
Oblique pace	7	123.4	53.2	43.1	84	162.8
Step angle	7	160°16'	10	6.2	152.9	167.7
External width	6	45.2	11.7	25.9	35.8	54.6
Footprint length	7	32.5	8.3	25.5	26.4	38.6
Footprint width	6	26.1	7.7	29.5	19.9	32.3
Length of digit II	7	8.5	2.3	27.1	6.8	10.2
Length of digit III	7	10.9	2.7	24.8	8.9	12.9
Length of digit IV	7	7.7	2.9	37.7	5.6	9.8
II^III	7	28°40'	13.1	45.7	19.0	38.4
III^IV	7	27°53'	9.5	34.0	20.9	34.9
II^IV	6	59°06'	22.9	38.7	40.8	77.4

possible explanation for this phenomenon: the standard deviation tends to increase as the value of the mean increases (which is, after all, the reason for computing the coefficient of variation). Paces and strides will nearly always have larger mean values than footprint lengths and widths, which could account for larger values of the standard deviation for step lengths compared to footprint dimensions (J. O. Farlow, personal communication, 2015). Although this is a heterogeneous association in terms of morphological diversity (three forms), the main dimensions of the footprints are not very different. The standard deviation of the digit lengths is very low (<11) because, although the trackmaker of SOPP 1 was a large iguanodontid, its toes are relatively short. The standard deviation values of the digit divergences are high, especially for total digital divergence II^IV, but this is normal even in homogeneous populations due to the high mobility of the toes in bipedal and tridactyl dinosaurs and to the effects of different substrate conditions.

The coefficient of variation is calculated as

$$\text{coefficient of variartion} = \frac{\sigma \times 100}{\mu},$$

where σ = standard deviation and μ = population (or assemblage) mean.

For the meaning of this statistical function, see Leonardi (1987a, 38, 53). The coefficient of variation is very high at Passagem das Pedras for both the stride and the oblique pace. Because the trackways are well impressed and generally very well preserved, this high value (>25%) indicates a strong heterogeneity of the association in terms of both trackmaker forms and their gaits. In fact, the coefficient of variation is high even for the *Moraesichnium barberenae* sample alone, with 5 measurable trackways among the sample of 7. For example, for the stride, the coefficient of variation is 44.0%.

The coefficient takes smaller values for the external width of trackway for two reasons:

1. The standard deviation is perforce lower for the above-described reason, but the mean is much lower because this is a parameter of much smaller magnitude than the stride.
2. The external width has a minimum limit while the stride has a much higher variability, as explained in greater detail below.

Nevertheless, the coefficient of variation of external width is here just moderately high, indicating a somewhat homogeneous association in terms of this parameter; in fact, it is an association of bipedal and adult dinosaurs.

The coefficient of variation for the step angle is extremely low, indicating a homogeneous population in terms of this parameter. It is normal that values of angles have lower coefficients because these angles are probably independent of the dimensions of the trackmaker, and the step angle has the aforesaid theoretical upper limit of 180°.

For the length and the width of the footprint, as well as for the digit lengths, values of the coefficient of variation are almost all close to 25%, for the reasons explained above in the discussion of the standard deviation of these parameters. In contrast, the coefficient for digital divergences is higher. In part, this may reflect the diversity of trackmaker kinds, but it also likely is due to high mobility of the toes in the theropods, as emphasized above. It is worth noting that values of the coefficient of variation are much higher for measurements of digit IV than the other toes, indicating that this digit, located on the outer (lateral) side of the foot, was the least important toe for the support and stability of the bipedal dinosaurs of this sample.

The confidence interval values for the sample confirm the heterogeneity of this association. Note the different behavior of the linear measures (very wide range of the confidence interval) and the angular measures (small range of the confidence interval).

Serrote do Letreiro—SOSL (table 5.54)

The means of the linear measures are very low compared to the means of whole sample, which points to the presence of a true population of small- and medium-size individuals of the same kind. Otherwise, the mean of the step angle is high because it is independent from the animal dimensions and points at a population of good runners.

The mean of the free length of digit IV is smaller than the free length of digit II, which is typical in this type of measurement method, as seen in the section on the data of footprints. The mean of the free length of digit III is higher than the means of each of the other two digits. The partial digital divergences show that the footprints are very symmetrical, with the extreme toes (II and IV) more or less equidistant from the median toe.

The standard deviations are almost always <10, pointing to a population sensu stricto, which is evident also at the ichnomorphological level. The relatively high value of the standard deviation for the stride depends mainly on the gait difference; in some cases there is a walking gait, in

Table 5.54 Summary Statistics from Serrote do Letreiro (SOSL, Population #4)

	N	M	σ_{n-1}	CV %	CI Lower	Upper
Stride	6	103.6	20.1	19.5	87.2	119.4
Oblique pace	6	52.6	10.4	19.8	44.3	60.9
Step angle	6	163°35'	11.1	6.8	154.7	172.5
External width	5	23.6	5	21.2	19.2	28
Footprint length	6	21	3.4	16.2	17.2	24.8
Footprint width	6	16.9	2	11.8	15.3	18.5
Length of digit II	6	6.3	0.8	12.7	5.7	6.9
Length of digit III	6	10	1.9	19	8.5	11.5
Length of digit IV	6	5.9	0.6	10.2	5.4	6.4
II^III	6	28°32'	12.6	44.2	18.4	38.6
III^IV	6	30°41'	3.1	10.1	28.2	33.2
II^IV	6	62°16'	16.5	26.5	49.1	75.5

others a running gait. The value of the digital divergence III^IV is smaller than that of the digital divergence II^III, which indicates a greater mobility of digit II.

Almost all the parameters have a coefficient of variation lower than 25%, confirming the sample homogeneity of the makers at the level of a Linnaean population. The fact that the coefficient of variation of a good number of parameters are near 20% depends on the dimensional heterogeneity of the sample. The highest values are in digital divergence and mainly in the digital divergence II^III, confirming the observation about the mobility of digit II, and in digital divergence II^IV. The relatively high value of this coefficient for the length of the phalangeal portion of digit III would point to it as the less important digit for foot stability, which seems to be uncommon and illogical in a theropod population.

The confidence interval values confirm the population homogeneity and the regularity of the trackways.

Fazenda Piau-Caiçara — SOCA (table 5.55)

The means of the stride, of the oblique pace, and of the step angle are very low, probably indicating, as observed in the first section of this chapter, that the individuals at this locality moved more slowly than dinosaurs of comparable size from the other localities. The means of the free length of digit IV are greater than for the free length of digit II. The means of the partial digital divergences are very similar.

The high values of the standard deviation of the trackway parameters of this sample point to the heterogeneity of the assemblage and are noteworthy, particularly in comparison to the comparable values for the Serrote do Letreiro sample, given the small sample size at this locality, which records data for a lesser number of animals. The fact is also confirmed by the high value of the standard deviation of the pace angle, which generally is lower.

Table 5.55 Summary Statistics from Piau-Caiçara (SOCA, Levels 4, 9, 13/2, 13/3)

	N	M	σ_{n-1}	CV 95%	CI	
					Lower	Upper
Stride	20	144.6	60.5	41.8	118.1	171.1
Oblique pace	20	86.6	26.8	30.9	74.9	98.3
Step angle	20	156°35'	17.4	11.1	149	164.2
External width	18	43.3	22.1	51	33.1	53.5
Footprint length	20	29.3	6.2	21.2	26.6	32
Footprint width	20	24.6	5.4	21.9	22.2	27
Length of II digit	19	9.3	2.8	30.1	8	10.6
Length of III digit	19	15	4	26.7	13.2	16.8
Length of IV digit	19	10	2.8	28	8.7	11.3
II^III	20	27°29'	10.3	37.5	23	32
II^IV	20	30°04'	9.6	31.9	25.9	34.3
II^IV	20	57°02'	15.3	26.8	50.3	63.7

The coefficient of variation is almost always high, which also confirms the heterogeneity of the sample, as the high values of coefficient of variation in this case cannot be attributed to the impression or to the preservation of the trackways, which are good quality; it probably cannot be attributed to the quality of the measurements either. The value for the step angle is lower than the coefficients of variation for the linear measures, as is normal. The relatively lower value of the coefficient of variation for the free length of digit III with relation to the extreme digits (II and IV) indicates that digit III is the most important, which in general is the more common condition in theropods. The coefficient of variation for digital divergences is high, but the three values of the partial and total digital divergences are well balanced between them.

The values of the confidence interval indicate heterogeneity of the association, which is reasonable inside a suborder (almost the entire sample belongs to theropods). These values do not reach the high values of the SOPP sample, where there are two different suborders of makers as well as very different gaits.

Engenho Novo—ANEN (table 5.56)

The values of the mean, standard deviation, and coefficient of variation for the ANEN sample are very similar to those for SOSL, and so the same remarks apply here. However, the values of the confidence interval are different, with larger intervals at Engenho Novo. This probably depends on the lesser number of individuals in Engenho Novo.

Serrote do Pimenta—SOES (table 5.57)

The values of the means are high, almost as large as those for Passagem das Pedras. This is because some of the individuals—some of the bipeds and 1 quadruped—were large, and because one of the dinosaurs was running. As with the localities already discussed, the mean FL is greater than

Table 5.56 Summary Statistics from Engenho Novo (ANEN, Theropod Tracks)

	N	M	σ_{n-1}	CV %	CI	
					Lower	Upper
Stride	6	138.9	62.1	44.7	89.2	188.6
Oblique pace	6	81.4	24.1	29.6	62.1	100.7
Step angle	6	159°28'	23.7	14.9	140.5	178.5
External width	6	28.2	12.2	43.3	18.4	38
Footprint length	6	21.8	3.9	17.9	18.7	24.9
Footprint width	6	17.4	4	23	14.2	20.6
Length of digit II	6	6.1	2	32.8	4.5	7.7
Length of digit III	6	11.6	1.7	14.7	10.2	13
Length of digit IV	6	6.7	2.2	32.8	4.9	8.5
II^III	6	24°38'	6.6	26.8	19.3	29.9
III^IV	6	29°01'	8.9	30.7	21.9	36.1
II^IV	6	53°39'	13.2	24.6	43.1	64.3

the mean FW because a large percentage of the sample (89%) consists of theropods.

The mean of the free lengths of digit IV is less than the mean of the free lengths of digit II. As usual, digital divarication II-III is less than the digital divarication III-IV—something particularly noticeable in SOES 11. The values of the standard deviations are high (exceeded only by those at Passagem das Pedras), indicating substantial variability in the size of the trackmakers.

The values of the coefficient of variations are similar to those of Engenho Novo, Piau-Caiçara, and Passagem das Pedras. They are higher than those of Serrote do Letreiro, which presents a less heterogeneous association. However, the coefficient of variation of the pace angulation (step angle) is low, indicating that this is a sample of very efficient

Table 5.57 Summary Statistics from Serrote do Pimenta (SOES, Theropods Tracks)

	N	M	σ_{n-1}	CV%	CI	
					Lower	Upper
Stride	9	208.9	99.1	47.4	144.2	273.6
Oblique pace	9	113.7	47.2	41.5	82.9	144.5
Step angle	9	161°39'	14.6	9	152.2	171.2
External width	8	53.7	16.6	30.9	42.2	65.2
Footprint length	9	33.8	8.2	24.3	28.4	39.2
Footprint width	9	27.7	8.8	31.8	22	33.4
Length of digit II	7	14.5	5.2	35.9	10.6	18.4
Length of digit III	8	18.7	5.8	31	14.7	22.7
Length of digit IV	9	13.4	3.7	27.6	11	15.8
II^III	9	22°58'	4.1	17.9	20.3	25.7
III^IV	9	28°49'	10.6	36.8	21.9	35.7
II^IV	9	51°40'	10.4	20.1	44.9	58.5

walkers and runners, including the quadrupedal ornithopod (*Caririchnium magnificum*).

The confidence interval is always wide.

Sequence Graphs

This kind of graph (Leonardi 1981d), characteristically used in ichnology, visually represents the gait of the trackmakers. In such graphs, the abscissa (horizontal axis) shows at periodic distance the successive cases of a parameter (e.g., the successive strides) in the same trackway. The ordinate (vertical axis) shows the values of the same parameter. The resulting curve gives a visual impression of a feature associated with trackmaker gait: its speed, acceleration, deceleration, cruising speeds, regularity, and any accidents. However, such curves *probably* show something about speed and acceleration but don't *necessarily* do so. What the curves show is changes in step length, without saying anything about the amount of time it took to take those steps. J. O. Farlow has watched walking emus (*Dromaius novaehollandiae*) suddenly stop, with the left and right foot marking off a pace, and hold that position for several minutes before starting to walk again. One would never know this had happened from looking at the trackway (J. O. Farlow, personal communication, 2015; 2018, 187–189). On similar topics, see also Bishop et al. (2017).

Figure 5.1 represents 9 trackways from the bed of the Peixe River between the localities of Passagem das Pedras (SOPP) and Poço do Motor (SOPM). The most important observation is that there is a marked difference in the manner of gait among 4 individuals of the same ichnospecies, *Moraesichnium barberenae*, whose trackways are recorded in the same rocky pavements, although in two contiguous levels. It is a very rare and interesting case. Trackways SOPP 2, 3, and 5 represent individuals of about the same size (FL between 35 and 38 cm), but the strides are completely different. The mean stride of SOPP 2 is 186.7 cm; SOPP 3 is 482.8; and SOPP 5 is 197.2 cm. Evidently, the trackmakers of SOPP 2 and 5 walked while SOPP 3 ran quickly, as shown by contrasting values of the stride length/footprint length (SL/FL) ratio: 5.6 (SOPP 2), 5.3 (SOPP 5), and 13.8 (SOPP 3), respectively. This is a rare demonstration of different gaits used by different individuals of the same size and, presumably, the same species; it further demonstrates that the parameters of the trackway, like the stride, depend only in part on static anatomical dimensions of the trackmaker (leg lengths). Dynamic features of limb kinematics, as they affect the capacity for running and jumping, have a very great influence.

Also, observe that individual SOPP 4, likely of the same species but smaller size, with footprints about 20 cm long (we lack complete data for this sequence due to the poor quality of the footprints, caused by destructive erosion), took distinctly longer strides (mean value of 244.6 cm) than those of SOPP 2 and 5, animals that were nearly twice the size of SOPP 4. This indicates that it proceeded at a faster speed. Trackways SOPP 2, 4, and 5 indicate a relatively regular speed. SOPP 5 seems to have been made by the slowest of these animals, with a rather tentative

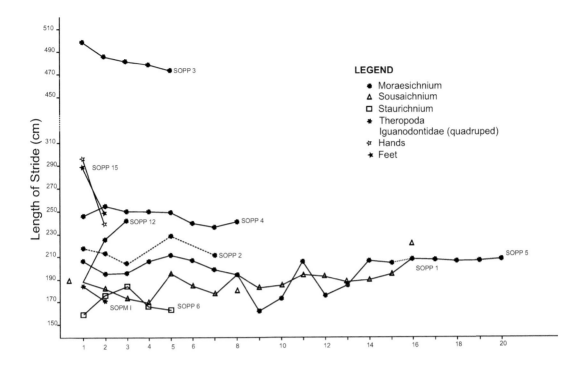

Figure 5.1. Sequence graphs of the strides of dinosaur trackways from the localities of Passagem das Pedras (SOPP) and Poço do Motor (SOPM).

and explorative pace, especially in the central part of the trackway, until the fourteenth footprint in this graph (footprint 63). In contrast, SOPP 3 was in a deceleration phase, indicating that values of the stride could have been >5 m prior to the first recorded footprint.

Little can be said about the other trackways. An uncommon quadruped trackway, SOPP 15, hardly identifiable because it is preserved as undertracks, was decelerating, as was SOPM 1, made by a large theropod with wide total digital divarication (II∧IV). Iguanodontid SOPP 6, *Staurichnium diogenis*, presents a short, irregular, and very slow trackway. The large iguanodontid (SOPP 1) *Sousaichnium pricei* shows a slow walking gait, with some hesitating irregularity in the middle part of the sequence but substantially uniform in terms of speed. This trackway has a distinctive characteristic: a certain sigmoidal increase or decrease in stride length, along with a slight serpentine swinging to and fro, more or less at about every third step, noticed visually in the trackway. The animal supported its body mainly on its hind-feet, both the left and the right, but impressed the sediment only with the right front foot.

Figure 5.2A represents the trackways from Serrote do Pimenta, all found in the same rocky surface. Again, there are differences of stride among an otherwise rather homogeneous association of large theropods. Thus SOES 17, with a SL/FL ratio of 11.6, indicates a much faster gait than the trackways SOES 6 (ratio 5.2), SOES 10 (7.4), SOES 11 (6.5), and SOES 12 (5.6), as observed in the diagram.

The same locality has a trackway of a quadrupedal iguanodontid (*Caririchnium magnificum* holotype, SOES 9) with a rather slow gait, as shown in figure 5.2, and a stride/hind-footprint length ratio of 5.0. Lower

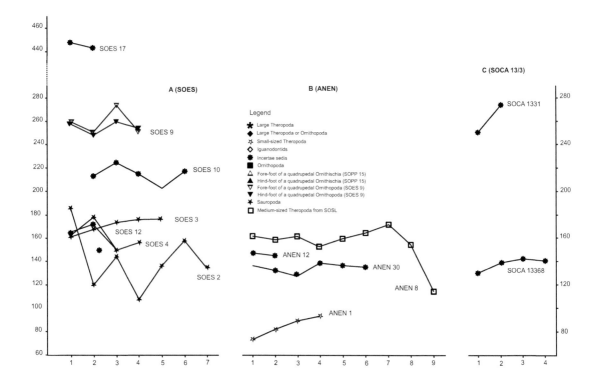

Figure 5.2. Sequence graphs of the gaits of dinosaur trackways from the localities of Serrote do Pimenta (*A*, SOES), Engenho Novo (*B*, ANEN), and Piau-Caiçara (*C*, SOCA).

in the same graph are data for some trackways of medium-size sauropods. The SL/FL ratios for these dinosaurs are clearly low: SOES 100 (formerly SOES 2) = 2.0; SOES 101 (formerly SOES 3) = 3.2; SOES 102 (formerly SOES 4) = 2.7. These animals advanced slowly, and some of them irregularly, in a herd, alongside other individuals whose footprints were discovered and excavated later and whose trackway patterns are not shown here. SOES 100 (formerly SOES 2) was the largest individual among the herd members and followed the other, smaller individuals; consequently, it had a trackway pattern more irregular than the others'.

Figure 5.2*B*, for locality Engenho Novo, shows a trackway (ANEN 8) made either by a bipedal dinosaur or perhaps by a quadruped whose feet totally overprinted tracks made by its hands, pointing to a slow gait with continuous deceleration in the final stretch. The trackmaker was either a relatively small sauropod or a large ornithopod. Despite plotting relatively high in the graph, its relatively long strides are simply a function of the large size of the animal; the dinosaur was in fact moving very slowly, with a SL/FL ratio of only 2.3. Figure 5.2*B* also shows two trackways made by large theropods (ANEN 12 and 30) that were walking with strides of fairly uniform length. Another trackway (ANEN 1) was made by a very small theropod that consistently increased its stride length; its mean SL/FL ratio is 5.4, but the ratio increases from 4.6 at the beginning of the trackway to 5.8 at its end. These values show acceleration.

Figure 5.2*C* for locality Piau-Caiçara (SOCA) presents the data for 2 theropods of different size and kind.

The histogram in figure 5.3A illustrates the frequency distribution of the values of the strides of all the measured dinosaurian trackways. The tracks of doubtful classification are included only in histogram A. The result is a unimodal polygon with positive skew (asymmetry). The mode is 162.5 cm, and the mean is approximately 168 cm—that is, to the right of the mode, as one would predict from the skewness. The positive skew is mainly due to the presence of 2 large theropods that took very long strides (implying a very fast gait). Theropods, mainly those of large size, have a dominant influence on the histogram because of their abundance. Almost half of the values of the stride (45.5%) are between 150 and 200 cm, and nearly 70% of the trackways have strides with values between 100 and 200 cm. The concentration of values is also indicated by the narrowly peaked, non-Gaussian (leptokurtic) shape of the frequency distribution. There are only 2 trackways with strides greater than 275 cm and only a single trackway with a stride less than 75 cm.

These histograms in figure 5.3, in their present form, are useful as summary statements about stride lengths of the dinosaurs at the various sites. To analyze the locomotion of the trackmakers in greater detail would require consideration of additional variables. Stride length is a function of the kind of trackmaker but also of the absolute size of the trackmaker—big animals usually take bigger steps than do small animals—as well as how quickly the animal is moving (J. O. Farlow, personal communication, 2015).

Figure 5.3B, with 62 cases (80.5% of the sample), represents the distribution of the length of the stride of all the theropod trackways in the Sousa basin. It shows a unimodal curve (mode = 187.5 cm and mean = 169.9 cm) with positive asymmetry strongly influenced by the presence of the running theropods. Between the values 150 and 200 cm, 45.2% of the cases are included, and 64.5% are included between 125 and 225 cm.

Figure 5.3C represents the distribution of the length of the stride of all the trackways attributed to ornithopods in the Sousa basin. The histogram, with only a few cases, shows as a result a substantially symmetrical,

Commentary on the Histograms of Figure 5.3

Figure 5.3. Histograms representing the distribution of the length of the stride of the measured dinosaurian trackways of the Sousa basin. A, Overall distribution of the length of the stride of all the measured dinosaurian trackways of the Sousa basin. B, Distribution of the length of the stride of the measured trackways attributed to theropods. C, Distribution of the length of the stride of the measured trackways attributed to ornithopods. D, Distribution of the length of the stride of the measured trackways attributed to sauropods.

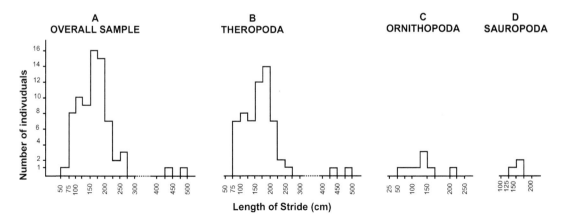

unimodal curve with a mode not so far from that of the histogram A (mode = 162.5 cm and mean = 161.9 cm).

Figure 5.3D is based on the 3 sauropod trackways with measurable strides from the Sousa basin. What data there are suggest a unimodal (mode = 162.5 cm) distribution with negative asymmetry. Six additional sauropod trackways from the same herd were later discovered but are not included in this histogram.

Commentary on the Histograms of Figure 5.4

Figure 5.4 shows histograms for the other main trackway parameters (all trackways: A, B, and C; theropod trackways only: D, E, and F).

The distribution in figure 5.4A, pertaining to the step angle, is quite different from the histograms of figure 5.3. At first glance, this might seem unlikely because the parameters are interdependent, and their values are taken from the same sample of trackways. The distribution for the step angle is strongly asymmetrical (negative skew) while that of the stride is only slightly asymmetrical (positive skew).

The stride length is a function of (1) the kind of trackmaker, (2) the size of the trackmaker, and (3) whether that animal moves with a slow or a rapid gait. Although there has to be some upper limit to the stride length of animals in the real world, this upper limit is not fixed to a particular value by anatomical and mathematical considerations inherent in the way it is measured. In contrast, when an animal is normally walking or running in a straight-line direction (without sudden changes of direction), the step angle is constrained to have a fixed upper limit of 180°. Without this logical upper limit, as the speed increased, the animal would start to cross its legs and possibly stumble. The values of the step angle reveal that the dinosaurs were structurally sophisticated in their locomotion, as one would also deduce from the biomechanics of the hind-limb skeleton. In bipedal dinosaurs, which make up the majority of this sample, the legs moved in a parasagittal manner. Thus 86.5% of the sample has values of the step angle >140°, and 48.6% >160°. This is mainly the case for the theropod trackways (fig. 5.4D), for which the mode of the frequency distribution is 175°, and the mean is 159°36'; the mean is less than the mode, as one would expect in a negatively skewed distribution.

In figure 5.4B, the oblique pace appears to have a rather broader distribution, especially in the left portion, when compared to the histogram for the strides (fig. 5.3). However, this is an artifact of the difference in abscissa interval widths in the two sets of histograms. The curve is almost symmetrical over the interval 20–150 cm but has a slight positive skew. The mode (95 cm), however, is to the right of the mean (91.3 cm). A large number (46.9%) of the values occur in the interval between 80 and 110 cm (n = 30) due to the dominance of theropod data cases (fig. 5.4E). The higher values correspond to the aforementioned runners.

Figure 5.4C illustrates the frequency distribution of values of the external width of the trackway. Values >100 cm correspond to sauropods, whose quadrupedal trackways are rather wide. The most frequent values

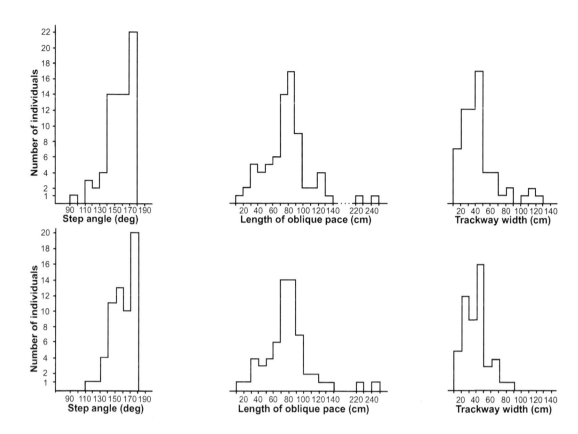

(65.1%) are concentrated between 20 and 50 cm, and almost all of them (88.9%) are between 10 and 70 cm. Values of the external width of the trackway usually cannot be less than the values of footprint width, because if this occurred, the animal would be crossing its legs. Consequently, the values are almost always constrained to values greater than the width of a small footprint and less than an upper limit defined by the combined widths of two large footprints from the same trackway, a left and a right, positioned side by side. Narrower trackways (proportional to the size of the individual) pertain to theropod trackways (fig. 5.3F).

Figure 5.4. Histograms showing the frequency distributions of values of the step angle (A, D), the oblique pace length (B, E), and the external width (C, F) of trackways of the Sousa basin. A–C are for the overall sample, including trackways of uncertain classification. D–F are only for trackways attributed to theropods.

Commentary on the Histograms of Figure 5.5

In figure 5.5, histograms A and B show the frequency distribution of footprint lengths for the total sample of measured footprints in trackways or isolated prints (fig. 5.5A) and for theropods alone (fig. 5.5B). The first histogram, based on 244 individuals, shows a positively skewed distribution, almost entirely due to the presence of sauropod footprints. Most of the sauropod hind-foot prints from the Rio do Peixe basins are 50 to 90 cm long, with an extreme case of 120 cm. The collected distribution for bipedal dinosaurs only—that is, theropods (the majority) and iguanodontids—would instead be more symmetrical, as seen in the histogram for theropods alone (fig. 5.5B). The size-frequency distribution for footprints of sauropods needs to be completely updated by inclusion of new data for a herd in the Riacho do Pique at Serrote do Letreiro (Carvalho 2000b),

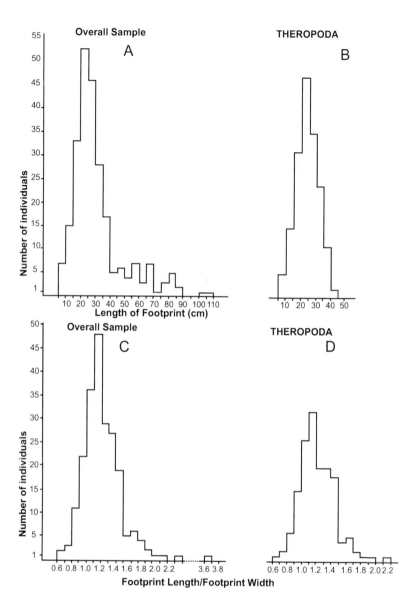

Figure 5.5. Histograms representing the distribution of the length of the footprint of: *A*, All the measurable individual tracks of the Rio do Peixe basins (in trackways [mean length] and as isolated footprints), including those of uncertain classification (n = 244). *B*, The same, only for theropods. *C–D*, Histograms representing the distribution of the ratio or index: length of the footprint/width of the footprint. *C*, All the measurable individuals of the Rio do Peixe basins (in trackways and as isolated footprints), including those of uncertain classification (n° = 244). *D*, The same, but for theropods only.

as well as on the basis of excavations in 1988 in the Serrote do Pimenta and the discovery of more material at Piau. However, we predict that the results would not be very different from what is reported here.

The central values of figure 5.5A are concentrated between 15 and 35 cm (65.6%) and the interval 20–30 cm encloses 40.6% of all the footprints. More than half of the footprints (52%) are in the interval of 20 to 35 cm. Analogous percentile values and higher are found in histogram B (theropods)—for instance, 80.6% of the footprints show values included in the interval between 15 and 35 cm.

Histograms C and D respectively show the numerical distribution of the footprint L/W ratio of all the footprints from the Rio do Peixe basins, and of theropod footprints only. The curve indicated in histogram C shows positive asymmetry but is more symmetrical than that of histogram

A. This is because sauropods show a footprint L/W index in histogram C close to that of theropods. In fact, in the Rio do Peixe ichnofauna, many theropod footprints are quite broad due to the high total digital divergence, in contrast to what is seen in many other dinosaurian ichnofaunas. The most notable value of the index is 1.0, which represents footprints whose length is equal to their width. Smaller values indicate footprints that are wider than they are long (17.4%), which represent, for the most part, theropods with the highest digital divarication (II^IV) or deformed sauropod footprints. Values greater than 1.0 indicate footprints that are longer than they are wide (82.6%). This latter situation is to be expected for an ichnoassociation of dinosaurs, mainly bipeds; however, it is quite different from expectations for an ichnoassociation of theropsids (therapsids + mammals). Thus, in the case of the almost contemporaneous ichnofauna of the Botucatu Formation (Lower Cretaceous of the Paraná basin, Brazil and Paraguay), there would be, with all probability, a large majority of footprints of the early mammal kind, with footprint FL/FW index values <1; there would be only a few of the dinosauroid kind with values >1 (Leonardi 1981d, 1989a, 1992, 1994; Leonardi and Carvalho 1999; Leonardi and Godoy 1980; Leonardi and Sarjeant 1986; Leonardi et al., 2007a, 2007b; Fernandes et al. 2011a, 2011b; Francischini et al. 2015; Leonardi and Carvalho 2020a, 2020b). Analogous considerations exposed about histogram C can be applied in histogram D, for theropods only.

Bivariate Relationships

Diagram A of figure 5.6 illustrates the relationship between stride and footprint length for all the trackways of the basins. Unsurprisingly, the two variables are positively correlated: big dinosaurs generally took longer steps than small dinosaurs. There is, however, a great deal of scatter in the relationship, without a single tight, well-defined axis. This is due in large part to the fact that data for more than one kind of trackmaker are plotted. Points associated with different trackmaker groups tend to plot in distinct regions of the graph, forming individualized polygons, each with its own trend axis.

Points for three sauropods from the same herd (SOES 100, 101, 102; other trackmakers in this group were found and excavated later but are not shown in this graph) cluster together near the top of the graph, indicating that these were the dinosaurs with the greatest footprint lengths. These big animals were of similar but not equal size. As big as they were, these dinosaurs had relatively short stride lengths.

A cluster of seven points, referring to the ornithopods, defines a long polygon nearly parallel to the axis of stride length (beneath the sauropod cluster); the rightmost point is associated with the hind-foot portion of the trackway of a quadruped attributed to iguanodontids (SOES 9). Another point associated with a long footprint length and stride, probably attributable to an ankylosaur or a quadrupedal ornithopod, is that of SOPP 15. Points associated with the forefoot prints of these trackways plot lower on the graph due to their smaller size.

Figure 5.6. Relationship between stride length and footprint length for all trackways of the Sousa basin, including those of uncertain classification. *A*, Log-transformed mean stride length vs. log-transformed mean footprint length. *B*, Log-transformed stride/footprint length ratio vs. log-transformed mean footprint length. In this and other graphs in this section, polygons are defined by connecting points around the periphery of data cases for tracks of the different categories of dinosaurs.

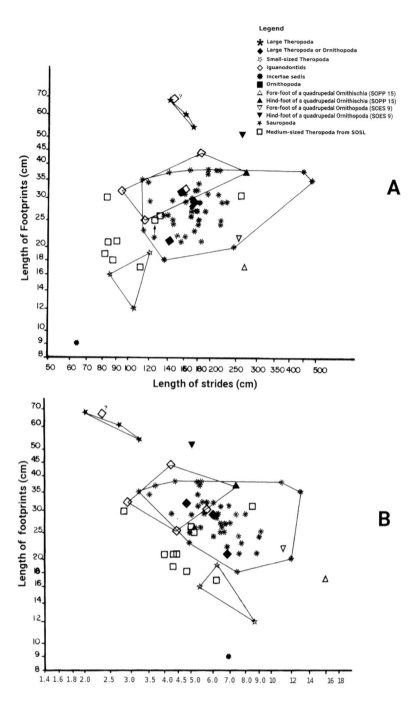

Mostly below but also overlapping the ornithopod cluster, a large polygon encloses points for theropods of mostly large size. The larger individuals in this cluster were presumably nearly or fully grown individuals, but there is a substantial size range of footprints, possibly indicating the presence of more than one species of trackmaker, or animals of different ontogenetic age. For example, the smallest individual, which plots at the bottom of the polygon, may have been younger than most. Most of the points inside this polygon group toward the left side of the graph, representing theropods that were walking slowly; only a few points are at the far right, representing animals that were taking progressively longer strides and so moving at faster speeds. The two right-most cases (trackways SOPP 3 and SOES 17) indicate dinosaurs that were running, with a suspended phase in which both feet were off the ground.

A group of only three points beneath the large-theropod cluster corresponds to trackways of small theropods with a long digit III. The three points admittedly are not particularly close together, and the cluster is fairly separated from the others due to the smaller size of these dinosaurs.

Some widely scattered points relate to a group of 10 small- to medium-size theropods from Serrote do Letreiro. The point represented by the coordinates 260, 3 (SOCA 1331), although based on an incomplete trackway, must probably belong to another group of theropods.

An interesting feature of the graph is that apart from the SOSL grouping (4), there are very few points for smaller theropods. Note, for example, the truncated left side of the polygon comprising points of the large theropod cluster, showing a general lack of smaller individuals with footprint morphologies that would put them in this group.

There is, however, one point for a very small dinosaur that plots close to the origin of the graph (with coordinates 62, 9). This trackway (SOCA 153) is of uncertain classification and labeled incertae sedis in the graph. It is positioned near the group of small theropod footprints.

The trackway with coordinates 148, 69 (ANEN 8) was initially thought, despite the lack of morphological details, to belong to ornithopods. However, diagram A places this trackway in the cluster of points for sauropods. The footprints in this trackway are large and rounded, with displacement rims.

To conclude with a general observation, points for sauropods seem to be distinct from the other groups because they clearly plot at the left but very high in the diagram, as would be expected for slow and heavy animals. In contrast, there is substantial overlap between the clusters for ornithopods and large theropods. Poor preservation and the trackways' lack of morphological details routinely make distinguishing between these two groups difficult. The long axes of the two clusters are almost parallel but do not coincide, indicating that the ornithopods had a proportionally shorter stride for a given footprint length.

Figure 5.6B shows these relationships in a slightly different manner, plotting the SL/FL ratio against FL. The sauropod trackways are even more clearly distinct from the others than in figure 5.6A, and the way trackway ANEN 8 plots among them increases the suspicion that this trackway could in fact have been made by a sauropod. Values for the forefeet portions of the quadruped trackways (SOES 9 and SOPP 15) are positioned well away from points for the corresponding hind-feet portions of the same trackways due to the large difference in size between the two kinds of footprint. As in figure 5.6A, large theropods and large ornithopods show a large amount of overlap. Points for the large theropods are less correlated than in figure 5.6A. The cluster of small theropod tracks is well separated from the large theropod cluster.

Even with this kind of bivariate diagram, we could not obtain complete separation of all the clusters of points. A major reason for this is that the large majority of the dinosaurs in the sample are bipedal animals that were nearly or fully grown individuals of not so different size, as one can see better in the histograms.

Diagram A of figure 5.7 studies the correlation between the length of stride and the index FL/FW. It is the least successful separator of groups of the bivariate relationships examined here; only the small theropods present a fairly distinct cluster. The points related to the medium-size theropods of Serrote do Letreiro plot alongside or overlap the small theropod track polygon.

Diagram B of figure 5.7 shows the correlation between the length of the footprints and the width of footprints, including isolated footprints as well as footprints in the trackways (mean value) in all the localities of the basin. All the points are arranged in a long elliptical cluster, indicating good correlation of the data, despite the fact that the data represent a rather heterogeneous ichnofauna.

The points are distributed in regions showing partial overlap. The points for sauropods form a narrow polygon, indicating a good correlation, except one point for a deformed footprint (coordinates 80, 22). The majority of the points are concentrated in the central area (values 52–88 cm for hind-foot print length; values 45–80 cm for the width). The cluster partly overlaps that of ornithopods, and two of the sauropod points intersect the cluster of large theropods. In general, the large size takes the sauropods to a clear dimensional distinction.

The long cluster of the ornithopods also indicates a good correlation of the two parameters. This last cluster and that of the large theropods show considerable overlap, but about two-thirds of the ornithopod points are outside of the larger region where the large theropod points are concentrated.

The large theropod polygon indicates, once again, a low level of correlation and, therefore, high variability. Although this may depend in part on the wide range of footprint sizes within this cluster, it probably also reflects a relatively high number of distinct forms of trackmakers. There is a strong concentration of points (~60%) in the central region, with footprint lengths between 25 and 35 cm and widths between 17 and 34 cm. The sample of this central area shows a tighter correlation between print length and width.

The small theropod points, which are more numerous in this diagram, are scattered throughout a relatively wide polygon, indicating low correlation and/or problems of identification/classification in discriminating between small and large theropod tracks. The points representing the medium-size theropods of the Serrote do Letreiro site are situated in the fringe of overlap between both groups. Despite the analysis presented in our four bivariate graphs, the nature of this population remains uncertain,

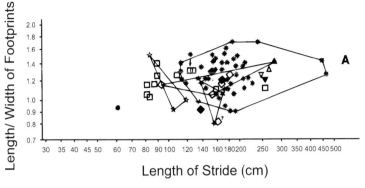

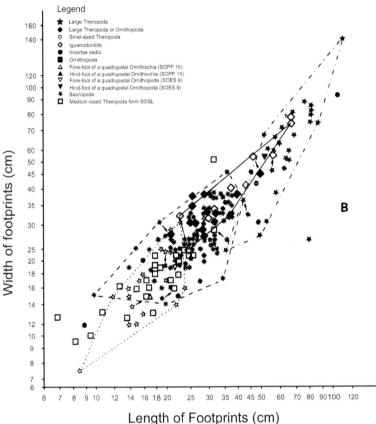

Figure 5.7. *A.* Relationship between mean length of the stride and the ratio mean length/mean width of the footprint, for all the trackways of the Sousa basin, including those of uncertain classification. The scale is logarithmic in both the axes. *B,* Correlation between the length and width of all the measurable individual tracks of the Sousa basin (trackways and isolated footprints), including those of uncertain classification. The scale is logarithmic in both the axes.

as it was from the morphologic and taxonomic point of view (Leonardi 1979b).

The slope of the sauropod relationship is steep, with its axis forming an inclination angle of approximately 65° with the abscissa. The corresponding axis for the small theropods is a bit less, with about 55° of dip (inclination angle). For the large theropods, the dip of the axis is around 50°. In these three groups, therefore, the data indicate that with increasing size, footprint width increases more rapidly than length. In the ornithopods, in contrast, the dip is around 45°, indicating that both parameters increase at the same rate.

Conclusions of the Statistical Analysis

· The heterogeneity of some ichnofaunas and the homogeneity of others are well emphasized by the study of the standard deviations and of the variability coefficient. The heterogeneity can depend on the presence of different forms (real heterogeneity) and/or of different gaits (biomechanical or behavior heterogeneity).

· A sample of the angular values of the step angle is always much more homogeneous than the values of the linear parameters of the same trackway (stride, oblique pace, etc.). In contrast, the angular values of the cross-angle of the footprint and the digital divarication are always rather heterogeneous. This last fact depends, without a doubt, on the mobility of the digits in the bipedal dinosaurs, especially in the theropods and on the different substrates.

· In some cases, the different localities show samples with evidence of different collective behavior: groups of both slower and faster dinosaurs, although these groups do not always seem to have been organized in packs or herds. Perhaps this diverse collective behavior must be attributed to the different geographic and environmental situations.

· The graphs are effective tools for the study of gait and of locomotor behavior. In the sample for Passagem das Pedras, for example, these graphs indicate, among other things, rather different gaits among individuals of same ichnospecies (on the same rocky surface, although on two contiguous levels).

· The study of several statistical functions and of graphs and diagrams allows us to confirm that the sauropods of Serrote do Pimenta marched in a herd.

· Examination of the bivariate graphs of two variables allowed better classification of some previously unidentified trackways and footprints. Even so, discriminating between footprints of large theropods and large ornithopods remains difficult. This is true even with graphs that compare correlate footprint shape indexes with linear data.

· Histograms of linear footprint dimensions generally show positive asymmetry. In contrast, the histogram of the step angle shows negative asymmetry because of the virtual upper limiting value of $180°$.

· All the studied dinosaurs show a sophisticated, efficient gait. The sauropods are less efficient than the bipedal forms.

· In a sample of footprints of bipedal dinosaurs, the print L/W ratio index is almost always greater than 1.0. In a sample of prints from Mesozoic synapids, the opposite would be true.

· The importance of the correlation between the stride and the length of the footprint was demonstrated. It defines the gait of the trackmakers as well as their efficiency as walkers or runners.

- The bivariate relationship graphs demonstrate the presence of a substantial diversity of large theropods. The other groups are more homogeneous.
- Examination of the correlation axis between two variables allows better resolution of the systematic groups of the trackmakers than the study of the corresponding clusters and polygons.
- We suggest a more detailed study in the future to examine the data relating to digits themselves and to their relationships, positions, and functional significance (see Lallensack et al. 2019).
- It is important to emphasize once again the large number of localities with dinosaur tracks in the Rio do Peixe basins (18 at the time when the present study of measurement and statistic was done; 37 as of 2021); of ichnofossiliferous levels (~96 in 2021); of individuals (535 in 1987); and of different forms. These numbers will greatly increase if a permanent team located in Sousa will continue systematic explorations and excavations in the Sousa, Pombal, and Uiraúna-Brejo das Freiras basins.
- The measurement of the numberless data contained in the tables of this volume gave prominence to many of the problems and limits stressed by Falkingham in his chapter on objective ichnologic methods (Falkingham et al. 2016) and proved—in advance—the wisdom of many of his suggestions.

We close this chapter by quoting and paraphrasing James O. Farlow's funny phrase in the "Final Thoughts" section of his gigantic and precious book *Noah's Ravens* (2018). While not achieving his level and his results, we can say, at the end of this long statistical section and at the conclusion of many measurements and tables of data laboriously obtained in the field, "We have used lots of statistics and tried very hard not to lie." We were enchanted to read this work and our single regret is that his book came out only after ours had been delivered to our publisher and major changes to our work were no longer allowed. But so life goes.

Table 5.1 Codes of the Main Sousa basin Ichnosites

CODE	MUNICIPALITY	LOCALITY	FORMATION
ANAC	São João do Rio do Peixe	Araçá de Cima	Sousa
ANAR	São João do Rio do Peixe	Aroeira	Sousa
ANDB	São João do Rio do Peixe	Barragem do Domício	Sousa
ANCA	São João do Rio do Peixe	Cabra Assada	Sousa
ANEN	São João do Rio do Peixe	Faz. Engenho Novo	Sousa
ANJU	São João do Rio do Peixe	Sítio Juazeirinho	Sousa
ANPV	São João do Rio do Peixe	Poço da Volta	Sousa
APTA	Aparecida	Tapera	Sousa
ANZO	São João do Rio do Peixe	Zoador	Sousa
APVR II	Aparecida	Várzea dos Ramos II	Sousa
APVR III	Aparecida	Várzea dos Ramos III	Sousa
SORO	Sousa	Riachão dos Oliveira	Sousa
SOAB	Sousa	Abreu	Sousa
SOBP	Sousa	Baixio do Padre	Sousa
SOCA	Sousa	Faz. Caiçara-Piau	Sousa
SOCV	Sousa	Curral Velho	Rio Piranhas
SOES	Sousa	Fazenda Estreito	Antenor Navarro
SOFB	Sousa	Floresta dos Borba	Antenor Navarro
SOFP	Sousa	Fazenda Paraíso	Rio Piranhas
SOLF I	Sousa	Lagoa do Forno I	Sousa-Rio Piranhas
SOLF II	Sousa	Lagoa do Forno II	Sousa-Rio Piranhas
SOMA	Sousa	Matadouro Municipal	Sousa
SOMD	Sousa	Mãe d'Água	Rio Piranhas
SOPA	Sousa	Piau II	Sousa
SOPE	Sousa	Pedregulho	Sousa
SOPI	Sousa	Piedade	Sousa
SOPM	Sousa	Poço do Motor	Sousa
SOPP	Sousa	Passagem das Pedras	Sousa
SORC	Sousa	Riacho do Cazé	Antenor Navarro
SOSA	Sousa	Sítio Saguim	Sousa
SOSL	Sousa	Serrote do Letreiro	Antenor Navarro
SOVR	Aparecida	Várzea dos Ramos	Sousa

Table 5.2 Classification of the Trackways and isolated Footprints, Rio do Peixe basins

Track	Footprint	Classification	Track	Footprint	Classification	Track	Footprint	Classification	Track	Footprint	Classification
ANAR	1	?GOr	SOCA	21	?LTh	SOCA	9991	LTh	SOCA	13244	MTh
ANBD	1	LTh	SOCA	22	?LTh	SOCA	13215	STh	SOCA	13245	IS
ANBD	2	LTh	SOCA	30	?GOr	SOCA	1323	STh	SOCA	13246	STh
ANBD	3	LTh	SOCA	33	STh	SOCA	13236	STh	SOCA	13247	?STh
ANCA	1	LTh	SOCA	30b	?MTh	SOCA	13499	GOr	SOCA	13248	MTh
ANCA	2	LTh	SOCA	31b	LTh	SOCA	13498	LTh/GOr	SOCA	13249	IS
ANCA	3	LTh	SOCA	32b	LTh	SOCA	132b	STh	SOCA	13250	MTh
ANCA	4	LTh	SOCA	41	IS	SOCA	1321	LTh	SOCA	13251	LTh
ANCA	5	LTh	SOCA	42	LTh	SOCA	1322	MTh	SOCA	13252	LTh/GOr
ANCA	5	LTh	SOCA	422	IS	SOCA	1323	MTh	SOCA	13253	MTh
ANCA	6	LTh	SOCA	43	GOr	SOCA	1324	MTh	SOCA	13254	IS
ANCA	7	GOr	SOCA	44	LTh	SOCA	1325	MTh	SOCA	13255	LTh
ANCA	8	LTh	SOCA	45	LTh	SOCA	1236	MTh	SOCA	13256	IS
ANCA	9	LTh	SOCA	452	LTh	SOCA	1327	IS	SOCA	13257	IS
ANCA	10	LTh	SOCA	46	GOr	SOCA	1328	IS	SOCA	13258	IS
ANCA	11	LTh	SOCA	47	MTh	SOCA	1329	IS	SOCA	13259	LTh
ANCA	12	LTh	SOCA	48	LTh	SOCA	13210	IS	SOCA	13261	MTh
ANCA	13	LTh	SOCA	49	LTh	SOCA	13211	MTh	SOCA	13262	LTh
ANEN	1	STh	SOCA	491	LTh/GOr	SOCA	13212	MTh	SOCA	13263	IS
ANEN	2	LTh	SOCA	492	MTh	SOCA	13214	STh	SOCA	13264	?LTh
ANEN	3	LTh	SOCA	493	LTh/GOr	SOCA	13215	MTh	SOCA	13265	MTh
ANEN	4	LTh	SOCA	52	IS	SOCA	13216	MTh	SOCA	13266	LTh
ANEN	5	LTh	SOCA	50b	MTh	SOCA	13217	LTh	SOCA	13267	IS
ANEN	6	IS	SOCA	51b	LTh	SOCA	13218	IS	SOCA	13268	LTh/GOr
ANEN	7	LTh	SOCA	54b	LTh/GOr	SOCA	13219	IS	SOCA	13269	MTh
ANEN	8	?GOr	SOCA	57b	LTh	SOCA	13220	LTh	SOCA	13270	LTh-Hs
ANEN	9	LTh	SOCA	59b	IS	SOCA	13221	MTh	SOCA	13271	LTh
ANEN	10	Sau	SOCA	61	LTh/GOr	SOCA	13222	MTh	SOCA	13272	LTh
ANEN	11	?LTh	SOCA	71	LTh	SOCA	13223	LTh	SOCA	13273	LTh
ANEN	12	LTh	SOCA	72	IS	SOCA	13224	MTh	SOCA	13274	LTh
ANEN	13	LTh	SOCA	73	LTh/GOr	SOCA	13225	MTh	SOCA	13275	?GOr
ANEN	14	LTh	SOCA	75	LTh	SOCA	13226	MTh	SOCA	13276	MTh
ANEN	15	LTh	SOCA	76	LTh	SOCA	13227	STh	SOCA	13277	IS
ANEN	16	LTh	SOCA	78	?GOr	SOCA	13228	MTh	SOCA	13278	LTh/Gor
ANEN	17	LTh	SOCA	79	IS	SOCA	13229	IS	SOCA	13279	IS
ANEN	18	LTh	SOCA	91	LTh	SOCA	13230	STh	SOCA	1331	MTh
ANEN	19	LTh	SOCA	92	LTh/GOr	SOCA	13231	STh	SOCA	1332	?LTh
ANEN	30	LTh	SOCA	94	LTh-Hs	SOCA	13232	MTh	SOCA	1333	LTh/GOr
ANEN	31	LTh	SOCA	97	LTh	SOCA	13233	IS	SOCA	1334	?STh
ANEN	32	LTh	SOCA	98	LTh	SOCA	13234	STh	SOCA	1335	LTh
ANEN	33	LTh	SOCA	99	LTh	SOCA	13235	IS	SOCA	1336	LTh
ANEN	34	LTh	SOCA	991	LTh/GOr	SOCA	13236	STh	SOCA	1337	LTh
ANEN	35	?LTh	SOCA	993	LTh	SOCA	13237	MTh	SOCA	1338	STh
ANJU	1	LTh	SOCA	9931	LTh	SOCA	13238	IS	SOCA	1339	IS
ANJU	2	LTh	SOCA	9932	LTh	SOCA	13239	?MTh	SOCA	13310	Sau
ANJU	3	LTh	SOCA	994	LTh	SOCA	13240	LTh	SOCA	13311	Sau
ANJU	4	LTh	SOCA	995	LTh	SOCA	13241	LTh	SOCA	13312	Sau
ANPV	1	LTh	SOCA	996	IS	SOCA	13242	STh	SOCA	13313	Sau
ANZO	1	MTh	SOCA	997	LTh-Hs	SOCA	13243	MTh	SOCA	13314	Sau

(continued on the next page)

Table 5.3 Classification of the Trackways and isolated Footprints, Rio do Peixe basins

Track	Footprint	Classification	Track	Footprint	Classification	Track	Footprint	Classification
SOCA	13315	IS	SOCA	13364	Sau	SOCA	1370	LTh
SOCA	13316	IS	SOCA	13366	STh	SOCA	140	IS
SOCA	13317	STh	SOCA	13367	Sau	SOCA	150	LTh/GOr
SOCA	13318	Sau	SOCA	13368	LTh	SOCA	151	LTh
SOCA	13319	MTh	SOCA	13369	LTh/GOr	SOCA	152	LTh
SOCA	13320	?STh	SOCA	13370	Sau	SOCA	153	IS
SOCA	13321	STh	SOCA	13371	IS	SOCA	154	STh
SOCA	13322	IS	SOCA	13372	?STh	SOCA	1601-1620	LTh-Hs
SOCA	13323	?Sau	SOCA	13373	IS	SOCA	1621	Or-Hs
SOCA	13324	Sau	SOCA	13374	IS	SOCA	1622	LTh-Hs
SOCA	13325	STh	SOCA	13375	STh	SOCA	1632	LTh
SOCA	13326	LTh	SOCA	13376	?STh	SOCA	1633-1655	LTh-Hs
SOCA	13327	Sau	SOCA	13377	Sau	SOCA	1656	LTh
SOCA	13328	Sau	SOCA	13378	STh	SOCA	171	?GOr
SOCA	13329	Sau	SOCA	13379	STh	SOCA	172	GOr
SOCA	13330	IS	SOCA	13380	STh	SOCA	173	LTh
SOCA	13331	IS	SOCA	13381	STh	SOCA	174	IS
SOCA	13332	STh	SOCA	13382	?STh	SOCA	175	LTh
SOCA	13333	STh	SOCA	13383	?STh	SOCA	176	LTh
SOCA	13334	Croc	SOCA	13384	?Sau	SOCV	1	LTh
SOCA	13335	Croc	SOCA	13385	Sau	SOCV	2	LTh
SOCA	13336	LTh	SOCA	13386	Sau	SOCV	3	STh
SOCA	13337	Sau	SOCA	13387	?STh	SOCV	4	IS
SOCA	13338	LTh	SOCA	13388	STh	SOCV	5	STh
SOCA	13339	Sau	SOCA	15389	?LTh	SOCV	6	LTh
SOCA	13340	Sau	SOCA	1341	Sau/GOr	SOCV	7	LTh
SOCA	13341	?Sau	SOCA	1342	Sau/GOr	SOCV	8	GOr
SOCA	13342	Sau	SOCA	1350	LTh	SOCV	9	GOr
SOCA	13343	Sau	SOCA	1350b	IS-Hs	SOCV	10	GOr
SOCA	133431	MTh	SOCA	1360	LTh	SOCV	11	LTh/GOr
SOCA	13344	LTh	SOCA	1361	LTh/GOr	SOCV	15	LTh
SOCA	13345	MTh	SOCA	1362	LTh	SOES	1	LTh
SOCA	13346	STh	SOCA	1363	LTh	SOES	2	Sau
SOCA	13347	STh	SOCA	1364/1	LTh	SOES	200	Sau
SOCA	13348	IS	SOCA	1364/2	LTh	SOES	201	Sau
SOCA	13349	STh	SOCA	1365	IS	SOES	202	Sau
SOCA	13350	MTh	SOCA	1366	LTh	SOES	3	Sau
SOCA	13351	STh	SOCA	1367	LTh	SOES	4	Sau
SOCA	13352	Sau	SOCA	1368	?LTh	SOES	5	Sau
SOCA	13353	STh	SOCA	1369	LTh	SOES	6	LTh
SOCA	13354	MTh	SOCA	13690	IS	SOES	7	Ank-q
SOCA	13355	STh	SOCA	13691	LTh/GOr	SOES	8	GOr
SOCA	13356	STh	SOCA	13692	?LTh	SOES	9	GOr-q
SOCA	13357	STh	SOCA	13693	?LTh	SOES	10	LTh
SOCA	13359	Sau	SOCA	13694	LTh	SOES	11	LTh
SOCA	13360	LTh/GOr	SOCA	13695	GOr	SOES	12	LTh
SOCA	13361	IS	SOCA	13696	LTh	SOES	15	Lac
SOCA	13362	IS	SOCA	13697	LTh	SOES	16	IS
SOCA	13363	Sau	SOCA	13698	IS	SOES	17	LTh

(continued on the next page)

Table 5.4 Classification of the Trackways and isolated Footprints, Rio do Peixe basins

Track	Footprint	Classification	Track	Footprint	Classification	Track	Footprint	Classification
SOES	18	LTh	SOPP	1 (-A)	GOr-q	SOSL	4997	MTh
SOES	20	MTh	SOPP	2 (-B)	LTh	SOSL	4998	MTh
SOES	21	?GOr	SOPP	3 (-C)	LTh	SOSL	4999	MTh
SOES	22	LTh	SOPP	4 (-D)	LTh	SOSL	49990	MTh
SOES	23	LTh	SOPP	5 (-E)	LTh	SOSL	49991	MTh
SOES	24	LTh	SOPP	6 (-F)	GOr	SOSL	49992	MTh
SOES	25	LTh	SOPP	7	LTh	SOSL	49993	MTh
SOES	26	?GOr	SOPP	7b	IS	SOSL	49994	MTh
SOES	27	LTh	SOPP	8	LTh	SOSL	49995	MTh
SOES	28	?LTh	SOPP	9	?LTh	SOSL	49996	MTh
SOES	29	?LTh	SOPP	9bis	IS	SOSL	49997	MTh
SOES	30	LTh	SOPP	iter	LTh	SOSL	5	MTh
SOES	31	?LTh	SOPP	10	LTh/GOr	SOSL	7	IS
SOES	32	LTh	SOPP	11	LTh	SOSL	8	GOr
SOES	33	LTh	SOPP	12	LTh	SOSL	9	LTh
SOES	34	LTh	SOPP	13	LTh	SOSL	10	IS
SOES	35	LTh	SOPP	14	LTh	SOVR	1-15	LTh
SOES	36	LTh	SOPP	15	IS-q	SOVR	16	?GOr
SOES	37	LTh	SOPP	16	IS	SOVR	17	LTh
SOES	38	?LTh	SOPP	17	LTh/GOr			
SOES	39	?LTh	SOSL	1	?GOr			
SOES	40	LTh	SOSL	2	MTh			
SOES	41	LTh	SOSL	3	IS			
SOES	42	LTh	SOSL	40	MTh			
SOES	43	IS	SOSL	41	MTh			
SOES	44	LTh	SOSL	42	MTh			
SOMA	1	LTh	SOSL	43	MTh			
SOMA	2	LTh	SOSL	44	MTh			
SOMA	3	?LTh	SOSL	45	MTh			
SOMA	4	IS	SOSL	46	MTh			
SOMD	1	GOr	SOSL	47	MTh			
SOMD	2	LTh	SOSL	48	MTh			
SOMD	3	IS	SOSL	49	MTh			
SOMD	4	IS	SOSL	491	MTh			
SOPE	1	LTh	SOSL	492	MTh			
SOPE	2	LTh	SOSL	493	MTh			
SOPI	1	LTh	SOSL	494	MTh			
SOPI	2	LTh	SOSL	495	MTh			
SOPI	3	GOr	SOSL	496	MTh			
SOPI	4	LTh/GOr	SOSL	497	MTh			
SOPI	5	LTh	SOSL	498	MTh			
SOPI	6	LTh	SOSL	499	MTh			
SOPI	7	IS	SOSL	4990	MTh			
SOPM	1	LTh	SOSL	4991	MTh			
SOPM	4	LTh	SOSL	4992	MTh			
SOPM	5	LTh	SOSL	4993	MTh			
SOPM	6	LTh	SOSL	4994	MTh			
SOPM	7	LTh	SOSL	4995	MTh			
SOPM	8	LTh	SOSL	4996	MTh			

Legend

? = uncertain classification
ANAR 1 = it is the code nº of a track
AN = municipality of Antenor Navarro
AR = locality Aroeira
1 = track#1 of the locality
Ank = Ankylosauria
B = bis (=twice)
STh = Small Theropoda with long digit III
Croc = Crocodilia
IS = Incertae Sedis
Lac = Lacertoid track
LTh = large Theropoda
GOr = Graviportal Ornithopoda
Sau = Sauropoda
Hs = Half-swimming
SOr = Small Ornithopoda
MTh = Medium-sized Theropoda
/ = or
q = Quadruped

Table 5.5 Parameters of the Trackways from São João do Rio do Peixe County

STRIDE

N°	ANBD 01	ANBD 02	ANBD 03	ANJU* 02	ANJU 03	ANJU 04	ANPV 01
1-3	156.5	150.0	173.0	192.3	97.8	212.5	173.5
2-4	-	-	181.5	-	121.0	-	-
3-5	-	-	-	-	144.3	-	-
n	1	1	2	1	3	1	1
min	-	-	173.0	-	97.8	-	-
max	-	-	181.5	-	144.3	-	-
M	-	-	175.3	-	121.0	-	-
σn-1	-	-	6.0	-	23.3	-	-

STEP ANGLE

N°	ANBD 01	ANBD 02	ANBD 03	ANJU* 02	ANJU 03	ANJU 04	ANPV 01
1-3	158°	-	162°	172°	134°	159°	179°
2-4	-	-	164°	-	141°	-	-
3-5	-	-	-	-	149°	-	-
n	1	0	2	1	3	1	1
min	-	-	162°	-	134°	-	-
max	-	-	164°	-	149°	-	-
M	-	-	163°	-	141°20'	-	-
σn-1	-	-	1.4	-	7.5	-	-

EXTERNAL WIDTH

N°	ANBD 01	ANBD 02	ANBD 03	ANJU* 02	ANJU 03	ANJU 04	ANPV 01
1-3	30.8	-	34.5	32.3	-	48.0	-
2-4	-	-	34.5	-	-	-	-
3-5	-	-	-	-	-	-	-
n	1	0	2	1	0	1	0
min	-	-	-	-	-	-	-
max	-	-	-	-	-	-	-
M	-	-	34.5	-	-	-	-
σn-1	-	-	0.0	-	-	-	-

INTERNAL WIDTH

N°	ANBD 01	ANBD 02	ANBD 03	ANJU* 02	ANJU 03	ANJU 04	ANPV 01
1-3	-	-	-	-19.3	-9.3	-	-
2-4	-	-	-	-	-	-	-
3-5	-	-	-	-6.5	-	-	-
n	0	0	0	1	1	0	-
min	-	-	-	-	-	-	-
max	-	-	-	-	-	-	-
M	-	-	-	-	-	-	-
σn-1	-	-	-	-	-	-	-

OBLIQUE PACE

N°	ANBD 01	ANBD 02	ANBD 03	ANJU* 02	ANJU 03	ANJU 04	ANPV 01
1-2	77.5	-	82.0	96.2	53.3	109.0	87.5
2-3	81.3	-	92.0	-	53.0	119.0	85.8
3-4	-	-	91.5	-	75.0	-	-
σn-1	-	-	-	-	74.8	-	-
n	2	0	3	1	4	2	2
min	77.5	-	82.0	-	53.0	109.0	85.8
max	81.3	-	92.0	-	75.0	119.0	87.5
M	79.4	-	88.5	-	64.1	114.0	86.7
σn-1	2.7	-	5.6	-	12.6	7.1	1.2

DIVERGENCE OF FOOTPRINTS FROM THE MIDLINE

N°	ANBD 01	ANBD 02	ANBD 03	ANJU* 02	ANJU 03	ANJU 04	ANPV 01
1	-11°30'	-	-8°	-20°30'	9°	-3°	0°
2	7°30'	-	2°	20°30'	-	-7°	0°
3	4°30'	-	-5°	-	1°	-	0°
4	-	-	-2°	-	-18°	-	-
5	-	-	-	-	-	-	-
n	3	0	4	2	3	2	3
min	-11°30'	-	-8°	-20°30'	-18°	-7°	0°
max	7°30'	-	2°	20°30'	9°	-3°	0°
M	0°10'	-	-3°30'	0°00'	-2°40'	-5°	0°
σn-1	10.2	-	4.3	29.0	13.9	2.8	0

GAUGE

N°	ANBD 01	ANBD 02	ANBD 03	ANJU* 02	ANJU 03	ANJU 04	ANPV 01
1-3	15.8	-	14.0	6.9	21.7	19.5	1.8
2-4	-	-	11.0	-	20.5	-	-
3-5	-	-	-	-	20.5	-	-
n	1	0	2	1	3	1	1
min	-	-	11.0	-	20.5	-	-
max	-	-	14.0	-	21.7	-	-
M	-	-	12.5	-	20.9	-	-
σn-1	-	-	2.1	-	0.7	-	-

* Data obtained by extrapolation

(continued on the next page)

Table 5.6 Parameters of the Footprints from São João do Rio do Peixe County

N°	LENGTH OF THE FOOTPRINT						
	ANBD 01	ANBD 02	ANBD 03	ANJU 02	ANJU 03	ANJU 04	ANPV 01
1	-	-	-	26.8	-	37.0	-
2	-	-	25.3	27.9	-	39.0	22.5
3	21.0	22.0	28.3	-	-	-	22.5
4	-	-	26.3	-	30.3	-	-
5	-	-	-	-	27.5	-	-
n	1	1	3	2	2	2	2
min	-	-	25.3	26.8	27.5	37.0	-
max	-	-	28.3	27.9	30.3	39.0	-
M	-	-	26.6	27.4	28.9	38.0	22.5
σn-1	-	-	1.5	0.8	2.0	1.4	0

	WIDTH OF THE FOOTPRINT						
	ANBD 01	ANBD 02	ANBD 03	ANJU 02	ANJU 03	ANJU 04	ANPV 01
1	-	-	-	27.9	-	28.5	-
2	14.5	-	21.3	24.3	-	27.8	21.0
3	14.0	17.8	22.0	-	-	-	18.8
4	-	-	-	-	25.5	-	-
5	-	-	-	-	23.8	-	-
n	2	1	2	2	2	2	2
min	14.0	-	21.3	24.3	23.8	27.8	18.8
max	14.5	-	22.0	27.9	25.5	28.5	21.0
M	14.3	-	21.7	26.1	24.7	28.2	19.9
σn-1	0.4	-	0.5	2.5	1.2	0.5	1.6

	CROSS-AXIS ANGLE						
	ANBD 01	ANBD 02	ANBD 03	ANJU 02	ANJU 03	ANJU 04	ANPV 01
1	-	-	101°	84°	96°	97°	-
2	93°	-	84°	93°	-	94°	58°
3	93°30'	104°	-	-	-	-	88°
4	-	-	-	-	103°	-	-
5	-	-	-	-	90°	-	-
n	2	1	2	2	3	2	2
min	93°	-	84°	84°	90°	94°	58°
max	93°30'	-	101°	93°	103°	97°	88°
M	93°15'	-	92°30'	88°30'	96°30'	95°30'	73°
σn-1	0.4	-	12.0	6.4	6.5	2.1	21.2

(continued on the next page)

Table 5.7 Parameters of the Footprints (DIGITS) from São João do Rio do Peixe

	FREE LENGTH OF DIGIT								
	ANBD 01			ANBD 02			ANBD 03		
N°	II	III	IV	II	III	IV	II	III	IV
1	-	13.0	9.5	-	-	-	-	15.5	10.0
2	10.3	-	10.3	-	-	-	9.3	15.0	8.0
3	8.0	12.0	11.0	7.5	12.3	6.8	7.3	15.8	14.0
4	-	-	-	-	-	-	-	11.5	8.0
5	-	-	-	-	-	-	-	-	-
n	2	2	3	1	1	1	2	4	4
min	8.0	12.0	9.5	-	-	-	7.3	11.5	8.0
max	10.3	13.0	11.0	-	-	-	9.3	15.8	14.0
M	9.2	12.5	10.3	-	-	-	8.3	14.5	10.0
σn-1	1.6	0.7	0.7	-	-	-	1.4	2.0	2.8

	LENGTH OF PHALANGEAL PORTION OF THE DIGIT								
	ANBD 01			ANBD 02			ANBD 03		
N°	II	III	IV	II	III	IV	II	III	IV
1	-	-	-	-	-	-	-	-	-
2	-	-	19.5	-	-	-	-	-	18.0
3	11.0	13.0	19.5	-	-	16.3	15.0	-	27.0
4	-	-	-	-	-	-	-	-	22.5
5	-	-	-	-	-	-	-	-	-
n	1	1	2	0	0	1	1	0	3
min	-	-	-	-	-	-	-	-	18.8
max	-	-	-	-	-	-	-	-	27.0
M	-	-	-	-	-	-	-	-	22.5
σn-1	-	-	-	-	-	-	-	-	4.5

	DIVARICATION OF DIGITS								
	II - III	III-IV	II - IV	II - III	III - IV	II - IV	II - III	III - IV	II - IV
1	-	21°30'	-	-	-	-	-	36°	-
2	24°	29°30'	53°30'	-	-	-	23°	35°	67°
3	9°	21°	29°30'	15°30'	46°	61°	31°	25°	67°
4	-	-	-	-	-	-	-	25°	-
5	-	-	-	-	-	-	-	-	-
n	2	3	2	1	1	1	2	4	2
min	9°	21°	29°30'	-	-	-	23°	25°	-
max	24°	29°30'	59°30'	-	-	-	31°	36°	-
M	16°30'	24°	41°30'	-	-	-	27°	30°30'	67°
σn-1	10.6	4.8	17.0	-	-	-	5.7	5.8	0

(continued on the next page)

Table 5.8 Parameters of the Footprints (DIGITS) from São João do Rio do Peixe

FREE LENGTH OF DIGIT

N°	ANJU 02 II	III	IV	ANJU 03 II	III	IV	ANJU 04 II	III	IV	ANPV 01 II	III	IV
1	11.4	15.7	9.3	-	-	-	14.5	25.0	18.0	-	-	-
2	10.8	18.5	9.6	-	-	-	10.3	15.0	6.5	12.0	16.8	9.8
3	-	-	-	-	-	-	-	-	-	9.3	14.0	8.5
4	-	-	-	4.8	14.0	9.8	-	-	-	-	-	-
5	-	-	-	11.0	17.0	8.5	-	-	-	-	-	-
n	2	2	2	2	2	2	2	2	2	2	2	2
min	10.8	15.7	9.3	4.8	14.0	8.5	10.3	15.0	6.5	9.3	14.0	8.5
max	11.4	18.5	9.6	11.0	17.0	9.8	14.5	25.0	18.0	12.0	16.8	9.8
M	11.1	17.1	9.5	7.9	15.5	9.2	12.4	20.0	12.3	10.7	15.4	9.2
σn-1	0.4	2.0	0.2	4.4	2.1	0.9	3.0	7.1	8.1	1.9	2.0	0.9

LENGTH OF PHALANGEAL PORTION OF THE DIGIT

N°	ANJU 02 II	III	IV	ANJU 03 II	III	IV	ANJU 04 II	III	IV	ANPV 01 II	III	IV
1	18.6	20.0	20.7	-	-	-	24.0	30.3	26.5	-	-	-
2	20.0	208	21.2	-	-	-	-	-	30.5	16.5	-	18.5
3	-	-	-	-	-	-	-	-	-	13.3	-	15.5
4	-	-	-	20.8	-	23.3	-	-	-	-	-	-
5	-	-	-	-	-	19.5	-	-	-	-	-	-
n	2	2	2	1	0	2	1	1	2	2	0	2
min	18.6	20.0	20.7	-	-	19.5	-	-	26.5	13.3	-	15.5
max	20.0	20.8	21.2	-	-	23.3	-	-	30.5	16.5	-	18.5
M	19.3	20.4	21.0	-	-	21.4	-	-	28.5	14.9	-	17.0
σn-1	1.0	0.6	0.4	-	-	2.7	-	-	2.8	2.3	-	2.1

DIVARICATION OF DIGITS

N°	II - III	III - IV	II - IV	II - III	III - IV	II - IV	II - III	III - IV	II - IV	II - III	III - IV	II - IV
1	40°	33°30'	74°	27°30'	38°	65°	19°30'	18°	38°	-	-	-
2	29°	35°30'	64°	-	-	-	27°	26°	53°	37°	47°	95°
3	-	-	-	-	-	-	-	-	-	12°	16°	28°
4	-	-	-	17°	23°	40°	-	-	-	-	-	-
5	-	-	-	33°	35°	60°30'	-	-	-	-	-	-
n	2	2	2	3	3	3	3	2	2	2	2	2
min	29°	33°30'	64°	17°	23°	40°	19°30'	18°	38°	12°	16°	28°
max	40°	35°30'	74°	33°	38°	65°	27°	26°	53°	37°	47°	95°
M	34°30'	34°30'	69°	25°50'	32°20'	55°10'	23°15'	22°	45°30'	24°30	31°30'	61°30'
σn-1	7.8	14	7.1	8.1	8.1	13.3	5.3	5.7	10.6	17.7	21.9	47.4

Table 5.9 Parameters of the Trackways from Engenho Novo (ANEN)

ANEN/N°	STRIDE							STEP ANGLE						
	01	03*	04*	08*	12	30	31*	01	03*	04*	08*	12	30	31*
1-3	72.8	191.8	114.2	160.7	146.9	136.2	220.4	122°	176°	114°	122°30'	161	174°30'	174
2-4	81.8	-	-	158.2	145.4	132.1	-	196°	-	-	125°	155	179	-
3-5	87.7	-	-	161.2	-	127.0	-	150°	-	-	130°30'	-	181	-
4-6	91.6	-	-	151.5	-	138.3	-	171°	-	-	114°	-	179	-
5-7	-	-	-	158.7	-	136.2	-	-	-	-	125°	-	169°30'	-
6-8	-	-	-	164.3	-	135.2	-	-	-	-	123°30'	-	169	-
7-9	-	-	-	171.4	-	-	-	-	-	-	134°	-	-	-
8-10	-	-	-	152.6	-	-	-	-	-	-	124°	-	-	-
9-11	-	-	-	114.3	-	-	-	-	-	-	76°	-	-	-
n	4	1	1	9	2	6	1	4	1	1	9	2	6	1
min	72.8	-	-	114.3	145.4	127.0	-	122°	-	-	76°	155	169°30'	-
max	91.6	-	-	171.4	146.9	138.3	-	196°	-	-	134°	161	181	-
M	83.5	-	-	154.7	146.2	134.2	-	159°30'	-	-	119°23'	158	175°20'	-
σn-1	8.2	-	-	16.3	1.1	4.1	-	31.4	-	-	17.2	4.2	5.2	-

	EXTERNAL WIDTH							INTERNAL WIDTH						
	01	03*	04*	08*	12	30	31*	01	03*	04*	08*	12	30	31*
1-3	31.5	21.4	49.5	124.0	35.7	19.9	24.5	-	-13.3	10.2	-	-9.2	-10.7	-18.4
2-4	7.7	-	-	124.0	-	15.3	-	-20.7	-	-	-42.3	-	-16.3	-
3-5	22.6	-	-	115.3	-	-	-	0.6	-	-	-42.3	-	-	-
4-6	19.9	-	-	115.8	-	-	-	-10.4	-	-	-26.5	-	-	-
5-7	-	-	-	108.7	-	-	-	-	-	-	-36.7	-	-	-
6-8	-	-	-	119.4	-	-	-	-	-	-	-35.7	-	-13.2	-
7-9	-	-	-	119.4	-	-	-	-	-	-	-15.8	-	-	-
8-10	-	-	-	-	-	-	-	-	-	-	-40.8	-	-	-
9-11	-	-	-	-	-	-	-	-	-	-	26.5	-	-	-
n	4	1	1	7	1	2	1	3	1	1	8	1	3	1
min	7.7	-	-	108.7	-	15.3	-	-20.7	-	-	-42.3	-	-16.3	-
max	31.5	-	-	124.0	-	19.9	-	0.6	-	-	26.5	-	-10.7	-
M	20.4	-	-	118.1	-	17.6	-	-10.2	-	-	-26.7	-	-13.4	-
σn-1	9.8	-	-	5.4	-	3.3	-	10.7	-	-	23..3	-	2.8	-

	OBLIQUE PACE							GAUGE						
	01	03*	04*	08*	12	30	31*	01	03*	04*	08*	12	30	31*
1-2	46.6	96.4	95.9	95.4	74.5	67.3	110.2	18.7	2.6	30.6	56.6	14.3	6.6	3.6
2-3	39.8	-	-	90.8	74.5	68.9	-	2.2	-	-	56.6	18.4	0.5	-
3-4	42.8	-	-	91.3	75.0	68.4	-	12.2	-	-	49.5	-	-	-
4-5	47.9	-	-	88.3	-	63.3	-	8.0	-	-	31.6	-	-	-
5-6	44.6	-	-	91.8	-	74.5	-	-	-	-	35.7	-	8.2	-
6-7	-	-	-	88.3	-	62.8	-	-	-	-	52.0	-	8.2	-
7-8	-	-	-	98.5	-	67.9	-	-	-	-	51.4	-	-	-
8-9	-	-	-	90.8	-	-	-	-	-	-	56.1	-	-	-
9-10	-	-	-	84.7	-	-	-	-	-	-	98.0	-	-	-
10-11	-	-	-	110.2	-	-	-	-	-	-	-	-	-	-
n	5	1	1	10	3	7	1	4	1	1	9	2	4	1
min	39.8	-	-	84.7	74.5	62.8	-	2.2	-	-	31.6	14.3	0.5	-
max	47.9	-	-	110.2	75.0	75.5	-	18.7	-	-	98.0	18.4	8.2	-
M	44.3	-	-	93.0	74.7	67.0	-	10.3	-	-	54.2	16.4	5.9	-
σn-1	3.2	-	-	7.1	0.3	3.9	-	7.0	-	-	18.8	2.9	3.7	-

* Data obtained by extrapolation

ANEN 01 is a crooked trackway

Table 5.10 Parameters of the Footprints (DIGITS) from São João do Rio do Peixe

ANEN/N°		DIVERGENCE OF FOOTPRINTS FROM THE MIDLINE					
	01	03*	04*	08*	12	30	31*
1	-	5°30'	0°	-	-5°	1°	-8°30'
2	15	5°30'	0°	-4°30'	-3°30'	-4°30'	8°30'
3	5	-	-	8°30'	-2°30'	9°30'	-
4	20	-	-	31°30'	11°30'	-4°30'	-
5	0	-	-	8°	-	-	-
6	15	-	-	46°30'	-	-19°	-
7	-	-	-	45°	-	-18°	-
8	-	-	-	43°	-	-13°	-
9	-	-	-	14°	-	-	-
10	-	-	-	-	-	-	-
11	-	-	-	60°	-	-	-
n	5	2	2	9	4	7	2
min	0	-	-	-	-5°	-18°	-8°30'
max	20	-	-	-	11°30'	9°30'	8°30'
M	11	5°30'	0°	-	0°07'	-6°55'	0°
σn-1	8.2	0	0	-	7.7	10.4	12.1

* Data obtained by extrapolation

ANEN 01 is a crooked trackway   (continued on the next page)

Table 5.11 Parameters of the Footprints from Engenho Novo (ANEN)

LENGTH OF THE FOOTPRINT

ANEN / N°	01	03	04	08	12	30	31
1	-	23.5	22.4	-	23.5	16.3	24.5
2	16.2	26.5	24.0	-	20.4	18.9	-
3	15.2	-	-	58.1	27.0	16.3	-
4	17.3	-	-	69.1	26.5	18.4	-
5	16.4	-	-	68.9	-	-	-
6	13.0	-	-	74.5	-	18.4	-
7	-	-	-	68.3	-	19.9	-
8	-	-	-	72.9	-	19.4	-
9	-	-	-	67.3	-	-	-
n	5	2	2	7	4	7	1
min	13.0	23.5	22.4	58.1	20.4	16.3	-
max	17.3	26.5	24.0	74.5	27.0	19.9	-
M	15.6	25.0	23.2	68.5	24.4	18.2	-
σn-1	1.6	2.1	1.1	5.3	3.1	1.4	-

FREE LENGTH OF DIGIT

ANEN / N°	01 II	01 III	01 IV	03 II	03 III	03 IV	04 II	04 III	04 IV	12 III	12 IV	30 III	30 IV	31 II	31 III	31 IV
1	-	-	5.6	-	14.3	10.2	12.2	4.6	5.6	13.3	8.2	11.2	5.1	9.2	13.3	8.7
2	2.1	7.2	2.4	-	10.7	5.1	11.7	5.6	4.1	10.2	5.6	11.2	5.1	-	-	8.1
3	4.8	11.1	8.4	-	-	-	-	5.1	5.6	13.3	-	10.2	7.7	-	-	-
4	4.4	8.8	2.2	-	-	-	-	8.2	-	13.3	7.1	10.2	7.7	-	-	-
5	2.3	8.2	2.2	-	-	-	-	-	-	-	-	-	-	-	-	-
6	2.1	7.7	6.2	-	-	-	-	-	-	10.7	6.6	10.7	3.6	-	-	-
7	-	-	-	-	-	-	-	-	-	10.2	6.1	10.2	5.1	-	-	-
n	5	5	5	-	2	-	2	2	2	4	3	5	5	1	1	2
min	2.1	7.2	2.2	-	10.7	-	11.7	4.6	4.1	10.2	5.6	10.2	3.6	-	-	8.1
max	4.8	11.1	8.4	-	14.3	-	12.2	5.6	8.2	13.3	8.2	11.2	7.7	-	-	8.7
M	3.1	8.6	4.3	-	12.5	-	12.0	5.8	5.8	13.3	6.5	10.7	5.8	-	-	8.4
σn-1	1.3	1.5	2.9	-	2.5	-	0.4	0.7	1.7	1.6	1.5	0.5	1.8	-	-	0.4

WIDTH OF THE FOOTPRINT

ANEN / N°	01	03	04	08	12	30	31
1	-	16.3	19.4	-	19.9	14.8	17.9
2	9.2	18.4	19.4	-	19.4	15.3	23.4
3	10.1	-	-	64.3	23.5	14.8	-
4	11.2	-	-	72.4	-	17.9	-
5	9.5	-	-	-	-	-	-
6	12.4	-	-	61.7	-	-	-
7	-	-	-	65.3	-	14.3	-
8	-	-	-	61.2	-	13.8	-
9	-	-	-	63.8	-	-	-
10	-	-	-	62.2	-	-	-
11	-	-	-	-	-	-	-
n	5	2	2	7	3	6	2
min	9.2	16.3	19.4	61.2	19.4	13.8	17.9
max	12.4	18.4	19.4	72.4	23.5	17.9	23.4
M	10.5	17.4	19.4	64.4	20.9	15.2	20.7
σn-1	1.3	1.5	0	3.8	2.2	1.4	3.9

LENGTH OF PHALANGEAL PORTION OF THE DIGIT

ANEN / N°	01 II	01 III	01 IV	03 III	03 IV	04 III	04 IV	12 III	12 IV	30 III	30 IV	31 III	31 IV
1	-	-	12.8	19.9	9.7	16.3	-	19.4	-	11.7	-	15.3	19.9
2	10.8	11.8	9.4	17.3	-	16.3	14.3	15.3	-	12.2	12.2	15.3	-
3	7.6	-	12.7	-	-	-	-	17.3	8.2	12.2	12.2	-	-
4	10.5	-	10.5	-	-	-	19.4	-	11.7	15.3	12.6	-	-
5	7.7	10.5	10.1	-	-	-	-	-	-	-	-	-	-
6	7.9	-	10.6	-	-	-	-	-	-	13.8	13.8	-	-
7	-	-	-	-	-	-	-	10.7	10.7	-	12.8	-	-
8	-	-	-	-	-	-	-	14.3	14.3	-	-	-	-
n	5	2	5	2	1	2	2	3	0	2	2	2	1
min	7.6	10.5	9.4	12.8	-	14.3	-	15.3	8.2	14.3	12.2	11.7	1
max	10.8	11.8	12.7	17.3	-	19.9	-	19.4	14.3	19.4	15.3	12.2	1
M	8.9	11.2	10.7	15.1	-	18.6	-	16.3	16.9	17.3	13.3	12.0	19.9
σn-1	1.6	0.9	1.2	3.2	-	1.8	-	0	3.6	2.1	1.3	0.4	-

Table 5.12 Parameters of the Trackways from Matadouro (SOMA)

SOMA/N°	STRIDE		STEP ANGLE	
	1	2*	1	2*
1-3	162.3	162.0	178°	156°
n	1	1	1	1

	EXTERNAL WIDTH		INTERNAL WIDTH	
	1	2*	1	2*
1-3	20.0	39.0	-20.0	-0.8
n	1	1	1	1

	GAUGE		OBLIQUE PACE	
	1	2*	1	2*
1-2	0.5	19.5	85.0	83.0
2-3	0.5	-	76.8	-
n	2	1	2	1
min	-	-	76.8	-
max	-	-	85.0	-
M	0.5	-	80.9	-
σn-1	0	-	5.8	-

	DIVERGENCE OF FOOTPRINTS FROM THE MIDLINE	
	1	2
1	10°30'	-
2	3°30'	-
3	-2°30'	-
n	3	-
min	-2°30'	-
max	10°30'	-
M	3°50'	-
σn-1	6.5	-

* Means obtained by extrapolation

(continued en the next page)

Table 5.13 Parameters of the Footprints from Matadouro (SOMA)

SOMA/N°	LENGTH OF THE FOOTPRINT		FREE LENGTH OF DIGIT					
	01	02	01			02		
			II	III	IV	II	III	IV
1	28.5	25.0	8.8	12.0	7.0	9.0	16.5	-
2	33.5	18.3	10.0	12.5	-	5.5	12.5	5.5
3	30.0	-	8.5	19.0	-	-	-	-
n	3	2	3	3	1	2	2	1
min	28.5	18.3	8.5	12.0	-	5.5	12.5	-
max	33.5	25.0	10.0	19.0	-	9.0	16.5	-
M	30.7	21.7	9.1	14.5	-	7.3	14.5	-
σn-1	2.6	4.7	0.8	3.9	-	2.5	2.8	-

	WIDTH OF THE FOOTPRINT		LENGTH OF THE PHALANGEAL PORTION OF THE DIGIT					
	01	02	01			02		
			II	III	IV	II	III	IV
1	22.0	20.3	23.0	-	18.5	-	-	15.0
2	20.5	20.0	20.5	-	17.5	-	-	-
3	21.0	-	20.0	21.5	15.5	-	-	-
n	3	2	3	1	3	0	0	1
min	20.5	20.0	20.0	-	15.5	-	-	-
max	22.0	20.3	23.0	-	18.5	-	-	-
M	21.2	20.2	21.2	-	17.2	-	-	-
σn-1	0.8	0.2	1.6	-	1.5	-	-	-

	CROSS-AXIS ANGLE		DIVARICATION OF DIGITS					
	01	02	II-III	III-IV	II-IV	II-III	III-IV	II-IV
1	74°30'	102°30'	41°30'	45°30'	87°	14°	31°	45°30'
2	111°	94°30'	24°30'	16°30'	41°30	34°	37°	71°
3	89°	-	26°	12°30'	43°	-	-	-
n	3	2	3	3	3	2	2	2
min	74°30'	94°30'	24°30'	12°30'	41°30	14°	31°	45°30'
max	111°	102°30'	41°30'	45°30'	87°	34°	37°	71°
M	91°30'	98°30'	30°40'	24°50'	57°10'	24°	34°	58°15'
σn-1	18.4	5.7	9.4	18.0	25.8	14.1	4.2	18.0

Table 5.14 Parameters of the Trackways from Piau-Caiçara (SOCA)

SOCA / N°

STRIDE

	42*	43	44	45*	46*	48*	493*	57	75*	91*	97	98*
1-3	182	86.5	201,5	193	113	177,5	155,5	134,6	229	167	152,5	181
2-4	-	100.5	-	-	-	-	-	-	-	-	123,5	-
3-5	-	-	-	-	-	-	-	-	-	-	-	-
4-6	-	-	-	-	-	-	-	-	-	-	74	-
n	1	2	1	1	1	1	1	-	-	1	3	1
min	-	82.5	-	-	-	-	-	-	-	-	74	-
max	-	100.5	-	-	-	-	-	-	-	-	152,5	-
M	-	93.5	-	-	-	-	-	-	-	-	116.7	-
σn-1	-	9.9	-	-	-	-	-	-	-	-	39.7	-

STEP ANGLE

	42*	43	44	45*	46*	48*	493*	57	75*	91*	97	98*
1-3	153°	139°	168°	172°	146°	172°	152°	-	164°	146°	123°	140°
2-4	-	206°	-	-	-	-	-	-	-	-	144°	-
3-5	-	-	-	-	-	-	-	-	-	-	-	-
4-6	-	-	-	-	-	-	-	-	-	-	-	-
n	1	2	1	1	1	1	1	0	1	1	2	1
min	-	139°	-	-	-	-	-	-	-	-	-	-
max	-	206°	-	-	-	-	-	-	-	-	-	-
M	-	172°30'	-	-	-	-	-	-	-	-	-	-
σn-1	-	47.3	-	-	-	-	-	-	-	-	-	-

OBLIQUE PACE

	42*	43	44	45*	46*	48*	493*	57	75*	91*	97	98*
1-2	94.0	46.0	115.4	97.0	59.0	86.5	79.0	-	115.8	88.0	94.5	96.3
2-3	-	46.5	94.0	-	-	-	-	-	-	-	78.5	-
3-4	-	56.8	-	-	-	-	-	-	-	-	68.0	-
4-5	-	-	-	-	-	-	-	-	-	-	-	-
5-6	-	-	-	-	-	-	-	-	-	-	-	-
n	1	3	2	1	1	1	1	0	1	1	3	1
min	-	46.0	94.0	-	-	-	-	-	-	-	68.0	-
max	-	56.8	115.4	-	-	-	-	-	-	-	94.5	-
M	-	49.8	104.7	-	-	-	-	-	-	-	80.3	-
σn-1	-	6.1	15.1	-	-	-	-	-	-	-	13.3	-

OBS: SOCA 43 is a crooked trackway
* data obtained by extrapolation
? Uncertain data

(continued on the next page)

Table 5.15 Parameters of the Trackways from Piau-Caiçara (SOCA)

EXTERNAL WIDTH

SOCA / N°	42*	43	44	45*	46*	48*	493*	57	75*	91*	97	98*
1-3	41.5	47.5	42.0	24.3	34.0	40.0	40.0	-	42.0	60.0	77.5	64.0
2-4	-	43	-	-	-	-	-	-	-	-	68.5	-
3-5	-	-	-	-	-	-	-	-	-	-	-	-
4-6	-	-	-	-	-	-	-	-	-	-	-	-
n	1	2	1	1	1	1	1	0	1	1	2	1
min	-	43	-	-	-	-	-	-	-	-	68.5	-
max	-	47.5	-	-	-	-	-	-	-	-	77.5	-
M	-	45.25	-	-	-	-	-	-	-	-	73.0	-
σn-1	-	3.18	-	-	-	-	-	-	-	-	8.4	-

GAUGE

SOCA / N°	42*	43	44	45*	46*	48*	493*	57	75*	91*	97	98
1-3	22.0	17.5	22.5	5.8	18.0	17.0	10.5	-	17.0	26.8	42.5	32.0
2-4	-	11.5	-	-	-	-	-	-	-	-	40.0	-
3-5	-	-	-	-	-	-	-	-	-	-	-	-
4-6	-	-	-	-	-	-	-	-	-	-	-	-
n	1	2	1	1	1	1	1	0	1	1	2	1
min	-	11.5	-	-	-	-	-	-	-	-	40.0	-
max	-	17.5	-	-	-	-	-	-	-	-	42.5	-
M	-	14.5	-	-	-	-	-	-	-	-	41.3	-
σn-1	-	4.2	-	-	-	-	-	-	-	-	1.8	-

INTERNAL WIDTH

SOCA / N°	42*	43	44	45*	46*	48*	493*	57	75*	91*	97	98*
1-3	5.8	-10.5	0.5	-14.3	-4.0	-5.5	-14.0	-	-8.0	-7.5	4.3	7.8
2-4	-	-40.5	-	-	-	-	-	-	-	-	3.8	-
3-5	-	-	-	-	-	-	-	-	-	-	-	-
4-6	-	-	-	-	-	-	-	-	-	-	-	-
n	1	2	1	1	1	1	1	0	1	1	2	1
min	-	-40.5	-	-	-	-	-	-	-	-	3.8	-
max	-	-10.5	-	-	-	-	-	-	-	-	4.3	-
M	-	-25.5	-	-	-	-	-	-	-	-	4.1	-
σn-1	-	21.2	-	-	-	-	-	-	-	-	0.4	-

DIVERGENCE OF FOOTPRINTS FROM THE MIDLINE

SOCA / N°	42*	43	44	45*	46*	48*	493*	57	75*	91*	97	98*
1	0°	12°	0°	0°	9°30'	3°	12°30'	-	0°	0°	0°	0°
2	0°	-12°	0°	0°	9°30'	3°	12°30'	-	0°	0°	-6°	0°
3	-	13°30'	-	-	-	-	-	-	-	-	-2°	-
4	-	7°30'	-	-	-	-	-	-	-	-	15°	-
5	-	-	-	-	-	-	-	-	-	-	8°	-
6	-	-	-	-	-	-	-	-	-	-	5	-
n	2	4	2	2	2	2	2	0	2	2	5	2
min	-	-12°	-	-	-	-	-	-	-	-	-2°	-
max	-	13°30'	-	-	-	-	-	-	-	-	15°	-
M	0°	5°30'	0°	0°	9°30'	3°	12°30'	-	0°	0°	3°	0°
σn-1	0	11.8	0	0	0	0	0	-	0	0	8.4	0

OBS: SOCA 43 is a crooked trackway
* data obtained by extrapolation
? Uncertain data

(continued on the next page)

Table 5.16 Parameters of the Trackways from Piau-Caiçara (SOCA)

STRIDE

SOCA / N°	99	991	993	1323	13215*	13236	1331	1336*	1337*	13368?	1350	150*	151	153*
1-3	186,5	172	212,5	103,1	83,8	127,5	250	82	143	129,5	181,3	141,8	183,8	61,8
2-4	175	-	-	-	-	113.8	273	-	-	139	176	-	-	-
3-5	-	-	-	-	-	-	-	-	-	142	-	-	-	-
4-6	-	-	-	-	-	-	-	-	-	140	-	-	-	-
n	2	1	1	1	1	2	2	1	1	4	2	1	1	1
min	175	-	-	-	-	113.8	250	-	-	129.5	176	-	-	-
max	186.5	-	-	-	-	127.5	273	-	-	142	181.3	-	-	-
M	180.8	-	-	-	-	120.7	261.5	-	-	137.6	178.7	-	-	-
σn-1	8.1	-	-	-	-	9.7	16.3	-	-	5.6	3.7	-	-	-

STEP ANGLE

	99	991	993	1323	13215*	13236	1331	1336*	1337*	13368?	1350	150*	151	153°
1-3	170°	170°39'	175°	168°	164°	128°	143°	150°	178°	148°	147°	166°	157°	170°
2-4	182°	-	-	-	-	128°30'	140°	-	-	144°	142°	-	-	-
3-5	-	-	-	-	-	-	-	-	-	132°	-	-	-	-
4-6	-	-	-	-	-	-	-	-	-	138°	-	-	-	-
n	2	1	1	1	1	2	2	1	1	4	2	1	1	1
min	170°	-	-	-	-	128°	140°	-	-	132°	142°	-	-	-
max	182°	-	-	-	-	128°'30	143°	-	-	148°	147°	-	-	-
M	176°	-	-	-	-	128°15'	141°30'	-	-	140°30'	144°30'	-	-	-
σn-1	8.5	-	-	-	-	0.4	2.1	-	-	7.0	3.5	-	-	-

OBLIQUE PACE

	99	991	993	1323	13215*	13236	1331	1336*	1337*	13368?	1350	150*	151	153*
1-2	94.0	78.0	100.8	51.9	85.6	70.6	120.0	85.0	144.0	69.0	84.8	137.7	97.3	31.2
2-3	93.0	90.8	112.5	53.1	-	65.5	144.0	-	-	67.0	104.0	-	90.5	-
3-4	106.0	-	-	-	-	-	147.0	-	-	79.5	81.8	-	-	-
4-5	-	-	-	-	-	-	-	-	-	76.0	-	-	-	-
5-6	-	-	-	-	-	-	-	-	-	75.5	-	-	-	-
n	3	2	2	2	1	2	3	1	1	5	3	1	2	1
min	93.0	78.0	100.8	51.9	-	-	120.0	-	-	67.0	81.8	-	90.5	-
max	106.0	90.8	112.5	53.1	-	-	147.0	-	-	79.5	104.0	-	97.3	-
M	97.7	84.4	106.7	52.5	-	30.6	137.0	-	-	73.4	90.2	-	93.9	-
σn-1	7.2	9.1	8.3	0.8	-	0	14.8	-	-	5.2	12.0	-	4.8	-

OBS: * data obtained by extrapolation

? Uncertain data

(continued on the next page)

Table 5.17 Parameters of the Trackways from Piau-Caiçara (SOCA)

SOCA / N°						EXTERNAL WIDTH								
	99	991	993	1323	13215*	13236	1331	1336*	1337*	13368?	1350	150*	151	153*
1-3	-	-	36.8	-	35.6	49.5	-	50.8	30.0	44.0	48.8	-	45.7	12.1
2-4	30.5	-	-	-	-	-	84.0	-	-	64.0	-	-	-	-
3-5	-	-	-	-	-	-	-	-	-	68.0	-	-	-	-
4-6	-	-	-	-	-	-	-	-	-	67.7	-	-	-	-
n	1	0	1	0	1	1	1	1	1	4	1	0	1	1
min	-	-	-	-	-	-	-	-	-	44.0	-	-	-	-
max	-	-	-	-	-	-	-	-	-	68.0	-	-	-	-
M	-	-	-	-	-	-	-	-	-	61.1	-	-	-	-
σn-1	-	-	-	-	-	-	-	-	-	11.5	-	-	-	-

						GAUGE								
	99	991	993	1323	13215*	13236	1331	1336*	1337*	13368?	1350	150*	151	153*
1-3	11.0	10.5	9.0	8.8	11.9	30.6	47.0	20.0	3.6	19.0	29.0	16.2	18.9	1.8
2-4	7.0	-	-	-	-	-	54.0	-	-	27.0	29.5	-	-	-
3-5	-	-	-	-	-	-	-	-	-	38.0	-	-	-	-
4-6	-	-	-	-	-	-	-	-	-	38.0	-	-	-	-
n	2	1	1	1	1	1	2	1	1	4	2	1	1	1
min	7.0	-	-	-	-	-	47.0	-	-	19.0	29.0	-	-	-
max	11.0	-	-	-	-	-	54.0	-	-	38.0	29.5	-	-	-
M	9.0	-	-	-	-	-	50.5	-	-	30.5	20.3	-	-	-
σn-1	2.8	-	-	-	-	-	4.9	-	-	9.3	0.4	-	-	-

						INTERNAL WIDTH								
	99	991	993	1323	13215*	13236	1331	1336*	1337*	13368?	1350	150*	151	153*
1-3	-16.0	-	-27.8	-11.9	1.3	10.0	14.0	-7.4	-23.4	-14.5	-2.5	-5.4	-1.4	-7.3
2-4	-28.0	-	-	-	-	-	21.0	-	-	-14.0	-	-	-	-
3-5	-	-	-	-	-	-	-	-	-	-5.0	-	-	-	-
4-6	-	-	-	-	-	-	-	-	-	-14.5	-	-	-	-
n	2	0	1	1	1	1	2	1	1	4	1	1	1	1
min	-28.0	-	-	-	-	-	14.0	-	-	-14.5	-	-	-	-
max	-16.0	-	-	-	-	-	21.0	-	-	-5.0	-	-	-	-
M	-22.0	-	-	-	-	-	17.5	-	-	-11.0	-	-	-	-
σn-1	8.5	-	-	-	-	-	4.9	-	-	4.6	-	-	-	-

						DIVERGENCE OF FOOTPRINTS FROM THE MIDLINE								
	99	991	993	1323	13215*	13236	1331	1336*	1337*	13368?	1350	150*	151	153*
1	16°30'	3°30'	2°	-12°	0°	-29°	-15°	0°	0°	-3°30'	-12°	4°30'	15°	0°
2	4°	-1°30'	32°	-3°	0°	-27°	9°	0°	0°	8°	0°	4°30'	11°	0°
3	13°	11°	-6°	7°	-	13°30'	-12°	-	-	3°	-13°	-	12°	-
4	39°	-	-	-	-	-	3°	-	-	0°	-	-	-	-
5	-	-	-	-	-	-	-	-	-	0°	-	-	-	-
6	-	-	-	-	-	-	-	-	-	4°30'	-	-	-	-
n	4	3	3	3	2	3	4	2	2	6	3	2	3	2
min	4°	-1°30'	-6°	-12°	-	-29°	-15°	-	-	-3°30'	-13°	-	11°	-
max	39°	11°	32°	7°	-	13°30'	9°	-	-	8°	0°	-	15°	-
M	18°07'	4°20'	9°20'	-2°30'	0°	-14°12'	-3°45'	0°	0°	2°	-8°20'	4°30'	12°42'	0°
σn-1	14.9	6.3	20.0	9.5	0	24.0	11.6	0	0	4.0	7.2	0	2.1	0

OBS: * data obtained by extrapolation
? Uncertain data
(continued on the next page)

Table 5.18 Parameters of the Footprints from Piau-Caiçara (SOCA)

SOCA/N°	42	43	44	45	46	48	493	57	75	91	97	98	99
					LENGTH OF THE FOOTPRINT								
1	-	35.3	-	29.8	21.5	-	29.8	27.5	28.5	38.3	35.5	34.0	-
2	29.0	-	31.7	35.0	18.0	26.5	35.0	-	-	-	35.5	34.0	30.0
3	-	30.0	30.0	-	-	-	-	24.2	-	-	31.5	-	31.5
4	-	31.5	-	-	-	-	-	-	-	-	34.0	-	33.5
5	-	-	-	-	-	-	-	-	-	-	-	-	-
6	-	-	-	-	-	-	-	-	-	-	32.0	-	-
n	1	3	2	2	2	1	2	2	1	1	5	2	3
min	-	30.0	30.0	29.8	18.0	-	29.8	24.2	-	-	31.5	-	30.0
max	-	35.3	31.7	35.0	21.5	-	35.0	27.5	-	-	35.5	-	33.5
M	-	32.3	30.9	32.4	19.8	-	32.4	25.8	-	-	33.7	34.0	31.7
σn-1	-	2.7	1.2	3.7	2.5	-	3.7	2.2	-	-	18.9	0	1.8

SOCA/N°	42	43	44	45	46	48	493	57	75	91	97	98	99
					WIDTH OF THE FOOTPRINT								
1	17.3	28.5	-	20.0	26.0	18.8	25.5	31.6	28.0	32.5	40.0	29.0	-
2	18.5	-	21.5	19.0	24.8	26.8	26.5	-	24.5	33.0	33.0	27.0	22.5
3	-	30.0	20.0	-	-	-	-	21.7	-	-	32.5	-	22.5
4	-	27.3	-	-	-	-	-	-	-	-	29.0	-	22.5
5	-	-	-	-	-	-	-	-	-	-	-	-	-
6	-	-	-	-	-	-	-	-	-	-	32.0	-	-
n	2	3	2	2	2	2	2	2	2	2	5	2	3
min	17.3	27.3	20.0	19.0	24.8	18.8	25.5	21.7	24.5	32.5	29.0	27.0	-
max	18.5	30.0	21.5	20.0	26.0	26.8	26.5	31.6	28.0	33.0	40.0	29.0	-
M	17.9	28.6	20.8	19.5	25.4	22.8	26.0	27.6	26.3	32.8	33.3	28.0	22.5
σn-1	0.8	1.4	1.1	0.7	0.8	5.7	0.7	7.0	2.5	0.4	4.1	1.4	0

SOCA/N°	42	43	44	45	46	48	493	57	75	91	97	98	99
					CROSS-AXIS ANGLE								
1	94°	92°	-	77°30'	90°	95°	82°	72°	69°30'	83°	81°30'	106°	63°
2	86°	-	96°	94°30'	97°	96°	103°	-	101°	95°	94°	102°	109°30'
3	-	106°30'	96°	-	-	-	-	81°	-	-	92°	-	78°
4	-	108°	-	-	-	-	-	-	-	-	78°	-	92°
5	-	-	-	-	-	-	-	-	-	-	-	-	-
6	-	-	-	-	-	-	-	-	-	-	85°	-	-
n	2	3	2	2	2	2	2	2	2	2	5	2	4
min	82°	92°	-	77°30'	90°	95°	82°	72°	69°	83°	78°	102°	63°
max	94°	108°	-	94°30'	97°	96°	103°	81°	101°	95°	94°	106°	109°30'
M	90°	102°10'	96°	86°	93°30'	95°30'	92°30'	76°30'	85°15'	89°	86°06'	104°	85°37'
σn-1	5.7	8.8	0	12.0	4.9	0.7	14.8	6.4	22.3	8.5	6.8	2.8	19.8

Table 5.19 Parameters of the Footprints from Piau-Caiçara (SOCA)

SOCA/N°

LENGTH OF THE FOOTPRINT

	991	993	1323	13215	13236	1331	1336	1337	13368?	1350	150	151	153
1	-	30.5	11.9	20.8	-	31.0	27.2	27.6	33.0	24.0	24.3	-	9.1
2	26.5	31.5	-	-	21.3	30.0	31.8	30.0	-	23.8	17.6	21.6	8.8
3	30.5	-	-	-	17.5	34.0	-	-	36.5	24.0	-	-	-
4	-	-	-	-	-	29.0	-	-	41.0	-	-	-	-
5	-	-	-	-	-	-	-	-	38.5	-	-	-	-
6	-	-	-	-	-	-	-	-	38.0	-	-	-	-
n	2	2	1	1	2	4	2	2	5	3	2	1	2
min	26.5	30.5	-	-	17.5	29.5	27.2	27.6	33.0	23.8	17.6	-	8.8
max	30.5	31.5	-	-	21.3	34.0	31.8	30.0	41.0	24.0	24.3	-	9.1
M	28.5	31.0	-	-	19.4	31.1	29.5	28.8	37.4	23.9	21.0	-	9.0
σ_{n-1}	2.8	0.7	-	-	2.9	2.0	3.3	1.7	2.9	0.1	4.7	-	0.2

WIDTH OF THE FOOTPRINT

	991	993	1323	13215	13236	1331	1336	1337	13368?	1350	150	151	153
1	23.3	28.3	12.5	20.4	19.6	27.0	28.6	26.6	23.0	23.0	23.0	29.7	9.4
2	-	29.5	-	-	21.5	-	28.4	26.6	-	18.3	-	18.9	10.0
3	-	-	13.5	-	17.5	28.0	-	-	28.5	24.0	-	21.6	-
4	-	-	-	-	-	28.5	-	-	32.0	-	-	-	-
5	-	-	-	-	-	-	-	-	30.0	-	-	-	-
6	-	-	-	-	-	-	-	-	32.5	-	-	-	-
n	1	2	2	1	3	4	2	2	5	3	1	3	2
min	-	28.0	12.5	-	17.5	27.0	28.4	26.6	23.0	18.3	-	18.9	9.4
max	-	29.5	13.5	-	21.5	28.5	28.6	26.6	32.5	24.0	-	29.7	10.0
M	-	28.7	13.0	-	19.5	27.8	28.5	26.6	31.1	21.8	-	23.4	9.7
σ_{n-1}	-	1.1	0.7	-	2.0	0.8	0.1	0	5.6	3.0	-	5.6	0.4

CROSS-AXIS ANGLE

	991	993	1323	13215	13236	1331	1336	1337	13368?	1350	150	151	153
1	115°	110°	74°	95°	119°	-	94°	90°	73°30'	95°	81°	85°	94°
2	91°30'	87°	-	-	128°	-	98°	70°	68°30'	96°30'	84°30'	112°	-
3	83°	78°	110°	-	97°	-	-	-	97°	77°	-	117°	-
4	-	-	-	-	-	-	-	-	91°	-	-	-	-
5	-	-	-	-	-	-	-	-	87°	-	-	-	-
6	-	-	-	-	-	-	-	-	89°	-	-	-	-
n	3	3	2	1	3	0	2	2	6	3	2	3	1
min	83°	78°	74°	-	97°	-	94°	70°	68°30'	77°	81°	85°	-
max	115°	110°	110°	-	128°	-	98°	90°	97°	96°30'	84°30'	117°	-
M	96°	91°40'	94°	-	114°40'	-	96°	80°	84°20'	89°30'	82°45'	104°40'	-
σ_{n-1}	16.6	16.4	25.5	-	15.9	-	2.8	14.1	11.0	10.9	2.3	17.21	-

(continued on the next page)

Table 5.20 Parameters of the Footprints (DIGITS) from Piau-Caiçara (SOCA)

FREE LENGTH OF DIGIT

SOCA / Nº	42			43			44			45			46		
	II	III	IV	II	III	IV	II	III	IV	II	III	IV	II	III	IV
1	6.0	13.0	6.5	6.3	12.5	6.0	-	-	-	11.3	18.0	10.0	7.0	10.0	3.5
2	7.5	14.5	8.5	-	-	-	10.5	17.5	11.5	8.5	18.0	13.5	2.8	8.5	-
3	-	-	-	5.5	12.5	8.3	9.0	15.5	9.5	-	-	-	-	-	-
4	-	-	-	9.0	16.0	10.0	-	-	-	-	-	-	-	-	-
5	-	-	-	-	-	-	-	-	-	-	-	-	-	-	-
6	-	-	-	-	-	-	-	-	-	-	-	-	-	-	-
n	2	2	2	3	3	3	2	2	2	2	2	2	2	2	1
min	6.0	13.0	6.5	5.5	12.5	10.0	9.0	15.5	9.5	8.5	-	10.0	2.8	8.5	-
max	7.5	14.5	8.5	9.0	16.0	8.3	10.5	17.5	11.5	11.3	-	13.5	7.0	10.0	-
M	6.8	13.8	7.5	6.9	13.7	8.1	9.8	16.5	10.5	9.9	18.0	11.8	4.9	9.3	-
σn-1	1.1	1.1	1.4	1.8	2.0	2.0	1.1	1.4	1.4	2.0	0	2.5	3.0	1.1	-

PHALANGEAL LENGTH

	42			43			44			45			46		
	II	III	IV	II	III	IV	II	III	IV	II	III	IV	II	III	IV
1	-	-	-	26.0	-	29.0	-	-	-	16.0	-	22.5	15.0	11.8	12.8
2	21.0	24.5	23.0	-	-	-	24.5	19.5	23.5	19.0	-	29.5	11.5	13.0	-
3	-	-	-	-	-	-	23.5	-	19.5	-	-	-	-	-	-
4	-	-	-	-	-	23.0	-	-	-	-	-	-	-	-	-
5	-	-	-	-	-	-	-	-	-	-	-	-	-	-	-
6	-	-	-	-	-	-	-	-	-	-	-	-	-	-	-
n	1	1	1	1	0	2	2	1	2	2	0	2	2	2	1
min	-	-	-	-	-	23.0	23.5	-	19.5	16.0	-	22.5	11.5	11.8	-
max	-	-	-	-	-	29.0	24.5	-	23.5	19.0	-	29.5	15.0	13.0	-
M	-	-	-	-	-	26.0	24.0	-	21.5	17.5	-	26.0	13.3	12.4	-
σn-1	-	-	-	-	-	4.2	0.7	-	2.8	2.1	-	4.9	2.5	0.8	-

DIVARICATION OF DIGITS

	II - III	III - IV	II - IV	II - III	III - IV	II - IV	II - III	III - IV	II - IV	II - III	III - IV	II - IV	II - III	III - IV	II - IV
1	36°	22°30'	59°	12°	34°30'	46°30'	-	-	-	12°	20°	31°	25°	23°	48°
2	7°	15°	21°	-	-	-	24°	29°	51°30'	15°30'	17°	32°30'	14°	-	-
3	-	-	-	20°30'	41°	61°30'	2°	31°	33°	-	-	-	-	-	-
4	-	-	-	37°30'	38°30'	80°	-	-	-	-	-	-	-	-	-
5	-	-	-	-	-	-	-	-	-	-	-	-	-	-	-
6	-	-	-	-	-	-	-	-	-	-	-	-	-	-	-
n	2	2	2	3	3	3	2	2	2	2	2	2	2	1	1
min	7°	15°	21°	12°	34°30'	46°30'	2°	29°	33°	12°	17°	31°	14°	-	-
max	36°	22°30'	58°	37°30'	41°	80°	24°	31°	51°30'	15°30'	20°	32°30'	25°	-	-
M	21°30'	18°45'	39°30'	23°20'	38°30'	62°40'	13°	30°	42°15'	13°48'	18°30'	31°48'	19°30'	-	-
σn-1	205	5.3	26.2	13.0	3.3	16.8	15.6	1.4	13.1	2.5	2.1	1.1	7.8	-	-

(continued on the next page)

Table 5.21 Parameters of the Footprints (DIGITS) from Piau-Caiçara (SOCA)

FREE LENGTH OF DIGIT

SOCA / N°	48			493			57			75			91		
	II	III	IV	II	III	IV	II	III	IV	II	III	IV	II	III	IV
1	7.0	15.0	-	13.0	16.5	9.5	14.2	17.5	9.2	13.0	17.5	12.5	13.5	20.0	12.5
2	9.5	13.0	9.0	12.0	20.8	15.5	5.8	13.3	5.8	8.0	-	6.0	9.0	-	8.0
3	-	-	-	-	-	-	-	-	-	-	-	-	-	-	-
4	-	-	-	-	-	-	-	-	-	-	-	-	-	-	-
5	-	-	-	-	-	-	-	-	-	-	-	-	-	-	-
6	-	-	-	-	-	-	-	-	-	-	-	-	-	-	-
n	2	2	1	2	2	2	2	2	2	2	1	2	2	1	2
min	7.0	13.0	-	12.0	16.6	9.5	5.8	13.3	5.8	8.0	-	6.0	9.0	-	8.0
max	9.5	15.0	-	13.0	20.8	15.5	4.2	17.5	9.4	13.0	-	12.5	13.5	-	12.5
M	8.3	14.0	-	12.5	18.7	12.5	10.0	15.4	7.5	10.5	-	9.3	11.3	-	10.3
σn-1	1.8	1.4	-	0.7	3.0	4.2	5.9	3.0	2.4	3.5	-	4.6	3.2	-	3.2

PHALANGEAL LENGTH

	48			493			57			75			91		
	II	III	IV	II	III	IV	II	III	IV	II	III	IV	II	III	IV
1	-	-	-	18.5	-	19.5	22.5	-	20.8	24.0	-	18.0	25.3	-	30.0
2	19.0	-	21.0	26.5	-	27.5	13.3	-	15.4	17.5	-	20.5	-	-	25.3
3	-	-	-	-	-	-	-	-	-	-	-	-	-	-	-
4	-	-	-	-	-	-	-	-	-	-	-	-	-	-	-
5	-	-	-	-	-	-	-	-	-	-	-	-	-	-	-
6	-	-	-	-	-	-	-	-	-	-	-	-	-	-	-
n	1	0	1	2	0	2	2	0	2	2	0	2	1	0	2
min	-	-	-	18.5	-	19.5	13.3	-	15.4	17.5	-	18.0	-	-	25.3
max	-	-	-	26.5	-	27.5	22.5	-	20.8	24.0	-	20.5	-	-	30.0
M	-	-	-	22.5	-	23.5	17.9	-	18.1	20.8	-	19.3	-	-	27.7
σn-1	-	-	-	5.7	-	5.7	6.5	-	3.8	4.6	-	1.8	-	-	3.3

DIVARICATION OF DIGITS

	II - III	III - IV	II - IV	II - III	III - IV	II - IV	II - III	III - IV	II - IV	II - III	III - IV	II - IV	II - III	III - IV	II - IV
1	18°	17°	34°30'	28°	10°	38°	38°	36°	74°	21°	47°	68°	40°	26°	65°30'
2	34°	29°	63°	16°	28°	43°	47°	36°	83°	22°	21°	43°	46°30'	21°30'	68°
3	-	-	-	-	-	-	-	-	-	-	-	-	-	-	-
4	-	-	-	-	-	-	-	-	-	-	-	-	-	-	-
5	-	-	-	-	-	-	-	-	-	-	-	-	-	-	-
6	-	-	-	-	-	-	-	-	-	-	-	-	-	-	-
n	2	2	2	2	2	2	2	2	2	2	2	2	2	2	2
min	18°	17°	34°30'	16°	10°	38°	38°	-	74°	21°	21°	43°	40°	21°30'	65°30'
max	34°	29°	63°	28°	28°	43°	47°	-	83°	22°	47°	68°	46°30'	26°	68°
M	26°	23°	48°15'	22°	19°	40°30'	42°30'	36°	78°30	21°30'	34°	55°30'	43°15'	23°48'	66°15'
σn-1	11.3	8.5	20.2	8.5	12.7	3.5	8.4	0	6.4	0.7	18.4	17.7	4.6	3.2	1.8

Table 5.22 Parameters of One Trackway from Fazenda Piedade (SOPI)

SOPI 1	STRIDE	STEP ANGLE
1-3	141.0	176°30'
n	1	1

	EXTERNAL WIDTH	INTERNAL WIDTH
1-3	19.0	-20.0
n	1	1
	OBLIQUE PACE	GAUGE
1-2	74.0	1.0
2-3	67.0	2.0
n	2	2
min	67.0	1.0
max	74.0	2.0
M	70.5	1.5
σn-1	4.9	0.7

	DIVERGENCE OF FOOTPRINTS FROM THE MIDLINE
1	8°
2	15°
3	9°
n	3
min	8°
max	15°
M	10°40'
σn-1	3.8

(continued on the next page)

Table 5.23 Parameters of the Footprints from Fazenda Piedade (SOPI)

SOPI 1	LENGTH OF FOOTPRINT	FREE LENGTH OF DIGIT		
		II	III	IV
1	26.5	10.0	16.0	9.0
2	25.5	-	18.0	11.0
3	27.5	9.0	15.0	8.5
n	3	2	3	3
min	25.5	9.0	15.0	8.5
max	27.0	10.0	18.0	11.0
M	26.3	9.5	16.3	9.5
$\sigma n-1$	0.8	0.7	1.5	1.3

	WIDTH OF THE FOOTPRINT	LENGTH OF THE PHALANGEAL PORTION OF THE DIGIT		
1	20.5	16.0	20.0	15.0
2	19.5	17.0	24.0	17.0
3	18.5	16.0	20.0	14.0
n	3	3	3	3
min	18.5	16.0	20.0	14.0
max	20.5	17.0	24.0	17.0
M	19.5	16.3	21.3	15.3
$\sigma n-1$	1.0	0.6	2.3	1.5

	CROSS-AXIS ANGLE	DIVARICATION OF DIGITS		
		II - III	III - IV	II - IV
1	93°	26°	28°	53°
2	112°	12°	40°	52°30'
3	111°30'	13°	39°	47°30'
n	3	3	3	3
min	93°	12°	28°	47°30'
max	112°	26°	40°	53°
M	105°30'	17°	35°40'	51°
$\sigma n-1$	10.8	7.8	6.7	3.0

Table 5.24 Parameters of the Trackways from Poço do Motor (SOPM)

SOPM/N°	STRIDE					STEP ANGLE				
	1	**4**	**5***	**6***	**7***	**1**	**4**	**5***	**6***	**7***
1-3	184.0	182.3	156.0	175.5	175.5	156	180	144	140	142
2-4	170.8	-	-	-	-	158	-	-	-	-
n	2	1	1	1	1	2	1	1	1	1
min	107.8	-	-	-	-	156	-	-	-	-
max	184.0	-	-	-	-	158	-	-	-	-
M	177.4	-	-	-	-	157	-	-	-	-
σn-1	9.3	-	-	-	-	1.4	-	-	-	-

	EXTERNAL WIDTH					INTERNAL WIDTH				
	1	**4**	**5***	**6***	**7***	**1**	**4**	**5***	**6***	**7***
1-3	-	-	42.0	55.5	47.5	-	-	7.0	10.0	10.0
2-4	-	-	-	-	-	-	-	-	-	-
n	0	0	1	1	1	0	0	1	1	1
min	-	-	-	-	-	-	-	-	-	-
max	-	-	-	-	-	-	-	-	-	-
M	-	-	-	-	-	-	-	-	-	-
σn-1	-	-	-	-	-	-	-	-	-	-

	OBLIQUE PACE					GAUGE				
	1	**4**	**5***	**6***	**7***	**1**	**4**	**5***	**6***	**7***
1-2	103.8	97.5	82.3	93.5	93.0	24.3	1.5	24.0	31.5	27.5
2-3	84.5	85.0	-	-	-	-	-	-	-	-
3-4	89.5	-	-	-	-	-	-	-	-	-
n	3	2	1	1	1	1	1	1	1	1
min	84.5	85.0	-	-	-	-	-	-	-	-
max	103.8	97.5	-	-	-	-	-	-	-	-
M	92.6	91.3	-	-	-	-	-	-	-	-
σn-1	10.0	8.8	-	-	-	-	-	-	-	-

	DIVERGENCE OF FOOTPRINTS FROM THE MIDLINE				
1	-8°	6°30'	0°	0°	-18°30'
2	8°30'	10°30'	0°	0°	12°30'
3	-	-	-	-	-
4	4°	-	-	-	-
n	3	2	2	2	2
min	-8°	6°30'	-	-	-18°30'
max	8°30'	10°30'	-	-	12°30'
M	1°30'	8°30'	0°	0°	-3°
σn-1	8.5	2.8	0	0	21.9

* Means obtained by extrapolation

(continued on the next page)

Table 5.25 Parameters of the Footprints from Poço do Motor (SOPM)

SOPM / N°	LENGTH OF THE FOOTPRINT				
	01	04	05	06	07
1	30.0	27.0	23.5	25.0	28.5
2	37.0	31.5	25.5	29.5	-
3	-	-	-	-	-
4	29.5	-	-	-	-
n	3	2	2	2	1
min	29.5	27.0	23.5	25.0	-
max	37.0	31.5	25.5	29.5	-
M	32.7	29.3	24.5	27.3	-
σn-1	4.2	3.2	1.4	3.2	-

	WIDTH OF THE FOOTPRINT				
	01	04	05	06	07
1	29.3	19.0	17.5	25.0	16.8
2	41.3	27.0	17.0	19.8	19.3
3	-	-	-	-	-
4	31.8	-	-	-	-
n	3	2	2	2	2
min	29.3	19.0	17.0	19.8	18.8
max	41.3	27.0	17.5	25.0	19.3
M	34.1	23.0	17.3	22.4	19.1
σn-1	6.3	5.7	0.4	3.7	0.4

	CROSS-AXIS ANGLE				
	01	04	05	06	07
1	93°	107°30'	81°	99°	101°30
2	-	79°	53°	95°	91°
3	-	-	-	-	-
4	98°	-	-	-	-
n	2	2	2	2	2
min	93°	79°	63°	95°	91°
max	98°	107°30'	81°	99°	101°30'
M	95°30'	93°15'	72°	97°	96°15'
σn-1	3.5	20.2	12.7	2.8	7.4

(continued on the next page)

Table 5.26 Parameters of the Footprints (DIGITS) from Poço do Motor (SOPM)

SOPM / N°

FREE LENGTH OF DIGIT

	01 II	01 III	01 IV	04 II	04 III	04 IV	05 II	05 III	05 IV	06 II	06 III	06 IV	07 II	07 III	07 IV
1	14.5	19.3	17.0	5.0	23.0	12.5	9.0	9.5	3.0	13.5	16.0	13.5	6.3	12.5	7.8
2	-	19.5	-	15.0	-	-	-	-	-	12.5	15.0	12.5	5.3	-	1.8
3	-	-	-	-	-	-	-	-	-	-	-	-	-	-	-
4	-	-	-	-	-	-	-	-	-	-	-	-	-	-	-
n	1	2	1	2	1	1	1	1	1	2	2	2	2	1	2
min	-	19.3	-	5.0	-	-	-	-	-	12.5	15.0	12.5	5.3	-	1.8
max	-	19.5	-	15.0	-	-	-	-	-	13.5	16.0	13.5	6.3	-	7.8
M	-	19.4	-	10.0	-	-	-	-	-	13.0	15.5	13.0	5.8	-	4.8
σn-1	-	0.1	-	7.1	-	-	-	-	-	0.7	0.7	0.7	0.7	-	4.2

LENGTH OF THE PHALANGEAL PORTION

	01 II	01 III	01 IV	04 II	04 III	04 IV	05 II	05 III	05 IV	06 II	06 III	06 IV	07 II	07 III	07 IV
1	-	-	26.5	-	21.0	18.3	11.1	12.0	10.5	-	-	18.0	-	-	12.5
2	-	-	-	-	-	-	-	-	-	-	-	-	-	-	-
3	-	-	-	-	-	-	-	-	-	-	-	-	-	-	-
4	-	-	-	-	-	-	-	-	-	-	-	-	-	-	-
n	0	0	1	-	1	1	1	1	1	-	-	1	-	-	1
min	-	-	-	-	-	-	-	-	-	-	-	-	-	-	-
max	-	-	-	-	-	-	-	-	-	-	-	-	-	-	-
M	-	-	-	-	-	-	-	-	-	-	-	-	-	-	-
σn-1	-	-	-	-	-	-	-	-	-	-	-	-	-	-	-

DIVARICATION OF DIGITS

	01 II-III	01 III-IV	01 II-IV	04 II-III	04 III-IV	04 II-IV	05 II-III	05 III-IV	05 II-IV	06 II-III	06 III-IV	06 II-IV	07 II-III	07 III-IV	07 II-IV
1	15°30'	34°	49°	25°30'	9°30'	53°	11°	3°	37°	34°	21°	26°	31°30'	38°30'	75°
2	53°	37°	83°	43°30'	31°	56°	34°	40°	43°	44°	23°	60°30'	44°	44°30'	82°
3	-	30°	-	-	-	-	-	-	-	-	-	-	-	-	-
4	-	-	-	-	-	-	-	-	-	-	-	-	-	-	-
n	2	3	2	2	2	2	2	2	2	2	2	2	2	2	2
min	15°30'	30°	49°	25°30'	9°30'	53°	11°	3°	37°	34°	21°	26°	31°30'	38°30'	75°
max	53°	37°	83°	43°30'	31°	56°	34°	40°	43°	44°	23°	60°30'	44°	44°30'	82°
M	34°15'	33°40'	66°	34°30'	20°15'	54°30'	22°30'	21°30'	40°	39°	22°	43°15'	37°45'	41°30'	78°30'
σn-1	26.5	3.5	24.0	12.7	15.2	2.1	16.3	26.2	4.2	7.1	1.4	24.4	8.8	4.2	4.9

Table 5.27 Parameters of the Trackways from Passagem das Pedras (SOPP)

SOPP/N°	STRIDE						
	1	2	3	4	6	12	13
1-3	188.7	217.0	499.0	244.5	158.6	186.5	169.0
2-4	180.7	213.0	483.0	254.0	175.7	225.0	-
3-5	173.4	203.0	481.0	249.0	183.6	241.0	-
4-6	169.4	-	478.0	248.5	166.4	-	-
5-7	195.2	227.5	473.0	249.0	163.3	-	-
6-8	183.9	-	-	238.0	-	-	-
7-9	177.4	210.0	-	234.5	-	-	-
8-10	195.2	-	-	239.0	-	-	-
9-11	182.3	-	-	-	-	-	-
10-12	183.9	-	-	-	-	-	-
11-13	193.6	-	-	-	-	-	-
12-14	191.9	-	-	-	-	-	-
13-15	187.1	-	-	-	-	-	-
14-16	187.1	-	-	-	-	-	-
15-17	192.8	-	-	-	-	-	-
16-18	204.9	-	-	-	-	-	-
n	16	5	5	8	5	3	1
min	169.4	203.0	473.0	234.5	158.6	182.0	-
max	204.9	227.5	499.0	254.0	183.6	241.0	-
M	186.7	214.1	482.8	244.6	169.5	217.5	-
σn-1	9.1	9.1	9.8	6.8	10.1	28.0	-

	EXTERNAL WIDTH						
	1	2	3	4	6	12	13
1-3	29.0	52.5	46.0	21.5	48.0	-	25.0
2-4	41.9	61.0	55.0	-	55.5	-	-
3-5	58.8	45.0	54.6	-	55.5	-	-
4-6	51.6	-	52.2	-	44.9	-	-
5-7	53.2	-	45.1	-	48.4	-	-
6-8	56.5	-	-	-	-	-	-
7-9	56.5	40.0	-	-	-	-	-
8-10	51.6	-	-	-	-	-	-
9-11	67.7	-	-	-	-	-	-
10-12	66.9	-	-	-	-	-	-
11-13	40.3	-	-	-	-	-	-
12-14	46.0	-	-	-	-	-	-
13-15	46.8	-	-	-	-	-	-
14-16	46.8	-	-	-	-	-	-
15-17	53.2	-	-	-	-	-	-
16-18	54.0	-	-	-	-	-	-
n	16	4	5	1	5	0	1
min	29.0	40.0	41.5	-	48.0	-	-
max	67.7	61.0	55.0	-	55.5	-	-
M	51.3	49.6	50.6	-	50.5	-	-
σn-1	9.7	9.2	4.7	-	4.8	-	-

(continued on the next page)

Table 5.28 Parameters of the Trackways from Passagem das Pedras (SOPP)

SOPP/N°	STEP ANGLE						
	1	2	3	4	6	12	13
1-3	161°	155°	171°30'	173°	146°	153°30'	175°
2-4	164°	150°30'	170°	171°	135°	146°30'	-
3-5	163°	158°	169°30'	163°	150°	144°30'	-
4-6	157°	-	170°30'	173°	155°	-	-
5-7	180°	-	173°	173°	150°	-	-
6-8	177°	-	-	173°30'	-	-	-
7-9	159°	167°	-	172°	-	-	-
8-10	169°	-	-	175°	-	-	-
9-11	158°	-	-	-	-	-	-
10-12	157°	-	-	-	-	-	-
11-13	171°	-	-	-	-	-	-
12-14	167°	-	-	-	-	-	-
13-15	166°	-	-	-	-	-	-
14-16	169°	-	-	-	-	-	-
15-17	159°	-	-	-	-	-	-
16-18	166°	-	-	-	-	-	-
n	16	4	5	8	5	3	1
min	157°	150°30'	169°30'	163°	135°	144°30'	-
max	180°	167°	173°	175°	155°	153°30'	-
M	165°11'	157°36'	170°54'	171°41'	147°12'	148°10'	-
σn-1	6.9	7.0	1.4	3.7	7.5	4.7	-

	INTERNAL WIDTH						
	1	2	3	4	6	12	13
1-3	-23.4	-10.0	-9.0	-4.0	-14.1	-	-16.0
2-4	-7.3	-10.5	-11.0	-	-1.2	-	-
3-5	-6.5	-10.5	-5.0	-	-13.3	-	-
4-6	-17.7	-	-13.0	-	-13.3	-	-
5-7	-18.5	-	-16.0	-	-7.0	-	-
6-8	-17.7	-	-	-	-	-	-
7-9	-14.2	-14.5	-	-	-	-	-
8-10	-29.0	-	-	-	-	-	-
9-11	-6.5	-	-	-	-	-	-
10-12	-25.0	-	-	-	-	-	-
11-13	-24.2	-	-	-	-	-	-
12-14	-17.7	-	-	-	-	-	-
13-15	-26.6	-	-	-	-	-	-
14-16	-11.3	-	-	-	-	-	-
15-17	-18.5	-	-	-	-	-	-
16-18	-32.3	-	-	-	-	-	-
n	16	4	5	1	5	0	1
min	-32.3	-14.0	-16.0	-	-14.1	-	-
max	-6.5	-10.0	-5.0	-	-1.2	-	-
M	-18.5	-11.3	-10.8	-	-9.8	-	-
σn-1	8.0	2.2	4.1	-	5.6	-	-

(continued on the next page)

Table 5.29 Parameters of the Trackways from Passagem das Pedras (SOPP)

SOPP/N°	OBLIQUE PACE						
	1	2	3	4	6	12	13
1-2	98.4	109.0	250.0	122.0	82.0	87.0	81.0
2-3	93.6	115.0	248.0	128.0	88.3	104.5	88.0
3-4	89.5	105.0	238.0	129.0	96.9	122.5	-
4-5	87.1	102.5	241.0	123.0	85.9	122.5	-
5-6	87.1	-	237.0	128.0	82.0	-	-
6-7	111.3	-	234.0	122.0	87.1	-	-
7-8	74.4	103.0	-	117.5	-	-	-
8-9	103.2	108.9	-	118.0	-	-	-
9-10	93.6	-	-	122.0	-	-	-
10-11	93.6	-	-	-	-	-	-
11-12	95.2	-	-	-	-	-	-
12-13	98.4	-	-	-	-	-	-
13-14	95.2	-	-	-	-	-	-
14-15	93.6	-	-	-	-	-	-
15-16	94.4	-	-	-	-	-	-
16-17	101.6	-	-	-	-	-	-
17-18	106.5	-	-	-	-	-	-
n	17	6	6	9	6	4	2
min	74.4	102.5	234.0	117.5	82.0	87.0	81.0
max	111.3	115.0	250.0	129.0	96.9	122.5	88.0
M	95.1	107.2	241.3	123.3	87.0	109.1	84.5
σn-1	8.4	4.7	6.4	4.2	5.5	17.0	5.0

	DIVERGENCE OF FOOTPRINTS FROM THE MIDLINE						
1	-4°	-7°	-3°	-4°	11°30'	-6°	10°
2	-8°	-7°	22°	-7°	-11°	-	-2°
3	-2°30'	0°	7°	3°30'	7°	-	6°
4	0°	-9°	1°30'	-	-5°	16°30'	-
5	4°	-4°	1°	16°	-3°	-	-
6	6°30'	-	7°	-	-5°	-	-
7	-6°30'	8°	-14°	6°	3°	-	-
8	-10°30'	-9°	-	-	-	-	-
9	2°	5°30'	-	-	-	-	-
10	-11°	6°	-	-	-	-	-
11	-5°30'	-	-	-	-	-	-
12	10°30'	-	-	-	-	-	-
13	4°30'	-	-	-	-	-	-
14	4°30'	-	-	-	-	-	-
15	5°	-	-	-	-	-	-
16	-20°30'	-	-	-	-	-	-
17	-4°30'	-	-	-	-	-	-
18	10°	-	-	-	-	-	-
n	18	9	7	5	7	2	3
min	-20°30'	-9°	-14°	-7°	-11°	-6°	-2°
max	10°30'	8°	22°	16°	11°30'	16°30'	10°
M	-1°26'	-1°50'	3°04'	2°54'	-0°21'	5°15'	4°40'
σn-1	8.2	6.9	11.0	9.0	7.2	15.9	6.1

(continued on the next page)

Table 5.30 Parameters of the Trackways from Passagem das Pedras (SOPP)

SOPP/N°	GAUGE						
	1	2	3	4	6	12	13
1-2	11.3	23.0	19.0	-	10.9	13.5	1.0
2-3	21.0	23.0	16.0	-	24.2	28.5	2.0
3-4	6.5	21.0	23.5	-	28.1	34.0	-
4-5	18.5	-	23.5	-	13.7	30.5	-
5-6	16.1	-	15.0	-	19.9	-	-
6-7	19.4	-	15.0	-	19.5	-	-
7-8	15.3	14.0	-	-	-	-	-
8-9	11.3	-	-	-	-	-	-
9-10	6.5	-	-	-	-	-	-
10-11	30.6	-	-	-	-	-	-
11-12	10.5	-	-	-	-	-	-
12-13	8.1	-	-	-	-	-	-
13-14	9.7	-	-	-	-	-	-
14-15	4.8	-	-	-	-	-	-
15-16	11.3	-	-	-	-	-	-
16-17	19.4	-	-	-	-	-	-
17-18	8.1	-	-	-	-	-	-
n	17	4	6	0	6	4	2
min	4.8	14.0	15.0	-	10.9	13.5	1.0
max	30.6	23.0	23.5	-	28.1	34.0	2.0
M	13.4	20.3	18.7	-	19.4	26.6	1.5
σn-1	6.7	4.3	4.0	-	6.4	9.0	0.7

(continued on the next page)

Table **5.33** Parameters of the Footprints (DIGITS) from Passagem das Pedras (SOPP)

FREE LENGTH OF DIGIT

SOPP/N°	01			02			03			04			06			12			13		
	II	III	IV	II	III	IV	II	III	IV	II	III	IV	II	III	IV	II	III	IV	II	III	IV
1	-	8.9	4.0	7.0	11.5	5.0	7.5	10.1	10.1	4.0	-	-	6.3	10.1	6.3	-	8.0	8.0	-	16.0	9.0
2	-	14.5	11.3	15.0	17.0	13.0	8.5	15.5	9.5	4.0	-	-	5.5	7.0	3.9	-	-	-	9.0	13.0	5.0
3	8.1	11.3	6.5	10.0	15.0	14.0	13.0	11.2	10.0	4.0	-	-	6.3	11.7	4.7	-	-	8.5	-	8.0	5.0
4	11.3	15.3	-	12.0	15.0	9.0	8.0	13.0	12.5	-	-	-	9.4	9.8	8.6	10.0	7.0	-	-	-	-
5	19.4	22.6	21.0	9.0	12.0	10.0	11.0	15.0	9.0	-	8.0	-	5.5	11.7	5.5	-	7.0	6.0	-	-	-
6	-	-	-	-	-	-	8.0	11.5	-	-	-	-	3.6	8.6	3.1	-	-	-	-	-	-
7	9.7	5.6	-	9.0	19.0	9.0	8.0	15.0	11.0	6.0	-	3.0	3.9	7.8	4.7	-	-	-	-	-	-
8	-	-	-	14.0	15.0	11.0	-														
9	-	-	-	8.0	13.0	8.0	-														
10	-	-	-	11.0	12.0	9.0	-														
11	8.1	-	4.0	-																	
12	8.1	17.7	18.5	-																	
13	-	-	-	-																	
14	-	-	-	-																	
15	-	-	-	-																	
16	-	-	-	-																	
17	9.7	8.1	8.1	-																	
18	9.7	11.3	12.1	-																	
n	8	9	8	9	9	9	7	7	6	4	1	1	5	7	4	1	3	3	1	3	3
min	8.1	5.6	4.0	7.0	11.0	5.0	7.5	10.1	9.0	4.0	-	-	3.9	9.0	3.1	-	7.0	6.0	-	8.0	5.0
max	19.4	22.6	21.0	15.0	19.0	14.0	13.0	15.5	12.5	5.0	-	-	9.4	11.7	6.3	-	8.0	8.5	-	16.0	9.0
M	10.5	12.8	10.7	10.6	14.3	9.8	9.1	13.0	10.4	4.5	-	-	6.1	9.5	4.7	-	7.3	7.5	-	12.3	6.3
σn-1	3.8	5.3	6.4	2.7	2.5	2.7	2.0	2.2	1.2	1.0	-	-	2.0	1.8	1.5	-	0.6	1.3	-	4.0	2.3

(continued on the next page)

Table 5.34 Parameters of the Footprints (DIGITS) from Passagem das Pedras (SOPP)

SOPP/N°	LENGTH OF THE PHALANGEAL PORTION OF THE DIGIT																				
	01			02			03			04			06			12			13		
	II	III	IV	II	III	IV	II	III	IV	II	III	IV	II	III	IV	II	III	IV	II	III	IV
1	-	-	-	16.0	26.0	11.0	-	-	-	-	-	-	14.8	18.0	25.8	-	-	23.0	-	-	16.0
2	-	-	-	-	-	-	-	-	-	-	-	-	14.8	18.0	10.2	-	-	-	16.0	18.0	8.0
3	-	-	-	-	-	-	-	-	-	-	-	-	-	-	-	-	-	27.5	-	17.0	9.0
4	-	-	-	-	-	-	-	-	-	-	-	-	15.6	18.0	16.4	28.0	-	-	-	-	-
5	-	-	-	-	-	-	-	-	-	-	-	-	10.9	21.9	18.8	-	-	18.0	-	-	-
6	24.2	-	-	-	-	-	-	-	-	-	-	-	-	17.2	12.5	-	-	-	-	-	-
7	-	-	-	-	-	-	-	-	-	-	-	-	-	-	-	-	-	-	-	-	-
8	-	-	-	-	-	-	-	-	-	-	-	-	-	-	-	-	-	-	-	-	-
9	-	-	-	-	-	-	-	-	-	-	-	-	-	-	-	-	-	-	-	-	-
10	15.3	22.6	22.6	-	-	-	-	-	-	-	-	-	-	-	-	-	-	-	-	-	-
11	-	-	-	-	-	-	-	-	-	-	-	-	-	-	-	-	-	-	-	-	-
12	-	-	-	-	-	-	-	-	-	-	-	-	-	-	-	-	-	-	-	-	-
13	16.1	16.1	16.1	-	-	-	-	-	-	-	-	-	-	-	-	-	-	-	-	-	-
14	16.1	16.9	17.7	-	-	-	-	-	-	-	-	-	-	-	-	-	-	-	-	-	-
15	19.4	16.1	17.7	-	-	-	-	-	-	-	-	-	-	-	-	-	-	-	-	-	-
16	15.3	21.0	21.0	-	-	-	-	-	-	-	-	-	-	-	-	-	-	-	-	-	-
17	-	-	-	-	-	-	-	-	-	-	-	-	-	-	-	-	-	-	-	-	-
18	-	-	-	-	-	-	-	-	-	-	-	-	-	-	-	-	-	-	-	-	-
n	6	5	5	1	1	1	0	0	0	0	0	0	4	5	5	1	0	3	1	2	3
min	15.3	16.1	16.1	-	-	-	-	-	-	-	-	-	10.9	17.2	10.2	-	-	18.0	-	17.0	8.0
max	24.2	22.6	22.6	-	-	-	-	-	-	-	-	-	15.6	21.9	25.8	-	-	27.5	-	18.0	16.0
M	17.7	18.5	19.0	-	-	-	-	-	-	-	-	-	14.0	18.6	16.7	-	-	22.8	-	17.5	11.0
σn-1	3.5	3.0	2.7	-	-	-	-	-	-	-	-	-	2.0	1.9	6.0	-	-	4.7	-	0.7	4.4

(continued on the next page)

Table 5.35 Parameters of the Footprints (DIGITS) from Passagem das Pedras (SOPP)

DIVARICATION OF DIGITS

SOPP/N°	01			02			03			04			06			12			13		
	II-III	III-IV	II-IV	II-III	III-IV	II-IV	II-III	III-IV	II-IV	II-III	III-IV	II-IV	II-III	III-IV	II-IV	II-III	III-IV	II-IV	II-III	III-IV	II-IV
1	19°	19°	-	30°	44°	74°	59°	54°30'	102°30'	19°	23°	44°	20°	18°	38°	-	26°	-	-	22°	-
2	-	27°	-	50°	36°	83°	57°	17°	73°30'	11°	16°30'	33°30'	13°	25°	40°	-	-	-	7°	23°	30°
3	26°30'	27°	53°	46°30'	23°	70°	56°	44°	100°30'	15°	-	-	17°30'	18°30'	36°	-	-	-	-	14°	-
4	16°	-	-	42°	37°	79°	46°	26°30'	75°	-	-	-	33°	21°	56°	-	-	-	-	-	-
5	12°30'	16°	29°	24°	34°	58°	57°30'	42°	100°30'	38°	31°	70°	15°	19°	35°	-	-	-	-	-	-
6	-	-	-	-	-	-	55°	-	-	-	-	-	19°	29°	49°	-	-	-	-	-	-
7	12°30'	-	-	38°	38°	76°	39°	41°30'	80°	41°	62°	108°	27°	27°	54°	-	11°	-	-	-	-
8	7°	13°30'	20°30'	32°	42°	73°	-	-	-	-	-	-	-	-	-	-	-	-	-	-	-
9	10°	17°	27°30'	35°	55°	88°	-	-	-	-	-	-	-	-	-	-	-	-	-	-	-
10	17°	11°	29°	50°30'	47°	97°	-	-	-	-	-	-	-	-	-	-	-	-	-	-	-
11	13°	20°	33°	-	-	-	-	-	-	-	-	-	-	-	-	-	-	-	-	-	-
12	2°	11°	13°30'	-	-	-	-	-	-	-	-	-	-	-	-	-	-	-	-	-	-
13	25°	11°	36°	-	-	-	-	-	-	-	-	-	-	-	-	-	-	-	-	-	-
14	15°	0°	17°	-	-	-	-	-	-	-	-	-	-	-	-	-	-	-	-	-	-
15	7°	-	0°	-	-	-	-	-	-	-	-	-	-	-	-	-	-	-	-	-	-
16	6°	8°	14°	-	-	-	-	-	-	-	-	-	-	-	-	-	-	-	-	-	-
17	5°	15°30'	21°30'	-	-	-	-	-	-	-	-	-	-	-	-	-	-	-	-	-	-
18	21°	14°30'	35°	-	-	-	-	-	-	-	-	-	-	-	-	-	-	-	-	-	-
n	15	14	13	9	9	9	7	6	6	5	4	4	7	7	7	-	2	0	1	3	1
min	2°	8°	13°30'	24°	23°	53°	39°	17°	73°30'	11°	16°30'	33°30'	13°	18°	35°	-	11°	-	7°	14°	30°
max	26°30'	27°	53°	50°30'	55°	97°	59°	54°30'	102°30'	41°	62°	108°	33°	29°	56°	-	26°	-	7°	23°	30°
M	13°	15°02'	25°18'	38°40'	39°33'	77°33'	52°47'	37°35'	88°40'	24°48'	33°07'	63°52'	20°38'	22°30'	44°	-	18°30'	-	7°	19°40'	30°
σn-1	7.2	7.1	13.1	9.3	9.0	11.1	7.4	13.5	13.9	13.8	20.1	33.2	7.0	4.5	8.8	-	10.6	-	-	4.9	-

Table 5.36 Parameters of One Trackway from Passagem das Pedras (SOPP 5)

N°	STRIDE	STEP ANGLE	EXTERNAL WIDTH	INTERNAL WIDTH	N°	GAUGE	OBLIQUE PACE	N°	DIVERGENCE OF FOOTPRINTS FROM THE MIDLINE
50-52	206.0	177°	34.0	-17.0	50-51	5.0	106.0	50	15°
51-53	194,5	163°	52.5	-16.5	51-52	10.0	100.5	51	9°
52-54	195.0	147°30'	62.5	-2.5	52-53	25.0	97.5	52	-18°
53-55	206.0	154°	65.0	-4.0	53-54	33.0	107.0	53	-18°
54-56	211.0	160°	48.0	-12.0	54-55	16.0	107.5	54	-2°
55-57	206.0	152°	52.5	-6.0	55-56	22.0	108.5	55	0°
56-58	197.5	162°	51.5	-6.0	56-57	24.5	105.0	56	-13°
57-59	195.0	168°	38.0	-19.0	57-58	8.0	97.0	57	16°
58-60	161.5	153°	54.0	-12.0	58-59	12.5	99.0	58	0°
59-61	172.5	145°	53.0	-2.0	59-60	28.0	69.0	59	-5°
60-62	204.5	160°30'	47.5	-9.0	60-61	21.0	112.0	60	27°
61-63	174.0	161°	42.0	-9.0	61-62	14.5	96.0	61	0°
62-64	183.0	163°30'	42.5	-15.0	62-63	17.5	80.0	62	0°
63-65	205.5	175°	36.5	-23.5	63-64	10.0	105.0	63	6°
64-66	203.5	175°30'	36.0	-20.5	64-65	4.0	102.0	64	28°
					65-66	9.0	103.0	65	10°
GAP					GAP	-	-	66	7°30'
76-78	206.0	168°30'	-	-	76-77	11.0	106.0	GAP	-
77-79	206.0	150°	-		77-78	10.0	103.0	76	4°
78-80	205.0	165°	-		78-79	9.0	105.5	77	7°30'
79-81	205.0	164°	-		79-80	17.0	101.5	78	21°
80-82	206.0	159°	-		80-81	26.0	107.0	79	2°
n	20	20	15	15	81-82	16.0	101.5	80	0°
min	161.5	145°	34.0	-23.5	n	22	22	81	9°
max	211.0	177°	65.0	-2.0	min	4.0	69.0	82	7°
M	197.2	161°10'	47.7	-11.6	max	33.0	112.0	n	24
σn-1	13.7	9.1	9.5	6.8	M	15.9	100.9	min	-18°
					σn-1	7.9	9.6	max	28°
								M	4°42'
								σn-1	11.8

Table 5.37 Parameters of One Trackway from Passagem das Pedras (SOPP 15)

SOPP/N°	STRIDE		STEP ANGLE		EXTERNAL WIDTH	INTERNAL WIDTH	INTERMANUS DISTANCE
	HAND	FOOT	HAND	FOOT			
1-3	296.3	288.8	174°30'	176°	34.0	-22.0	-21.0
2-4	237.5	248.0	173°30'	179°	31.5	-25.0	-13.5
n	2	2	2	2	2	2	2
min	237.5	248.0	173°30'	176°	31.5	-25.0	-21.0
max	296.3	288.8	174°30'	179°	34	-22.0	-13.5
M	266.9	268.4	174°	177°30'	32.8	-23.5	-17.3
σn-1	41.3	28.8	0.7	1.5	0.8	2.1	5.3

	GAUGE		OBLIQUE PACE			DISTANCE BETWEEN MANUS AND PES	DIVERGENCE OF FOOTPRINTS FROM THE MIDLINE	
	HAND	FOOT	HAND	FOOT			HAND	FOOT
1-2	8.8	5.3	168.5	155.8	1	17.8	10°30'	15°
2-3	8.5	3.8	128.5	133.5	2	30.5	-18	-1°
3-4	3.8	2.8	109.5	115.3	3	25.5	31°	23°
n	3	3	3	3	4	19.8	44°	29°
min	3.8	2.8	109.5	115.3	n	4	4	4
max	8.8	5.3	168.5	155.8	min	17.8	18°	-1°
M	7.0	4.0	135.5	134.9	max	30.5	44°	29°
σn-1	2.8	1.3	30.1	20.3	M	23.4	16°52'	16°30'
					σn-1	5.7	23.4	13.0

Table 5.38 Parameters of the Trackways from Serrote do Letreiro (SOSL)

SOSL/N°	STRIDE								STEP ANGLE							
	40*	41	42	43*	44	4999*	5	8	40*	41	42	43*	44	4999*	5	8
1-3	108.6	80.5	127.3	87.0	189.0	89.9	90.0	158.0	150°	158°30'	165°	172°	180°	156°	151°	-
2-4	-	84.3	128.7	-	138.0	-	-	-	-	156°	165°	-	174°	-	-	-
3-5	-	79.3	-	-	111.4	-	-	-	-	160°30'	-	-	164°	-	-	-
4-6	-	-	-	-	105.3	-	-	-	-	-	-	-	157°	-	-	-
5-7	-	-	-	-	120.3	-	-	-	-	-	-	-	170°30'	-	-	-
6-8	-	-	-	-	114.2	-	-	-	-	-	-	-	188°	-	-	-
7-9	-	-	-	-	107.2	-	-	-	-	-	-	-	211°	-	-	-
8-10	-	-	-	-	114.7	-	-	-	-	-	-	-	196°	-	-	-
n	1	3	2	1	8	1	1	1	1	3	2	1	8	1	1	0
min	-	79.6	127.3	-	105.3	-	-	-	-	156°	-	-	157°	-	-	-
max	-	84.3	128.7	-	189.1	-	-	-	-	160°30'	-	-	211°	-	-	-
M	-	81.5	128.0	-	125.0	-	-	-	-	158°30'	165°	-	180°03'	-	-	-
σn-1	-	2.5	1.0	-	27.8	-	-	-	-	2.2	0	-	17.7	-	-	-

	EXTERNAL WIDTH								INTERNAL WIDTH							
	40*	41	42	43*	44	4999*	5	8	40*	41	42	43*	44	4999*	5	8
1-3	29.6	23.4	26.2	17.3	-	-	28.6	-	1.9	-8.9	-10.5	-14.0	-	-	-5.0	-
2-4	-	25.3	26.6	-	-	-	-	-	-	-11.2	-13.1	-	-	-	-	-
3-5	-	25.3	-	-	-	-	-	-	-	-	-	-	-	-	-	-
4-6	-	-	-	-	26.2	-	-	-	-	-	-	-	-	-	-	-
5-7	-	-	-	-	26.2	-	-	-	-	-	-	-	-	-	-	-
6-8	-	-	-	-	19.2	-	-	-	-	-	-	-	-	-	-	-
7-9	-	-	-	-	12.7	-	-	-	-	-	-	-	-	-	-	-
8-10	-	-	-	-	15.4	-	-	-	-	-	-	-	-	-	-	-
n	1	3	2	1	5	0	1	0	1	2	2	1	0	0	1	0
min	-	23.4	26.2	-	26.2	-	-	-	-	-11.2	-13.1	-	-	-	-	-
max	-	25.3	26.6	-	12.7	-	-	-	-	-8.9	-10.5	-	-	-	-	-
M	-	24.7	26.4	-	19.9	-	-	-	-	-10.1	-11.8	-	-	-	-	-
σn-1	-	1.1	0.3	-	6.2	-	-	-	-	1.6	1.8	-	-	-	-	-

* Data obtained by extrapolation

(continued on the next page)

Table 5.39 Parameters of the Trackways from Serrote do Letreiro (SOSL)

SOSL/N°	OBLIQUE PACE								GAUGE							
	40*	41	42	43*	44	4999*	5	8	40*	41	42	43*	44	4999*	5	8
1-2	56.2	41.2	59.5	44.0	110.4	46.3	47.1	-	14.7	9.4	9.4	3.7	3.9	-	11.4	-
2-3	-	41.2	69.3	-	77.7	-	46.4	-	-	9.4	7.9	-	3.2	-	-	-
3-4	-	44.9	60.4	-	60.9	-	-	-	-	8.9	-	-	12.2	-	-	-
4-5	-	35.6	-	-	52.4	-	-	-	-	-	-	-	12.2	-	-	-
5-6	-	-	-	-	54.8	-	-	-	-	-	-	-	8.9	-	-	-
6-7	-	-	-	-	66.5	-	-	-	-	-	-	-	9.4	-	-	-
7-8	-	-	-	-	48.2	-	-	-	-	-	-	-	7.3	-	-	-
8-9	-	-	-	-	60.9	-	-	-	-	-	-	-	13.9	-	-	-
9-10	-	-	-	-	55.7	-	-	-	-	-	-	-	-	-	-	-
n	1	4	4	1	9	1	2	0	1	3	2	1	8	0	1	0
min	-	35.6	59.5	-	48.2	-	46.4	-	-	8.9	7.9	-	3.2	-	-	-
max	-	44.9	69.3	-	110.4	-	47.1	-	-	9.4	9.4	-	13.9	-	-	-
M	-	40.7	63.1	-	65.3	-	46.8	-	-	9.2	8.7	-	8.9	-	-	-
σn-1	-	3.8	5.4	-	19.0	-	0.5	-	-	0.3	1.1	-	3.9	-	-	-

* Data obtained by extrapolation

(continued on the next page)

Table 5.40 Parameters of the Trackways from Serrote do Letreiro (SOSL)

SOSL/N°	DIVERGENCE OF FOOTPRINTS FROM THE MIDLINE							
	40*	41	42	43*	44	4999*	5	8
1	0°	-3°	-18°30'	-4°30'	5°	-9°30'	-9°30'	-
2	0°	13°	-2°	-4°30'	-	-9°30'	0°	-
3	-	-6°	-10°30'	-	6°	-	11°	-
4	-	-6°	-17°30'	-	-	-	-	-
5	-	6°30'	-	-	0°	-	-	-
6	-	-	-	-	-	-	-	-
7	-	-	-	-	-	-	-	-
8	-	11°	-	-	-7°30'	-	-	-
9	-	-	-	-	4°	-	-	-
10	-	-	-	-	13°	-	-	-
n	2	6	4	2	6	2	3	0
min	-	-6°	-18°30'	-	-7°30'	-	-9°30'	-
max	-	13°	-2°	-	13°	-	11°	-
M	0°	2°45'	-12°06'	-4°30'	3°25'	-9°30'	0°30'	-
σn-1	0	8.4	7.6	0	6.8	0	10.3	-

* Data obtained by extrapolation

(continued on the next page)

Table 5.41 Parameters of the Footprints from Serrote do Letreiro (SOSL)

SOSL/N°	LENGTH OF THE FOOTPRINT							
	40	41	42	43	44	4999	05	08
1	18.2	-	27.6	17.3	25.2	21.1	21.4	76.6
2	16.6	18.7	26.9	19.2	-	20.6	21.4	-
3	-	17.1	23.7	-	30.1	-	18.6	59.8
4	-	19.3	24.3	-	-	-	-	-
5	-	19.3	-	-	20.0	-	-	-
6	-	-	-	-	18.5	-	-	-
7	-	-	-	-	24.9	-	-	-
8	-	20.8	-	-	27.2	-	-	-
9	-	-	-	-	25.8	-	-	-
10	-	-	-	-	25.8	-	-	-
n	2	5	4	2	8	2	3	2
min	16.6	17.1	23.7	17.3	18.5	20.6	18.6	59.8
max	18.2	20.8	27.6	19.2	30.1	21.1	21.4	76.6
M	17.4	19.0	25.6	18.3	24.7	20.5	20.5	68.2
$\sigma n\text{-}1$	1.1	1.3	1.9	1.3	3.8	1.6	1.6	11.9

	WIDTH OF THE FOOTPRINT							
	40	41	42	43	44	4999	05	08
1	15.9	-	18.0	16.8	18.7	16.8	16.4	75.4
2	11.9	16.1	19.2	15.0	-	-	11.4	-
3	-	16.9	20.4	-	-	-	16.4	62.2
4	-	16.7	19.9	-	-	-	-	-
5	-	16.1	-	-	16.3	-	-	-
6	-	-	-	-	-	-	-	-
7	-	-	-	-	-	-	-	-
8	-	16.5	-	-	21.5	-	-	-
9	-	-	-	-	-	-	-	-
10	-	-	-	-	-	-	-	-
n	2	5	4	2	3	1	3	2
min	11.9	16.1	18.0	15.0	16.3	-	11.4	62.2
max	15.9	16.9	20.4	16.8	21.5	-	16.4	75.4
M	13.9	16.5	19.4	15.9	18.8	-	16.7	68.8
$\sigma n\text{-}1$	2.8	0.4	1.0	1.3	2.6	-	2.9	9.3

	CROSS-AXIS ANGLE							
	40	41	42	43	44	4999	05	08
1	84°	-	100°30'	73°	94°	86°	94°30'	86°
2	-	87°	80°	77°	-	-	78°	-
3	-	79°	77°	-	-	-	77°	-
4	-	-	-	-	-	-	-	-
5	-	77°	-	-	99°30'	-	-	-
6	-	-	-	-	-	-	-	-
7	-	-	-	-	-	-	-	-
8	-	78°	-	-	93°30'	-	-	-
9	-	-	-	-	84°	-	-	-
10	-	-	-	-	75°	-	-	-
n	1	4	3	2	5	1	3	1
min	-	77°	77°	73°	75°	-	77°	-
max	-	87°	100°30'	77°	99°30'	-	94°30'	-
M	-	80°15'	85°48'	75°	89°12'	-	83°10'	-
$\sigma n\text{-}1$	-	4.6	12.8	2.8	9.7	-	9.8	-

(continued on the next page)

Table 5.42 Parameters of the Footprints (DIGITS) from Serrote do Letreiro (SOSL)

SOSL/N°	40			41			42			43		
	FREE LENGTH OF DIGIT											
	II	III	IV	II	III	IV	II	III	IV	II	III	IV
1	5.6	10.3	7.1	-	7.5	5.6	-	12.0	7.2	5.6	9.4	6.5
2	-	5.6	3.7	6.7	8.2	3.5	9.1	15.1	8.1	6.1	10.3	5.6
3	-	-	-	6.0	8.2	3.9	8.7	12.5	5.4	-	-	-
4	-	-	-	-	6.8	4.9	5.1	11.1	-	-	-	-
5	-	-	-	4.4	8.9	5.2	-	-	-	-	-	-
6	-	-	-	-	-	-	-	-	-	-	-	-
7	-	-	-	-	-	-	-	-	-	-	-	-
8	-	-	-	6.6	11.2	7.5	-	-	-	-	-	-
9	-	-	-	-	-	-	-	-	-	-	-	-
10	-	-	-	-	-	-	-	-	-	-	-	-
n	1	2	2	4	6	6	3	4	3	2	2	2
min	-	5.6	3.7	4.4	6.8	3.5	5.1	11.1	5.4	5.6	9.4	5.6
max	-	10.3	7.1	6.7	11.2	7.5	9.1	15.1	8.1	6.1	10.3	6.5
M	-	8.0	5.4	5.9	8.5	5.1	7.6	12.7	6.9	5.9	9.9	6.1
σn-1	-	3.3	2.4	1.1	1.5	1.4	2.2	1.7	1.4	0.4	0.6	0.6

SOSL/N°	40			41			42			43		
	LENGTH OF THE PHALANGEAL PORTION											
	II	III	IV	II	III	IV	II	III	IV	II	III	IV
1	12.2	-	10.8	-	-	-	-	-	-	-	-	15.0
2	-	-	-	11.2	-	13.1	12.7	-	20.6	-	14.5	15.9
3	-	-	-	10.3	-	12.4	11.2	-	18.7	-	-	-
4	-	-	-	-	-	-	12.2	-	-	-	-	-
5	-	-	-	10.3	-	15.0	-	-	-	-	-	-
6	-	-	-	-	-	-	-	-	-	-	-	-
7	-	-	-	-	-	-	-	-	-	-	-	-
8	-	-	-	12.2	15.9	18.7	-	-	-	-	-	-
9	-	-	-	-	-	-	-	-	-	-	-	-
10	-	-	-	-	-	-	-	-	-	-	-	-
n	1	0	1	4	1	4	3	0	2	0	1	2
min	-	-	-	10.3	-	12.4	11.2	-	18.7	-	-	15.0
max	-	-	-	12.2	-	18.7	12.7	-	20.6	-	-	15.9
M	-	-	-	11.0	-	14.8	12.0	-	19.7	-	-	15.5
σn-1	-	-	-	0.9	-	2.8	0.7	-	1.3	-	-	0.6

SOSL/N°	40			41			42			43		
	DIVARICATION OF DIGITS											
	II-III	III-IV	II-IV	II-III	III-IV	II-IV	II-III	III-IV	II-IV	II-III	III-IV	II-IV
1	44°30'	44°	89°	-	41°30'	-	-	46°	-	49°	32°	81°
2	-	17°30'	-	9°	18°30'	27°30'	26°30'	25°	51°	38°	27°	65°
3	-	-	-	52°	16°30'	78°	46°	31°30'	78°	-	-	-
4	-	-	-	-	35°	-	4°30'	-	-	-	-	-
5	-	-	-	29°	26°	55°	-	-	-	-	-	-
6	-	-	-	-	-	-	-	-	-	-	-	-
7	-	-	-	-	-	-	-	-	-	-	-	-
8	-	-	-	3°	29°	32°	-	-	-	-	-	-
9	-	-	-	-	-	-	-	-	-	-	-	-
10	-	-	-	-	-	-	-	-	-	-	-	-
n	1	2	1	4	6	4	3	3	2	2	2	2
min	-	17°30'	-	3°	16°30'	32°	4°30'	25°	51°	38°	27°	65°
max	-	44°	-	52°	41°30'	48°	46°	46°	78°	49°	32°	81°
M	-	30°48'	-	23°15'	27°45'	50°37'	25°42'	34°12'	64°30'	43°30'	29°30'	73°
σn-1	-	18.6	-	22.2	9.6	20.7	20.8	10.8	19.1	7.8	3.5	11.3

(continued on the next page)

Table 5.43 Parameters of the Footprints (DIGITS) from Serrote do Letreiro (SOSL)

FREE LENGTH OF DIGIT

SOSL/N°	44			4999			05			08		
	II	III	IV	II	III	IV	II	III	IV	II	III	IV
1	8.3	10.5	8.4	5.6	9.4	6.1	8.6	12.1	8.6	26.3	25.1	19.1
2	-	-	-	-	8.4	-	6.4	10.0	7.1	-	-	-
3	-	14.0	6.7	-	-	-	7.1	10.0	5.7	-	-	-
4	-	-	-	-	-	-	-	-	-	-	-	-
5	6.1	11.8	6.5	-	-	-	-	-	-	-	-	-
6	-	-	7.2	-	-	-	-	-	-	-	-	-
7	-	-	3.1	-	-	-	-	-	-	-	-	-
8	6.6	12.0	5.6	-	-	-	-	-	-	-	-	-
9	-	11.2	5.6	-	-	-	-	-	-	-	-	-
10	-	11.2	4.7	-	-	-	-	-	-	-	-	-
n	3	6	8	1	2	1	3	3	3	1	1	1
min	6.1	10.5	3.1	-	8.4	-	6.4	10.0	5.7	-	-	-
max	8.3	14.0	8.4	-	9.4	-	8.6	12.1	8.6	-	-	-
M	7.0	11.8	6.0	-	8.9	-	7.4	10.7	7.1	-	-	-
σn-1	1.2	1.2	1.6	-	0.7	-	1.1	1.2	1.5	-	-	-

LENGTH OF THE PHALANGEAL PORTION

	44			4999			05			08		
	II	III	IV	II	III	IV	II	III	IV	II	III	IV
1	14.0	-	15.4	10.3	-	18.7	12.6	12.6	17.1	74.2	-	75.4
2	-	-	-	-	-	-	10.0	13.6	17.6	-	-	-
3	-	-	21.9	-	-	-	9.3	12.1	15.7	-	-	-
4	-	-	-	-	-	-	-	-	-	-	-	-
5	14.5	-	14.0	-	-	-	-	-	-	-	-	-
6	-	-	16.4	-	-	-	-	-	-	-	-	-
7	-	-	-	-	-	-	-	-	-	-	-	-
8	21.5	-	20.6	-	-	-	-	-	-	-	-	-
9	-	-	-	-	-	-	-	-	-	-	-	-
10	-	-	20.6	-	-	-	-	-	-	-	-	-
n	3	0	6	1	0	1	3	3	3	1	0	1
min	14.0	-	14.0	-	-	-	9.3	12.1	15.7	-	-	-
max	21.5	-	21.9	-	-	-	12.6	13.6	17.6	-	-	-
M	16.7	-	18.2	-	-	-	10.6	12.8	16.8	-	-	-
σn-1	4.2	-	3.3	-	-	-	1.7	0.8	1.0	-	-	-

DIVARICATION OF DIGITS

	44			4999			05			08		
	II-III	III-IV	II-IV	II-III	III-IV	II-IV	II-III	III-IV	II-IV	II-III	III-IV	II-IV
1	12°	18°	29°	14°	34°30'	48°30'	37°	24°30'	60°30'	35°	35°	70°
2	-	-	-	-	-	-	33°	26°	59°	-	-	-
3	-	26°30'	-	-	-	-	37°30'	35°	72°30'	-	-	-
4	-	-	-	-	-	-	-	-	-	-	-	-
5	25°	34°	59°	-	-	-	-	-	-	-	-	-
6	-	-	-	-	-	-	-	-	-	-	-	-
7	-	-	-	-	-	-	-	-	-	-	-	-
8	24°	32°30'	56°	-	-	-	-	-	-	-	-	-
9	-	37°	-	-	-	-	-	-	-	-	-	-
10	-	18°	-	-	-	-	-	-	-	-	-	-
n	3	6	3	1	1	1	3	3	3	1	1	1
min	12°	18°	29°	-	-	-	33°	24°30'	59°	-	-	-
max	25°	37°	59°	-	-	-	37°30'	35°	72°30'	-	-	-
M	20°18'	27°20'	48°	-	-	-	35°50'	28°30'	64°	-	-	-
σn-1	7.2	8.2	16.5	-	-	-	2.5	5.7	7.4	-	-	-

Table 5.44 Parameters of the Trackways from Serrote do Pimenta (SOES)

SOES/Nº					STRIDE						
	1	2	3	4	6	10	11	12	17	18	22*
1-3	112.8	181.5	160.1	161.7	195.7	-	-	163.5	448.0	172.2	124.3
2-4	-	118.8	166.7	176.6	-	212.6	-	172.2	442.5	-	-
3-5	-	143.6	173.3	148.5	-	223.5	-	149.3	-	-	-
4-6	-	105.6	174.9	155.1	-	214.7	-	-	-	-	-
5-7	-	135.3	174.9	-	-	200.6	-	-	-	-	-
6-8	-	156.8	-	-	-	215.8	-	-	-	-	-
7-9	-	132.0	-	-	-	-	197.3	-	-	-	-
n	1	7	5	4	1	5	1	3	2	1	1
min	-	118.8	160.1	148.5	-	200.6	-	149.3	442.5	-	-
max	-	181.5	174.9	176.6	-	223.5	-	172.2	448.0	-	-
M	-	139.1	170.0	160.5	-	213.4	-	161.7	445.3	-	-
σn-1	-	24.9	6.5	12.0	-	8.3	-	11.6	3.9	-	-

					EXTERNAL WIDTH						
	1	2	3	4	6	10	11	12	17	18	22*
1-3	36.0	128.7	112.2	105.6	58.1	-	65.4	36.5	46.7	33.8	62.2
2-4	-	151.8	108.9	110.6	-	-	-	51.8	-	-	-
3-5	-	135.3	113.9	115.5	-	-	-	-	-	-	-
4-6	-	148.5	113.9	115.5	-	-	-	-	-	-	-
5-7	-	127.1	94.1	-	-	-	-	-	-	-	-
6-8	-	118.8	-	-	-	-	-	-	-	-	-
n	1	6	5	4	1	0	1	2	1	1	1
min	-	118.8	94.1	105.6	-	-	-	36.5	-	-	-
max	-	151.8	113.9	115.5	-	-	-	51.8	-	-	-
M	-	125.0	108.6	111.8	-	-	-	44.2	-	-	-
σn-1	-	12.9	8.4	4.7	-	-	-	10.8	-	-	-

					OBLIQUE PACE						
	1	2	3	4	6	10	11	12	17	18	22*
1-2	58.0	108.9	94.1	75.9	102.6	-	105.7	80.7	222.9	86.7	130.8
2-3	54.8	112.2	97.4	108.9	102.9	107.9	-	85.0	226.7	88.3	-
3-4	-	108.9	99.0	92.4	-	103.5	-	92.7	218.0	-	-
4-5	-	92.4	108.9	92.4	-	120.4	-	65.4	-	-	-
5-6	-	82.5	92.4	82.5	-	94.8	-	-	-	-	-
6-7	-	94.1	99.0	-	-	106.3	-	-	-	-	-
7-8	-	102.3	-	-	-	109.0	106.3	-	-	-	-
8-9	-	75.9	-	-	-	-	94.8	-	-	-	-
N	2	8	6	5	2	6	3	4	3	2	1
min	54.8	75.9	92.4	82.5	102.6	94.8	94.8	65.4	218.0	86.7	-
max	58.0	112.2	108.9	108.9	102.9	120.4	106.3	92.7	226.7	88.3	-
M	56.4	97.5	98.5	90.4	102.8	107.0	102.3	81.0	222.5	87.5	-
σn-1	2.3	12.6	5.8	12.5	0.2	8.3	6.5	11.5	4.4	1.1	-

					DIVERGENCE OF FOOTPRNTS FROM THE MIDLINE						
	1	2	3	4	6	10	11	12	17	18	22*
1	5°	17°	7°	-17°	20°30'	-	5°30'	-14°30'	1°	7°	0°
2	5°	-22°	-9°	11°	-11°	0°	-18°	-	8°30'	-	0°
3	-	-21°	4°30'	-7°30'	-7°	6°	-	-13°30'	-	4°30'	-
4	-	26°30'	18°	13°30'	-	-	-	-12°	-	-	-
5	-	-19°	13°	-10°	-	7°	-	-	-	-	-
6	-	15°	8°	17°30'	-	12°30'	-	-	-	-	-
7	-	-14°	13°	-	-	-7°	-26°	-	-	-	-
8	-	-22°	-	-	-	10°	-22°	-	-	-	-
N	2	8	7	6	3	6	4	3	2	2	2
min	-	-22°	-9°	-17°	-11°	-7°	-26°	-14°30'	1°	4°30'	-
max	-	26°30'	18°	17°30'	20°30'	12°30'	5°30'	-12°	8°30'	7°	-
M	5°	-4°56'	7°74'	1°15'	0°5'	4°45'	-15°07'	-13°20'	4°45'	5°45'	0°
σn-1	0	20.7	8.7	14.5	17.1	7.1	14.1	1.3	5.3	1.8	0

(continued on the next page)

Table 5.45 Parameters of the Trackways from Serrote do Pimenta (SOES)

SOES/N°						STEP ANGLE					
	1	2	3	4	6	10	11	12	17	18	22
1-3	180°	110°	115°	119°	147°30'	-	-	163°	172°30'	162°	144°
2-4	-	66°	115°	122°	-	179°	-	152°	177°30'	-	-
3-5	-	93°	111°	118°	-	179°	-	138°30'	-	-	-
4-6	-	74°	121°	126°	-	183°	-	-	-	-	-
5-7	-	98°30'	132°	-	-	180°	-	-	-	-	-
6-8	-	110°30'	-	-	-	178°	-	-	-	-	-
7-9	-	100°	-	-	-	-	168°	-	-	-	-
n	1	7	5	4	1	5	1	3	2	1	1
min	-	66°	111°	118°	-	178°	-	138°30'	172°30'	-	-
max	-	110°30'	132°	126°	-	183°	-	163°	177°30'	-	-
M	-	93°8'	118°48'	121°15'	-	179°48'	-	151°10'	175°	-	-
σn-1	-	17.1	8.2	3.6	-	1.9	-	12.3	3.5	-	-

						INTERNAL WIDTH					
	1	2	3	4	6	10	11	12	17	18	22
1-3	-17.3	8.3	-6.6	-11.6	-7.7	-	-	-	-5.8	-6.5	25.1
2-4		21.5	-6.6	-11.6	-	-	-	-	-	-	-
3-5		-9.9	-5.0	-3.3	-	-	-	-	-	-	-
4-6		-9.9	-9.9	-23.1	-	-	-	-	-	-	-
5-7		-6.6	-11.6	-	-	-	-	-	-	-	-
6-8		-3.3	-	-	-	-	-26.1	-	-	-	-
n	1	6	5	4	1	0	1	0	1	1	1
min		-9.9	-11.6	-23.1	-	-	-	-	-	-	-
max		21.5	-5.0	-3.3	-	-	-	-	-	-	-
M		0	-7.9	-12.4	-	-	-	-	-	-	-
σn-1		12.5	2.7	8.1	-	-	-	-	-	-	-

						GAUGE					
	1	2	3	4	6	10	11	12	17	18	22
1-2	-	66.0	56.1	49.5	22.4	-	-	14.7	20.7	16.1	39.6
2-3	3.5	62.7	46.2	38.0	29.8	3.8	-	29.4	8.2	-	-
3-4	-	77.6	57.8	56.1	-	2.2	-	29.4	-	-	-
4-5	-	59.4	59.4	52.8	-	3.8	-	-	-	-	-
5-6	-	59.4	39.6	24.8	-	3.8	-	-	-	-	-
6-7	-	52.8	36.3	-	-	1.1	-	-	-	-	-
7-8	-	56.1	-	-	-	-	-	-	-	-	-
8-9	-	52.8	-	-	-	-	34.9	-	-	-	-
N	1	8	6	5	2	5	1	3	2	1	1
min	-	52.8	36.3	24.8	22.4	1.1	-	14.7	8.2	-	-
max	-	77.6	59.4	56.1	29.8	3.8	-	29.4	20.7	-	-
M	-	60.9	49.2	44.2	26.1	2.9	-	24.5	14.5	-	-
σn-1	-	8.2	9.9	12.8	5.2	1.2	-	8.5	8.8	-	-

(continued on the next page)

Table 5.46 Parameters of the Footprints from Serrote do Pimenta (SOES)

SOES/N°	LENGTH OF THE FOOTPRINT										
	01	02	03	04	06	10	11	12	17	18	22
1	-	71.0	-	-	37.1	-	29.1	30.1	41.5	-	-
2	36.0	67.7	56.1	62.7	41.3	24.3	38.4	-	36.6	27.7	25.1
3	34.0	-	54.5	56.1	34.3	-	-	27.4	36.0	-	-
4	-	72.6	64.4	56.1	-	-	-	-	39.6	-	-
5	-	69.3	56.1	61.1	-	-	-	-	-	-	-
6	-	72.6	46.2	64.4	-	32.3	-	-	-	-	-
7	-	59.4	46.2	-	-	-	27.6	-	-	-	-
8	-	62.7	-	-	-	29.8	26.9	-	-	-	-
n	2	7	6	5	3	3	4	2	4	1	1
min	34.0	59.4	46.2	56.1	34.3	24.3	26.9	27.4	36.0	-	-
max	36.0	72.6	64.4	64.4	41.3	32.3	38.4	30.1	41.5	-	-
M	35.0	67.9	53.9	60.1	37.6	28.8	30.5	28.8	38.4	-	-
σn-1	1.4	5.1	5.1	3.8	3.5	4.1	5.3	1.9	2.6	-	-

	WIDTH OF THE FOOTPRINT										
	01	02	03	04	06	10	11	12	17	18	22
1	-	59.4	-	-	29.1	-	35.4	21.8	27.3	-	21.0
2	25.0	59.4	52.8	52.8	30.1	22.6	34.2	-	-	19.8	17.7
3	26.0	46.2	51.2	52.8	29.4	22.1	-	21.5	-	20.7	-
4	-	72.6	51.2	57.8	-	-	-	-	-	-	-
5	-	56.1	56.1	59.4	-	-	-	-	-	-	-
6	-	69.3	49.5	46.2	-	27.8	-	-	-	-	-
7	-	49.5	46.2	-	-	-	31.1	-	-	-	-
8	-	52.8	-	-	-	24.3	32.9	-	-	-	-
n	2	8	6	5	3	4	4	2	1	2	2
min	25.0	46.2	46.2	46.2	29.1	22.1	31.1	21.5	-	19.8	17.7
max	26.0	72.6	56.1	59.4	30.1	27.8	35.4	21.8	-	20.7	21.0
M	25.5	58.2	51.2	53.8	29.5	24.2	33.4	21.7	-	20.3	19.4
σn-1	0.7	9.2	3.3	5.2	0.5	2.6	1.8	0.2	-	0.6	2.3

	CROSS-AXIS ANGLE										
	01	02	03	04	06	10	11	12	17	18	22
1	-	-	-	-	90°	-	93°	90°30'	91°	-	115°30'
2	102°	-	-	-	93°	99°30'	99°	-	84°	-	114°
3	87°	-	-	-	71°	102°	-	98°	-	91°30'	-
4	-	-	-	-	-	-	-	93°	-	-	-
5	-	-	-	-	-	-	-	-	-	-	-
6	-	-	-	-	-	97°	-	-	-	-	-
7	-	-	-	-	-	-	101°	-	-	-	-
8	-	-	-	-	-	96°	90°	-	-	-	-
n	2	0	0	0	3	4	4	3	2	1	2
min	87°	-	-	-	71°	96°	90°	90°30'	84°	-	114°
max	102°	-	-	-	93°	102°	101°	98°	91°	-	115°30'
M	94°30'	-	-	-	84°40'	98°37'	95°45'	93°50'	87°30'	-	114°45'
σn-1	10.6	-	-	-	11.9	2.7	5.1	3.8	4.9	-	11.1

(continued on the next page)

Table 5.47 Parameters of the Footprints (DIGITS) from Serrote do Pimenta (SOES)

SOES/N°	01			02			03			04		
FREE LENGTH OF DIGIT												
	II	III	IV	II	III	IV	II	III	IV	II	III	IV
1	-	-	-	-	-	-	-	-	-	-	-	-
2	10.8	-	17.5	-	-	-	-	-	-	-	-	-
3	9.5	-	-	-	-	-	-	-	-	-	-	-
4	-	-	-	-	-	-	-	-	-	-	-	-
5	-	-	-	-	-	-	-	-	-	-	-	-
6	-	-	-	-	-	-	-	-	-	-	-	-
7	-	-	-	-	-	-	-	-	-	-	-	-
8	-	-	-	-	-	-	-	-	-	-	-	-
n	2	0	1	0	0	0	0	0	0	0	0	0
min	9.5	-	-	-	-	-	-	-	-	-	-	-
max	10.8	-	-	-	-	-	-	-	-	-	-	-
M	10.2	-	-	-	-	-	-	-	-	-	-	-
σn-1	0.9	-	-	-	-	-	-	-	-	-	-	-

	01			02			03			04		
LENGTH OF THE PHALANGEAL PORTION												
	II	III	IV	II	III	IV	II	III	IV	II	III	IV
1	-	18.5	-	-	-	-	-	-	-	-	-	-
2	20.0	22.3	25.5	-	-	-	-	-	-	-	-	-
3	-	24.0	19.5	-	-	-	-	-	-	-	-	-
4	-	-	-	-	-	-	-	-	-	-	-	-
5	-	-	-	-	-	-	-	-	-	-	-	-
6	-	-	-	-	-	-	-	-	-	-	-	-
7	-	-	-	-	-	-	-	-	-	-	-	-
8	-	-	-	-	-	-	-	-	-	-	-	-
n	1	3	2	0	0	0	0	0	0	0	0	0
min	-	18.5	19.5	-	-	-	-	-	-	-	-	-
max	-	24.0	25.5	-	-	-	-	-	-	-	-	-
M	-	21.6	22.5	-	-	-	-	-	-	-	-	-
σn-1	-	2.8	4.2	-	-	-	-	-	-	-	-	-

	II-III	III-IV	II-IV	II-III	III-IV	II-IV	II-III	III-IV	II-IV	II-III	III-IV	II-IV
DIVARICATION OF DIGITS												
1	-	-	-	-	-	-	-	-	-	-	-	-
2	20°	20°30'	40°30'	-	-	-	-	-	-	-	-	-
3	27°	17°	44°	-	-	-	-	-	-	-	-	-
4	-	-	-	-	-	-	-	-	-	-	-	-
5	-	-	-	-	-	-	-	-	-	-	-	-
6	-	-	-	-	-	-	-	-	-	-	-	-
7	-	-	-	-	-	-	-	-	-	-	-	-
8	-	-	-	-	-	-	-	-	-	-	-	-
n	2	2	2	0	0	0	0	0	0	0	0	0
min	20°	17°	40°30'	-	-	-	-	-	-	-	-	-
max	27°	20°30'	44°	-	-	-	-	-	-	-	-	-
M	23°30'	18°45'	42°15'	-	-	-	-	-	-	-	-	-
σn-1	4.9	2.5	2.5	-	-	-	-	-	-	-	-	-

(continued on the next page)

Table 5.48 Parameters of the Footprints (DIGITS) from Serrote do Pimenta (SOES)

SOES/N°	FREE LENGTH OF DIGIT											
	06			10			11			12		
	II	III	IV	II	III	IV	II	III	IV	II	III	IV
1	17.5	25.2	16.8	-	-	-	16.9	18.7	14.5	-	-	16.6
2	14.7	22.8	11.2	-	13.1	9.8	15.7	22.2	16.7	-	-	11.7
3	-	16.1	14.7	11.3	-	9.8	-	-	-	-	-	-
4	-	-	-	-	-	-	-	-	-	-	-	12.0
5	-	-	-	-	-	-	-	-	11.4	-	-	-
6	-	-	-	16.7	23.1	16.4	-	-	-	-	-	-
7	-	-	-	-	-	-	14.4	20.5	12.0	-	-	-
8	-	-	-	8.2	18.1	11.9	19.0	19.6	13.1	-	-	-
n	2	3	3	3	3	4	4	4	4	0	0	3
min	14.7	16.1	11.2	8.2	13.1	9.8	14.4	18.7	12.0	-	-	11.7
max	17.5	25.2	16.8	16.7	23.1	16.4	19.0	22.2	16.7	-	-	16.6
M	16.1	21.4	14.2	12.1	18.1	12.0	16.5	20.3	14.1	-	-	13.4
σn-1	2.0	4.7	2.8	4.3	5.0	3.1	2.0	1.5	2.0	-	-	2.7

	LENGTH OF THE PHALANGEAL PORTION											
	06			10			11			12		
	II	III	IV	II	III	IV	II	III	IV	II	III	IV
1	26.6	-	-	-	-	-	24.0	-	24.5	18.0	21.3	24.0
2	21.7	-	-	-	-	15.8	20.7	25.1	30.5	-	-	-
3	-	-	18.2	-	-	22.3	-	-	-	-	-	24.0
4	-	-	-	-	-	-	-	-	-	-	-	-
5	-	-	-	-	-	-	-	-	-	-	-	-
6	-	-	-	24.5	-	24.5	-	-	-	-	-	-
7	-	-	-	-	-	-	20.7	-	16.9	-	-	-
8	-	-	-	20.7	23.4	15.3	22.9	24.0	17.4	-	-	-
n	2	0	1	2	1	4	4	2	4	1	1	2
min	21.7	-	-	20.7	-	15.3	20.7	24.0	16.9	-	-	-
max	26.6	-	-	24.5	-	24.5	24.0	25.1	30.5	-	-	-
M	24.2	-	-	22.6	-	19.5	22.1	24.6	22.3	-	-	24.0
σn-1	3.5	-	-	2.7	-	4.6	1.7	0.8	6.5	-	-	0

	DIVARICATION OF DIGITS											
	II-III	III-IV	II-IV	II-III	III-IV	II-IV	II-III	III-IV	II-IV	II-III	III-IV	II-IV
1	21°	31°30'	52°30'	-	-	-	19°30'	52°	71°30'	27°	24°30'	52°30'
2	27°30'	28°	56°	35°	14°	49°	17°	43°30'	61°	-	-	-
3	20°	28°	47°	19°	59°	78°	-	-	-	20°30'	31°	52°
4	-	-	-	-	-	-	-	-	-	10°30'	35°	45°
5	-	-	-	-	37°	-	-	-	-	-	-	-
6	-	-	-	32°	28°	60°	-	-	-	-	-	-
7	-	-	-	-	18°	-	14°30'	58°30'	73°	-	-	-
8	-	-	-	39°30'	27°	67°	27°	57°	83°	-	-	-
n	3	3	3	4	6	4	4	4	4	3	3	3
min	20°	28°	47°	19°	14°	49°	14°30'	43°30	61°	10°30'	24°30'	45°
max	27°30'	31°30'	56°	39°30'	59°	78°	27°	58°30'	83°	27/	35°	52°30'
M	22°50'	29°10'	51°50'	31°22'	30°30'	63°30'	19°30'	52°45'	72°07'	19°20'	30°10'	49°50'
σn-1	4.1	2.0	4.5	8.8	16.1	12.2	5.4	6.8	9.0	8.3	5.3	4.2

(continued on the next page)

Table 5.49 Parameters of the Footprints (DIGITS) from Serrote do Pimenta (SOES)

FREE LENGTH OF DIGIT

SOES/N°	17			18			22		
	II	III	IV	II	III	IV	II	III	IV
1	16.9	26.8	16.6	-	-	-	9.5	-	-
2	-	22.5	11.7	-	-	-	4.5	14.8	12.5
3	-	-	-	-	6.5	6.5	-	-	-
4	-	-	12.0	-	-	-	-	-	-
5	-	-	-	-	-	-	-	-	-
6	-	-	-	-	-	-	-	-	-
7	-	-	-	-	-	-	-	-	-
8	-	-	-	-	-	-	-	-	-
n	1	2	3	0	1	1	2	1	1
min	-	22.3	11.7	-	-	-	4.5	-	-
max	-	26.8	16.6	-	-	-	9.5	-	-
M	-	24.6	13.4	-	-	-	7.0	-	-
σn-1	-	3.2	2.7	-	-	-	3.5	-	-

LENGTH OF THE PHALANGEAL PORTION

SOES/N°	17			18			22		
	II	III	IV	II	III	IV	II	III	IV
1	30.5	28.9	29.4	-	-	-	20.2	-	-
2	-	-	25.6	-	-	-	16.4	-	18.0
3	-	-	-	-	-	-	-	-	-
4	-	-	28.9	-	-	-	-	-	-
5	-	-	-	-	-	-	-	-	-
6	-	-	-	-	-	-	-	-	-
7	-	-	-	-	-	-	-	-	-
8	-	-	-	-	-	-	-	-	-
n	1	1	3	0	0	0	2	0	1
min	-	-	25.6	-	-	-	16.4	-	-
max	-	-	29.4	-	-	-	20.2	-	-
M	-	-	28.0	-	-	-	18.4	-	-
σn-1	-	-	2.1	-	-	-	2.8	-	-

DIVARICATION OF DIGITS

	II-III	III-IV	II-IV	II-III	III-IV	II-IV	II-III	III-IV	II-IV
1	25°	17°30'	42°	26°	26°30'	52°30'	18°30'	25°	43°
2	-	14°30	-	-	-	-	22°	18°	39°
3	-	-	-	-	33°	-	-	-	-
4	-	-	-	-	-	-	-	-	-
5	-	-	-	-	-	-	-	-	-
6	-	-	-	-	-	-	-	-	-
7	-	-	-	-	-	-	-	-	-
8	-	-	-	-	-	-	-	-	-
n	1	2	1	1	2	1	2	2	2
min	-	14°30'	-	-	26°30'	-	18°30'	18°	39°
max	-	17°30'	-	-	33°	-	22°	25°	43°
M	-	16°	-	-	29°45'	-	20°15'	21°30'	41°
σn-1	-	2.1	-	-	4.6	-	2.5	4.9	2.8

Table 5.50 Parameters of one Trackway from Serrote do Pimenta (SOES 9)

SOES/N°	STRIDE		STEP ANGLE		EXTERNAL WIDTH	INTERNAL WIDTH	INTERMANUS DISTANCE
	HAND	FOOT	HAND	FOOT			
1-3	257.5	259.0	165°	143°	87.0	-10.5	-8.3
2-4	247.5	248.5	176°	148°	80.5	-11.5	-23.0
3-5	259.0	273.0	176°30'	148°	82.0	-11.5	-22.3
4-6	253.0	250.3	162°	151°	82.0	-12.3	-1.8
n	4	4	4	4	4	4	4
min	247.5	248.5	162°	143°	80.5	-12.3	-23.0
max	259.0	273.0	176°30'	151°	87.0	-10.5	-1.8
M	254.3	257.7	169°52'	147°30'	82.9	-11.5	-13.9
σn-1	5.2	11.2	7.5	x.3	2.8	0.7	10.5

	GAUGE		OBLIQUE PACE	
	HAND	FOOT	HAND	FOOT
1-2	17.5	36.0	134.8	144.0
2-3	18.0	35.0	124.3	128.0
3-4	2.5	32.5	122.8	132.5
4-5	1.0	34.5	135.5	149.5
5-6	39.3	34.5	124.0	111.3
n	5	5	5	5
min	1.0	32.5	122.8	111.3
max	39.3	36.0	135.5	149.5
M	15.7	34.5	128.3	133.1
σn-1	15.5	1.3	6.3	14.9

	DISTANCE BETWEEN MANUS AND PES	DIVERGENCE OF FOOTPRINTS FROM THE MIDLINE	
		HAND	FOOT
1	51.0	-25°	-24°
2	50.0	14°	-15°
3	46.5	13°30'	-16°30'
4	45.0	36°	-10°
5	52.5	-63°	-19°
6	48.5	19°	-3°
n	6	6	6
min	45.0	-63°	-24°
max	52.5	36°	-3°
M	48.9	-0°55'	-14°35'
σn-1	2.8	36.4	7.3

(continued on the next page)

Table 5.51 Parameters of the Footprints from Serrote do Pimenta (SOES 9)

SOES/Nº	LENGTH		FREE LENGTH OF DIGIT		
				FOOT	
	HAND	FOOT	II	III	IV
1	26.5	54.5	22.0	-	21.0
2	-	52.5	-	25.0	-
3	25.0	46.0	25.5	25.5	17.5
4	16.5	50.3	19.5	23.0	22.0
5	16.8	56.0	11.8	14.0	7.0
6	23.0	52.0	35.0	25.0	29.5
n	6	6	5	5	5
min	16.5	46.0	11.8	14.0	7.0
max	26.5	56.0	35.0	25.5	29.5
M	21.6	51.9	22.8	22.5	19.4
σn-1	4.7	3.5	8.5	4.8	8.2

	WIDTH		LENGTH OF THE PHALANGEAL PORTION		
1	16.0	51.3	-	29.3	-
2	23.8	49.0	-	-	-
3	25.5	46.3	-	-	-
4	11.5	50.3	-	-	-
5	10.5	43.0	-	-	-
6	16.3	47.0	-	-	-
n	6	6	0	1	0
min	10.5	43.0	-	-	-
max	23.8	51.3	-	-	-
M	17.3	47.8	-	-	-
σn-1	6.2	3.0	-	-	-

	CROSS-AXIS ANGLE		DIVARICATION OF DIGITS		
			II-III	III-IV	II-IV
1	100°	-	21°	24°	46°
2	96°	-	14°	31°	45°
3	98°	-	17°	35°	53°
4	91°	-	17°	36°	53°
5	79°	-	33°	25°	58°
6	92°	-	12°	34°	45°
n	6	0	6	6	6
min	79°	-	12°	24°	45°
max	100°	-	33°	36°	58°
M	92°30'	-	19°	30°30'	50°
σn-1	7.5	-	7.5	5.2	5.4

Table 5.52 Parameters of the Footprints from Curral Velho (SOCV) and Mãe D'água (SOMD) isolated Footprints

SOCV/N°	LENGTH	FREE LENGTH OF DIGIT		
		II	III	IV
1	25.6	-	-	-
2	29.2	-	-	-
3	17	4.1	12.9	-
4	45.8	16.8	19.5	8.4
5	20.0	8.1	11.1	3.3
6	30.8			
7	32.2	8.5	20.0	12.2
8	49.8	5.8	13.2	-
9	56.3	10.5	21.6	14.7
10	38.1	11.4	16.7	10.5
11	28.6	5.8	12.3	6.2
15	29.0	7.7	13.5	7.7

SOCV/N°	WIDTH	LENGTH OF THE PHALANGEAL PORTION		
		II	III	IV
1	27.8	19.3	20.0	25.6
2	28.8	23.5	21.2	23.5
3	-	8.9	-	-
4	41.1	32.1	-	37.4
5	17.0	14.4	-	13.0
6	21.9	19.2	20.0	26.2
7	23.7	15.6	-	22.6
8	40.5	-	-	-
9	47.9	-	-	-
10	35.2	-	-	-
11	24.1	-	-	17.6
15	25.2	-	-	-

SOMD/N°	LENGTH	FREE LENGTH OF DIGIT		
		II	III	IV
1	35.7	-	19.6	10.0
2	32.7	-	-	-
3	48.3	10.4	30.4	-
4	48.7	11.7	20.7	12.2

SOMD/N°	WIDTH	LENGTH OF THE PHALANGEAL PORTION		
		II	III	IV
1	-	-	-	-
2	19.3	14.7	25.7	18.0
3	37.4	-	-	-
4	26.1	-	-	-

SOMD/N°	CROSS-AXIS ANGLE	DIVARICATION OF DIGITS		
		II-III	III-IV	II-IV
1	-	-	33°	-
2	-	6°30´	18°30´	24°30´
3	-	26°30´	26°	53°
4	-	39°	35°30´	75°

SOCV/N°	CROSS-AXIS ANGLE	DIVARICATION OF DIGITS		
		II-III	III-IV	II-IV
1	-	40°	12°	52°
2	-	37°30´	22°30´	60°
3	-	20°30´	-	-
4	66°	51°	0°	51°
5	-	19°30´	15°30´	35°
6	-	23°	35°	58°
7	110°	9°	43°	52°
8	-	23°	41°	63°
9	-	14°	30°	44°
10	-	15°30´	15°30´	31°
11	-	11°30´	16°30´	28°
15	-	20°	37°	57°

The Trackmakers of the Ichnofaunas of the Rio Do Peixe Basins

It is not always easy to attribute a track to a trackmaker, and this is particularly true for the ichnofaunas of northeastern Brazil. In South America, unfortunately for our study, Cretaceous dinosaurian faunas are best known and studied from Argentina (a long way from Sousa!) rather than from Brazil. The only dinosaurian notable fauna (as known from bones) from the Neocomian terrains of South America is that of the La Amarga Formation (and correlative formations) in the Neuquén basin, Patagonia, Argentina.

In Brazil, dinosaurs are more abundant in the south and southeast than in the northeast and in Triassic and Upper Cretaceous terrains rather than those of the Lower Cretaceous. The other South American countries are very poor in dinosaur skeletons, probably because there are only a few or no vertebrate paleontologists searching for those bones. As for dinosaurs of the Early Cretaceous (especially the Neocomian) more generally, they are poorly known not only in South America but the world over, in comparison with the rich dinosaurian faunas of the Middle and Late Jurassic and the Late Cretaceous.

South American dinosaurs are mostly very different from those of the northern continents, with the exception of Hadrosauria, Ankylosauria (Nodosauridae), and some Ceratopsia that invaded this continent toward the end of the Cretaceous, coming from North and Central America. This dissimilarity corresponds to the very probable geographic and biogeographic isolation of South American and, more generally, Gondwanan dinosaur faunas from those of northern continents (Laurasia) during a large part of the Jurassic and almost all of the Cretaceous, an interesting phenomenon of endemism (Bonaparte 1986, 2007). Consequently, there is a large difference between South American and Laurasian dinosaur faunas and a notable affinity between South American dinosaur faunas and those of the other Gondwanan plates: Africa, Madagascar, India, Antarctica, and Australia (fig. 6.1). Toward the end of the Cretaceous, about 10 million years before the K/T boundary, contact was reestablished between North and South America in the Caribbean belt, and there was an interchange between those plates. As a result, titanosaurids and South American theropods expanded their ranges northward and North American theropods, hadrosaurs, and other ornithopods (Rozadilla et al. 2016), ankylosaurs, and ceratopsids invaded the South American continent (Bonaparte 2007). However, one should note the interesting results of Dunhill et al. (2016) on the sound hypothesis that the separation of the continents surely made the faunal exchanges among them, right up

Figure 6.1. An environmental reconstruction of the meandering fluvial system with a wide floodplain, where perennial and temporary lakes were developed, corresponding to the depositional environment of the Sousa Formation. The reconstruction includes the more common dinosaurs represented by the local ichnofauna: theropods, probably different forms of abelisaurids, and large and medium-size titanosaurid sauropods. The mesofauna is represented by notosuchid crocodiles and ararypemid aquatic turtles. (Art by Deverson Silva (Pepi).)

to the end of the Cretaceous, more difficult, but it did not completely prevent them.

Another difficulty in attributing South American tracks to their track-makers, and in comparing them to those of other continents, especially Laurasia, is that in general both bones and tracks of dinosaurs are found in continental terrains, where precise dating is often very difficult. This is because correlation of continental formations and faunas with the classic stratigraphic and chronological units instituted for the mainly marine sequence of Europe is always problematic. Furthermore, radiometric dating requires specific kinds of rock for analysis and is not at all possible for the sedimentary rocks of the Rio do Peixe basins.

Last but not least, reasonably complete foot skeletons are rarely found, and, particularly for theropods, the feet are usually conservative in structure—not just the feet but also the overall body plan, apart from the skull (Hendrickx et al. 2015). As a consequence, the resolution at which different ichnomorphotypes can be associated with pedal skeletons of particular zoological taxa is not very great. For example, at Rio do Peixe, many theropod tracks have a wide total digital divarication (between digits II-IV), which is an interesting characteristic; however, it is difficult to interpret the systematic significance of this observation in the absence of good foot skeletons.

These problems notwithstanding, the tracks described above can be attributed, with some probability and confidence, to the following groups from South America, Africa, and other Gondwanan plates.

Theropods

Large Theropods

Theropod classification is especially complicated, and it is particularly difficult to attribute theropod tracks to their makers. This group includes several distinct clades and subclades—more than twenty at the level of family and superfamily, according to Hendrickx et al. (2015). However, because they are found in the Lower Cretaceous terrains of South America, many large theropod tracks of the Rio do Peixe basins very probably pertain mainly to medium-size and large theropods (5–9 m long; Hendrickx et al. 2015) of the family Abelisauridae Bonaparte and Novas, 1985 (fig. 6.2; probably part of the Ceratosauria). This clade was definitely recorded from the Aptian-Albian to Maastrichtian. Moreover, bones of a small abelisauroid, a large abelisaurid, and a probably medium to large megalosaurid tetanuran were recently recorded from the type locality of the Bajada Colorada Formation (Lower Cretaceous, Berriasian and Valanginian), Neuquén, Argentina (Canale et al. 2016), but perhaps this clade was already present in the Aalenian-Bajocian (Middle Jurassic) of Patagonia (Hendrickx et al. 2015).

Among the abelisaurids from the Cenomanian of South America are the genera *Xenotarsosaurus* Martínez, Giménez, Rodríguez and Bochatey, 1986, an abelisaurid with a very interesting tarsal articulation (Barreal Formation; Ocho Hermanos, province of Chubut, Argentina); *Ekriksinatosaurus* Calvo, Rubilar-Rogers and Moreno, 2004 (Candelero

Figure 6.2. A panorama of the Sousa basin. Two abelisaurids observe, from the Precambrian hills outside the Early Cretaceous graben, a small herd of sauropods. Art by Ariel Milani Martine.

Formation, Añelo, Province of Neuquén, Argentina); and *Ilokelesia* Coria and Salgado, 1998 (Huincul Formation, Aguada Grande, Plaza Huincul, Province of Neuquén, Argentina).

Abelisaurids may not have been the only theropod trackmakers of the Rio do Peixe basins, however. Some of the large theropod tracks are possibly attributed to large and very large (8–17m long; Hendrickx et al. 2015) Spinosauridae Stromer, 1915 sensu Sereno et al. (1998), the sister group of Megalosauridae. These theropods fed on fish, dinosaurs, and pterosaurs. Outside of South America, spinosaurids are recorded in Niger in Africa and the Barremian of England and Portugal (Hendrickx et al. 2015).

For South America itself, from the Aptian-Albian terrains (more probably Albian; Araripe basin, State of Ceará, northeastern Brazil; Romualdo Formation) comes the genus *Irritator* Martill et al., 1996, classified as Maniraptora by Martill and coauthors, but more probably assigned to Spinosauridae (according to Kellner 1996; Sues et al. 2002, and Bonaparte 2007). Unfortunately, this form is represented only by a partial skull. Another probable spinosaurid is *Angaturama limai* Kellner and Campos, 1996, from the same Araripe basin and the same stratigraphic position (Ceará, Brazil) as *Irritator changelleri*, and now considered a junior synonym thereof (Charig and Milner 1997; Sues et al. 2002).

Medeiros (2006) recognized spinosaurid teeth, comparable in size to those of the enormous Albian-Cenomanian North African form *Spinosaurus*, from northeastern Brazil (Cajual Island, Bahia de São Marcos, Maranhão; Alcântara Formation, Eocenomanian). From the same basin (São Luis) and stratigraphic position comes the spinosaurid *Oxalaia* (Kellner et al. 2011).

Another group of candidates for makers of large Brazilian theropod tracks are the Carcharodontosauridae, medium-size to very large allosauroid theropods (6–14 m long; Hendrickx et al. 2015). Outside of South America, they have been seen in the Kimmeridgian-Tithonian of Tanzania, the Barremian of Spain, the Aptian-Albian of North America, the Cenomanian-?Santonian of north Africa, and the Turonian of China (Hendrickx et al. 2015). South American forms include the colossal *Giganotosaurus* Coria and Salgado, 1995 (Candeleros Formation, El Chocón, Neuquén, Argentina), whose feet are unfortunately unknown; *Mapusaurus* Coria and Currie, 2006 (at least 7 individuals; Huincul Formation, Las Cortaderas, Plaza Huincul, Province of Neuquén, Argentina); and the huge *Tyrannotitan* Novas et al., 2005 (2 individuals; Cerro Barcino Formation, Estancia La Juanita, Paso de Indios, Province of Chubut, Argentina). Allosauridae sensu stricto must probably be excluded as possible trackmakers of the large theropod tracks of the Rio do Peixe basins, as they are strictly Late Jurassic in age.

Although we can make plausible guesses about the clades of theropods that might have been responsible for footprints in the Rio do Peixe basins, we cannot go into any details about which species or genera the actual trackmakers were. We do not know any form of large theropod from the Neocomian terrains of the South American continent; all we have are some isolated and up to now unidentified teeth from the La Amarga Formation. Recently, however, Canale et al. (2016) described new theropod materials from the type locality of the Bajada Colorada Formation (Lower Cretaceous, Berriasian-Valanginian; Neuquén, Argentina), which include a small abelisauroid, a large abelisaurid, and a probably medium-to-large megalosaurid tetanuran.

In any case, the great diversity of large theropods in the Cretaceous of South America (in terms of the number of the genera and families) may explain the corresponding diversity of the large theropod footprint morphotypes in the Rio do Peixe basins. Something analogous can probably be said for the footprints of small theropods and for the footprints of sauropods.

Small Theropods, Some with a Long Digit III

Small- and medium-size theropod footprints with digit III substantially longer than digits II and IV (and often sinuous) were classically attributed to the Coelurosauria (see, e.g., Haubold 1971), as they were then understood (von Huene 1914). Today, however, the makers of this kind of morphotype, recognized during the nineteenth and twentieth centuries mainly on the basis of Triassic and Lower Jurassic tracks of the Laurasian

continents (i.e., those pertaining to the *Grallator-Eubrontes* plexus), are no longer considered to be Coelurosauria but very probably Ceratosauria.

Small theropod tracks of this kind, or very similar, are found in the Rio do Peixe basins, although they are rather rare (fig. 6.3). These morphotypes can be related here to some South American (and more broadly Gondwanan) theropod families that occupied, in South America, ecological niches analogous to those filled by several small and medium-size theropods in Laurasia. Prime examples of such southern small- and medium-size theropods are the abelisauroid Noasauridae Bonaparte and Powell, 1980 and Velocisauridae Bonaparte, 1991. Noasauridae are recorded in the Barremian-Early Aptian of Argentina, although they are better known in the latter part of the Gondwanan Cretaceous (Hendrickx et al. 2015).

From the Neocomian comes the genus *Ligabueino* Bonaparte, 1996 (Abelisauridae or, more probably, Noasauridae; La Amarga Formation, Province of Neuquén, Argentina). Unfortunately, it is impossible to compare the Brazilian tracks with the foot skeleton of this form, of which just two phalanges are recorded.

Of particular interest is the genus *Santanaraptor* Kellner, 1999, from the Aptian, which comes from the Chapada do Araripe (state of Ceará, northeastern Brazil; Romualdo Member of the Santana Formation; probably Albian), just some 200 km WSW of the Rio do Peixe basins. Equally interesting, its metatarsus III is notably stouter and longer than metatarsals II and IV. Mainly because of this common characteristic, Bonaparte (2007) attributes *Santanaraptor* to the family Velocisauridae Bonaparte, 1991, together with the Senonian (late Cretaceous) genus *Velocisaurus* Bonaparte, 1991, from the upper Cretaceous of Patagonia (Bajo de la Carpa Formation, Neuquén, province of Neuquén, Argentina). *Velocisaurus* is a Senonian theropod; however, its feet would perfectly fit some small theropod tracks of the Rio do Peixe basins.

From the same member and formation as *Santanaraptor* also comes *Mirischia* Naish, Martill and Frey, 2004, which is probably a basal coelurosaur (Bonaparte 2007), pertaining to the Compsognathidae, according to Hendrickx et al. (2015). Its feet are unfortunately unknown. South American coelurosaurs from Cenomanian terrains include the small-bodied genus "*Aniksosaurus*" Martínez and Novas, 1997, which probably is a basal coelurosaur but waits for a formal institution; it is based on 5

Figure 6.3. Small- or medium-size theropod with digit III substantially longer than digits II and IV (and often sinuous) pertaining to the Ceratosauria, perhaps to the Noasauridae. Their tracks are found in the Rio do Peixe basins, although they are rather rare. Art by Ariel Milani Martine.

incomplete individuals from the Bajo Barreal Formation, south of the province of Chubut, Argentina, and the fine small form *Buitreraptor* Makovicky, Apesteguía and Agnolín, 2005 (Early Dromeosauridae; La Buitrera, province of Río Negro, Argentina).

Sauropods

South American Sauropods: Neocomian

Sauropod tracks of the Rio do Peixe basins were perhaps impressed by Dicraeosauridae, Rebbachisauridae, or early Titanosauriformes (figs. 6.1, 6.4). Among the Dicraeosauridae is the genus *Amargasaurus* Salgado and Bonaparte, 1991, from the La Amarga Formation, Neocomian of the province of Neuquén, Argentina, with its characteristic very long neural spines in the cervical vertebrae. More recently (Canale et al. 2016), bones of diplodocid and dicraeosaurid sauropods were recorded from the type locality of the Bajada Colorada Formation (Lower Cretaceous, Berriasian and Valanginian), Neuquén, Argentina.

For the Rebbachisauridae, note *Rayososaurus* Bonaparte, 1996 (Rayoso Formation, Neocomian, Neuquén, Argentina). Also, dicraeosaurid and diplodocid sauropod material was recently recorded (Canale et al. 2016) from Bajada Colorada Formation (type locality, Lower Cretaceous, Berriasian-Valanginian).

The set of three articulated caudal vertebrae and two isolated chevrons, whose characteristics allow attribution to the basal Titanosauriformes, recently found in the Uiraúna-Brejo das Freiras (Triunfo) basin, points to the basal Titanosauriformes as possible trackmakers of some of the sauropod tracks from the Sousa basin.

The unique formally described dinosaur species from Rio do Peixe basins is *Triunfosaurus leonardii* Carvalho, Salgado, Lindoso, Araújo, Nogueira and Agnelo, 2017 (fig. 6.5). It was found in the Rio Piranhas Formation (Triunfo basin) and allows new perspectives into the comprehension of paleobiogeographical and temporal distribution of the titanosaur sauropods (Carvalho et al. 2017). Titanosaurs are common in Upper Cretaceous rocks of Brazil and Argentina, but the Early Cretaceous age of this species shows the first steps of titanosaur evolution throughout Gondwanaland.

Considering that the size of some sauropod footprints is compatible with that of *Triunfosaurus leonardii*, we include this species among the possible trackmakers of the Rio do Peixe basins.

South American Sauropods: Aptian-Albian

Other South American sauropods of the Early Cretaceous, although more recent than the Neocomian, are *Zapalasaurus* Salgado, Carvalho, and Garrido 2006 (early Aptian; La Amarga Formation; Neuquén, Argentina; ?Diplodocoidea); *Amazonsaurus* Carvalho, Santos Avilla, and Salgado, 2003 (?Diplodocoidea; Itapecuru Formation, Aptian, Maranhão, northeastern Brazil); *Ligabuesaurus* Bonaparte, Gonzales Riga, and Apesteguía, 2006 (Ligabuesauridae, Titanosauriformes, Lohan Cura Formation, Aptian-Albian, Neuquén, Argentina); *Agustinia* Bonaparte,

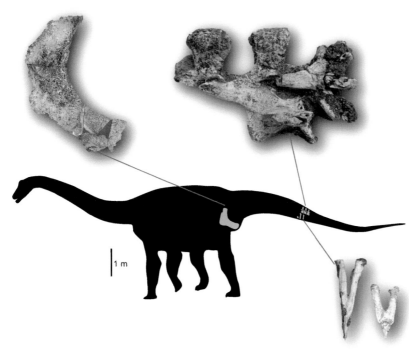

Figure 6.4. Herds of sauropods, represented by assemblages of at least 7–15 parallel trackways, are rather frequent in the Sousa basin. Indeed, the social or gregarious behavior is here represented almost only by sauropods. Art by Ariel Milani Martine.

Figure 6.5. The discovery of a dinosaur specimen in the Rio Piranhas Formation of the Triunfo basin (*Triunfosaurus leonardii*) opens new perspectives into the comprehension of paleogeographical and temporal distribution of the titanosaur sauropods, as well as in the interpretation of the sauropod tracks in the Sousa basin.

1 m

Figure 6.6. The giant *Argentinosaurus*, with its enormous feet, can explain the very large sauropod tracks (diameter up to 120 cm) found in the Sousa basin, especially in level 13/3 at Caiçara-Piau farms. This photo was taken in an exhibition of Argentinian dinosaurs in Padua (Italy).

1999 (Agustiniidae, Lohan Cura Formation, Aptian-Albian, Neuquén, Argentina).

South American Sauropods: Cenomanian

Included here are *Limaysaurus* Calvo and Salgado, 1995 (Salgado et al. 2005) (Rebbachisauridae, Cenomanian, Candeleros Formation, Neuquén, Argentina); *Andesaurus* Bonaparte and Calvo, 1991 (Titanosauridae; Cenomanian, Candeleros Formation, Neuquén, Argentina); the gigantic *Argentinosaurus* Bonaparte and Coria, 1993 (Titanosauridae; Cenomanian, Huincul Formation, Neuquén, Argentina; fig. 6.6); *Epachthosaurus* Powell, 1986 (Titanosauridae; Cenomanian, Bajo Barreal Formation, province of Chubut, Argentina); *Chubutisaurus* Del Corro, 1975 (Titanosauriformes; Cenomanian, Cerro Barcino Formation, province of Chubut, Argentina).

The morphology of the hand-foot set from Serrote do Pimenta (SOES 7, Antenor Navarro Formation), discovered by Leonardi in 1979 (Leonardi 1982, 1994), indicates that it probably belongs to an ankylosaur, most likely a nodosaurid (fig. 6.7), similar to little *Minmi* Molnar, 1980 from Queensland, Australia (Albian). The quadrupedal trackway (undertrack) from Passagem das Pedras (SOPP 15, Sousa Formation) could also, perhaps, belong to this group. Possible ankylosaur trackways were already known from Bolivia at Toro-Toro, Potosí (two parallel trackways of *Ligabueichnium bolivianum* Leonardi, 1984, classified by him as possible ankylosaurs; see also Bonaparte et al. 1993; Meyer et al. 2018, 2019). Many more (at least 7 long trackways) were discovered more recently (1994) by geologist José Hugo Heymann and by paleontologist Mario Suárez Riglos (1995) and were initially studied there on the big wall at Cal Orck'o quarry, Sucre (1998), by Christian Meyer, Martin Lockley, and Leonardi (Lockley et al. 2002; Meyer et al. 2018, 2019). The Bolivian trackways are, however, found in terrains of the Santonian-Maastrichtian (Upper Cretaceous), and skeletal remains of this group are very rare and doubtful in South America. The Bolivian prints were presumably made by Nodosauridae immigrants from North America sometime during the last ten million years of the Cretaceous.

Ankylosauria

Figure 6.7. A hand-foot set of a quadrupedal dinosaur of medium-large size found at Serrote do Pimenta site (SOES 7) was attributed to ankylosaurians, probably nodosaurids. It was the first ankylosaur track reported from South America. Art by Ariel Milani Martine.

There are no skeletons or isolated bones of large, graviportal ornithopods in the record of the Lower Cretaceous of South America, or in general, before the immigration of North American hadrosaurids in the Campanian or Early Maastrichtian. This probably means only that such fossils have yet to be found because the ornithopod tracks from the Rio do Peixe basins prove that such plant-eating dinosaurs were present, at least in northeastern Brazil (fig. 6.8).

Ornithopoda

Figure 6.8. The ichnofauna of the Rio do Peixe basins contains the tracks of 38 graviportal ornithopod iguanodontids, some of them bipedal, some semiquadrupedal or quadrupedal. Art by Ariel Milani Martine.

Figure 6.9. Some rare small ornithopod tracks from Rio do Peixe basins may belong to juveniles of large ornithopods or to small-size endemic ornithopods, perhaps dryosaurids. Art by Ariel Milani Martine.

Indeed, the large trackways and isolated footprints with three rounded hooves, like *Caririchnium*, *Sousaichnium*, and *Staurichnium*, as well as the short trackway attributable to *Iguanodonichnus* at Serrote do Letreiro, are surely attributable to medium-size to large graviportal iguanodontian ornithopods—similar, for example, to *Ouranosaurus* Taquet, 1976 from the Aptian-Albian of Niger or *Muttaburrasaurus* Bartholomai and Molnar, 1981 from the Albian of Queensland. Such ornithopod trackmakers would quite surely have been endemic to Gondwanaland. *Ouranosaurus* would be especially interesting as a maker of some ornithopod tracks of the Sousa basin, if northeastern Brazil were really "out of Africa" (Maisey 2011).

According to Díaz-Martínez et al. (2015, 40) *Caririchnium magnificum* tracks from Brazil could have been made by a basal iguanodontid, maybe something like an Ankylopollexia and/or a Styracosterna. However, bones of these clades have not been found up to now in the Lower Cretaceous of South America.

As for the rare small ornithopod tracks from Rio do Peixe, they may belong to juveniles of large ornithopods like the ones described above or, more likely, to some form of small-size endemic ornithopods—perhaps dryosaurids (fig. 6.9), already known from South American lower Cretaceous terrains (Coria and Salgado 1996).

Crocodyliformes

The batrachopodid hand-foot set at Piau is poorly impressed but could be attributed to Crocodyliformes. The more recently published crocodylian traces from Tapera are attributed (possibly) to Mesoeucrocodylia (Campos et al. 2010). The trackways of Crocodyliforme are generally rare. One of the few known ichnospecies is *Angolaichnus adamanticus* from the Early Cretaceous of Angola (Mateus et al. 2017). Another interesting specimen is from Bajocian, Middle Jurassic terrains of Central Iran (Abbassi 2015). And in the Uiraúna-Brejo das Freiras basin, one femur of a small notosuchian crocodile was found (Carvalho and Nobre 2001; fig. 3.25A).

The innumerable very small half-swimming tracks at Piau-Caiçara are very probably nearly all attributable to turtles (Leonardi and Carvalho 2000; Leonardi and Santos 2006; Dentzien-Dias et al. 2010), most likely members of the Araripemydae Gaffney and Meylan 1991 (fig. 6.10).

Chelonia

Figure 6.10. The main element of the mesofauna in the Rio do Peixe basins is represented by numerous small tracks very probably imprinted in the substrate by swimming turtles, perhaps of the family Araripemydae. Art by Ariel Milani Martine.

Legend

?	uncertain classification
1	track#1 of the ichnosite
Ank	Ankylosauria
STh	Small Theropoda with long digit III
MTh	Medium-sized Theropoda
LTh	Large Theropoda
GOr	Graviportal Ornithopoda
Sau	Sauropoda
IS	Incertae sedis (uncertain classif.)
/	or

Table 7.1 Classification of the Sousa basin Trackways.

ANBD	1	LTh	SOES	1	LTh
ANBD	2	LTh	SOES	100	Sau
ANBD	3	LTh	SOES	101	Sau
ANEN	1	STh	SOES	102	Sau
ANEN	3	LTh	SOES	103	Sau
ANEN	4	LTh	SOES	104	Sau
ANEN	8	GOr	SOES	105	Sau
ANEN	12	LTh	SOES	106	Sau
ANEN	30	IS	SOES	107	Sau
ANEN	31	LTh	SOES	6	LTh
ANJU	2	LTh	SOES	9	GOr
ANJU	3	LTh	SOES	10	LTh
ANJU	4	LTh	SOES	11	LTh
ANPV	1	LTh	SOES	12	LTh
SOCA	42	LTh	SOES	17	LTh
SOCA	43	GOr	SOES	18	LTh
SOCA	44	LTh	SOES	22	LTh
SOCA	46	GOr	SOMA	1	LTh
SOCA	46	GOr	SOMA	2	LTh
SOCA	48	LTh	SOPI	1	LTh
SOCA	493	LTh/GOr	SOPM	1	LTh
SOCA	57	LTh	SOPM	4	LTh
SOCA	75	LTh	SOPM	5	LTh
SOCA	97	LTh	SOPM	6	LTh
SOCA	98	LTh	SOPM	7	LTh
SOCA	99	LTh	SOPP	1	GOr
SOCA	991	LTh/GOr	SOPP	2	LTh
SOCA	993	LTh	SOPP	3	LTh
SOCA	1323	STh	SOPP	4	LTh
SOCA	13215	MTh	SOPP	5	GOr
SOCA	13236	STh	SOPP	12	LTh
SOCA	1331	MTh	SOPP	15	?Ank
SOCA	1336	LTh	SOSL	40	MTh
SOCA	1337	LTh	SOSL	42	MTh
SOCA	13368	LTh	SOSL	43	MTh
SOCA	1350	LTh	SOSL	44	MTh
SOCA	150	LTh/GOr	SOSL	4999	MTh
SOCA	151	LTh	SOSL	5	MTh
SOCA	153	IS			

Behavior of the Rio Do Peixe Basins Dinosaurs

7

We ordinarily think of biosedimentary traces as ephemeral: one says "to write in sand," meaning to vote to oblivion, but as we have seen, footprints can become more permanent fossils. Paleoichnology is not limited to description, illustration, and classification of traces in para-Linnaean taxa and, when possible, attributing them to Linnaean taxa.

Paleoichnology also is the primary and "unrivaled" (Gatesy and Ellis 2016) tool for making inferences about the behavior of the trackmakers: "Whilst bones present a static view of extinct animals, fossil footprints are a direct record of the activity and motion of the track maker" (Falkingham et al. 2020, 1). What stances and gaits did they use, at what levels of activity did they progress, in what directions, and for what reasons? Did they move by themselves or as members of social groups? What do trackway patterns possibly reveal about the body structure of the trackmakers, their diets, and even their metabolic physiology? (Leonardi 2011; Lockley and Meyer 1999; Falkingham et al. 2016a, 2020; Gatesy and Ellis 2016). Ichnology becomes for us a true dinosaur registry office, with its ichnological database and census studies (see, e.g., Xing et al. 2019).

In fact, tracks constitute a remarkable reservoir of data about gait, speed estimation, posture, glenoacetabular distance, limb and foot kinematics, ethology, and metabolism (Leonardi 1984b, 1984c; 1994, 129–30; 1997; 2011; Citton et al. 2015a; Romano and Citton 2020). A list of abbreviated terms used in graphs and tables may be found at the beginning of chapter 4.

Speeds

A relatively small number of trackways (75) from the Rio do Peixe basins allowed estimation of trackmaker speeds (tables 7.1, 7.2), following the methods of Alexander (1989) as modified by Thulborn (1990).

The distribution of calculated trackmaker speeds shows marked positive asymmetry. This is, however, a mixed sample, composed as it is of trackways of sauropods, ornithopods, and (mainly) theropods (table 7.2). The calculated speed of 58 of these trackways (78.67% of the sample) is between 3 and 7 km/h (fig. 7.1). These animals were therefore moving with a walking gait and speed. At the far left of the histogram are 7 trackways with a very slow calculated speed (≤ 2 km/h; 9.33%); 4 of these are sauropods. To the far right of the central (modal) block of columns, nine (12%) trackways indicate a speed between 8 and 23 km/h. Of these, 5 (6.67%) have estimated speeds of 7–10 km/h; another 4 (5.33%) are scattered over a range between 12.8 and 22 km/h. These last 4 trackways

Table 7.2 Estimated speeds of the makers of Sousa basin dinosaur trackways.

Code of the track	classif.	FL	xSL	h=4FL	h=yFL	Vt	Vc h=4FL	Vc h=yFL	SL/FL	SL/h h=4FL	SL/h h=yFL
SOPM 1	LTh	32.2	177.4	128.8	157.8	4.3-4.7	5.5	4.2	5.5	1.4	1.1
SOPM 4	LTh	29.3	182.3	117.2	143.6	4.1-4.5	6.4	5.0	6.2	1.5	1.3
SOPM 5	LTh	24.5	156	98	110.2	2.8-4	6.0	5.3	6.7	1.6	1.4
SOPM 6	LTh	27.3	175.5	109.2	133.8	4-4.4	6.6	5.1	6.4	1.6	1.3
SOPM 7	LTh	28.5	175.5	114	139.6	4.1-4.5	6.1	4.9	6.2	1.5	1.3
SOPP 1	GOr	44.1	186.7	176.4	260.2	4.9-5.8	4.2	2.6	4.2	1.1	0.7
SOPP 2	LTh	37.7	214.1	150.8	184.7	4.6-5	6.1	5.4	5.7	1.4	1.2
SOPP 3	LTh	35	482.8	140	171.5	4.5-4.9	26.2	20.7	13.8	3.4	2.8
SOPP 4	LTh	20	244.6	80	98	3.5-3.8	16.3	12.8	12.2	2.0	2.5
SOPP 5	LTh	29.9	197.2	119.6	146.5	4.2-4.5	7.0	5.6	6.6	1.6	1.3
SOPP 6	GOr	29.9	169.6	119.6	176.4	4.2-4.9	5.4	3.5	5.7	1.4	1.0
SOPP 12	LTh	24.2	217.5	96.8	108.9	3.8-4	10.7	9.2	9	2.2	2.0
SOPP 13	LTh	26	189	104	127.4	3.9-4.3	7.7	6.1	7.3	1.8	1.5
SOCA 42	LTh	29	182	115	142.1	4.1-4.5	4.5	6.0	6.3	1.6	1.3
SOCA 43	GOr	32.3	93.5	129.2	190.6	4.3-5.1	1.8	1.2	2.9	0.7	0.5
SOCA 44	LTh	30.9	201.5	123.6	151.4	4.2-4.6	7.0	5.6	6;5	1.6	1.3
SOCA 45	LTh	32.4	193	129.6	158.8	4.3-4.7	6.2	4.9	5.9	1.5	1.2
SOCA 46	GOr	19.8	113	79.2	95	3.5-3.8	4.4	3.6	5.7	1.4	1.2
SOCA 48	LTh	26.5	177.5	106	129.8	4-4.3	6.8	5.4	6.7	1.7	1.4
SOCA 493	LTh or GOr	32.4	155.5	129.6	175	4.3-4.9	4.3	3.0	4.8	1.2	0.9
SOCA 57	LTh	25.8	134.6	103.2	126.4	3.9-4.3	4.5	3.5	5.2	1.3	1.1
SOCA 75	LTh	28.5	229	114	139.6	4.1-4.5	9.7	7.5	8	2	1.6
SOCA 91	LTh	38.3	167	153.2	187.7	4.6-5	4.0	3.2	4.4	1.1	0.9
SOCA 97	LTh	33.7	116.7	134.8	165.1	4.4-4.8	2.6	2.0	3.5	0.9	0.7
SOCA 98	LTh	34	181	136	166;6	4.4-4.8	5.3	4.2	5.3	1.3	1.1
SOCA 99	LTh	31.7	180.8	126.8	155.3	4.3-4.7	5.7	4.5	5.7	1.4	1.2
SOCA 991	LTh or GOr	28.5	172	114	154	4.1-4.6	6.0	4.2	6.0	1.5	1.1
SOCA 993	LTh	31	212.5	124	151.9	4.2-4.6	7.7	6.0	6.8	1.7	1.4
SOCA 1323	STh	11.9	103.1	47.6	53.5	2.8-3	7.1	6.2	8.7	2.2	1.9
SOCA 13215	MTh	20.8	83.8	83.2	93.6	3.6-3.8	2.6	2.2	4.0	1.0	1.9
SOCA 13236	STh	19.4	120.1	77.6	87.3	3.5-3.6	5.1	4.5	6.2	1.5	1.4
SOCA 1331	MTh	31.1	261.5	124.4	152.39	4.2-4.6	10.9	8.5	8.4	2.1	1.7
SOCA 1336	MTh	29.5	82	118	144.55	4.1-4.5	1.6	1.3	2.8	0.7	0.6
SOCA 1337	LTh	28.8	143	115.2	141.1	4.1-4.5	4.3	3.4	5.0	1.2	1.0
SOCA 13368	LTh	37.4	137.6	149.6	183.3	4.6-5	3.0	2.4	3.7	0.9	0.7
SOCA 1350	LTh	23.9	178.7	95.6	107.5	3.8-4	7.8	6.8	7.5	1.9	1.7
SOCA 150	LTh or GOr	21	141.8	84	97.6	3.6-3.8	6.2	5.2	6.7	1.7	1.4
SOCA 151	LTh	21.6	183.8	86.4	97.2	3.6-3.8	9.3	8.0	8.5	2.1	1.9
ANBD 1	LTh	21	156.5	84	94.5	3.6-3.8	7.3	6.3	7.4	1.9	1.6
ANBD 2	LTh	22	150	88	99	3.7-3.9	6.4	5.6	6.8	1.9	1.5

Code of the track	classif.	FL	xSL	h=4FL	h=yFL	Vt	Vc h=4FL	Vc h=yFL	SL/FL	SL/h h=4FL	SL/h h=yFL
ANBD 3	LTh	26.6	175.3	106.4	130.3	4-4.3	6.7	5.3	6.6	1.6	1.3
ANEN 1	STh	15.6	83.5	62.4	70.2	3.2-3.3	3.6	3.1	5.3	1.4	1.2
ANEN 3	LTh	25	191.8	100	122.5	3.9-4.2	8.4	6.6	7.7	1.9	1.6
ANEN 4	LTh	23.2	114.2	92.8	104.4	3.7-3.9	3.8	3.3	4.9	1.2	1.1
ANEN 8	Sau or / LTh	68.5	134.2	274	404.1	5.9-7	1.4	0.9	2.0	0.5	0.3
ANEN 12	LTh	24.4	146.2	97.6	109.8	3.8-4	5.4	4.7	6.0	1.5	1.3
ANEN 30	LTh	18;2	134.2	72.8	81.9	3.4-3.6	6.6	5.8	7.4	1.8	1.6
ANEN 31	LTh	24.5	220.4	98	110.2	3.8-4	10.7	9.3	9.0	2.2	2.0
ANJU 2	LTh	27.4	192.3	109.6	134.3	4-4.4	7.4	5.9	7.0	1.7	1.4
ANJU 3	LTh	28.9	121	115.6	141.6	4.1-4.5	3.2	2.6	4.2	1.0	0.8
ANJU 4	LTh	38	212.5	152	186.2	4.6-5	6.0	4.7	5.6	1.4	1.1
ANPV 1	LTh	22.5	173.5	90	101.2	3.7-3.9	7.9	6.9	7.7	1.9	1.7
SOPI 1	LTh	26.3	141	106	128.9	4-4.3	4.6	3.7	5.4	1.3	1.1
SOMA 1	LTh	20.7	162.3	122.8	150.4	4.2-4.6	4.9	3.9	5.3	1.3	1.1
SOMA 2	LTh	21.6	162	86.4	97.2	3.6-3.8	7.5	6.6	7.5	1.9	1.7
SOSL 40	LTh	17.4	108.6	69.6	78.3	3.3-3.5	4.9	4.5	6.2	1.6	1.4
SOSL 41	LTh	19	81.5	76	85.5	3.4-3.6	2.7	2.4	4.3	1.1	0.9
SOSL 42	LTh	25.6	128	102.4	115.2	3.9-4.1	4.1	3.6	5.0	1.2	1.1
SOSL 43	LTh	18.3	87	73.2	82.3	3.4-3.6	3.2	2.8	4.7	1.2	1.1
SOSL 44	LTh	24.7	125	98.8	111.1	3.8-4	4.1	3.6	5.0	1.3	1.1
SOSL 4999	LTh	20.9	89.9	83.6	91	3.6-3.8	2.9	2.5	4.3	1.1	0.9
SOSL 5	LTh	20.5	90	82	92.2	3.6-3.7	3.0	2.9	4.4	1.1	1.0
SOSL 8	GOr	68.2	158	272.8	402.4	6-6.9	1.8	1.2	2.3	0.6	0.4
SOES 1	LTh	35	475	140	171.5	4.5-4.9	25.5	22	13.6	3.4	2.8
SOES 100	Sau	67.9	139.1	271.6	400.6	5.9-6.9	1.5	0.9	2.0	0.5	0.3
SOES 101	Sau	53.9	170	215.6	318	5.3-6.3	2.8	1.8	3.2	0.8	0.5
SOES 102	Sau	60.1	160.5	240.4	354.6	5.6-6.6	2.2	1.7	2.7	0.7	0.5
SOES 6	LTh	37.6	195.7	150.4	184.2	4.6-5	5.2	4.1	5.2	1.3	1.1
SOES 9	GOr	51.9	254.3	207.6	306.2	5.3-6.2	5.6	3.6	4.9	1.2	0.8
SOES 10	LTh	28.8	213.4	115.2	141.1	4.1-4.5	8.5	6.6	7.4	1.8	1.5
SOES 11	LTh	30.5	197.3	122	149.4	4.2-4.6	6.9	5.5	6.5	1.6	1.3
SOES 12	LTh	28.8	161;7	115.2	141.1	4.1-4.5	7.4	4.2	5.6	1.4	1.1
SOES 17	LTh	38.4	445.3	153.6	188.2	4.6-5	24.9	16.3	11.6	2.9	2.4
SOES 18	LTh	27.7	172.2	110.8	135.7	4-4.4	2.7	2.3	6.2	1.5	1.3
SOES 22	LTh	25.1	124.3	100.4	123	3.9-4.2	2.8	2.2	5.0	1.2	1.0

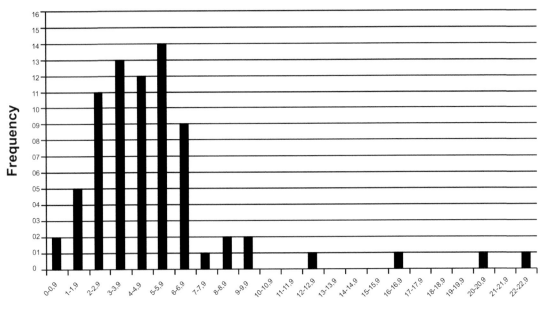

Calculated Speed (km/h)

Figure 7.1. Distribution of values of calculated speed (V$_c$, where h = yFL; km/h) of all the measured trackways of the Rio do Peixe basins. N = 75.

(SOES 1, SOES 17, SOPP 3, and SOPP 4), all assigned to medium to large theropods, represent the fastest runners of the Rio do Peixe ichnofauna (fig. 7.2).

The mean calculated speed (V$_c$, where h = yFL) for the medium and large theropods (n = 62 individuals, including SOCA 493, 991, 150; 82.67% of all the trackways of this sample) is 5.57 km/h; for the small theropods alone (n = 3 individuals, 4% of all the trackways of this sample) the mean is 4.6 km/h; for the iguanodontids, combining data for both bipeds and quadrupeds (n = 6 individuals; 8% of all the trackways), it is 2.62 km/h; for the few sauropod trackways (n = 4 [including ANEN 8], 5.33% of all the trackways) it is 2.47 km/h.

The average calculated speed for the tracks of the Sousa Formation alone (n = 55; 73.33% of all the trackways) is 5.23 km/h; for the sample from the Antenor Navarro Formation trackways (n = 20; 26.66% of all the trackways), the average value of V$_c$ is 4.72% km/h. A histogram of the distribution of the values of the theoretical or predicted speeds (V$_t$, where h = 4FL) of all the complete and measured trackways of the Rio do Peixe basins would be even more irregular than the histogram of the V$_c$ values (fig. 7.1). This is probably because this is a mixed sample containing sauropod, ornithopod, and mainly theropod trackways, showing an absolute and relative difference in the footprint dimensions that is rather large. This fact seems to demonstrate that the idea of calculating h by multiplying FL by 4 (h = 4FL) is not an accurate method but a rule of thumb (Thulborn 1990, 251).

Trackway data (fig. 7.3) allow comparison of the stride/hip height ratio (SL/h; N = 75) for trackways of the Rio do Peixe basins. The distribution

Figure 7.2. Social behavior of some theropods from the Sousa Formation. The tracks of some of these theropods indicate high velocity during chasing activities. (Art by Deverson Silva [Pepi].)

again indicates positive asymmetry, once again mainly due to the 4 running theropod trackways described in our earlier discussion of trackmaker speed estimates. On the basis of SL/h, three gaits can be recognized. Where SL/h < 2 (0–1.9), the dinosaurs are presumed to have been walking; values of 2 ≤ SL/h ≤ 2.9 point to slow running; and values of SL/h > 2.9 would represent a running gait. Nearly two-thirds of the studied trackways (49 individuals, 65.33% of the sample) have SL/h between 1 and 1.5, indicating a very slow walking gait (table 7.2, final column); 69 individuals (94.6% of the sample) have SL/h less than 1.9, indicating a walking gait; only 6 individuals, all of them identified as theropods (5.4% of the sample) point to a slow running gait (2 < SL/h < 2.9); and none of the trackmakers shows a true running gait (x > 2.9), although a few trackways are very near to the value 3. It may be that the SL/h ratio understates the velocity of the quicker trackmakers; certainly, in our everyday experience, speeds like those estimated for trackways SOES 17 (16.3 km/h), SOPP 3 (20.7 km/h), SOES 1 (22 km/h), and perhaps also SOPP 4 (12.8 km/h) would seem to qualify as running.

As for particular groups of trackmakers, the few sauropod trackways in the sample show SL/h between 0.3 and 0.5, and the ornithopod trackways have 0.5 ≤ SL/h ≤ 1.2; thus, all of the trackways attributed to plant eaters take values of SL/h ≤ 1.2, indicating rather or very slow speeds. The theropod trackways, in contrast, show values of SL/h in the range of 0.6 ≤ SL/h ≤ 2.8.

Speed estimates predicted or calculated from fossil trackways must be regarded with appropriate caution (Lockley and Meyer 1999), but, that caveat notwithstanding, the dinosaurs of the Rio do Peixe basins generally walked and only very rarely ran. The quadrupeds always moved slowly or even very slowly. The bipeds, theropods included, did not run very often or very quickly either; the maximum estimated speed we found here is about 22 km/h.

We are sure this is not just the case in the Rio do Peixe: as a matter of fact, this is a very common situation, almost general to the whole world and along the whole nonavian dinosaurs' time (Leonardi and Mietto 2000, for Early Jurassic). A recent statement on dinosaurs' low speed can be found, for example, in Xing et al. 2014a. There, for the sauropods, as for some diverse assemblages from the Feitianshan Formation, Lower Cretaceous of Sichuan Province, southwest China, the maximum speed value is 3.17 km/h; for ornithopods, it's 6.01 km/h. Another record, from the Cretaceous Neungju Group in the quarry at Seoyu-ri, Hwasun County, Jeollanam-do, South Korea, shows 1,000 or so dinosaur footprints and trackways, mainly theropodian (88%); this points to speeds between 3.3 and 20.5 km/h, with the small theropods often low running or running and the large theropods just walking (Huh et al. 2006). This is by no means surprising. If you look at large animals today, they usually walk slowly and run only when they must (J. O. Farlow, personal communication, 2015; see also Lockley and Meyer 1999).

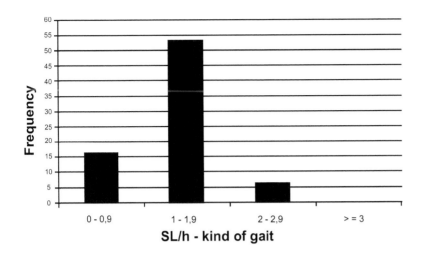

Figure 7.3. Distribution of values of the stride/hip height ratio (SL/h), where hip height (h) is estimated by the formula h = yFL (table 7.2), for all of the trackways of the Rio do Peixe basins measured to date.

This corresponds to Tony Thulborn's (1990, 264) statement wherein he concludes, from the observation and measurements of several hundred trackways, that at least bipedal dinosaurs preferred walking and that trackways showing a trotting or running gait are quite uncommon. See also Wade (1989), Molnar and Farlow (1990), and Lockley and Meyer (1999). However, see, for example, Moratalla et al. (2017), on the probable presence of a trackway of a trotting theropod.

Today there are many lines of evidence (Bakker 1986a, 1986b, 1972; Spotilla et al. 1973; Ricqles 1974; Ostrom 1980; Ricqles 1969, 1974, 1984; Farlow 1976, 1980; Farlow et al. 1995; Ruben et al. 1996; Reid 1997; Seebacher 2003; Padian et al. 2004; O'Connor and Claessens 2005; Sereno et al. 2008; Zheng et al. 2009; Pontzer et al. 2009) for high metabolic rates, locomotor costs, or endothermy for dinosaurs. Besides the histology and growth of the bony tissues, these lines of evidence include the presence of postcranial skeletal pneumaticity, mainly in the vertebrae; the apparent absence of nasal turbinates; a new vision on dinosaur posture, environment, and ecology; experimental comparisons between dinosaurs and currently existent animals; and the discovery of several dinosaur bodies showing feathers or filaments (cf. Pontzer et al. 2009).

However, the dinosaurs' high rate of metabolism is not so evidently supported in the known paleoichnological record, and it ought to be better discussed on the basis of many detailed and statistic works on their trackways (see Molnar and Farlow 1990, 218, fig. 7.5), rather than on the basis of some isolated, selected, biased, and/or uncontrolled data. When the latter happens, it leads to the huge racing dinosaurs of Bakker (1986a, 1986b) and Paul (1987a, 1987b). (for this concept, see also Molnar and Farlow 1990).

Gaits

Dinosaur tracks (trackways as well as isolated footprints) of the Rio do Peixe basins made by bipeds (at least 417 individual tracks, or ~86.33% of the identified tracks) greatly outnumber those attributed to quadrupeds (about 66 individuals, or ~13.66% of the identified tracks). The ratio of biped tracks to quadruped tracks is thus 6.32:1.

Among the individual tracks made by quadrupeds, 59 are attributed to sauropods (89.39% of the plant-eaters), 5 to quadrupedal graviportal ornithopods (7.57%), 1 to an ankylosaur (1.52%), and 1 to a small quadrupedal ornithischian (1.52%). However, about 30 other isolated ornithopod footprints might actually pertain to quadrupedal or semiquadrupedal trackmakers. The bipeds comprise 329 large theropods (78.90% of the bipeds), 31 small theropods (7.43%), 5 small theropods of another kind (1.20%), 16 indeterminate theropods (3.84%), 34 probably bipedal ornithopods of medium to large size (8.15%), and 2 small ornithopods (0.48%).

All the trackways are rather narrow, attesting to a fully erect stance, also in the quadrupeds; however, they are evidently narrower for the bipeds, where the step angle can reach 180° or, in some rare cases, more than this—that is, above this limit. This rather rare phenomenon occurs sometimes in quite straight trackways, when the maker slightly crossed its feet because of the high speed and when turning. As J. O. Farlow observed in the emu (*Dromaius novaehollandiae*), 180° will usually be the upper limit of the pace angle, unless the beast makes a sudden change of direction such that a foot comes down on the opposite side of the trackway midline—for example, a right foot comes down to the left of the trackway midline (J. O. Farlow, personal communication, 2015; Farlow 2018, 163–164).

Both the quadruped and biped trackways usually indicate a walking gait, sometimes very slow. Only the bipeds present cases of a (nearly?) running gait (fig. 7.2). No hopping or galloping gaits are represented at the Rio do Peixe basins. The sprawling gait is probably represented only by the lacertoid footprint SOES 15 at Serrote do Pimenta, but the complete trackway is unfortunately lacking.

One special gait is half-swimming (Leonardi 1987a), first described by Coombs (1980). In this kind of progression, the trackmakers swam in rather shallow water in temporary lakes, perhaps while engaged in fishing, and impressed in the sediment only the distal portions of the toes—the claws or, in one case, the hooves. The morphology of theropod footprints interpreted as showing this manner of locomotion is very characteristic: the tip of the claw mark of digit III is triangular and straight; in contrast, the tips of the claws of digits II and IV are curved because they are impressed with an outward and rearward rotational movement (figs. 4.30, 4.34C, 4.47B, 4.47E, 4.51–4.55). The half-swimming tracks permit calculation of the water depth at the place and time the tracks were made because the water level would have been at about the height of the articulation of the femur in the acetabulum (= h). It is very probable that at least some dinosaurs could swim in deeper water, but in such cases, they obviously would not have generally touched the bottom to produce tracks. However, as shown by Xing et al. (2016), the assemblage of sauropod tracks from the Lower Cretaceous Hekou Group of Gansu

Province, northern China, characterized by the preservation of only the pes claw traces, were interpreted as having been left by individuals that were walking, not buoyant or swimming. These tracks were explained as the result of animals moving on a soft mud-silt substrate, projecting their claws deeply and so registering their traces on an underlying sand layer where they gained more grip during progression.

In passing, we are not applying the ichnogeneric name *Charachichnos* Whyte and Romano, 2001 (see Milner and Lockley 2016) to the half-swimming tracks (or swim tracks) because it seems inappropriate to institute and use names referring to animal behavior; otherwise, one should, for example, name the trackways of running animals *Runningichnos* and those of walking animals *Walkingichnos* (Leonardi and Carvalho 2020b; however, on this topic, see also Belvedere et al. 2018).

In the Sousa basin, there are half-swimming dinosaur tracks at a number of sites, including Engenho Novo (ANEN, 3 theropods; fig. 4.8). Piau-Caiçara (SOCA) also has several such footprints (figs. 4.27, 4.32, 4.39, 4.55); among these are a cluster of about 40 footprints and short trackways in level 16 (figs. 4.34, 4.39, 4.47B, 4.51–4.55) that are very typical and attributable to middle-size to large theropods, very similar to those illustrated by Coombs (1980). Additional examples of such footprints include SOCA 57 bis, 941, 997, 9994, 13270, and 175; Piau II (SOPU; 2 probable half-swimming footprints attributable to theropods); Várzea dos Ramos (SOVR; 1 half-swimming footprint attributable to theropods); Várzea dos Ramos III (APVR III; 4 half-swimming theropod tracks); and Lagoa do Forno I (SOLF I; a partial theropod footprint that appears to have been left by a swimming animal).

Altogether, there are about 60 theropod half-swimming tracks and a single probable ornithopod half-swimming track (SOCA 1621; figs. 4.51, 4.52) in our basins. From this record, it appears that among the dinosaurs at Sousa, nearly only the theropods are known to have swum or, at least, half-swum. J. O. Farlow, however, thinks Coombs's interpretation of what the Dinosaur State Park (Rocky Hill, Connecticut) "swimming" trackmakers were doing remains controversial (personal communication, 2015).

Many other tetradactyl and pentadactyl half-swimming footprints were not made by dinosaurs but rather by other, smaller reptiles, very probably turtles (figs. 4.27–4.30, 4.55) and, less frequently, lizardlike reptiles. Examples occur in the rocky pavement of level SOCA 9; the tracks SOCA 52, 59 bis, and 1359 bis; and some turtle tracks at Poço do Motor site (SOPM, level 9). Fazenda Piedade SOPI has 1 large half-swimming track with a lacertoid pattern that is otherwise unclassifiable (fig. 4.59C). It is puzzling that we did not find ordinary walking tracks of turtles or lizardlike reptiles corresponding to the inferred half-swimming tracks of such reptiles.

Graviportal ornithopods present another special gait. In the Sousa basin, these dinosaurs were evidently sometimes quadrupedal (SOES 9), although they carried their weight almost entirely on the hind-limbs; the forefoot tracks are considerably smaller and very shallow, as if the animal had touched the ground with its hands but had not put much weight on them. Other ornithopod trackways seem to be bipedal, but this is probably an artifact of the poor condition of the rocky surface. This is probably the case of the *Caririchnium magnificum* trackway (SOBP 1; fig. 4.11) at Baixio do Padre, for example; another example is ANEN 8 (figs. 4.3, 4.4, 4.7), where the trackway appears to be that of a biped, but the trackmaker was probably quadrupedal, with total overlap of pedal and manual prints.

On the other hand, the trackways referred to as *Staurichnium diogenis* or *Staurichnium* sp. (probably a nomen dubium) seem to be consistently bipedal. For example, the holotype—the first to be found, and the better specimen—is almost surely the trackway of a true biped. This is because the rock layer in which it is impressed is a fine siltstone whose surface is very smooth and well preserved, and yet no forefoot prints are visible, and complete overlap of forefoot and hind-foot tracks is very improbable. As a matter of fact, in such a good substrate, forefoot prints would be seen, although overlapped by the hind-feet tracks.

One special case is SOPP 1, the very first trackway discovered at Sousa. This is not the tracks of a biped, as it was characterized in the first paper by Leonardi (1979a) on this subject but rather reflects semibipedal (or semiquadrupedal) motion, with strong heteropody—large pedes and very small manus (figs. 4.69A, 4.69C, 4.69D). The trackmaker impressed just the right manus into the sediment, and only very slightly. The small manus prints are oval or elliptical in shape; very probably they represent the impression of just the tips of some fingers. Perhaps this was an individual with an injured or even missing left forelimb. For limping behavior of a trackmaker—another ornithopod in the case—that does not affect the dinosaur's speed, see Razzolini et al. 2016; generally speaking, see also Lockley and Meyer (1999).

Another odd feature of this trackway (and its midline) is its characteristically undulatory pattern: the footprints run 3 by 3, periodically with more on the left or more on the right. In contrast, dinosaur trackways in our basins are generally quite straight (e.g., the main trackways at the Passagem das Pedras site, fig. 4.66) because the trackmakers were going from one place to another in a purposeful manner. There are, however, some uncommon curved trackways, such as the small theropod ANEN 1 (figs. 4.3, 4.4A, 4.7) and ornithopod trackway SOCA 43 (fig. 4.22). This prevalence of rectilinear trackways is normal all over the world, with few exceptions. A rare or perhaps unique case of a U-turn has been recorded for a fine sauropod trackway within the Early Cretaceous Zhaojue dinosaur tracksite at the Sichuan Province, China (Xing et al. 2015a).

A kind of movement that is not represented in the Rio do Peixe basins is that of herbivorous dinosaurs grazing on the herb or shrub stratum of the area's vegetation and consequently showing an extremely slow gait, with very short strides and step angles, imprinting their footprints several times in a linear meter, as pasturing cows and grazing deer do today. This is not an astonishing matter because this absence is apparent in trackways around the world. Perhaps this is because patches of habitat with a dense enough plant cover to warrant such grazing would also be unlikely to record footprints.

One striking aspect of the dinosaurian gaits is their generally harmonic regularity (see figs. 5.1 and 5.2 in chap. 5). Of course, this depends a lot on constant substrate conditions in terms of both the shape and arrangement of the footprints: several trackways do show large differences among footprints in the same trackway. Such is the case for trackways ANEN 1 (fig. 4.7), ANJU 4, SOMA 1 (figs. 4.10, 4.12), SOCA 1336 (fig. 4.44), and SOCA 48 (figs. 4.24B, 4.24F, 4.26). That being the case, one can conclude that instituting ichnotaxa on the basis of isolated footprints is in vain, as earlier described for the trackway ANEN 1, attributed to a small theropod in chapter 4.

This kind of phenomenon was very recently and usefully demonstrated by Razzolini et al. (2014) for the El Frontal dinosaur tracksite (Early Cretaceous, Cameros basin, Soria, Spain), and by Ishigaki (2010) for a number of theropod trackways (Upper Cretaceous of Mongolia) with deep and/or elongated footprints because of the laborious locomotion of the trackmakers on soft and deep substrate.

For more on this topic, see Tracks and Substrate in chapter 3.

Bearings

In the Sousa basin, there are 386 trackways and single footprints (the latter just theropods) from which the trackmakers' directions of travel were measured. The travel directions of the dinosaurs at Piau-Caiçara were studied in detail (Godoy and Leonardi 1985; and see above) and by themselves constitute 195 trackways or isolated footprints (50.52% of the general sample of the Sousa basin).

We will now examine all the classifiable and measurable tetrapod ichnological material from the Rio do Peixe Group of the Sousa basin and its three constituent formations: the Sousa, Antenor Navarro, and Rio Piranhas. Our analysis includes data for the 40 half-swimming theropod tracks from Piau-Caiçara, level 16, but excludes data from the innumerable small half-swimming footprints attributed mainly to chelonians and other members of the vertebrate mesofauna. This is a large sample, based on 386 individual dinosaurs (table 7.3).

A rose diagram of these data (fig. 7.4) results in a rather tetramodal pattern, with two main modes in the NNE and SW quadrants, and two secondary modes in the other two quadrants. However, this pattern is not as clear as in the Piau-Caiçara ichnofauna (fig. 4.57). There seems, then, to be four groups of trackmakers (table 7.5).

The group of tracks concentrated in the northeastern quadrant (as well as two adjacent sectors of 10° each to the left of due north [0°], such that it is more convenient to call this a north-northeast sector) has a mode of 19 (in sector 341°–350°) and a mean of 9.83 occurrences per sector of 10° and spreads across a total sector of 90° (341°–70° in a clockwise direction; 25% out of 360°). This is a sample of 118 individuals (32.69% of the total

Table 7.3 Bearings of the tracks of the Rio do Peixe basins.

SOUSA FORMATION						A. NAV. FORMATION	
site	bearing in degrees	site	bearing in degrees	site	bearing in degrees	site	bearing in degrees
ANDB	24	SOMA	12	SOVR	61	SOSL	8
ANDB	336	SOMA	176	SOVR	63	SOSL	11
ANDB	345	SOMA	231	SOVR	65	SOSL	13
ANEN	360	SOMA	354	SOVR	77	SOSL	18
ANEN	183	SOPE	44	SOVR	83	SOSL	21
ANEN	89	SOPI	6	SOVR	88	SOSL	32
ANEN	280	SOPI	22	SOVR	104	SOSL	47
ANEN	240	SOPI	26	SOVR	115	SOSL	102
ANEN	168	SOPI	123	SOVR	120	SOSL	105
ANEN	322	SOPI	188	SOVR	120	SOSL	109
ANEN	124	SOPI	191	SOVR	133	SOSL	112
ANEN	156	SOPM	44	SOVR	180	SOSL	112
ANEN	145	SOPM	130	SOVR	180	SOSL	113
ANEN	88	SOPM	135	SOVR	180	SOSL	115
ANEN	57	SOPM	140	SOVR	192	SOSL	115
ANEN	181	SOPM	199	SOVR	208	SOSL	117
ANEN	228	SOPP	17	SOVR	214	SOSL	117
ANEN	358	SOPP	37	SOVR	221	SOSL	119
ANEN	356	SOPP	67	SOVR	221	SOSL	127
ANEN	356	SOPP	65	SOVR	221	SOSL	130
ANEN	65	SOPP	88	SOVR	227	SOSL	130
ANEN	220	SOPP	99	SOVR	231	SOSL	139
ANJU	17	SOPP	132	SOVR	234	SOSL	194
ANJU	134	SOPP	164	SOVR	234	SOSL	217
ANJU	300	SOPP	172	SOVR	235	SOSL	221
ANZO	225	SOPP	187	SOVR	237	SOSL	223
SOMA	12	SOPP	198	SOVR	238	SOSL	229
SOMA	176	SOPP	192	SOVR	240	SOSL	236
SOMA	231	SOPP	242	SOVR	244	SOSL	229
SOMA	354	SOPP	265	SOVR	248	SOSL	236
SOPE	44	SOPP	287	SOVR	250	SOSL	286
SOPI	6	SOPP	314	SOVR	250	SOSL	288
SOPI	22	SOPP	349	SOVR	254	SOSL	296
SOPI	26	SOPP	357	SOVR	255	SOSL	300
SOPI	123	SOPP	353	SOVR	266	SOSL	305
SOPI	188	SOSA	56	SOVR	270	SOSL	318
SOPI	191	SOSA	208	SOVR	306	SOSL	349
SOPM	44	SOSA	225	SOVR	308	SOSL	350
SOPM	130	SOSA	328	SOVR	319	SOSL	353
SOPM	135	SOVR	4	SOVR	322	SOES	4
SOPM	140	SOVR	4	SOVR	326	SOES	79
SOPM	199	SOVR	12	SOVR	338	SOES	240

| SOUSA FORMATION | | | | | | A. NAV. FORMATION | |
site	bearing in degrees	site	bearing in degrees	site	bearing in degrees	site	bearing in degrees
SOPP	17	SOVR	18	SOVR	342	SOES	328
SOPP	37	SOVR	22	SOVR	342	**RIO PIRANHAS Fm.**	
SOPP	67	SOVR	22	SOVR	346	site	bearing in degrees
SOPP	65	SOVR	22	SOVR	348		
SOPP	88	SOVR	24			SOFP	45
SOPP	99	SOVR	50			SOFP	72
SOPP	132	SOVR	60				
SOPP	164	SOVR	61			total	191

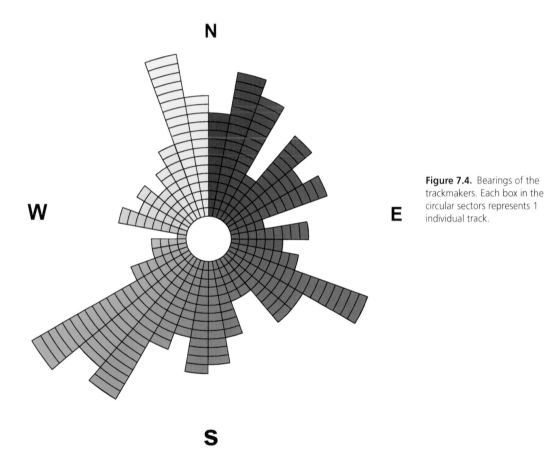

N

W

E

S

Figure 7.4. Bearings of the trackmakers. Each box in the circular sectors represents 1 individual track.

sample). This large sector therefore represents a large proportion of the dinosaurian ichnofauna, indicating one of the two main travel corridors, undoubtedly a preferred one.

The southwest (and adjacent sectors) group has a mode of 21 in sector 231°–240° and a mean of 9.83 occurrences per sector and occupies a total sector of 90° (161°–250°; 25% out of 360°). By chance, as with the

Table 7.4 Bearings of the Trackways at Piau–Caiçara

Disk Sectors of 10°	Frequency	Disk Sectors of 10°	Frequency
1°–10°	7	181–190	7
11°–20°	9	191–200	4
21–30	7	201–210	9
31–40	4	211–220	16
41–50	8	221–230	7
51–60	7	231–240	10
61–70	5	241–250	2
71–80	1	251–260	2
81–90	4	261–270	1
91–100	5	271–280	0
101–110	4	281–290	5
111–120	6	291–300	3
121–130	4	301–310	4
131–140	4	311–320	2
141–150	4	321–330	2
151–160	4	331–340	7
161–170	7	341–350	11
171–180	7	351–360	6

**Total 195
trackways**

north-northeast block, the sample is based on 118 individuals (32.69% of the total number of dinosaurs) and again no doubt represents a preferred travel corridor. The 30° sector between 211°–240° (8.33% of the 360° total) alone represents 57 individuals (15.79%).

The southeast group has mode 17 in sector 111°–120°, a mean of 11.25 occurrences per sector; occupies a total sector of 40° (111°–140°; 11.11%); and presents a sample of 45 individuals (12.47%). This is yet another preferred corridor.

The northwest group is the least likely to represent a preferred sector of travel. Its mode is only 8 in sector 281°–290°; the mean (average) is 7 cases per sector; and the group is spread over a sector of 30° (281°–310°; 8.33%). It constitutes a sample of 21 individuals (5.82%).

The four compass sectors recording the lowest numbers of trackmaker travel directions correspond to 71°–100°, 141°–160°, 251°–280°, and 311°–340°, adding up to a total sector of 110° (30.55% out of 360°), with a total of 59 occurrences (16.34% of the individuals).

The concentration of dinosaur travel directions in the NE and SW quadrants, and especially the modal sector that shifts the north-northeast block to the northwest of due north, depends mainly on the large sample of 40 half-swimming theropod tracks from level 16 of Piau-Caiçara, which shows a strong NNE–SSE trend.

Apart from the Piau-Caiçara level 16 sample, the data show a primary NE–SW corridor along which two large groups of dinosaurs moved in opposite directions. The north-northeast and the southwest groups of tracks by chance have the same number of individual trackmakers (N

Table 7.5 Characteristics of figure 7.4.

Block ▶	N-NE	interval	SE	interval	SW	interval	NW	interval
Sector of the block	341°-70°	71°-100°	101°-140°	141°-160°	161°-250°	251°-280°	281°-310°	311°-340°
N of individuals	118	19	45	10	118	9	21	21
% of the fauna	32.69%	5.26%	12.47%	2.77%	32.69%	2.49%	5.82%	5.82%
N of sectors of 10°	12	3	4	2	9	3	3	3
N of degrees	90°	30°	40°	20°	90°	30°	30°	30°
Degrees rate (360°)	25%	8.33%	11.11%	5.55%	25%	8.33%	8.33%	8.33%
Mean of individuals per sector of 10°	9.83	6.33	11.25	5	9.83	3	7	7
Mode	19	-	17	-	21	-	8	-

= 118 for both), but the group proceeding in a southwesterly direction is more clearly defined, as demonstrated by the fact that this latter group has the largest mode (N = 21; disk sector 231°–240°) in the entire Sousa basin sample. The northeast and southwest groups together account for 180° (50% out of 360°) of compass and include 236 individuals (65.37% of the individuals).

In addition to this main bipolar concentration of travel directions, there is a secondary but rather well-defined concentration of opposite directions, positioned around the axis 101°–140° and 281°–310°. In this case, there is a clear predominance of the group heading southeast. The SE and NW sectors occupy together 70° (19.44%) and together account for 66 individuals (18.28% of the individuals).

A tetramodal pattern of preferred travel directions is also obtained if data are limited to tracks from the Sousa Formation alone (318 individuals; 88.09% of the total sample). The same is true for the Antenor Navarro Formation alone (41 individuals; 11.36%); but in this formation the main concentration of tracks is heading southeast.

Both of the calculated bearings of tracks from the Rio Piranhas Formation (2 individuals; 0.55%) fall in the northeastern sector.

As previously described for the Piau-Caiçara tracks (table 7.4), and also true (but with less evidence) for tracks of the Sousa basin more generally, the majority of the tracks are parallel or almost parallel to the crests of ripple marks, where the latter are present. In turn, the dominant orientation of the shorelines, as indicated by these crests, is frequently parallel to the regional faults that created the Sousa basin. It is evident, therefore, that the movements of the dinosaurs were strongly influenced by the local and regional landscape, in particular by the lakes (shorelines) and, indirectly, the regional tectonic texture. Consequently, the dinosaurs traveled along preferred corridors (Leonardi 1989a, 1989b).

Stance

All our trackways, including those of the sauropods, are rather narrow, attesting to a fully erect trackmaker stance. All of the sauropods were obviously quadrupedal, and none of their trackways is manus-only at the

Rio do Peixe basins, such as those found in Morocco (Ishigaki 1989), at Cal Orck'o quarry near Sucre (Bolivia; Lockley et al. 2002), or in the CertainTeed gypsum mine near the Briar Site, in Howard County, Arkansas (Platt et al. 2018). Note, incidentally, that the Morocco "hand-only trackways" were first interpreted (Ishigaki 1989) as imprinted by swimming sauropods; today, however, it seems more likely that it is a case of preservation: they would be just undertrackways where the prints of the forefeet, whose area is lesser than that of the hind-feet, were imprinted, and the hind-feet were not (Lockley and Rice 1980; Ishigaki and Matsumoto 2009).

The theropods, large and small, were obviously all bipedal (Molnar and Farlow 1990, 213, as for the large theropods), with very narrow trackways, in contrast to the old model of large theropods giving a fair show of Cossack dance (Molnar and Farlow 1990; Wade 1989). The ornithopods were variously bipedal, quadrupedal, or, in one case (SOPP 1), semiquadrupedal.

In a few instances, the mark of the tail is perhaps preserved: SOPP 16 (fig. 4.79D), SOSL 49997, and perhaps also SOCA 13361 and 13362 (figs. 4.20, 4.36). The rareness of tail impressions is usual for dinosaur trackways. It is therefore evident that almost all the dinosaurs of the Rio do Peixe basins, bipedal as well as quadrupedal (like almost all the dinosaurs all over the world), kept their tails clear of the ground. The bipeds did so to counterbalance the dinosaur's trunk weight, and their dorsal column was probably held almost horizontally (Molnar and Farlow 1990), with the result that their barycenter was placed over the feet (fig. 7.2). The quadrupeds probably did so to avoid abrasion of the skin from friction with the ground, and to keep the tail ready as a weapon against possible attackers (fig. 7.5).

Cases of impression of a dinosaur's body in the substrate, in which the animal sat or wallowed, are unknown in the Rio do Peixe basins and are rare worldwide (Lockley et al. 2003a). One well-known instance of "sitting" prints is the *Anomoepus* tracks from the Early Jurassic of eastern North America (Lull 1953) and of Europe (Avanzini et al. 2001c). A candidate for a dinosaur wallow comes from sandstone of the Corda Formation (Lower Cretaceous) along the Tocantins River in Itaguatins, state of Tocantins, Brazil (Leonardi 1980e); the cited paper mistakenly attributed trackways to ornithischians that the same author later assigned to sauropods, probably titanosaurids. Adjacent to the 7 sauropod trackways was a structure, several meters in diameter, that could be interpreted as a place where a great dinosaur had splashed about in the mud.

Social Behavior

The majority of the dinosaurs of the Rio do Peixe basins were solitary animals, except for sauropods, which nearly always moved in herds. Such behavior is evidenced by groupings of parallel sauropod trackways of at least 7–15 individuals (Leonardi, 1989a, 1989b, 1994; Carvalho 2000b; Leonardi and Santos 2006); the number of animals in these herds could

Figure 7.5. Raindrop prints associated with the tracks show the effects of the rainy season. Theropods and sauropods were probably widely distributed in the floodplains of the central Sousa basin. (Art by Deverson Silva [Pepi].)

well have been greater, because some trackways were likely destroyed by erosion, and some have yet to be uncovered (fig. 7.5). Theropods and ornithopods, in contrast, usually moved as single animals. There are, however, three exceptions: the population of small and medium-size theropods at Serrote do Letreiro (figs. 4.90, 4.91, 4.93); the assemblage of 5 long-heeled theropods of the ichnogenus *Moraesichnium* at Passagem das Pedras (figs. 4.71, 4.74, 4.78); and perhaps, but just perhaps, the 30 or so theropods of Piau-Caiçara at level 13/2 (fig. 4.35).

Here, ornithopods were always solitary animals, *Caririchnium* tracks included. This is a surprising observation for Rio do Peixe ornithopods, because *Caririchnium* and other trackways of graviportal ornithopods from tracksites of comparable age around the world commonly occur in parallel or subparallel groupings (e.g., in South Korea: Lockley et al. 2006; Paik et al. 2020).

Indeed, for South America more generally, gregariousness seems limited almost exclusively to sauropods, and these dinosaurs nearly always proceeded in herds, exactly as in the Rio do Peixe ichnosites at Piau-Caiçara, Serrote do Letreiro, Serrote do Pimenta, and Floresta dos Borba. Occurrences of isolated footprints or trackways of sauropods are rare, most probably due to lack of exposures sufficiently extensive to record passages of groups or, more rarely, pairs of animals (Sucre, Bolivia; Meyer et al. 1999; Lockley et al. 2002).

It is doubtful that theropods proceeded in gregarious groups on this continent, with the exception of the above-mentioned cases at Paraíba, two cases at Toro-Toro (Potosí, Bolivia; Leonardi 1984b; Meyer et al. 2019), at Bajada Colorada Formation, Argentina (Canale et al. 2016), and outside South America at Broome Sandstone, Australia (Salisbury et al. 2017). The other theropod tracks on this continent would point to lone hunters rather than to pack hunters (Molnar and Farlow 1990, 212; see also Paik et al. 2020).

South American ornithopods likewise seem always to have been antisocial loners. In contrast, on other continents there are important examples indicating frequent gregariousness not just for sauropods but also for theropods and ornithopods. A striking set of examples comes from La Rioja, Spain, a region blessed with abundant dinosaurian ichnofaunas from many splendid Lower Cretaceous (Berriasian to Aptian) outcrops. The footprint assemblages are similar in age and composition to those of the Rio do Peixe basins. However, they are different regarding the gregarious behavior. At La Rioja, the cases of gregariousness are numerous. As a matter of fact, García-Ortiz and Pérez-Lorente (2014, 113) points to twenty-eight records of gregariousness in that megatracksite, among them nine of sauropods, twelve of theropods, seven of ornithopods, and one group of unknown bipedal dinosaurs. The same author also presents a very interesting synthesis on two main different kinds of gregarious behavior—accumulations of trackways and parallel trackways—and describes how these different styles are recorded for the diverse dinosaur clades (see also Pérez-Lorente 2015).

There is no concrete reason to hypothesize dinosaurian migrations in our area. However, the possibility of that phenomenon cannot be excluded, whether in terms of regular seasonal or occasional migrations. For example, some paleontologists admitted this phenomenon for the dinosaurs of very high latitudes; it was first suggested for dinosaur tracks from the Spitsbergen Islands (Lapparent, 1962; Lockley and Meyer 1999), northern Canada, and Alaska (Lockley and Meyer 1999); from the Antarctic continent; and from southern Australia (Rich and Rich 1989). These dinosaurs would have passed the winter in lower latitudes and in milder climates, then they would gather in the springtime to migrate toward the zones of higher latitudes—especially the herbivores, to feed on the energy-rich fodder of the northern (or southern, in the northern continents) woods, of the steppes, and of the tundra of those times; the carnivores, perhaps, would follow the herds of herbivores. It happens this way today with caribou, followed by their predators (Rich and Rich 1989). Most workers, however, no longer think that seasonal migrations from polar regions to lower latitudes and back took place. It now seems that the polar dinosaur faunas were endemic to high latitudes (see Fiorillo and Gangloff 2001; Bell and Snively 2008; Godefroit et al. 2009; Chinsamy et al. 2012; Fiorillo and Tykoski 2014). See especially Woodward et al. (2011), who propose that early rapid growth (demonstrated by cyclical suspensions in bone growth as well as high growth rates early in life) are preadaptations that could explain the dinosaurs' successful exploitation of a hostile high-latitude ecosystem.

The hypothesis of the large-scale dinosaurian migrations may be valid for movements among different habitats. Huge numbers of large ungulates routinely move large distances across the savannas and woodlands of tropical Africa, following the rains and consequent growth of fresh fodder (Currie 1992; see also Fricke et al., 2011). It is also possible that every herd of large herbivorous dinosaurs had one ample territory of its own for a cyclical migration, since some reptiles, such as crocodilians (Béland and Russell 1980), know and practice territoriality.

Navarrete et al. (2014) speculated about the possibility of migrations of dinosaurs (Camarillas Formation, ~130.6–128.4 mya, Barremian age) in Teruel, Spain, along coastal regions, which are always important routes for migration; see the dinosaur migration route along the western beaches of the western Interior Sea (United States) during Late Jurassic–Early Cretaceous. Dinosaur migrations, just like present-day animal migrations, were probably aroused by the search for food (Lockley and Hunt 1995; Fricke et al. 2011).

As a matter of fact, starting from the chemical analysis of tooth enamel of the sauropod *Camarasaurus*, Fricke, Hencecroth and Hoerner (2011) proposed that dinosaurs walked—or, better, migrated—hundreds of miles in Late Jurassic North America, more exactly in what are today Utah and Wyoming. The sauropod herds would have accomplished a lowland–upland periodical and seasonal migration of about 200 miles or about 300 km, searching for food and water availability. They would leave a floodplain area at the beginning of the summer, the dry season, when fodder began to run out, and would reach highlands that were probably cooler and wetter in summer. They would come back to the lowlands at the beginning of winter, when that region would be satisfactorily supplied with water and fodder.

Fricke et al. (2011) arrived at this result by examining and comparing ratios of oxygen isotopes in the *Camarasaurus* teeth enamel with oxygen isotopes found in layers of lowland rocks. As dinosaurs' teeth were replaced every few months, each tooth is a unique record of what the animal had drunk during the tooth's short life span. Lowland sediment and *Camarasaurus* teeth show clearly different oxygen-isotope ratios, pointing to the fact that many teeth had formed somewhere else. The ratios in the teeth were akin to what one would expect from teeth grown in the highlands, drinking highland water. Fricke and colleagues noted that on the African (Tanzania and Kenya) Serengeti plain the large mammal herds do a wet season-dry season migration of the very same kind. Perhaps here also the large sauropods would be followed in their seasonal migration by theropods like *Allosaurus*.

The shores of water bodies (Lockley and Hunt 1995; see also Navarrete et al. 2014) such as those inferred for our basins, would seem logical routes for such migrations. Indeed, fossil footprint assemblages provide evidence that dinosaurs could migrate or disperse along corridors in some very difficult and improbable terrains. For example, the rich ichnofaunas of the Calcari Grigi di Noriglio Formation (Lower Jurassic, Hettangian-Sinemurian; Rovereto, Italy) indicate that these animals passed from the African Plate to a Bahamas-like archipelago of muddy and sandy islands and islets (carbonate platforms) and detrital carbonate bars that would one day constitute Italy (Leonardi 1996, 1998, 2000, 2008; Leonardi and Lanzinger 1992; Avanzini et al. 1995, 2000b, 2000c, 2001d; Leonardi and Mietto 2000; Avanzini and Leonardi 2002; Avanzini et al. 2003; Nicosia et al. 2005). The same is more or less true for the Triassic, Late Tithonian-Late Cenomanian, and Coniacian-Maastrichtian (Leonardi and Leghissa 1990; Citton et al. 2015b) and the Aptian-Albian (Leonardi and Teruzzi 1993; Dal Sasso and Maganuco 2011; Dal Sasso et al. 2016) of the same Italian region.

Because some of the dinosaurs of the Rio do Peixe basins seem to have been gregarious, it is possible that they engaged in migrations, which is commonly a social and collective behavior. The region of the Rio do Peixe could have been abandoned by herbivorous dinosaurs (and perhaps subsequently by their predators) during intervals of severe, prolonged drought, as these huge plant eaters wandered in search of richer feeding grounds—perhaps in other, contemporaneous sedimentary basins of northeastern Brazil. Of course, it is also possible that the dinosaurs of our basins didn't migrate but rather only opportunistically expanded the territories over which they ranged. In the Early Cretaceous, they might even have wandered off as far as Africa, which, then, was not so far away (about 450 km).

The Dinosaur Community

8

The paleoichnologic record is generally biased toward preservation of footprints of large animals. This partly reflects the fact that small animals leave their footprints only if the sediment is fine-grain, smooth, moist, and very plastic (fig. 8.1). It also is due, however, to the fact that many researchers in Mesozoic terrains are satisfied with discovering and surveying the larger—and for them more interesting and fascinating—dinosaur footprints. They do not kneel down on the ground, as one ought to do, to examine rock surfaces more carefully, particularly when the surfaces are lit by oblique, low-angle light. In rocks of the Rio do Peixe Group, and especially in the fine-grain, very smooth-surfaced sedimentary rocks of the Sousa Formation, tracks of small animals, if present, should be preserved and readily found. In fact, the rock surfaces were carefully examined, inch by inch, at all of the outcrops, but without great success: the mesoichnofauna constitutes, as earlier described, only two sets of batrachopodid prints, crocodilian traces, one lacertoid footprint, and a large number of chelonian half-swimming tracks. The tetrapod community of the Rio do Peixe basins, as represented by footprints, is mainly a dinosaurian assemblage, and this is a mystery. There ought to be more mesofauna—but the tracks are just not there. A similar situation is described by Molnar and Farlow (1990) on the dinosaur association and on the rarity of mesofauna representatives at Cleveland-Lloyd Quarry in the Morrison Formation of central Utah. Especially at the Rio do Peixe ichnosites, there ought to be avian tracks, which are so common and highly diverse in North America and in localities of South Korea and, recently, also described in China (Ignotornidae from the Lower Cretaceous of Donghai County, Jiangsu Province; Xing et al. 2017a)—but there are none.

Small outcrops obviously cannot preserve footprints of all the taxa living in the surrounding region. However, the 37 sites of the Rio do Peixe basins crop out across quite a wide area, sampling a variety of ancient environments. Consequently, it is possible that the footprint assemblages as a whole do a reasonable job of sampling the composition of the regional fauna, at least of the larger animals.

As previously described, the number of individual plant-eating dinosaurs represented by tracks in the Rio do Peixe basins sums up to 101 (21% of the identified tracks); the number of individuals of large theropods (most or all of which presumably were carnivores) comes to 381 (79% of the

Dinosaurs and Mesofauna

Plant Eaters and Meat Eaters

Figure 8.1. Environmental reconstruction of Piau-Caiçara (SOCA) level 9. A turtle hauls itself out of the water of a temporary lake, onto a mud-bank marked by numerous tracks made by its fellow chelonians when the water was higher. In the background, an ornithopod crosses the already exposed, drying mud of the lakebed. Painting by Ariel Milani Martine.

Figure 8.2. *(facing)* Diagram illustrating the relative size and abundance of meat-eating versus plant-eating dinosaurs from the most trampled track levels at Piau-Caiçara farm, based on interpretation of the various footprints and trackways from each level. Drawing by Leonardi, redrawn from Leonardi (1984b).

identified tracks). The ratio of herbivorous/carnivorous individuals in this dinosaurian ichnofauna is 1/3.77 (fig. 8.2).

In the Sousa Formation, which represents a paleoenvironment of temporary lakes and their margins, there are tracks of 36 plant eaters (10.84% of the identified tracks) and 296 meateaters (89.16% of the identified tracks). The ratio of herbivorous/carnivorous individuals in this dinosaurian ichnofauna is 1/8.22.

The corresponding numbers and percentages for the sediments of marginal fans of the combined Antenor Navarro and Rio Piranhas Formations come to 64 for the plant eaters (42.95% of the identified tracks) and 85 for the meat eaters (57.05% of the identified tracks). The ratio of herbivorous/carnivorous individuals in these dinosaurian ichnofaunas is thus 1/1.33—a noticeable difference from the ichnofauna of the Sousa Formation.

These data differ from the theoretical ecological model of a great prevalence of plant eaters and a small number of predators. The abundance of theropod footprints against the plant eater footprints could be caused by a different level of activity of their respective trackmakers and/or by the grouping of certain taxa in particular environments.

Moreover, the prevalence of theropod tracks is a characteristic of many tracksites all over the world (Schult and Farlow 1992; Moreno et

CARNIVOROUS	HERBIVOROUS
2	
3-3 bis	
4	
5 bis	
7	
9-9 bis	
10	
13/2	
13/3	
13/4-5	
13/6	
15	
16	

al. 2011; Wagensommer et al. 2012; Farlow et al. 2015), and especially in South America (Leonardi 1989a, 1994). See, for example, the case of Quebrada Chacarillas, Tarapacá region, Chile. At Chacarilla, Lower Cretaceous, the theropod tracks are 80% of the ichnofauna of the third site of Quebrada Chacarillas, and the ornithopod tracks are 20% (Soto-Acuña et al. 2015).

Another example is the ichnofauna from the Cenomanian-age units of the Kem Kem beds of southeastern Morocco, in which, especially in the upper unit, theropod tracks are very abundant, in contrast with the rareness of sauropod and ornithischian tracks. The same happens in this area for the record of bones and teeth (Ibrahim et al. 2014). An important example is the case of the completely theropod-dominated ichnofauna at the Linjiang site, Jiaguan Formation, Lower Cretaceous of Guizhou Province, China (Xing et al. 2018); there is also the ichnosite from the Cretaceous Neungju Group in a quarry in Seoyu-ri, Hwasun County, Jeollanam-do, South Korea, where approximately 1,500 well-preserved dinosaur footprints, mainly theropod (88%), including 61 trackways, have been excavated and studied (Huh et al. 2006).

This Korean ichnosite of Hwasun County, with its great abundance of theropod tracks of different kinds, allows us to have a different perception of the global Korean ichnological record (Huh et al. 2006, 134). The makers of these tracks were theropods, ornithopods, and sauropods, but the presence of theropod footprints is statistically very high compared to other dinosaur sites in South Korea. In fact, the theropod tracks in this important ichnosite of Hwasun correspond to 88% of the entire number of footprints of the site. There are abundant theropod tracks of <40 cm; and most of the ichnological material of the site belongs to small theropods.

In contrast, other important Korean ichnosites are dominated by the tracks of plant-eating dinosaurs, that is, of ornithopods and sauropods, while the trackways of theropods are rather uncommon (Lockley et al. 2006, 73, 88, fig. 6, and table 1). In the ichnosite of Hwasun County, small theropod footprints and trackways are quite predominant (Huh et al. 2006): at the L1 level, for example, among 216 footprints, 205 are theropodian; among that 205, 199 are shorter than 40 cm (Huh et al. 2006). At the L2 level, about 750 dinosaur footprints occur in a dense concentration. Theropod footprints comprise 73% (61% shorter and 12% longer than 40 cm; Huh et al. 2006). Many of them were very probably imprinted by ornithomimosaurs (Huh et al. 2006), and so many—probably most—could be omnivores or herbivores. Only the makers of the large tracks from this locality were surely carnivorous, and so they belonged to the upper tier of the food chain.

Still another example is the Lavini di Marco (Rovereto, Trento, Italy) site from the Hettangian-Sinemurian Calcari Grigi di Noriglio Formation, where theropod tracks represented (as of the year 2000) 57% of all the individual tracks (Leonardi and Mietto 2000). An important theropod-dominated ichnofauna is also found in the Reuchenette Formation

(Kimmeridgian) in the Swiss Jura Mountains, northwest Switzerland (Castanera et al. 2018).

The relative abundance of herbivores and carnivores in all of these ichnofaunal samples differs dramatically from the expected relative abundance in a large-vertebrate community, in which large numbers of large plant eaters should be preyed upon by a much smaller number of predators (Molnar and Farlow 1990). Several explanations, not mutually exclusive, could account for this discrepancy.

The most obvious possibility is misidentification of the trackmakers. There may be a bias on the part of ichnologists: many researchers (among them Leonardi) will probably identify tridactyl footprints of bad quality as theropod by default; it is possible that fewer workers would default to an ornithopod identification of dubious footprints. Indeed, distinguishing theropod from ornithopod footprints is not always easy (Demathieu 1990; Farlow and Lockley 1993; Moratalla et al. 1988a, 1992; Leonardi et al. 1987a, 1987b, 1987c; Farlow et al. 2012a, 2012b, 2013). See also this volume's statistical chapter on this problem, especially graphs 6 and 7 and the correspondent comments.

A second possibility is that many of the footprints attributed to theropods were indeed made by members of that clade, but by herbivorous or omnivorous rather than strictly carnivorous forms. This is particularly likely for small to midsize theropods. Members of several smaller-bodied theropod clades are now thought to have been primarily herbivores (Holtz 2012). This possibility should be kept in mind in ichnofaunas where a substantial proportion of the footprints attributed to theropods is relatively small.

Even large and undoubtedly carnivorous theropods were considerably smaller animals than most sauropods, though, which brings us to a third explanation that could account for the discrepancy in the relative abundance of putative carnivore as opposed to herbivore tracks in many dinosaurian ichnofaunas. In the Rio do Peixe track assemblages, theropod footprints are always much smaller than those of sauropods. In dinosaur faunas where the large-herbivore component of the community was dominated by sauropods, the predator:prey biomass ratio would have been considerably less than the predator:prey numerical abundance ratio. One sauropod carcass would have fed a large number of theropods!

That brings us to the final possible explanation for the overabundance of theropod footprints as opposed to those of plant eaters, and the one favored here: differences in behavior, including activity levels, of carnivorous as opposed to herbivorous trackmakers (Molnar and Farlow 1990). An imbalance of herbivores and carnivores can develop at times in special subenvironments (and facies), and especially in the places where sediments are likely to be deposited. Even if, as reptiles, dinosaurs had to drink less than mammals, they nevertheless probably had to drink. If the Early Cretaceous dry season in the Rio do Peixe basins (especially the region of deposition of the Sousa Formation) was characterized by

Figure 8.3. Life restoration of dinosaurs of the Sousa basin, as at the locality of Piau-Caiçara farm. The carnivores ought to be outnumbered by the herbivores in the community, but this does not reflect on the tracks. Painting by Ariel Milani Martine.

rather few waterholes, these would have been spots where animals tended to congregate; the paths leading to them would have been more crowded than most places. This would have been especially true of the shores of temporary lakes during periodic lowering of water levels. Modern African large carnivores will haunt such areas, waiting or actively looking for prey, because herbivores living in the hinterland must eventually come for water (Béland and Russell 1980). The greater abundance of theropod tracks in the sediments of the Sousa Formation, and the high number of plant eater tracks in the marginal fans, would correspond well to this situation.

Something else to keep in mind about modern large herbivores and droughts: it isn't necessarily the lack of water that kills the beasts. An animal can go only so far away from a source of water; it must be able to return to that waterhole before it succumbs to thirst. As the drought worsens, the distance an animal can wander away from the diminishing water sources becomes less and less. Eventually the herbivores are forced to stay near the water, and because of this they consume all the available vegetation and starve before they die of thirst (J. O. Farlow, personal communication, 2015. See also Carpenter 1987).

However, for this explanation to account for the greater abundance of carnivore tracks in the Rio do Peixe group outcrops, it would be necessary to show that in the ancient basins there was a marked variety of subenvironments and that the area of the Sousa Formation track-bearing outcrops, found today mainly on the bed of the actual Peixe River, were, for some ancient reason, a more abundant source of fresh water than other habitats on the landscape. Furthermore, a greater availability of water could presumably allow more luxuriant vegetation, which could attract an abundance of herbivores. All of this is, however, only hypothetical, because the plant fossils are rather rare in these rocks.

Finally, if, as earlier suggested, the carnivorous dinosaurs were more active than the large herbivores (Leonardi 1989a, 1994; Farlow 2001; Farlow et al. 2015), covering more ground in a single day as they searched for food, they might have made many more footprints than the large plant eaters (fig. 8.3).

The overall ichnofauna of the Rio do Peixe basins contains, as mentioned previously, tracks of 381 theropods, 59 sauropods (altogether, 440 saurischian individuals), 38 graviportal and 2 small ornithopods (altogether, 40 ornithopods), 1 ankylosaur, 1 small quadrupedal ornithischian (altogether, 42 ornithischian individuals), and at least 53 unclassifiable dinosaurian tracks (fig. 8.4).

There are only a few saurischian-dominated ichnosites in the Rio do Peixe ichnofaunas, the kind with just theropod and sauropod tracks (like at Baixio do Padre, Engenho Novo, Fazenda Paraíso, Lagoa do Forno I, Riacho do Cazé, and Varzea dos Ramos I; 17.65% of the thirty-four localities bearing clearly classifiable tracks). This is in contrast with many classic ichnolocalities and some recently discovered ones (González Riga et al. 2015; Xing, et al. 2016d; Platt et al. 2018).

There is a large number of exclusively theropodian ichnosites that also record Saurischia-dominated ichnofaunas (Abreu*, Baleia*, Barragem do Domício*, Juazeirinho, Lagoa do Forno II*, Matadouro, Pedregulho*, Piau II*, Poçinho*, Poço do Motor, Poço da Volta*, Varzea dos Ramos II* and III, Zoador*; 47.06% of the sites). However, many of these theropod-dominated ichnosites are, as a matter of fact, less important sites, with only 1–3 individual tracks; these are recorded in the list above with an asterisk (*). This phenomenon is not surprising because theropod individual tracks correspond to 79% of the classifiable individual tracks of these basins. Finally, one locality is characterized by the presence of just 1 sauropod trackway (Aroeira; 2.94% of the ichnosites).

Some localities show tracks of both theropods and ornithopods (Cabra Assada, Curral Velho, Fazenda Piedade, and Sítio Saguim; 11.76% of

Saurischia versus Ornithischia

Figure 8.4. An environmental reconstruction of the floodplain corresponding to the depositional environment of the Sousa Formation. A quadrupedal ornithopod produces its trackway using a pacing gait as a local form of pterosaur flies overhead. Art by Ariel Milani Martine.

the ichnosites). Other sites record more abundant and mixed ichnofaunas: Floresta dos Borba (theropods, sauropods, and ornithopods); Mãe d'Água (theropods, sauropods, and ornithopods); Passagem das Pedras (theropods, sauropods, ornithopods, and a quadrupedal ornithischian); Piau-Caiçara (theropods, sauropods, and ornithopods; crocodiles and chelonians); Serrote do Letreiro (theropods, sauropods, and ornithopods); Serrote do Pimenta (theropods, sauropods, ornithopods, one ankilosaur, and a lacertoid reptile); and Tapera (theropods, sauropods, ornithopods, and crocodiles). These mixed ichnosites correspond to 20.59% of the considered 34 ichnosites.

At the Rio do Peixe basins, there are no ornithischian-dominated ichnofaunas, like those from the Lower Cretaceous Jiaguan Formation (Barremian–Albian) of Qijiang, south-central China (Xing et al. 2015b), the Eumeralla Formation (Aptian-Albian) of Australia, and the Gladstone Formation (Barremian) of Canada, and in a number of known ichnosites from the United States and eastern Asia.

It is obviously difficult to distinguish footprints of adults of small-bodied dinosaur species from young individuals of dinosaur species that as adults would reach a large size. In some fossil footprint assemblages (Peabody 1948), there are series of tracks of different sizes but similar shapes that plausibly comprise an ontogenetic series of individuals of the same species in different phases of growth, from hatchlings to adults. This interpretation is easier to make when dealing with footprints of pentadactyl and quadrupedal animals (e.g., chirotheroids), in which a large number of morphological characteristics (reference points or landmarks) can be compared. The situation is more difficult with tracks of bipedal, tridactyl dinosaurs because in the absence of impressions of the forefeet and the absence of the inner and outer toes of the hind-foot (digits I and V), there are far fewer characteristics for analysis.

Consequently, little can be said about the age-class makeup of the makers of our ichnofauna of the Rio do Peixe. The only footprint in our basin that is almost certainly that of a juvenile is isolated tridactyl track SOPP 18, which is the smallest dinosaur track discovered in our study (footprint length = 5.6 cm). There are no other very small dinosaurian individuals (hind-foot prints shorter than 12 cm). It is possible, of course, that this is an artifact of preservation. It is possible that dinosaurs of very small size were simply not heavy enough to leave footprints because of the relative compactness of the substrate.

The sample of footprints of theropods of small and medium size from level A of the Serrote do Letreiro (SOSL 4–5; fig. 8.5), representing 37 individuals, is a plausible candidate for an ontogenetic series. The trackmakers impressed footprints ranging in length from 16.6 to 30.1 cm, showing a high standard deviation. This association of prints of smaller (younger?) theropods (e.g., SOSL 492, 499, and 4993) with those of larger individuals (adults of the same species?) is remarkably rare in the Rio do Peixe basins. Even here, however, there are no truly juvenile (equivalents of a *pullus* in modern gamefowl) individuals.

SOSL is, however, unusual. More typically in our tracksites, a large proportion of the individual dinosaurs represented by footprints are animals of close to the same size, whether theropods or sauropods.

If the relative scarcity of small-dinosaur footprints in our ichnoassemblages really does reflect a small proportion of very small animals in the dinosaur populations, this might indicate very heavy mortality on the part of very young individuals (Leonardi 1981c), as in some living reptiles (e.g., turtles and crocodilians). Small juveniles may frequently have been eaten by larger predators or otherwise have met their death before many of them could make footprints, as was surely the case for the cute *Scipionyx samniticus* Dal Sasso and Signore, 1998 (Leonardi and Teruzzi 1993; Dal Sasso and Signore 1998; Dal Sasso 2004; Dal Sasso and Manganuco 2011). Larger individuals, with quite lower indexes of death rate, may have dominated the assemblages (Farlow et al. 2010, 409–10).

Figure 8.5. Serrote do Letreiro (SOSL). More images of the footprint assemblage at subsite 4. *A*, Trackway SOSL 42 (moving upward from the bottom), SOSL 41 (crossing from right to left), and other scattered footprints (SOSL 40/2, 46, 48, 4493, etc.). *B*, The crossing of trackways 42 and 40. *C*, Trackway 41 (moving upward from the bottom), crossed by trackways SOSL 42 and 44, with petroglyphs. *D*, Another view of the same prints. Graphic scales in centimeters.

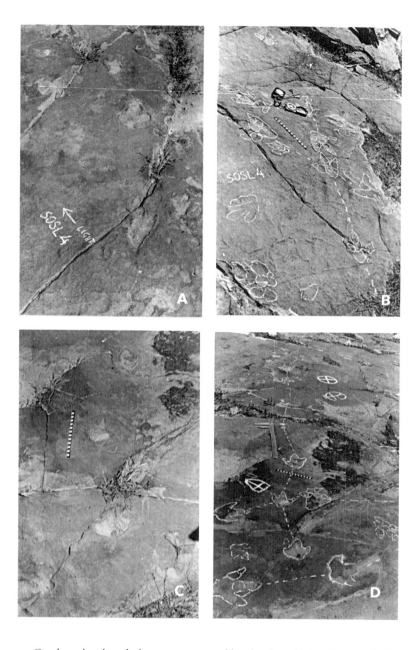

On the other hand, there were several kinds of small-size theropods. As a matter of fact, comparison with present-day tetrapod predators indicates that similar forms coexisting in the same environments as a rule reduced interspecific competition for prey by regular spacing of body dimensions. When coexisting predatory species are not so different in size, they usually present morphological differences related to prey capture—that is, different behavior. If predators are very similar in dimensions, morphology, and behavior, competition is avoided by habitat partition (Farlow and Planka 2002, 21; see also Farlow et al. 2010).

It is impossible to distinguish between tracks of males and females of extinct animals, and particularly dinosaurs. Indeed, among other things, dinosaurian feet do not have those "bizarre structures" (Padian and Horner 2010) that, if present, could be interpreted as qualitative features of sexual dimorphism.

However, it seems now possible, at least cautiously and tentatively, to identify sexual dimorphism in dinosaur skeletons of some clades. Standard sexual dimorphism (when males are larger and present peculiar characteristics) seems to be at least probable for lambeosaurinae

Males and Females

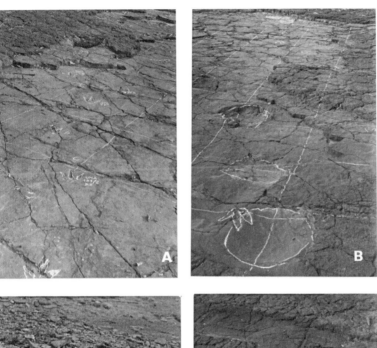

Figure 8.6. Piau-Caiçara (SOCA). Four views of the great sloping surface of level 13 at subsite 13/3. The surface was gridded in square meters for purpose of mapping. *A*, Theropod trackway SOCA 1331 and surrounding tracks. *B*, Sauropod undertracks, one of which was trampled on by a theropod footprint; at the top of the same photo one can see another theropod trackway. *C*, Sauropod tracks associated with theropod trackway 1331 and other theropod tracks. *D*, Bioturbation of the surface by the crossing of a sauropod herd.

hadrosaurs and neoceratopsids (especially Chasmosaurinae; cf. Lehmand 1990), and reverse sexual dimorphism (when the opposite happens) seems possible in some clades (Dodson and Currie 1990; Sampson and Ryan 1997; Chapman et al. 1997). On the other hand, other recent authors (e,g, Maiorino et al. 2015) do not agree with the sexual dimorphism in Proto-ceratopsidae and are not very favorable for the other above-mentioned dinosaur clades, estimating that the alleged dimorphism could depend, instead, on ontogenetic or normal intraspecific variations.

Furthermore, it seems now possible, in certain cases, to recognize nonavian female dinosaur individuals by analyzing some special tissues of their skeletons. As a matter of fact, fossil tissue interpreted as medul-lary bone—a temporary deposit of tissue along the walls of the medullary cavity, which some female birds form as a storage organ for calcium (and other minerals) just before egg formation and laying (Dacke et al. 1993)—was discovered recently along the medullary cavity of the long bones of *Tyrannosaurus rex* by Schweitzer et al. (2005). This recent and important discovery furnishes a concrete tool of gender distinction in dinosaurs. An analogous discovery was made three years later by Lee and Werning (2008) in *Allosaurus* and *Tenontosaurus*, demonstrating the presence of medullary bone not only in theropods (avian theropods included) but also in Ornithischia (Lee et al. 2013). Unfortunately, this discovery does not help in classifying the sex of a dinosaur on the basis of its tracks.

Plant Eater and Predator Interaction

The trackways in the Rio do Peixe basins are, as a rule, completely linear, with no indications of interactions between trackmakers; these dinosaurs were just passing through. Although some tracks of large plant eaters are associated with those of meat eating dinosaurs, sometimes heading more or less in the same direction or crossing one another, there is no suggestion that the carnivores were pursuing the herbivores. The closest approach to such an interaction is seen in occurrences in which theropod footprints (e.g., in Riacho do Pique in the Serrote do Letreiro site and perhaps at Piau-Caiçara, level 13/3, fig. 8.6) were impressed in the dis-placement rims of sauropod footprints while the sediment was still plastic. In such cases, it is difficult to think that the predators were unaware of the sauropod herd, considering their probable well-developed olfactory perception (Molnar and Farlow 1990)—but there is no evidence of actual interaction (fig. 8.6B, 8.6C).

Invertebrate Trails and Traces

9

Invertebrate ichnofossils were found only in the Sousa basin, in the Antenor Navarro and Sousa Formations.

In the Antenor Navarro Formation, the invertebrate ichnological material comes from Serrote do Letreiro, close to the northern border of the basin. The stratigraphic succession (fig. 9.1) is characterized by successive tabular bodies of conglomeratic sandstones, coarse sandstones, and centimetric levels of fine sandstones and siltstones, showing a clear fining upward in three depositional cycles (levels A, B, and C). These have sedimentary structures characterized by trough-cross and plane-parallel

Antenor Navarro
Formation

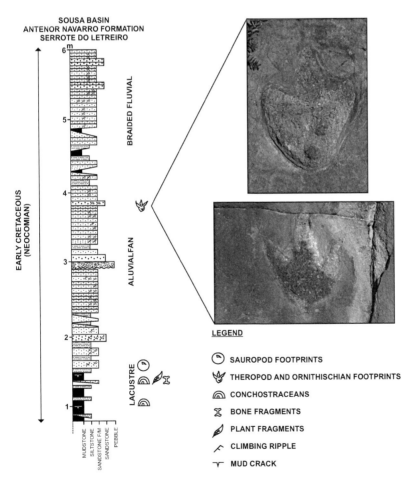

Figure 9.1. Stratigraphic section of the Antenor Navarro Formation at Serrote do Letreiro, Sousa basin. The levels A, B, and C correspond respectively to lacustrine, alluvial fan, and braided fluvial environments. (From Carvalho 1989).

LEGEND

- Ⓢ SAUROPOD FOOTPRINTS
- ☘ THEROPOD AND ORNITHISCHIAN FOOTPRINTS
- Ⓐ CONCHOSTRACEANS
- ⵥ BONE FRAGMENTS
- ⬙ PLANT FRAGMENTS
- ⋏ CLIMBING RIPPLE
- ⊤ MUD CRACK

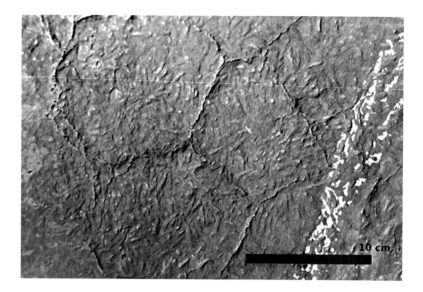

stratifications as well as desiccation cracks that can be observed on level C. Both invertebrate and vertebrate (dinosaur footprints) ichnofossils appear at the top of levels A and C, in the fine sandstones' and sandy siltstones' layers (Carvalho 2000b; Fernandes and Carvalho 2001).

The invertebrate ichnofossils from level A of the Antenor Navarro Formation in Serrote do Letreiro can be found in the finest portions of poorly sorted sandstone (fig. 9.2). They are on the same surface and do not display phobotaxis. The largest tracks can reach 1.5 m in length and have an average width of 1.0 cm. They usually do not display any kind of internal structure, but in some cases it is possible to observe underdeveloped menisci. However, at this level, the tracks without internal structures and filled with a fine sandstone similar to the one in the surrounding matrix predominate.

At level C, invertebrate trails seem to correspond to feeding (Fodinichnia) and locomotion (Repichnia) traces. The former were identified as belonging to the ichnogenus *Taenidium*. The granulometry of the sediments that fill these burrows is lower than the one of the surrounding matrix, but they are still very fine quartz sandstones. Another feeding structure present are branching sinuous tracks with highly complex meanders, possibly corresponding to ichnogenus *Phycodes*. Unlike the former, these ichnites appear over a clay film, in which contraction cracks with an extremely regular polygonal shape have been developed. The tracks overlap the cracks, which shows that they were formed during subaerial exposure of the sediments. Level C also contains straight and slightly sinuous tracks up to 80 cm in length. They are preserved in concave epirelief, without any kind of filling. Their distribution, however, is restricted and not significant. It is possible that these tracks correspond to types of *Taenidium* isp., in which the bad preservation or the recent exposure of the layers prevented the survival of their respective fillings.

Countless invertebrate ichnofossils, distributed in different stratigraphic levels, are especially preserved in the Sousa Formation. Their description, ethological classification, and importance are described below.

Circular structures are present in the Pedregulho district (Sousa County). They have an average diameter of 0.4 cm and are probably arranged in pairs along the bedding surface of an argillite. They are slightly darker than the brown matrix and are randomly distributed, having a probable two-by-two association along the stratification plane (MN 5.608-I). This material probably represents the extremities of a vertical U tube. Annelids can create this type of structure, which is attributed to the ichnogenus *Arenicolites*. It is ethologically classified as Domichnia (Carvalho 1989).

The structures related to organisms that feed on organic detritus found in the substrate (Fodinichnia)—such as those produced by recent arthropods and annelids—are common at Fazenda Caiçara-Piau (Sousa municipality, Sousa basin; MN 5.612-I—MN Museu Nacional, Universidade Federal do Rio de Janeiro). They have long and horizontal nonbranched shapes, some of which display internal consecutive divisions that represent menisci (MN 5.612-I.A). Their extremities can be rounded. Their cross section is oval-shaped, with the longer axis aligned in parallel with the stratification plane. Round shapes have an inflection of about 90°, with similar structural details to the straight forms. There are no overlaps. These structures are widely spaced along the stratification surface. They appear in a reddish-brown siltic argillite and are different from it due to their pronounced relief. Length varies between 2.0 and 6.0 cm and width between 0.5 and 1.0 cm. Straight tubes with a convex epirelief (MN 5.612—I.D) are also considered feeding structures (Fodinichnia) and are not limited to one stratification surface, thus having a partial and irregular distribution over the bedding planes. Their cross section is oval-shaped. They display a tracery and intermittent meniscoid structure on some portions of their length. When present, consecutive menisci appear at 0.1 cm intervals. Visible length along a bedding plane has a maximum value of 4.0 cm, and the average diameter of their cross section is 1.0 cm. The size and morphology of these structures suggest that their origin is related to the activity of terrestrial arthropods—probably insects.

The locomotion ichnofossils (Repichnia) found at Fazenda Caiçara-Piau are straight tracks, distributed over the bedding surface and preserved as a convex epirelief (Carvalho 1989). They are randomly distributed. Their cross section is oval-shaped, their width varies between 0.2–0.3 cm, and their maximum length is 3.5 cm (MN 5.612—I.C). The origin of these biogenic structures could be related to Oligochaeta annelids or terrestrial arthropods. There is intense bioturbation in some levels, with the association of different types of ichnofossils, such as MN 5.613—I.A, which is a long, slightly curved structure that displays no branching. This form's filling is different from the surrounding matrix; it is a clay, fine sandstone of the same color as the clay matrix (reddish-brown). It is 8.5

cm long, and its average width is 0.8 cm. It was probably produced by an arthropod, possibly an insect.

There are also ichnofossils related to the rest of the organisms (Cubichnia). In sample MN 5.613—I. B, there are straight and curved forms with a median crest. The tracks are intermittent and in some cases distributed successively in a diagonal distribution pattern. The extremities are V shaped, with the vertices turned to the interior, toward the central crest. The grooves are shallow, and their surface does not display any ornamentation. Preservation occurs as concave epirelief (epichnia). The bilobed track can have a total width of 0.6–0.7 cm, with the better-preserved forms displaying grooves with 0.3 cm of width and a median crest of 0.1 cm. Its length varies between 1.0 and 1.7 cm. The fact that it is not a long, continuous structure suggests that the organism that produced this ichnofossil did not have a grazing habit (Pasichnia). This being the case, annelids can be discarded. The organism had a natatorial habit and only occasionally came into contact with the substrate. This kind of ichnofossil could represent a resting moment of the animal at the bottom, like the movement performed by extant aquatic crustaceans (Carvalho 1989).

Other forms of undetermined nature from Fazenda Caiçara-Piau have round, hemispherical shapes in a convex epirelief. They occur throughout the bedding surface. Its average diameter is around 0.5 cm. It has the same color as the siltic-clayey matrix (reddish-brown), distinguished only by its prominent relief (MN5.612—I.B). There are also slightly curved tracks with unclear morphological traits, sometimes displayed as a central depression that creates two symmetrically positioned lobes, with no internal meniscoid structures (MN 5.617—I). These are preserved in convex epirelief and are slightly distinct from the clay matrix. The cross section is oval-shaped. Maximum length is around 5.0 cm, and the average width is 0.7 cm.

In Fazenda Piedade (Sousa municipality) there are straight, discontinuous burrows that display no segmentation or menisci and are randomly distributed over the bedding surface in different sedimentary levels. They are not bifurcated and sometimes overlap. The external borders of the tracks are parallel, and their extremities are rounded. The granulometry of the filling material of these tracks (siltstone) is larger than that of the surrounding material (argillite). The length of these ichnofossils varies between 1.5 and 3.5 cm. The average width of the tracks is 0.3 cm. To recognize the producer of these structures is problematic. This type of ichnofossil can be identified ethologically as feeding structures—that is, Fodinichnia. Tentatively, these are reported here as resulting from the ingestion of sediments by annelids.

The tracks ethologically identified as Fodinichnia are the most commonly found in the deposits of the Sousa Formation. In Fazenda Piau (Sousa County), there are straight or slightly sinuous, smooth tracks preserved in a convex epirelief on a surface with symmetrical ripple marks (MN 5.624—I). They do not have an internal structure, and their filling is similar to the surrounding matrix. The tracks are randomly distributed and do not display phobotaxis. Length varies between 0.1 cm and 5.0 cm, with a maximum width of 0.1 cm. The level where these tracks occur is intensely bioturbed. The morphological pattern of this ichnofossil suggests it was produced by Oligochaeta annelids. Sample MN 5.625—I.A also has straight tracks in a convex epirelief that are classified ethologically as Fodinichnia. They have parallel longitudinal striations along their surface. The filling sediment is similar to the matrix. They are randomly distributed and are surrounded by several millimetric tracks. Maximum length is 4.0 cm, and average width is 0.2 cm. Its origin is probably related to fecal matter expelled by freshwater invertebrates. Ichnogenus *Taenidium* was also identified at this site. There were straight tracks 1.5 cm long and 0.3 cm in width that were preserved in a convex epirelief, filled with the same material as the surrounding matrix and displaying regularly distributed menisci (MN 5.625—I.B). This ichnofossil was produced by a sediment eater organism, such as Oligochaeta. The ingestion of sediments and their posterior elimination during the animal's locomotion would lead to a pattern of successive menisci (Carvalho 1989).

At Fazenda Piau, there are invertebrate ichnofossils ambiguously identified as ethological structures of Domichnia. These are round, randomly distributed structures in a convex epirelief. The average diameter is 0.5 cm. The filling sediment is the same as the matrix. Greater concentrations can occur in certain points, which lead to the coalescence of these structures into misshapen masses of clay sediment.

In Saguim (Sousa County), there are also Fodinichnia structures (MN 5.627—I). They are long and randomly distributed. These tracks interlace, which in many cases leads to their fusion, and that can create a bifurcated aspect. The large number of overlapping tracks creates a peculiar mosaic pattern. Each track has an average length varying between 0.5 and 2.0 cm and an average width of 0.3 cm. They do not have any internal structure, and their filling is comprised of material that is slightly lighter than the matrix. These are ichnofossils related to sediment-eating organisms, and the large number of tracks indicates a high level of nutrients in the substrate (Carvalho 1989).

Paleoecological Interpretation of the Invertebrate Ichnofossils

The ichnofossils of Antenor Navarro Formation, such as the long, straight, or sinuous tracks with internal meniscus that allowed the identification of *Taenidium* isp., have been attributed to lumbriciform soft-bodied invertebrates (Oligochaeta annelids) with exoskeletons (especially arthropods such as conchostracans and insects), as well as vertebrates like fishes and reptiles (Buatois and Mángano 1996; Carvalho and Fernandes 2007), which actively rework the usually wet and semiconsolidated continental sediments. Conchostracans' activity is particularly intense at the bottom and shore of shallow lakes that undergo periodic desiccation. The annelids can operate both in the sediment at the bottom of lakes and in regions where the sediment is highly humid. On the other hand, insects rework the sediment on lakeshores and riverbanks as well as the drier deposits away from bodies of water. Not only is *Taenidium* considered a feeding ichnite, but it has also been interpreted as a simple locomotion structure (Graham and Pollard 1982). When lumbriciform organisms moved within the sediment, they ingested fine-grain clastic material to obtain the necessary nutrients for their metabolism, forming burrows with internal meniscoid structures due to sediment packing. This movement was performed through a peculiar mechanism: a posterior anchorage prevented the animal from moving backward when the anterior region of the body was projected forward, and another anterior or terminal anchorage allowed the posterior portion of the body to move forward (Heinberg 1974). If interpreting *Taenidium* as a feeding structure, the menisci would result from two alternating sediments, in which the packages that are lithologically similar to the matrix would correspond to undigested sediment; and the finer granulation menisci would have a coprolithic origin, resulting from the deposition of fecal remains by organisms such as annelids (D'Alessandro and Bromley 1987). The locomotion of arthropods that actively burrow could also create menisci through the transport of sediments over their surface, from the frontal region to the posterior empty space created by the movement of the organism, with the consecutive pressing of the sediments. This, however, does not seem to be the case of the tracks of Serrote do Letreiro, which have an interstratal feature (Fernandes and Carvalho 2001).

The ichnological association of the Antenor Navarro Formation resembles the *Scoyenia* ichnofacies, with a low ichnodiversity, sometimes monoichnospecific, comprised mainly of habitation, feeding, and locomotion ichnites, represented by vertical burrows, meniscoid tracks, vertebrate tracks, and root imprints. Among the sedimentary structures, mud cracks are a special feature. It reflects a subaerial environment. Serrote do Letreiro deposits have a fluvial origin, having been accumulated during torrential regimens close to the source areas and forming an alluvial fan system.

Unlike floodplains, fluvial environments dominated by channels are characterized by high energy and fluctuations of the sedimentation and erosion rates, resulting in an environment that is not favorable for the formation and preservation of biogenic structures. Furthermore, lack of nutrients prevents the maintenance of a diversified fauna, which can lead to a poor ichnofossiliferous record. At Serrote do Letreiro, the ichnites attributed to invertebrate activity are found in finer sediments accumulated in low energy conditions, on top of sandy bars and channel banks. The preserved tracks suggest that the sediments were firm and closely packed when the organisms moved through them, allowing their preservation. These conditions usually occur during periodic interruptions in sedimentation, when the sandy bars of the channel are exposed due to fluctuations in discharge. Biogenic activity is more common during the low water influx (Carvalho 1989).

The ichnocoenosis of invertebrates from the Sousa Formation is found in a sedimentary succession characterized by alternating siltstones and

Figure 9.3. Burrows in fine siliciclastic rocks observed in the Sousa Formation. They are indicative of feeding activities of invertebrates in the context of a floodplain area.

argillites, intercalated by centimetric levels of siltic sandstones and clayey sandstones. There can be occurrences of symmetrical and asymmetrical waves, contraction cracks, and flaser and linsen structures. The convolute lamination is associated with sites where tetrapod footprints are abundant (dinoturbation). There are feeding (Fodinichnia), crawling (Repichnia), resting (Cubichnia), and habitation (Domichnia) structures. The identified ichnogenera are ambiguously attributed to *Arenicolites* and *Taenidium*. The size and general morphology of the several Fodinichnia structures are varied, and their origins can be related to the activity of freshwater and terrestrial invertebrates, such as annelids, coleopters, orthopters, hymenopters, and larvae of several other insect orders. It is possible that the levels in which the feeding structures are abundant may indicate high nutrient content, which would allow maintenance of a considerable amount of benthic infauna and epifauna (fig. 9.3).

Repichnia and Cubichnia ichnofossils are less common. Repichnia are straight, nonbranching tracks preserved in a convex epirelief in the bedding plane. The maximum length of these structures is 8.5 cm, and the levels in which they occur represent time intervals of sedimentation interruption, creating omission surfaces—that is, sufficiently long periods of nondeposition that allow intense bioturbation. Cubichnia have an unusual aspect: dimorphic traits with V-shaped extremities in which the vertices are turned to the inside, toward a central crest. The tracks are discontinuous and suggest that the organism that generated them had a natatorial habit, only sporadically having contact with the substrate; one example could be a crustacean.

There are two extremely distinct types of habitation structures (Domichnia). One is represented by small, circular areas (average diameter of 0.4 cm) probably distributed in pairs and which would represent the extremities of a vertical U-shaped tube created by annelids. The other type, ambiguously considered a habitation, is comprised of small round masses with an average diameter of 0.5 cm, that when accreted create misshapen masses of clay sediment.

There are also some ichnofossils that were not ethologically classified. These are bilobed tracks and hemispherical masses distributed over the bedding plane. They are distinct from the matrix because they are preserved in a convex epirelief.

The ichnological assemblage of the Sousa Formation is similar to others already described in the literature, such as those from the Fleming Fjord (Triassic, Greenland; Bromley and Asgaard 1979), Diligencia (Miocene, California; Squires and Advocate 1984), and Duchesne River (Eocene, Utah; D'Alessandro et al. 1987) Formations, and in extant environments (Ratcliffe and Fagerstrom 1980). The invertebrate ichnocoenosis of the Sousa Formation is related to quiet lacustrine conditions. The absence of ichnofossils in certain stratigraphic levels is related to occasional water fluxes that would disturb the clay bottom, preventing the record of benthic biogenic activity. These lakes would usually be very shallow, sometimes developing in floodplains, and the intermittent water influx would lead to constant change in its borders. This would create a cyclic deposition responsible for the repetitive colonization by sediment-eating organisms, which would permanently inhabit the water body. The depositional conditions probably occurred in an environment that was very rich in nutrients, allowing the maintenance of an abundant and diverse benthic community (Carvalho 1989).

Localities Visited without Vertebrate Ichnological Results

10

The farms and localities of Saguim de Cima, Várzea da Jurema, Tabuleiro, Catolé da Piedade (WNW of Sousa, in the Sousa Formation); Pau d'Arco (SE of Sousa, in the Rio Piranhas Formation); the entire bed of the Peixe River between the Caiçara-Piau site and the Passagem das Pedras site, as well as between the Poço do Motor and Lagoa dos Patos; Murumbica, Lagoa dos Estrelas, and all the surfaces W of the road to Pereiros at Serrote do Letreiro; Benção de Deus; São Gonçalo, Guandu, Paquetá, Matumbe, and Curral; Várzea da Novena; Forno Velho, Triângulo, Vaca Morta, Saco, Xique-xique, Merejo, Riacho Chipo, Passarinho, Boa Esperança, Santo Antônio, Recanto, and Prazeres; and all the Antenor Navarro Formation rocky surfaces between Estreito and westward to Logradouro, José Lourenço, and Lagoinha.

The Municipality of Sousa

From west to east, the farms and localities of Singapura, Malhada da Pedra, Catingueiro, Caraíbas, Barros, Caldeirão, Caxias, Boa Vista, Santo Antônio, Bairro da Ponte, Rancho do Jacob, Outra Banda, Poço Cercado, Olho d'Água, Carnaúba, Sozinho, Várzea da Serrinha, Tentação, Poço Cercado, Mulunguzinho, Araçá de Baixo, Pavão, Juazeirinho dos Nogueira, Barragem do Açude dos Nogueira, Juazeiro da Emboscada, Poço da Lama, Rio Novo, Poço do Cantão, São Luiz, Vera Cruz, Riacho da Caçimba Nova, Baixio, Catolé, Foveiro, Bálsamo, Alto do Riacho, Serrote, Carnaúba, Areias, Lagoa do Mel, Barra de São Bento, Umari, and Livramento, Viração II.

The Municipality of São João do Rio do Peixe

The farms and localities at Acauã, Barra da Motuca, Riacho do Boi morto, Riacho do Gado. NB: There is a theropod footprint on a flagstone of the Sousa Formation in a sidewalk in the main street of the small town of Aparecida, seen by Leonardi in 2003.

The Municipality of Aparecida

The Municipality of Uiraúna	The farms and localities of Siriema, Heroismo, Açude das Areias, Açudes dos Caboclos, Morada, Pereiros, Fazenda Rio do Peixe, Alto dos Gomes, Extrema, Geraldo, Serrote da Pedra do Cipó, Tigre, Ipueiras, Lograddouro, Turim, Várzea do Cavalo, Curupati, and Varginha; and the neighborhoods of Uiraúna.
The Municipality of Pombal	From west to east, the farms and localities of Paissandu, Caieira, Boa Vista, Bezerro, Várzea Comprida, Malhada da Vaca, and Caraíbas.

Protections Acts

11

The ichnosites of the Rio do Peixe basins deserve protection (Schobbenhaus et al. 2002; Leonardi and Carvalho 2002; Carvalho et al. 2013; Siqueira et al. 2011; Santos et al. 2015, 2019; for a more global approach, cf. Alcalá et al. 2016). The main distribution area of dinosaur footprints at Passagem das Pedras (Ilha Farm), in Sousa County, is nowadays a natural park. In December 1992, through a state act (Diário Oficial do Estado da Paraíba, Decreto no 14.833, December 20, 1992), this area was defined as Dinosaur Valley Natural Monument (Monumento Natural Vale dos Dinossauros; fig. 11.1).

Figure 11.1 The occurrence of dinosaur footprints at Passagem das Pedras (Ilha Farm), in Sousa County, is now a natural park: the Dinosaur Valley Natural Monument (Monumento Natural Vale dos Dinossauros).

Figure 11.2. The use of symbols and images of dinosaurs and their tracks was enthusiastically adopted by the municipality, giving a new identity to the region.

In 1996, an agreement regarding a development plan was established among the Ministry of the Environment, the Paraíba state government (SUDEMA), and the Sousa Municipal City Hall (Convênio MMA/ PNMA/PED no 96 CV00030/96). The financial investment was approximately US$800,000.

Over the last forty-five years, the regular presence of researchers and undergraduate and graduate students, and the media's diffusion of news about the existence of dinosaur tracks in Sousa County, helped the local residents to change their perception of the relevance of the scientific research in the region. During this time span, ongoing paleontological discoveries captured the popular imagination. There was a progressive influence of the vertebrate ichnology on the social, sportive, and local commerce. Symbols and images of dinosaurs and their tracks became emblematic of the municipality, arousing a new identity for the region, which previously had been known mainly for its cotton agriculture.

Eventually, the ichnological information was progressively introduced into the population's everyday life, as evidenced by the inclusion of

Figure 11.3. *(left)* A canal cut to change the main course of the Peixe River. The intent is to protect the stratigraphic levels with footprints that previously suffered erosion during floods.

Figure 11.4. *(below)* Visitor center at the Dinosaur Valley Natural Monument. The exhibition area features thematic dioramas, fossils, a video room, and documents pertaining to the history of the research in the area.

symbolic elements in the community routine (fig. 11.2). This allowed the establishment of the Dinosaur Valley Natural Monument (locally the "natural park" or simply "Dinosaur Valley") and invited paleontological tourism in Sousa County—an ecological endeavor in which the tourist desires a unique experience, and the main motivation is knowledge. It is an important activity for the economic development of regions far from the great industrial centers and the traditional tourist attractions.

Figure 11.5. The natural park Vale dos Dinossauros at Passagem das Pedras (SOPP), Sousa Formation, Sousa. *A*, Carvalho and students at the main gate. *B*, The footbridge over the Peixe River and the marginal forest. *C*, Fiberglass models of dinosaurs, produced by our lab in Sousa. *D*, The visitor center. *E*, The landscape along the Peixe River, showing one of the footbridges and a small dam that thus far has prevented flooding of the trackways.

It is possible to preserve the area's ichnological heritage through public politics, increasing tourism in the region, local museum development, emphasis of areas with scientific and cultural importance, support of scientific research, and engagement of the local people through increased valuation of the ichnofossils as objects that are important to the economy (Carvalho and Leonardi 2007; Fernandes and Carvalho 2007; Santos 2014: Santos et al. 2015).

The protection of this ichnofossiliferous site comprises

· Changing the Peixe River main course. This aims to protect the stratigraphic levels with footprints that have been eroded during

flooding periods. A secondary artificial channel as well as bridges over it and the river channel were constructed (fig. 11.3).

· Planting of native vegetation at channel borders and neighboring areas of the Peixe River.
· Construction of an access road to the natural monument from federal highway BR-391.
· Addition of concrete footbridges over the trackways to avoid the direct contact between visitors and the fossil footprints.
· Creation of a visitor center with a 222-m² area with an exposition area with thematic dioramas, fossils, a video room, and documents concerning the history of research in the area. It also features a library, a souvenir shop, and a snack bar (fig. 11.4).

Presently, this is one of the best-preserved paleontological sites in Brazil. The area is now a tourist complex, with trained staff to guide visitors and to protect the paleontological site. However, some factors strongly contribute to the vulnerability of the ichnological sites (figs. 11.5, 11.6)—mainly, natural weathering and anthropic action. Siqueira et al. (2011) considered that the sites with tracks and footprints present a medium-to-high degree of vulnerability. The most vulnerable sites are Pedregulho, Piau/Caiçara, Várzea dos Ramos, Lagoa dos Patos, Cabra Assada, and Matadouro.

12

Although this book is the fruit of some four decades of fieldwork (1975–2021), we hope it does not represent the end of research on the Rio do Peixe ichnofaunas. There remains a great deal of field, lab, and brain work to be done. There are important local ichnofaunas that were discovered and surveyed in only a cursory manner, without thorough study, such as those of Floresta dos Borba (SOFB) and Tapera (APTA), with their innumerable tracks. There are also some small minor ichnofaunas that were seen and quickly surveyed by Leonardi many years ago that deserve a new survey and revision. Baixio do Padre (SOBP), discovered in January 2008, is worth a good deal of digging and a complete survey. At Piau (SOCA), the numerous layers upstream of the dam ought to be surveyed in a very dry year, when the reservoir is completely empty, to extend the site and add new rock layers, surely rich in tracks, to the twenty-five ichnofossiliferous levels of the site. Abreu (SOAB) and Araçá de Cima (SOAC) ought to be surveyed. Aroeira (ANAR) also deserves another look.

More exploration is needed in the three basins of the Rio do Peixe Group. The Sousa basin was not fully surveyed west of the town of São João do Rio do Peixe to the western border of the basin. The Uiraúna-Brejo das Freiras basin, not very rich in tracks, was explored very quickly, and its northwestern extremity was not examined. The southern part of the Pombal basin, where outcrops are generally rare, also was not completely surveyed. Along with promising sites that may be found during these explorations, new ichnofossiliferous sites can be expected to be discovered even in known areas of the basins that have been intensely explored.

It is worth noting that every year, during the rainy season, the Rio do Peixe and its tributaries (and to a lesser extent the Piranhas River) may uncover new layer surfaces and footprints. However, the waters that aid discovery may also destroy previously exposed tracks that detach from their substrates and are eroded. Tracks close to roads and human activity also may be damaged or ruined. This at-risk ichnological material should be taken up and moved out of harm's way and placed with scientific institutions for conservation and study. The Agência Nacional de Mineração also needs to be part of the initiative to curate and preserve the paleontological heritage.

Excavations should be continued at Passagem das Pedras and at the other main localities, always under the supervision of a specialist. One of the goals of future paleoichnologist activity in the Rio do Peixe basins should be continuing thorough documentation and study of entire track-bearing rocky pavements, both those already known as well as those newly appearing, using more advanced methods—improving and revising previous ichnological studies with modern techniques like laser scanning, geometric photogrammetry, three-dimensional modeling, use of drones (Belvedere et al. 2012; Breithaupt and Matthews 2012; Gatesy and Ellis 2012; Lužar-Oberiter et al. 2017), and other techniques that will arise in the future.

It must be recognized that over time many of the track-bearing surfaces will surely be destroyed by natural erosion, especially in the Sousa Formation, where the tracks crop out in the beds of rivers. All of this documentation—vintage-type and modern type—could eventually be utilized to digitally reconstruct the sites and the whole basins as they were prior to excavation and prior to erosion (Falkingham et al. 2014).

The authors gladly express their total agreement with the policies of protecting and valorizing the main locality of Passagem das Pedras by means of the Dinosaur Valley Natural Monument in Sousa. Apart from its cultural and scientific significance, this site has great economic and social value as a stimulus to tourism, which in turn enhances employment opportunities for local residents, which is so important and necessary for that region. We also express our wish that other sites (especially Serrote do Pimenta and Serrote do Letreiro) be fenced, protected, and valorized, eventually following the detailed and concrete proposals and maps prepared in collaboration with the Departamento Nacional de Produção Mineral-D.N.P.M. and Conselho Nacional de Desenvolvimento científico e tecnológico-CNPq and officially presented to the competent authorities (the mayors of Sousa and of São João do Rio do Peixe) in the 1980s by Leonardi. We recommend that the few fossil footprints that survive at Piau-Caiçara (almost all of the best rocky exposures were unfortunately destroyed by seasonal floods) be collected, following suggestions by Leonardi in the 1980s through 2016, under the supervision of a paleontologist and reposited in an institution, preferably the visitor center of the Dinosaur Valley Natural Monument at Sousa. Keep in mind that the so-called Dinosaur Valley, as characterized in the original proposal, is not solely the Passagem das Pedras (SOPP) site. As splendidly as this site has been developed, it is only one of four proposed modules in the project. Serrote do Pimenta at the Estreito farm (SOES), Serrote do Letreiro (SOSL) in the municipality of Sousa, and Engenho Novo (ANEN) in the municipality of São João do Rio do Peixe, a locality that has regrettably been destroyed, were also suggested as candidates for protection and development.

Last but by no means least, it is very important that there be a resident vertebrate paleontologist at Sousa, responsible for ongoing systematic survey, study, documentation, and protection of the paleontological resources of the Dinosaur Valley Natural Monument and the four basins.

References

Abbassi, N., H. Alimohammadian, S. Shakeri, S. Broumand, and A. Broumand. 2018. "Aptian Small Dinosaur Footprints from the Tirgan Formation, Kopet-Dagh Region, Northeastern Iran." In *Fossil Record*, 6, edited by S. G. Lucas and R. M. Sullivan. *New Mexico Museum of Natural History and Science Bulletin* 80, 5–14.

Abbassi, N., S. D'Orazi Porchetti, A. Wagensommer, and M. G. Dehnavi. 2015. "Dinosaur and crocodylomorph footprints from the Hojedk Formation (Bajocian, Middle Jurassic) of north Kerman, central Iran." *Italian Journal of Geosciences*, 134 (1): 86–94.

Abbassi, N., and S. Madanipour. 2014. "Dinosaur tracks from Jurassic Shemshak Group in Central Alborz Mountains, Northern Iran." *Geologica Carpathica* 65(2): 99–115.

Albuquerque, J. P. T. 1970. Inventário hidrogeológico do Nordeste, folha n°15- Jaguaribe SE. Recife, SUDENE, Divisão de Documentação, Brazil (SUDENE—Hidrogeologia, 32).

Alcalá, L., M. G. Lockley, A. Cabos, L. Mampel and R. Royo-Torres. 2016. "Evaluating the Dinosaur Track Record: An Integrative Approach to Understanding the Regional and Global Distribution, Scientific Importance, Preservation, and Management of Tracksites." In *Dinosaur Tracks: The Next Steps*, edited by P. L. Falkingham, D. Marty, and A. Richter, 100–16. Bloomington: Indiana University Press.

Alexander, R. McNeill. 1989. *Dynamics of Dinosaurs and Other Extinct Giants*. New York: Columbia University Press.

Almeida, F. F. M., and Y. Hasui. 1984. *O Pré-Cambriano do Brasil*. São Paulo: Edgard Blücher.

Amaral, C. A. 1983. Projetos Mapas Metalogenéticos e de Previsão de Recursos Minerais. Folha SB 24-Z-A Sousa, Volumes I, II (textos e mapas) CPRM (Superintendência Regional de Recife), Brazil, 1–13.

ANP. Brasil Round 9, 2008. Rio do Peixe Basin. Nona Rodada de Licitações. Cid Queiroz Fontes. Bid Area Department.

Antonelli M., M. Avanzini, M. Belvedere, M. Bernardi, P. Ceoloni, P. Citton, M. A. Conti, et al., 2020. "Updated Italian Vertebrate Ichnology Reference List." 233–65 In *Tetrapod Ichnology in Italy: The State of the Art*, edited by M. Romano and P. Citton. *Journal of Mediterranean Earth Sciences* 12 (2020).

Arai, M. 2006. "Revisão estratigráfica do Cretáceo Inferior das Bacias Interiores do Nordeste do Brasil." *Geociências* 25(1): 7–15.

Araújo, R. E. B., F. H. R. Bezerra, F. C. C. Nogueira, F. Balsamo, B. R. B. M. Carvalho, J. A. B. Souza, J. C. D. Sanglard, and C. C. M. Alanny. 2018. Basement control on fault formation and deformation band damage zone evolution in the Rio do Peixe Basin, Brazil. Tectonophysics 745: 117–131.

Assis, J. F. P., J. B. Macambira, and G. Leonardi. 2010. "Dinossauros terópodes do Ribeirão das Lajes, primeiro registro fóssil da Formação Sambaíba (Neotriássico-Eojurássico), Bacia do Parnaíba: Fortaleza dos Nogueiras, Maranhão-Brasil." Congresso Brasileiro de Geologia 45, Belém—PA, Brazil.

Avanzini, M. 1998. "Anatomy of a Footprint: Bioturbation as a Key to Understanding Dinosaur Walk Dynamics." *Ichnos* 6: 129–39.

Avanzini, M., M. Campolongo, G. Leonardi, and R. Tomasoni. 2001a. "Tracce di dinosauri nel Giurassico inferiore della Valle del Sarca (Italia Nord-Orientale)." *Studi Trentini di Scienze Naturali—Acta Geologica* 76: 167–82.

Avanzini, M., P. Ceoloni, M. A. Conti, G. Leonardi, R. Manni, N. Mariotti, P. Mietto, et al. 2001b. "Permian and Triassic Tetrapod Ichnofaunal Units of Northern Italy, Potential Contribution to Continental Biochronology." In Permian Continental Deposits of Europe and Other Areas. Regional Reports and Correlations, edited by G. Cassinis. Natura Bresciana, Monografia 25: 89–107.

Avanzini, M., F. M. Dalla Vecchia, V. De Zanche, P. Gianolla, P. Mietto, N. Preto, and G. Roghi. 2000a. "Aspetti Stratigrafici Relativi Alla Presenza Di Tetrapodi Nelle Piattaforme Carbonatiche Mesozoiche Del Sudalpino. Crisi biologiche, radiazioni adattative e dinamica delle piattaforme carbonatiche." *Accademia Nazionale di Scienze Lettere e Arti di Modena, Collana di Studi* 21(2000): 15–20.

Avanzini, M., S. Frisia, E. Keppens, G. Leonardi, M. Rinaldo, and K. Van den Driessche. 1995. Sedimentology and Diagenesis of a Dinosaur Tracksite: New Perspectives on Early Jurassic Palaeogeography of the Southalpine; 10–11 in Bathurst Meeting of Carbonate Sedimentologists, 10. Abstract volume for Talks and Posters. Egham, Surrey, 2–5 July 1995. Royal Holloway University of London, UK.

Avanzini, M., G. Gierlinski, and G. Leonardi. 2001c. "First Report of Sitting *Anomoepus* Tracks in European Lower Jurassic (Lavini di Marco Site—Northern Italy)." *Rivista Italiana di Paleontologia e Stratigrafia, Milano* 107(1): 131–36.

Avanzini, M., and G. Leonardi. 1993. "I dinosauri dei Lavini di Marco e i grandi vertebrati fossili del Trentino-Alto Adige." *Natura Alpina* 44(3): 1–14.

———. 1999. "Prima segnalazione di orme di un dinosauro accucciato (*Anomoepus*) nel Giurassico inferiore europeo"; 63–65 in Geoitalia 1999. Bellaria (Rimini), 20–23 settembre 1999, Forum Italiano Di Scienze Della Terra, 2, Riassunti 1. Italy.

———. 2002. *Isochirotherium inferni* ichnosp. n. in the Illyrian (Late Anisian, Middle Triassic) of Adige Valley (Bolzano—Italy). *Bollettino della Società Paleontologica Italiana* 41(1): 41–50.

Avanzini, M., G. Leonardi, D. Masetti, and P. Mietto. 2000b. "Conclusioni." In *Dinosauri in Italia. Le orme giurassiche dei Lavini di Marco (Trentino) e gli altri resti fossili italiani*, edited by G. Leonardi, and P. Mietto, 393–98. Pisa-Roma, Italy: Accademia Editoriale.

Avanzini, M., G. Leonardi, and P. Mietto. 2003. "*Lavinipes cheminii* Ichnogen., Ichnosp. nov., a Possible Sauropodomorph Track from the Lower Jurassic of Italian Alps." *Ichnos* 10: 179–93.

Avanzini, M., G. Leonardi, P. Mietto, and G. Roghi. 2000c. "Icnofaune dinosauriane nel Giurassico inferiore della Piattaforma di Trento; aspetti stratigrafici, paleoambientali e paleogeografici." Ottantesima Riunione Estiva Della Società Geologica Italiana. Riassunti delle Comunicazioni orali e Poster Trieste, 6–8 settembre 2000, Italy, 42–44.

Avanzini, M., G. Leonardi, and R. Tomasoni. 2001d. "Enigmatic Dinosaur Trackways from the Lower Jurassic (Pliensbachian) of the Sarca Valley, Northeast Italy." *Ichnos* 8: 235–42.

Azevedo, S. A. K. 1993. "Novas pegadas de dinossauros em Sousa, estado da Paraíba, Brasil." *Anais da Academia Brasileira de Ciências* 65(3): 279–83.

Bakker, R. T. 1972. "Anatomical and Ecological Evidence of Endothermy in Dinosaurs." *Nature* 238: 81–85.

———. 1986a (1993, 7th printing). *The Dinosaur Heresies*. New York: Zebra Books, Kensington.

———. 1986b. "The Return of the Dancing Dinosaurs." In *Dinosaurs Past and Present*, edited by S. J. Czerkas and E. C. Olson, vol. 1, 38–69. Seattle: Washington University Press.

Barbosa, W. V., I. B. Silva, R. C. Santos, A. C. Santos, C. A. C. Pimentel, V. A. Nóbrega, and J. M. Mabesoone. 1986. Revisão geológica da parte oriental da sub-bacia de Souza (Bacia do Rio do Peixe, Paraíba); 1: 308–320 in Congresso Brasileiro de Geologia, 34, Anais, Goiânia, 1986, Goiânia, Sociedade Brasileira de Geologia, Goiás, Brazil.

Bartholomai, A., and R. E. Molnar. 1981. "Muttaburrasaurus, a New Iguanodontid (Ornithischia: Ornithopoda) Dinosaur from the Lower Cretaceous of Queensland." *Memoirs of the Queensland Museum* 20: 319–49.

Béland, P., and D. A. Russell. 1978. "Paleoecology of Dinosaur Provincial Park (Cretaceous), Alberta, Interpreted from the Distribution of Articulated Vertebrate Remains." *Canadian Journal of Earth Sciences* 15(6): 1012–24.

———. 1980. Paléoécologie des dinosaures de la Formation Oldman (Crétacé supérieur Canada). *Mémoires de la Société géologique de France* N.S. 139: 15–18.

Bell, P. R., and E. Snively. 2008. "Polar dinosaurs on parade: a review of dinosaur migration." *Alcheringa: An Australian Journal of Palaeontology* 32(3): 271–284.

Belvedere, M., A. Baucon, S. Furin, P. Mietto, F. Felletti, and G. Muttoni. 2012. "The Impact of the Digital Trend on Ichnology: Ichnobase." In A. Richter and M. Reich (eds.). Dinosaur Tracks 2011. An International Symposium. Obernkirchen, April 14–17, 2011. Göttingen, Göttingen Universitätsdrucke, 2012, Germany.

Belvedere, M., M. R. Bennett, D. Marty, M. Burdka, S. C. Reynolds, and R. Bakirov. 2018. "Stat-tracks and Mediotypes: Powerful Tools for Modern Ichnology Based on 3D Models." *PeerJ* 6:e4247. https://doi.org/10.7717/peerj.4247.

Belvedere, M. and J. O. Farlow. 2016. "A Numerical Scale for Quantifying the Quality of Preservation of Vertebrate Tracks: In *Dinosaur Tracks: The Next Steps*, edited by P. L. Falkingham, D. Marty, and A. Richter, 92–98; chap. 6. Bloomington: Indiana University Press.

Belvedere M., P. Mietto, and S. Ishigaki. 2010. A Late Jurassic diverse ichnocoenosis from the siliciclastic Iouaridène Formation (Central High Atlas, Morocco). Geological Quarterly 54 (3): 367–380.

Bertozzo, F., F. M. Dalla Vecchia, and M. Fabbri. 2017. "The Venice Specimen of Ouranosaurus nigeriensis (Dinosauria, Ornithopoda)." *PeerJ* 5:e3403. https://doi.org/10.7717/peerj.3403.

Beurlen, K. 1967a. "Estratigrafia da faixa sedimentar costeira Recife-João Pessoa." *Boletim da Sociedade Brasileira de Geologia* 16(1): 43–54.

———. 1967b. "A estrutura geológica do Nordeste do Brasil" Congresso Brasileiro de Geologia, 21, pp. 150–158, Curitiba, Sociedade Brasileira de Geologia, Curitiba, Paraná, Brazil.

———. 1971. "Bacias sedimentares no bloco brasileiro." *Estudos Sedimentológicos* 1(2): 7–31.

Beurlen, K., and J. M. Mabesoone. 1969. "Bacias Cretáceas intracontinentais do Nordeste do Brasil." *Notícias Geomorfológicas* 9(18): 19–34.

Bishop, P. J., C. J. Clemente, R. E. Weems, D. F. Graham, L. P. Lamas and J. R. Hutchinson. 2017. "Using Step Width to Compare Locomotor Biomechanics between Extinct, Non-avian Theropod Dinosaurs and Modern Obligate Bipeds." *Journal of the Royal Society Interface* 14: 20170276. http://doi.org/10.1098/rsif.2017.0276.

Blalock, H. M. 1969. *Statistica per la ricerca sociale*. Bologna, Società editrice il Mulino, Italia, 731 pp.

Boa Nova, F. P. 1940. "Águas termais de Brejo das Freiras." *Mineração e Metalurgia* 5(28): 176–77.

Böhme, A., U. Stratmann, M. Wiggenhagen, T. van der Lubbe, and A. Richter. 2009. "New Tracks on the Rock: Parallel Trackways of a New Type of Iguanodontipus-Caririchnium-like Morphology from the Lower Cretaceous Sandstones of Obernkirchen, Northern Germany." *Journal of Vertebrate Paleontology* 29(3): 66A.

Boletim do Paleo Nordeste, Recife—Pernambuco, 9 a 10 de Dezembro de 2004, 3.

Boletim de Geociências da Petrobras. 2007. Cartas estratigráficas. Petrobras, Rio de Janeiro 15(2).

Bonaparte, J. F. 1986. "History of the Terrestrial Cretaceous Vertebrates of Gondwana," 2: 63–95, in Congreso Argentino de Paleontología Y Bioestratigrafía, 4. Mendoza, Argentina.

———. 1996. *Dinosaurios de América del Sur*. Museo Argentino de Ciencias Naturales "Bernardino Rivadavia," Buenos Aires, Argentina.

———. 2007. *Dinosaurios y pterosaurios de América del Sur*. Albatros, Buenos Aires, Argentina.

Bonaparte J. F., E. H. Colbert, P. Currie, A. de Ricqles, Z. Kielan-Jaworowska, G. Leonardi, N. Morello, and P. Taquet. 1984. *Sulle orme dei dinosauri*. Erizzo, Venezia (Esplorazioni e ricerche, IX), 2nd ed. 335 pp.

Bonaparte, J. F., and R. A. Coria. 1993. "Un nuevo y gigantesco saurópodo titanosaurio de la formación Río Limay (Albiano-Cenomaniano) de la provincia de Neuquén, Argentina." *Ameghiniana* 30: 271–83.

Bonaparte, J. F., B. Gonzales Riga, and S. Apesteguía. 2006. "*Ligabuesaurus leanzai* (Dinosauria, Sauropoda) a New Titanosaur from the Lohan Cura Formation (Aptian, Lower Cretaceous) of Neuquén, Patagonia, Argentina." *Cretaceous Research* 27: 364–76.

Bonaparte J. F., and F. E. Novas. 1985. "*Abelisaurus camahuensis* n.g., n.sp., Carnosauria del Cretácico tardio de Patagonia." *Ameghiniana* 21(2–4): 259–65.

Bonaparte, J. F., and J. E. Powell. 1980. "A Continental Assemblage of Tetrapods from the Upper Cretaceous Beds of El Brete, NW Argentina (Sauropoda Coelurosauria Carnosauria Aves)." *Mémoires de la Société géologique de France*, N.S. 39: 19–28.

Boutakiout, M., M. Hadri, J. Nouri, I. Díaz-Martínez, and F. Pérez-Lorente. 2009. Rastrilladas de icnitas terópodas gigantes del Jurásico Superior (Sinclinal de Iouaridène, Marruecos). *Revista Española de Paleontología* 24 (1), 31-46.

Branner, J. C. 1919. "Outlines of the Geology of Brazil to Accompany the Geologic Map of Brazil." *Bulletin of the Geological Society of America* 30(1919): 189–338.

Braun, O. P. G. 1966. Estratigrafia dos sedimentos da parte interior da região Nordeste do Brasil. Rio de Janeiro, DNPM/DGM. Boletim 236.

———. 1969. Geologia da Bacia do Rio do Peixe, Nordeste do Brasil. Recife, Departamento Nacional da Produção Mineral/ Divisão de Geologia e Mineralogia, Brazil Relatório Interno.

———. 1970. Geologia da bacia do Rio do Peixe, Nordeste do Brasil, 208–09. in Congresso Brasileiro de Geologia, 24, Brasília (DF), Resumo das Conferências e Comunicações. Sociedade

Brasileira de Geologia/Núcleo Centro-Oeste. Boletim Especial, 1, Brazil.

Breithaupt, B. H., and N. A. Matthews. 2012. "Neoichnology and Photogrammetric Ichnology to Interpret Theropod Community Dynamics." In A. Richter, and M. Reich (eds.), Dinosaur Tracks 2011. An International Symposium. Obernkirchen, April 14–17, 2011. Göttingen, Göttingen Universitätsdrucke, 2012, Germany.

Brett-Surman, M. K., T. R. Holtz, and J. O. Farlow, eds. 2012. The Complete Dinosaur. 2nd ed. Bloomington: Indiana University Press.

Brito, I. M. 1975. As bacias sedimentares do Nordeste do Brasil. Rio de Janeiro, UFRJ/Departamento de Geologia, vol.1, pp. 1–107.

Brodrick, H. 1909. "Note on footprint cast on the Inferior Oolite near Whitby, Yorks." Proc. Liverpool Geol. Soc. 10: 327–35.

Bromley, R., and U. Asgaard. 1979. "Triassic Freshwater Ichnocoenoses from Carlsberg Fjord, East Greenland." Palaeogeography, Palaeoclimatology, Palaeoecology 28: 39–80.

Bronner, G. and G. Demathieu. 1977. "Premières traces de reptiles archosauriens dans le Trias autochtone des Aiguilles Rouges (Col des Corbeaux, Vieil Emosson, Valais Suisse). Conséquences paléogéographiques et chronostratigraphiques." Comptes Rendus de l'Académie des Sciences, Paris 285(D): 649–52.

Browne, G. H. 2009. "First New Zealand Record of Probable Dinosaur Footprints from the Late Cretaceous North Cape Formation, Northwest Nelson." New Zealand Journal of Geology and Geophysics 52: 367–77.

Brum, P. A. R., and J. C. Souza. 1985. "Níveis de nutrientes minerais para gado, em lagoas ('Baías' e 'Salinas') no pantanal sul-Mato-Grossense." Pesquisa Agropecuária Brasileira Brasília 29(12): 1451–54.

Buatois, L. A., and M. G. Mángano. 1996. "Icnología de ambientes continentales: problemas y perspectivas." Reunión Argentina de Icnología 1(4): 5–30. Asociación Paleontológica Argentina, Publicación Especial, Argentina.

Calvo, J. O. 1991. "Huellas de dinosaurios en la Formación Rio Limay (Albiano-Cenomaniano?), Picún Leufú, Provincia de Neuquén, República Argentina. (Ornithischia-Saurischia: Sauropoda-Theropoda)." Ameghiniana 28: 241–58.

Calvo, J. O., and J. F. Bonaparte. 1991. "Andesaurus delgadoi gen. et sp. nov. (Saurischia, Sauropoda), dinosaurio Titanosauridae de la Formación Río Limay (Albiano-Cenomaniano), Neuquén, Argentina." Ameghiniana 28: 303–10.

Calvo, J. O., R. Rubilar-Rogers, and K. Moreno. 2004. "A New Abelisauridae (Dinosauria: Theropoda) from Northwest Patagonia." Ameghiniana 41: 555–63.

Calvo, J. O., and L. Salgado. 1995. "Rebbachisaurus tessonei a New Sauropoda from the Albian-Cenomanian of Argentina: New Evidence of the Origin of the Diplodocidae." Gaia, 11: 13–33.

Campos, H. B. N. 2004. "A New Archosaur Tracksite from the Early Cretacous of Sousa Basin, Northeastern Brasil." Paleo (2004): 3, Reunião Anual Regional da Sociedade Brasileira de Paleontologia, Resumos Recife, Pernambuco, Brazil.

———. 2005. "A New Archosaur Tracksite from the Early Cretaceous of Sousa Basin, Northeastern Brasil." Paleontologia em Destaque 49: 39.

Campos, H. B. N., R. C. Silva, and J. Milàn. 2010. "Traces of a Large Crocodylian from the Lower Cretaceous. Sousa Formation, Brazil." In Crocodyle Tracks and Traces, edited by J. Milàn, S. G. Lucas, M. G. Lockley, and J. A. Spielmann. New Mexico Museum of Natural History and Science, Bulletin 51: 109–14.

Campos, M. B., A. J. Braga, A. P. G. Souza, E. M. Silva, E. M. Franca, J. B. Medeiros, and M. Freitas. 1974. Projeto Rio Jaguaribe. Relatório de Fotointerpretação. Sureg, RE 1974, vol. 1, CPRM. [Unpublished].

Canale, J. I., S. Apesteguía, P. A. Gallina, F. A. Gianechini, and A. Haluza. 2016. "The Oldest Theropods from the Neuquén Basin: Predatory Dinosaur Diversity from the Bajada Colorada Formation (Lower Cretaceous: Berriasian–Valanginian), Neuquén, Argentina." Cretaceous Research (2016). https://doi.org/10.1016/j.cretres.2016.11.010.

Carpenter, K. 1987. "Paleoecological Significance of Droughts during the Late Cretaceous of the Western Interior." In 4th Simposium Mesozoic Terrestrial Ecosystems, edited by P. J. Currie and E. H. Koster, 42–47. Drumheller, Alberta, Canada: Tyrrell Museum of Paleontology.

Carvalho, I. S. 1989. Icnocenoses continentais: bacias de Sousa, Uiraúna-Brejo das Freiras e Mangabeira. MS thesis, Universidade Federal do Rio de Janeiro, Rio de Janeiro, Brazil [unpublished].

———. 1993. Os conchostráceos fósseis da bacias interiores do Nordeste do Brasil. DSc thesis, Universidade Federal do Rio de Janeiro, Rio de Janeiro, Brazil [unpublished].

———. 1994a. "As ocorrências de icnofósseis de vertebrados na bacia de São Luís, Cretáceo Superior, Estado do Maranhão," pp. 119–22, in Simpósio Sobre o Cretáceo do Brasil, 3, Boletim, Rio Claro, São Paulo, UNESP, Brazil.

———. 1994b. Contexto tafonômico das pegadas de terópodes da praia da Baronesa (Cenomaniano, bacia de São Luís); 3: 211–12 in Congresso Brasileiro de Geologia, 38, Boletim de Resumos Expandidos Camboriú. Santa Catarina, Sociedade Brasileira de Geologia, Santa Catarina, Brazil.

———. 1994c. "Candidodon: um crocodilo com heterodontia (Notosuchia, Cretáceo Inferior)." Anais da Academia Brasileira de Ciências, 66(3): 331–46

———. 1996a. "As pegadas de dinossauros da bacia de Uiraúna-Brejo das Freiras (Cretáceo Inferior, estado da Paraíba)," Simpósio Sobre o Cretáceo do Brasil, 4: 115–21, Boletim, Rio Claro, São Paulo, UNESP, Brazil.

———. 1996b. "Paleogeographic distribution of esthereliidean conchostracans on the Cretaceous rift interior basins of Northeastern Brazil," 7: 387–89, in Congresso Brasileiro de Geologia, 39, Anais, Salvador, Bahia, Brazil.

———. 2000a. "Geological Environments of Dinosaur Footprints in the Intracratonic Basins of Northeast Brazil during the Early Cretaceous Opening of the South Atlantic." Cretaceous Research 21(2000): 255–67.

———. 2000b. "Huellas de saurópodos Eocretácicas de la cuenca de Sousa (Serrote do Letreiro, Estado da Paraíba, Brasil)." Ameghiniana 37(3): 353–62.

———. 2004a. "Bacias cretáceas interiores do Nordeste." Fundação Paleontológica Phoenix 70: 1–4.

———. 2004b. "Dinosaur Footprints from Northeastern Brazil: Taphonomy and Environmental Setting." Ichnos 11: 311–21.

———. 2009. "Os conchostráceos das bacias interiores do Nordeste do Brasil e o quimismo das águas continentais do Cretáceo," 106 in Congresso Brasileiro de Paleontologia, 21, Atas, Belém, Sociedade Brasileira de Paleontologia, Belém, Pará, Brazil.

Carvalho, I. S., H. I. Araújo-Júnior, F. C. C., Nogueira, J. A. Soares, L. Salgado, R. M. Lindoso, and G. Leonardi. 2016. Taphonomic and Paleoenvironmental Aspects of the Lower Cretaceous Rio Piranhas Formation (Triunfo Basin, Northeastern Brazil) Based on Faciological and Paleontological Data. In Congresso Brasileiro de Geologia, 48°, FIERGS, Porto Alegre, RS, Brazil. Volume: Abstracts and programs.

Carvalho, I. S., L. S. Avilla, and L. Salgado. 2003. "Amazonsaurus maranhensis (Sauropoda, Diplodocoidea) from the Lower

Cretaceous (Aptian-Albian) of Brazil." *Cretaceous Research* 24: 697–713.

Carvalho, I. S., L. Borghi, and G. Leonardi. 2013. "Preservation of Dinosaur Tracks Induced by Microbial Mats in the Sousa Basin (Lower Cretaceous), Brazil." *Cretaceous Research* 44(2013): 112–21.

Carvalho, I. S. and D. A. Campos. 1988. "Um mamífero triconodonte do Cretáceo Inferior do Maranhão, Brasil." *Anais da Academia Brasileira de Ciências* 60(4): 437–46.

Carvalho, I. S., and M. G. P. Carvalho. 1990. O significado paleoambiental dos conchostráceos da Bacia de Sousa; 329–33 in Simpósio Sobre a Bacia do Araripe e Bacias Interiores do Nordeste, Crato, 1. D. A. Campos, M. S. S. Viana, P. M. Brito, and G. Beurlen, G. (eds.). Crato, Ceará, Sociedade Brasileira de Paleontologia, Brazil.

Carvalho, I. S., and A. C. S. Fernandes. 1992. Os icnofósseis da Bacia de Mangabeira, Cretáceo do Ceará; 105–06 in: Simpósio Sobre as Bacias Cretácicas Brasileiras, 2, Boletim de Resumos Expandidos, Rio Claro, 1992. São Paulo, UNESP, Brazil

———, eds. 2007. Icnologia. Sociedade Brasileira de Geologia, Série Textos, 3. São Paulo, Brazil, 178 pp.

Carvalho, I. S. and R. A. Gonçalves. 1994. "Pegadas de dinossauros neocretáceas da Formação Itapecuru, bacia de São Luís (Maranhão, Brasil)." *Anais da Academia Brasileira de Ciências* 66(3): 279–92.

Carvalho, I. S., and G. Leonardi. 1992. "Geologia das bacias de Pombal, Sousa, Uiraúna-Brejo das Freiras e Vertentes (Nordeste do Brasil)." *Anais da Academia Brasileira de Ciências* 64: 231–52.

———. 2007. "The Dinosaur Valley Natural Monument: Dinosaur Tracks from Rio do Peixe Basins (Lower Cretaceous, Brazil)," 51 in *Reunión Argentina de Icnología*, 5, Y Reunión de Icnología del Mercosur, 3. Ushuaia, Argentina, 28–30 Marzo, 2007, Ushuaia, Argentina.

———. 2020. Fossil footprints as biosedimentary structures for paleoenvironmental interpretation: Examples from Gondwana. *Journal of South American Earth Sciences*, https://doi.org/10.1016/j.jsames.2020.102936.

Carvalho, I. S., G. Leonardi, A. M. Rios-Netto, L. Borghi, A. P. Freitas, J. A. Andrade, and F. I. Freitas. 2020. Dinosaur Trampling from the Aptian of Araripe Basin, Brazil, as Tools for Stratigraphic Correlation. *Cretaceous Research* https://doi.org/10.3301/IJG.2020.24.

Carvalho, I. S., G. Leonardi, and W. F. S. Santos. 2013. Vale dos Dinossauros: a relevância das pegadas fósseis da Bacia de Sousa como patrimônio geológico (Abstract); 2 in Geobr Heritage, Simpósio Brasileiro de Patrimônio Geológico, 2, Anais, Ouro Preto, Minas Gerais, 24 a 28 de setembro de 2013, Brazil.

Carvalho, I. S., J. C. Mendes, and T. Costa. 2013. "The Role of Fracturing and Mineralogical Alteration of Basement Gneiss in the Oil Exhsudation in the Sousa Basin (Lower Cretaceous), Northeastern Brasil." *Journal of South American Earth Sciences* 47(2013): 47–54.

Carvalho, I. S., and P. H. Nobre. 2001. "Um Crocodylomorpha (?Notosuchia) da Bacia de Uiraúna (Cretáceo Inferior), Nordeste do Brasil." *Revista Brasileira de Paleontologia* 2: 123–24.

Carvalho, I. S., F. C. C. Nogueira, J. A. Soares, L. Salgado, and G. Leonardi. 2014. Dinosauria da Formação Antenor Navarro, Bacia de Uiraúna-Brejo das Freiras; 278 in Congresso Brasileiro de Geologia, 47, Anais, Salvador, Bahia, Brazil.

Carvalho, I. S., A. M. Rios-Netto, L. Borghi, A. P. Freitas, G. Leonardi, J. A. Andrade, and F. I. Freitas. 2019. Dinosaur Trampling from the Rio da Bateira Formation–Lower Cretaceous of Araripe Basin, Brazil: 21. In: Fialho, P. & Silva, R. (eds.)

Livro de Resumos. Paleo Fall Meeting 2019. Universidade de Evora, Portugal.

Carvalho, I. S., L. Salgado, R. M. Lindoso, J. R. Araújo Jr., F. C. C. Nogueira, and J. Soares. 2017. "A new basal titanosaur (Dinosauria, Sauropoda) from the Lower Cretaceous of Brazil." *Journal of South American Earth Sciences* 75: 74–84.

Carvalho, I. S., M. S. S. Viana, and M. F. Lima Filho. 1993a. "Bacia de Cedro: a icnofauna cretácica de vertebrados." *Anais da Academia Brasileira de Ciências* 65: 459–60.

———. 1993b. "Os icnofósseis de vertebrados da bacia do Araripe (Cretáceo Inferior, Ceará–Brasil)." *Anais da Academia Brasileira de Ciências* 65: 459.

———. 1994. Dinossauros do Siluriano: um anacronismo cronogeológico nas bacias interiores do Nordeste?; 3: 213–214 in Congresso Brasileiro de Geologia, 38, Boletim de Resumos Expandidos, Sociedade Brasileira de Geologia, Camboriú, Santa Catarina, Brazil.

Casamiquela, R. M., and A. Fasola. 1968. Sobre pisadas de dinosaurios del Cretácico Inferior de Colchagua (Chile), Departamento de Geología, Universidad de Chile, Santiago 30(1968): 1–24.

Castanera, D., M. Belvedere, D. Marty, G. Paratte, M. Lapaire-Cattin, C. Lovis, and C. A. Meyer. 2018. "A Walk in the Maze: Variation in Late Jurassic Tridactyl Dinosaur Tracks—A Case Study from the Late Jurassic of the Swiss Jura Mountains (NW Switzerland)." *PeerJ Preprints*. https://doi.org/10.7287/peerj.preprints.3506v1.

Castanera, D., J. Colmenar, V. Sauqué, and J. I. Canudo. 2015. "Geometric Morphometric Analysis Applied to Theropod Tracks from the Lower Cretaceous (Berriasian) of Spain." *Palaeontology* 58(1): 183–200.

Castro, D. L., D. C. Oliveira, and R. M. G. Castelo Branco. 2007. "On the Tectonics of the Neocomian Rio do Peixe Rift Basin, NE Brazil: Lessons from Gravity, Magnetics, and Radiometric Data." *Journal of South American Earth Sciences* 24: 184–202.

Cavalcanti, D. F. 1947. Pegadas de Dinosáurios no Rio do Peixe, Paraíba. Ceará, Secretaria de Agricultura e Obras Públicas, Boletim 1(1): 45–49.

Chafetz, H. S., and C. Buczynski. 1992. "Bacterially Induced Lithification of Microbial Mats." *Palaios* 7: 277–93.

Chapman, R. E. 1990. "Shape Analysis in the Study of Dinosaur Morphology." In *Dinosaur Systematics: Perspectives and Approaches*, edited by K. Carpenter and P. J. Currie, 21–42. New York: Cambridge University Press.

Chapman, R. E., D. Weishampel, G. Hunt, and D. Rasskin-Gutman. 1997. "Sexual Dimorphism in Dinosaurs." *Dinofest International Proceedings*: 83–93.

Charig, A. J., and A. C. Milner. 1997. "*Baryonyx walkeri*, a Fish-Eating Dinosaur from the Wealden of Surrey." *Bulletin of the Natural History Museum* Geology Series, 53: 11–70.

Chinsamy, A., D. B. Thomas, A. R. Tumarkin-Deratzian, and A. R. Fiorillo. 2012. "Hadrosaurs were perennial polar residents." *The Anatomical Record: Advances in Integrative Anatomy and Evolutionary Biology*, 265(4): 610–14.

Citton, P., U. Nicosia, I. Nicolosi, R. Carluccio, and M. Romano. 2015a. "Elongated Theropod Tracks from the Cretaceous Apenninic Carbonate Platform of Southern Latium (Central Italy)." *Palaeontologia Electronica* 18.3.49A: 1–12.

Citton, P., U. Nicosia, and E. Sacchi. 2015b. "Updating and Reinterpreting the Dinosaur Track Record of Italy." *Palaeogeography, Palaeoclimatology, Palaeoecology* 439: 117–25.

Citton, P., M. Romano, R. Carluccio, F. A. Caracciolo, I. Nicolosi, U. Nicosia, E. Sacchi, G. Speranza, and F. Speranza. 2017. "The First Dinosaur Tracksite from Abruzzi (Monte Cagno,

Central Apennines, Italy)." *Cretaceous Research* 73(2017): 47–59.

Cladellas, M. S. C., and J.-V. Llopis. 1971. "Icnitas de reptiles Mesozóicos en la Provincia de Logroño." *Acta Geológica Hispánica* 6(5): 139–42.

Cohen, A., M. Lockley, J. Halfpenny, and A. E. Michel. 1991. "Modern Vertebrate Track Taphonomy at Lake Manyara." *Palaios* 6: 371–89.

Conti, M.A., G. Leonardi, N. Mariotti, and U. Nicosia. 1975. "Tetrapod Footprints, Fishes and Molluscs from the Middle Permian of the Dolomites (N. Italy)." *Memorie Geopaleontologiche dell'Università di Ferrara* 3(II-1): 139–150.

———. 1977. "Tetrapod Footprints of the 'Val Gardena Sandstone' (North Italy). Their Paleontological, Stratigraphic and Paleoenvironmental Meaning." *Palaeontografia italica* 70 (N.S. 40): 91.

Conti, M. A., M. Morsilli, U. Nicosia, E. Sacchi, V. Savino, A. Wagensommer, L. Di Maggio, and P. Gianolla. 2005. "Jurassic Dinosaur Footprints from Southern Italy: Footprints as Indicators of Constraints in Paleogeographic Interpretation." *Palaios* 20: 534–50.

Coombs, W. P. 1980. "Swimming Ability of Carnivorous Dinosaurs." *Science* 207: 1198–1200.

Córdoba, V. C., A. F. Antunes, E. F. Jardim de Sá, A. N. Silva, D. C. Sousa, and F. A. P. L. Lins. 2008. "Análise estratigráfica e estrutural da Bacia do Rio do Peixe, Nordeste do Brasil: integração a partir do levantamento sísmico pioneiro 0295_RIO_DO_PEIXE_2D." *Boletim de Geociências da Petrobras, Rio de Janeiro* 16(1): 53–68.

Coria, R. A., and L. Salgado. 1996. "A Basal Iguanodontian (Ornithischia: Ornithopoda) from the Late Cretaceous of South America." *Journal of Vertebrate Paleontology* 16(3): 445–57.

Costa, S. A. G. 1969. Geologia da bacia do Rio do Peixe—PB: área 2—Região Leste de Sousa. Relatório de Graduação, Recife, 69 pp. [Unpublished]

Costa, W. D. 1964. "Nota preliminar da geologia da Bacia do Rio do Peixe." *Boletim Geologia* 4: 47–50. Recife: Universidade de Pernambuco.

Costa-Pérez, M., J. J. Moratalla, and J. Marugán-Lobón. 2019. "Studying Bipedal Dinosaur Trackways Using Geometric Morphometrics." *Palaeontologia Electronica* 22.3. pvc_3: 1–13. https://doi.org/10.26879/980.

Crandall, R. 1910. "Geografia, geologia, suprimento d'água, transportes e açudagem nos estados orientais do norte do Brasil: Ceará, Rio Grande do Norte e Paraíba." *Inspetoria de Obras Contra as Secas* 4 (série I): 1–131.

Currie, P. J. 1992. "Le Migrazioni dei Dinosauri." In *Dinosauri. Passato Presente e Futuro*, edited by B. Preiss and R. Silverberg, 199–211. Milano, Italy: Mondadori.

Dacke, C. G., S. Arkle, D. J. Cook, I. M. Wormstone, S. Jones, M. Zaidi, and Z. A. Bascal. 1993. "Medullary Bone and Avian Calcium Regulation." *Journal of Experimental Biology*, 184: 63–88.

Dai, H., L. Xing, D. Marty, J. Zhang, W.S. Persons IV, H. Hu, and F. Wang. 2015. "Microbially-Induced Sedimentary Wrinkle Structures and Possible Impact of Microbial Mats for the Enhanced Preservation of Dinosaur Tracks from the Lower Cretaceous Jiaguan Formation near Qijiang (Chongqing, China)." *Cretaceous Research* 53(2015): 98–109.

D'Alessandro, A., and R. G. Bromley. 1987. "Meniscate Trace Fossils and the Muensteria-Taenidium Problem." *Palaeontology* 30(4): 743–63.

D'Alessandro, A., A. A. Ekdale, and M. D. Picard. 1987. "Trace Fossils in Fluvial Deposits of Duchesne River Formation (Eocene) Uinta Basin, Utah." *Paleogeography, Palaeoclimatology, Palaeoecology* 61: 285–301.

Dal Sasso, C. 2004. *Dinosaurs of Italy*. Bloomington: Indiana University Press.

Dal Sasso, C., and S. Maganuco. 2011. "Scipionyx samniticus (Theropoda: Compsognathidae) from the Lower Cretaceous of Italy: Osteology, ontogenetic assessment, phylogeny, soft tissue anatomy, taphonomy, and palaeobiology." Memorie della Società italiana di Scienze naturali e del Museo civico di Storia naturale di Milano XXXVII, I.

Dal Sasso, C., G. Pierangelini, F. Famiani, A. Cau, and U. Nicosia. 2016. "First Sauropod Bones from Italy Offer New Insights on the Radiation of Titanosauria between Africa and Europe." *Cretaceous Research* 64(2016): 88–109.

Dal Sasso, C. and M. Signore. 1998. "Exceptional soft-tissue preservation in a theropod dinosaur from Italy." *Nature*, 392: 383–87.

Dantas, J. R. A. 1974. Carta geológica do Brasil ao milionésimo. Folha Jaguaribe (SB. 24) e Folha Fortaleza (AS. 24). Texto Explicativo. Brasília, DNPM/MME, Brazil.

Dantas, J. R. A., and J. A. L. Caula. 1982. "Estratigrafia e geotectônica," in SERM/CDRM. Mapa geológico do Estado da Paraíba: texto explicativo. João Pessoa, SERM/CDRM, Brasil.

Da Rosa, A. A. S. 1996. "Palaeogeografia e proveniência dos arenitos cretácicos da sequência pré-rifte das bacias interiores do Nordeste do Brasil." Master's Dissertation, Mestrado em Geologia—Área de Concentração: Geologia Sedimentar, Universidade do Vale do Rio dos Sinos, São Leopoldo, Brasil [Unpublished].

Da Rosa, A. A. S., and A. J. V. Garcia. 1993. "Petrologia das unidades terrígenas inferiores da sequência pré-rift das bacias interiores do nordeste do Brasil: palaeocorrentes e análise de proveniência." *Acta Geologica Leopoldensia* 38: 133–141.

———. 2000. "Palaeobiogeographic Aspects of Northeast Brasilian Basins during the Berriasian before the Break Up of Gondwana." *Cretaceous Research* 21: 221–39.

Demathieu, G. 1970. *Les empreintes de pas de vertébrés du Trias de la bordure nord-est du Massif Central*. Cahiers de Paléontologie. Paris: Éditions du Centre national de le recherche scientifique.

Demathieu, G. R. 1990. "Problems in Discrimination of Tridactyl Dinosaur Footprints, Exemplified by the Hettangian Trackways, the Causses, France." *Ichnos* 1(2): 97–110.

Dentzien-Dias, P. C., A. E. Q. Figueiredo, F. L. Pinheiro and C. L. Schultz. 2010. "Primeira evidência icnológica de um tetrápode natante no Membro Crato (Cretáceo Inferior), Formação Santana (Bacia do Araripe, Nordeste do Brasil)." *Revista Brasileira de Paleontologia* 13(3): 1–3, Setembro/Dezembro 2010.

Departamento Nacional da Produção Mineral, Brazil, Boletim 236.

Díaz-Martínez I., X. Pereda-Suberbiola, F. Pérez-Lorente, and J. I. Canudo. 2015. "Ichnotaxonomic Review of Large Ornithopod Dinosaur Tracks: Temporal and Geographic Implications." *PLoS ONE* 10(2): e0115477. https://doi.org/10.1371/journal.pone.0115477.

Difley, R. L. and A. A. Ekdale. 2019. Footprints of Utah's Last Dinosaurs: Track Beds in the Upper Cretaceous (Maastrichtian) North Horn Formation of the Wasatch Plateau, Central Utah. *Palaios* 17(4): 327–346.

Dodson, P. 1990. Counting dinosaurs: How many kinds were there? *Proceedings of the National Academy of Sciences of the USA* 87: 7608–7612.

Dodson, P. and P. J. Currie, 1990. "Neoceratopsia." In *The Dinosauria*, edited by D. Weishampel, P. Dodson, and H. Osmólska, 593–618. Berkeley: University of California Press.

Dodson, P., C. A. Forster, and S. D. Sampson. 2004. "Ceratopsidae." In *The Dinosauria*, 2nd ed., edited by D. Weishampel, P. Dodson, and H. Osmólska, 494–513. Berkeley: University of California Press.

Dunhill, A. M., J. Bestwick, H. Narey, J. and Sciberra. 2016. "Dinosaur Biogeographical Structure and Mesozoic Continental Fragmentation: A Network-Based Approach." *Journal of Biogeography* 43(9):1691–1704.

Dupraz, C., and P. T. Visscher. 2005. "Microbial Lithification in Marine Stromatolites and Hypersaline Mats." *Trends in Microbiology* 13: 429–38.

Dupraz, C., P. T. Visscher, L. K. Baumgartner, and R. P. Reid. 2004. "Microbe-mineral Interactions: Early Carbonate Precipitation in a Hypersaline Lake (Eleuthera Island, Bahamas)." *Sedimentology* 51: 745–65.

Esteves, F. A. 1988. Fundamentos de Limnologia. Rio de Janeiro, Interciência, Brasil.

Falkingham, P. L. 2016. "Applying Objective Methods to Subjective Tracks Outlines." In *Dinosaur Tracks: The Next Steps*, edited by P. L. Falkingham, D. Marty, and A. Richter, 72–80, chap. 4. Bloomington: Indiana University Press.

Falkingham, P. L., K. T. Bates, M. Avanzini, M. Bennett, E. M. Bordy, B. H. Breithaupt, D. Castanera, P. Citton, I. Diaz-Martinez, J. O. Farlow, et al. 2018. A Standard Protocol for Documenting Modern and Fossil Ichnological Data. *Palaeontology, Frontiers In Palaeontology* 2018, pp. 1–12.

Falkingham, P. L., K. T. Bates, and J. O. Farlow. 2014b. "Historical Photogrammetry: Bird's Paluxy River Dinosaur Chase Sequence Digitally Reconstructed as It Was Prior to Excavation 70 Years Ago." *Public Library of Science ONE* 9(4): e93247. https://doi.org/10.1371/journal.pone.0093247.

Falkingham, P. L., Bates, K. T., Margetts, L., Manning, P. L. 2014a. The "Goldilocks" effect: preservation bias in vertebrate track assemblages. *Journal of the Royal Society Interface* 8(61): 1142-1154.

Falkingham, P. L., and S. M. Gatesy. 2014. "The Birth of a Dinosaur Footprint: Subsurface 3D Motion Reconstruction and Discrete Element Simulation Reveal Track Ontogeny." *Proceedings of the National Academy of Sciences of the USA Early Edition* 111(51): 18279–84.

Falkingham P. L., L. Margetts, and P. L. Manning. 2010. "Fossil Vertebrate Tracks as Paleopenetrometers: Confounding Effects of Foot Morphology." *Palaios* 25: 356–60.

Falkingham, P. L., D. Marty, and A. Richter. 2016a. "Introduction." In *Dinosaur Tracks: The Next Steps*, edited by P. L. Falkingham, D. Marty, and A. Richter, 2–27. Bloomington: Indiana University Press.

———, eds. 2016b. *Dinosaur Tracks: The Next Steps*. Bloomington: Indiana University Press.

Falkingham, P. L., M. L. Turner, and S. M. Gatesy. 2020. Constructing and testing hypotheses of dinosaur foot motions from fossil tracks using digitization and simulation. *Palaeontology* 2020, doi: 10.1111/pala.12502.

Farlow, J. O. 2018. *Noah's Ravens: Interpreting the makers of tridactyl dinosaur footprints*. Indiana University Press, Bloomington.

Farlow, J. O. 1976. "A Consideration of the Trophic Dynamics of the Late Cretaceous Large Dinosaurs Community (Oldman Formation)." *Ecology* 57(5): 841–57.

———. 1980. "Predator/prey biomass ratios, community food webs and dinosaur physiology." In *A Cold Look at the Warm-Blooded Dinosaurs*, edited by R. D. K. Thomas and E. C. Olson, 55–83. Boulder, CO: Westview.

———. 1989. "Ostrich Footprints and Trackways: Implication for Dinosaur Ichnology." In *Dinosaur Tracks and Traces*, edited by

D. D. Gillette and M. G. Lockley, 243–48. New York: Cambridge University Press.

———. 1990. "Dinosaur Energetics and Thermal Biology." In *The Dinosauria*, edited by D. Weishampel, and H. Osmólska, 43–55. Berkeley: University of California Press.

———. 2001. "*Acrocanthosaurus* and the Maker of Comanchean Large-Theropod Footprints." In *Mesozoic Vertebrate Life*, edited by D. Tanke and K. Carpenter, 408–27. Bloomington: Indiana University Press.

———. 2015. "Hindlimb Proportions and Relative Stride Length in Bipedal and Potentially Bipedal Dinosaurs." *Geological Society of America Abstracts with Programs* 47(5): 36.

Farlow, J. O., K. T. Bates, R. M. Bonem, B. F. Dattilo, P. L. Falkingham, R. Gildner, J. Jacene, et al. 2015. "Dinosaur Footprints from the Glen Rose Formation (Paluxy River, Dinosaur Valley State Park, Somervell County, Texas)." In *Early- and Mid-Cretaceous Archosaur Localities of North-Central Texas*, edited by C. Noto, 50. Guidebook for the field trip held October 13, 2015. SVP 2015 Meeting. Dallas, Texas.

Farlow, J. O., R. E. Chapman, B. Breithaupt, and N. Matthews. 2012a. "The Scientific Study of Dinosaur Footprints." In *The Complete Dinosaur*, 2nd edition, edited by M. K. Brett-Surman, T. R. Holtz, and J. O. Farlow, 713–59. Bloomington: Indiana University Press.

Farlow, J. O., I. D. Coroian, and J. R. Foster. 2010. "Giants on the Landscape: Modelling the Abundance of Megaherbivorous Dinosaurs of the Morrison Formation (Late Jurassic, western USA)." *Historical Biology: An International Journal of Paleobiology* 22(4): 403–29.

Farlow, J. O. and P. Dodson. 1975. "The Behavioral Significance of Frill and Horn Morphology in Ceratopsian Dinosaurs." *Evolution* 29: 353–61.

Farlow, J. O., P. Dodson, and A. Chinsamy. 1995. "Dinosaur Biology." *Annu Rev Ecol Syst.* 26: 445–71.

Farlow, J. O., T. R. Holtz, T. H. Worthy, and R. E. Chapman. 2013. "Feet of the Fierce (and Not So Fierce): Pedal Proportions in Large Theropods, Other Non-avian Dinosaurs, and Large Ground Birds." In *Tyrannosaurid Paleobiology*, edited by J. M. Parrish, R. E. Molnar, P. J. Currie, and E. B. Koppelhus, 88–132. Bloomington: Indiana University Press.

Farlow, J. O., and M. G. Lockley. 1993. "An Osteometric Approach to the Identification of the Makers of Early Mesozoic Tridactyl Dinosaur Footprints." *New Mexico Museum Natural History Science Bulletin* 3: 123–31.

Farlow, J. O., M. O'Brien, G. J. Kuban, B. F. Dattilo, K. T. Bates, P. L. Falkingham P. L., L. Piñuela, et al. 2012b. "Dinosaur Tracksites of the Paluxy River Valley (Glen Rose Formation, Lower Cretaceous), Dinosaur Valley State Park, Somervell County, Texas." *Jornadas Internacionales Sobre Paleontología de Dinosaurios y su Entorno*, 5: 41–69, Actas Salas de los Infantes, Burgos, Spain.

Farlow J. O., J. G. Pittman, and J. M. Hawthorne. 1989. "*Brontopodus birdi* Lower Cretaceous Sauropod Footprints from the U.S. Gulf Coastal Plain." In *Dinosaur Tracks and Traces*, edited by D. D. Gillette and M. G. Lockley, 371–94. New York: Cambridge University Press.

Farlow, J. O., and E. R. Planka. 2002. "Body Size Overlap, Habitat Partitioning and Living Space Requirements of Terrestrial Vertebrate Predators: Implications for the Paleoecology of Large Theropod Dinosaurs." *Historical Biology: An International Journal of Paleobiology* 16(1): 21–40.

Farlow, J. O., N. J. Robinson, C. J. Kumagai, F. V. Paladino, P. L. Falkingham, R. M. Elsey, and A. J. Martin. 2017. "Trackways of the American Crocodile (*Crocodylus acutus*) in Northwestern

Costa Rica: Implications for Crocodylian Ichnology." *Ichnos* (July 31, 2017): 30–65. https://doi.org/10.1080/10420940.2017.13 50856.

Farlow, J. O., E. R. Schachner, J. C. Sarrazin, H. Klein, and P. J. Currie. 2014. "Pedal Proportions of *Poposaurus gracilis*: Convergence and Divergence in the Feet of Archosaurs." *Anatomical Record: Advances in Integrative Anatomy and Evolutionary Biology*, 297(6): 1022–46.

Fernandes, A. C. S., and I. S. Carvalho. 2001. "Icnofósseis de invertebrados da Bacia de Sousa (Estado da Paraíba, Brasil): a localidade de Serrote do Letreiro." *Simpósios Sobre a Bacia do Araripe e Bacias Interiores do Nordeste*, 1: 147–155, 1 e 2. Novembro de 1997. Crato—Ceará. Comunicações 2001. Coleção Chapada do Araripe, Brazil.

———. 2007. As pegadas de dinossauros da Bacia do Rio do Peixe: elementos de transformação cultural em Sousa, Paraíba—Brasil; p. 57 in Reunión Argentina de Icnologia, 5, Y Reunión de Icnología del Mercosur, 3, Resúmenes, Laboratório de Geología Andina CADIC—CONICET, Argentina.

Fernandes, M. A., A. M. Ghilardi, I. S. Carvalho, and G. Leonardi. 2011a. Pegadas de dinossauros Theropoda do paleodeserto Botucatu (Jurássico Superior-Cretáceo Inferior) da Bacia do Paraná; 4: 609–621 in I.S. Carvalho, N. K. Srivastava, O. Strohschoen Jr., and C.C. Lana (eds.), Paleontologia: Cenários de Vida. Editora Interciência, Rio de Janeiro, Brazil.

———. 2011b. Pegadas de dinossauros Theropoda do paleodeserto Botucatu (Jurássico Superior-Cretáceo Inferior) da Bacia do Paraná; p.1 in Congresso Brasileiro de Paleontologia, 22, Natal, Rio Grande do Norte, 2011, Brazil.

Ferraz, A., ed. 2004. Além do Rio. Uma fotografia da paisagem urbana. Edição comemorativa do sesquicentenário 1854–2004—Sousa—Paraíba. Sousa, AGT Produções, 2004.

Ferrusquía-Villafranca, I., S. P. Applegate, and L. Espinosa-Arrubarrena. 1978. "Rocas Volcanosedimentarias Mesozoicas y Huellas de Dinosaurios en la Región Suroccidental Pacífica de México." *Revista de la Universidad Nacional Autónoma de México, Instituto de Geología* 2(2): 150–62.

———. 1980. "Las Huellas más australes de Dinosaurios en Norteamérica y su significación geobiológica." 1, Actas: 249–63 in 1er Congreso Latinoamericano de Paleontología, Buenos Aires, 1978, Argentina.

Fiorillo, A. R., and R. A. Gangloff. 2001."Theropod Teeth From The Prince Creek Formation (Cretaceous) Of Northern Alaska, With Speculations On Arctic Dinosaur Paleoecology." *Journal of Vertebrate Paleontology*, 20(4): 675–682.

Fiorillo, A. R., and R. S. Tykoski. 2014. "A Diminutive New Tyrannosaur From The Top Of The World." *PLoS ONE* 9(3): e91287. https://doi.org/10.1371/journal.pone.0091287.

Francischini, H., P. C. Dentzien-Dias, M. A. Fernandes, and C. L. Schultz. 2015. "Dinosaur Ichnofauna of the Upper Jurassic/Lower Cretaceous of the Paraná Basin (Brazil and Uruguay)." *Journal of South American Earth Sciences* 63(2015): 180–90.

Fricke, H. C., J. Hencecroth, and M. E. Hoerner. 2011. "Lowland–Upland Migration of Sauropod Dinosaurs During the Late Jurassic Epoch." *Nature* 480: 513–15.

Fujita, M., Y.-N. Lee, Y. Azuma, and D. Li. 2012. "Unusual Tridactyl Trackways with Tail Traces from the Lower Cretaceous Hekou Group, Gansu Province, China." *Palaios* 2012 (27): 560–70.

García, A. J., and A. Wilbert. 1994. "Palaeogeography Evolution of Mesozoic Pre-rift Sequences in Coastal and Interior Basins of Northeastern Brasil." In *Pangea: Global Environments and Resources*, edited by A. F. Embry, B. Beauchamp, and D.J. Glass. *Memoir of Canadian Society of Petroleum Geology* 17: 123–30.

García-Ortiz, E., and F. Pérez-Lorente. 2014. "Palaeoecological Inferences about Dinosaur Gregarious Behaviour Based on the Study of Tracksites from La Rioja Area in the Cameros Basin (Lower Cretaceous, Spain)." *Journal of Iberian Geology* 40(1): 113–27.

Gatesy, S. M., and R. G. Ellis. 2012. Tracks as 3-D Particle Trajectories; 23 in A. Richter and M. Reich (eds.), Dinosaur Tracks 2011. An International Symposium. Obernkirchen, April 14–17, 2011. Göttingen, Göttingen Universitätsdrucke, 2012, Germany.

———. 2016. "Beyond Surfaces: A Particle-Based Perspective on Track Formation." In *Dinosaur Tracks: The Next Steps*, edited by P. L. Falkingham, D. Marty, and A. Richter, 82–91, chap. 5. Bloomington: Indiana University Press.

Ghilardi, A. M., T. Aureliano, R. R. C. Duque, M. A. Fernandes, A. M. F. Barreto, and A. Chinsamy. 2016. "A New Titanosaur from the Lower Cretaceous of Brazil." *Cretaceous Research* 16 (2016): 16–24. https://doi.org/10.1016/j.cretres.2016.07.001.

Ghilardi, A. M., T. Aureliano, R. R. C. Duque, E. V. Oliveira, and A. M. F. Barreto. 2014. "An Early Cretaceous Dinosaur from Sousa Basin, Brazil," in International Paleontological Congress, p. 327, 4, Abstract Volume, Mendoza, Argentina.

Gierlinski, G. D., P. Menducki, K. Janiszewska, I. Wicik, and A. Boczarowski. 2009. "A Preliminary Report on Dinosaur Track Assemblages from the Middle Jurassic of the Imilchil Area, Morocco." *Geological Quarterly* 53(4): 477–82.

Gierlinski, G. D., I. Ploch, E. Gawor-Biedawa, and G. Niedzwiedzki. 2008. "The First Evidence of Dinosaur Tracks in the Upper Cretaceous of Poland." *Oryctos* 8: 107–13.

Godefroit, P., L., S. Golovneva, G. Shchepetov, G. García, and P. Alekseev. 2009. "The Last Polar Dinosaurs: High Diversity Of Latest Cretaceous Arctic Dinosaurs in Russia." *Naturwissenschaften*, 96: 495–50.

Godoy, L. C., and G. Leonardi. 1985. Direções e comportamento dos dinossauros da localidade de Piau-Caiçara, Sousa, Paraíba (Brasil), Formação Sousa (Cretáceo Inferior); 2: 65–73 in Brasil, Departamento Nacional da Produção Mineral. Coletânea de Trabalhos Paleontológicos. Série "Geologia", 27, Seção Paleontologia e Estratigrafia, Brazil.

Golonka, J., M. I. Ross, and C. R. Scotese. 1994. "Phanerozoic Palaeogeographic and Palaeoclimatic Modeling Maps." In *Pangea: Global Environments and Resources*, edited by A. F. Embry, B. Beauchamp, and D. J. Glass. *Canadian Society of Petroleum Geology, Memoir* 17: 1–47.

González Riga, B. J., L. D. Ortiz David, M. B. Tomaselli, C. R. A. Candeiro, J. P. Coria, and M. Prámparo. 2015. "Sauropod and Theropod Dinosaur Tracks from the Upper Cretaceous of Mendoza (Argentina): Trackmakers and Anatomical Evidences." *Journal of South American Earth Sciences* 61: 134–41.

Graham, J. R., and J. E. Pollard. 1982. "Occurrence of the Trace Fossil *Beaconites antarcticus* in the Lower Carboniferous Fluviatile Rocks of County Mayo, Ireland." *Palaeogeography, Palaeoclimatology, Palaeoecology* 38: 257–68.

Hasiotis, S. T., B. F. Platt, D. I. Hembree, and M. J. Everhart. 2007. "The Trace-Fossil Record of Vertebrates." In *Trace Fossils Concepts, Problems, Prospects*, edited by W. Miller III, 196–218. Amsterdam: Elsevier.

Haubold, H. 1971. Ichnia Amphibiorum et Reptiliorum Fossilium. Handbuch der Paläoherpetologie, pt. 18. Stuttgart, Germany and Portland, USA. Gustav Fischer, vii + 124 pp.

———. 2012. Dinosaur footprints: taxonomy, phantom taxa and the relation between core and surface investigation; 187 pp. in A. Richter and M. Reich eds., Dinosaur Tracks 2011. An International Symposium. Obernkirchen, April 14–17, 2011. Göttingen, Göttingen Universitätsdrucke, 2012, Germany.

Heinberg, C. 1974. A dynamic model for a meniscus filled tunnel (Ancorichnus n. ichnogen.) from the Jurassic Pecten Sandstone of Milne Land, East Greenland. Kobenhavn, Gronlands Geologiske Undersogelse, 20 pp. (Rapport 62).

Henderson, D. M. 2017. "The First Evidence of Iguanodontids (Dinosauria: Ornithischia) in Alberta, Canada—A Fossil Footprint from the Early Cretaceous." *Cretaceous Research* 76 (2017): 28–33. https://doi.org/10.1016/j.cretres.2017.04.015.

Hendrickx, C., S. A. Hartman, and O. Mateus. 2015. "An Overview of Non-Avian Theropod Discoveries and Classification." *PalArch's Journal of Vertebrate Palaeontology* 12(1): 1–73.

Hernández-Medrano, N., C. P. Arribas, F. Pérez-Lorente, and R. Sesma. 2014. "Icnitas terópodas, ornitópodas y de pterosaurio en la formación Aguilar del Río Alhama. Grupo de Oncala, cuenca de Cameros." *Zubía* 32: 33–71.

Hitchcock, E. 1845. "An Attempt to Name, Classify, and Describe the Animals that Made the Fossil Footmarks of New England." Proceedings of the Assocciation of American Geologists and Naultralists, New Haven, 6th Meeting: 23–25.

———. 1858. *Ichnology of New England*. Boston: White.

Holtz, T. R. 2012. "Theropods." In *The Complete Dinosaur*, 2nd ed., edited by M. K. Brett-Surman, T. R. Holtz, and J. O. Farlow, 347–78. Bloomington: Indiana University Press.

Hornung, J. J., A. Böhme, and M. Reich. 2012a. The type material of the theropod ichnotaxon "Bueckeburgichnus" maximus Kuhn, 1958—reconsidered; 187 pp. in Dinosaur Tracks 2011. An International Symposium. A. Richter and M. Reich (eds.), Obernkirchen, April 14–17, 2011. Göttingen, Göttingen Universitätsdrucke, 2012, Germany.

Hornung, J. J., A. Böhme, N. Schlüter, and M. Reich. 2016. "Diversity, Ontogeny, or Both? A Morphometric Approach to Iguanodontian Ornithopod (Dinosauria: Ornithischia) Track Assemblages from the Berriasian (Lower Cretaceous) of Northwestern Germany." In *Dinosaur Tracks: The Next Steps*, edited by P. L. Falkingham, D. Marty, and A. Richter, : 202–25, chap. 12. Bloomington: Indiana University Press.

Hornung, J. J., H.-V. Karl, and M. Reich. 2012b. A well-preserved isolated turtle footprint from the lowermost Cretaceous (Berriasian) of northern Germany. in Dinosaur Tracks 2011; 187 pp. An International Symposium. Richter, A. and Reich, M. (eds.), Obernkirchen, April 14–17, 2011. Göttingen, Göttingen Universitätsdrucke, 2012, Germany.

Huene, F. R., von. 1914. Das natürliche System der Saurischia. Zentralblatt für Mineralogie, Geologie, und Paläontologie B: 154–58.

———. 1931. Verschiedene mesozoische Wirbeltierreste aus Südamerika. Neues Jahrbuch Mineralogie und Geologie B 66: 181–98.

Huh, M., I. S. Paik, M. G. Lockley, K. G. Hwang, B. S. Kim, and S. K. Kwak. 2006. "Well-Preserved Theropod Tracks from the Upper Cretaceous of Hwasun County, Southwestern South Korea, and Their Paleobiological Implications." *Cretaceous Research* 27 (2006): 123–38.

IBGE. 2010. Instituto Brasileiro de Geografia e Estatística, Censo 2010. Governo do Brasil, Brazil. http://censo2010.ibge.gov.br.

Ibrahim, N., D. J. Varricchio, P. C. Sereno, J. A. Wilson, D. B. Dutheil, *D. M. Martill, L. Baidder, and S. Zouhn.* 2014. "Dinosaur Footprints and Other Ichnofauna from the Cretaceous Kem Kem Beds of Morocco." *PLoS ONE* 9(3): e90751, http://doi.org/10.1371/journal.pone.0090751.

Iemini, J. A. 2009. Fácies orgânicas de uma sucessão sedimentar cretácea da Bacia de Sousa, PB, Brasil. Programa de Pós-Graduação em Geologia, Universidade Federal do Rio de Janeiro. Dissertação de Mestrado, Brazil [Unpublished].

Ishigaki, S. 1989. "Footprints of Swimming Sauropods from Morocco." In *Dinosaur Tracks and Traces*, edited by D. D. Gillette and M. G. Lockley, 83–86. New York: Cambridge University Press.

———. 2010. "Theropod Trampled Bedding Plane with Laboring Trackways from the Upper Cretaceous Abdrant Nuru Fossil Site, Mongolia." *Hayashibara Museum of Natural Sciences Research Bulletin* 3 (2010): 133–41.

Ishigaki, S., and Y. Matsumoto. 2009. "Re-examination of Manus-Only and Manus-Dominated Sauropod Trackways from Morocco." *Geological Quarterly* 53(4): 441–48.

Ishigaki, S., M. Watabe, Kh. Tsogtbaatar, and M. Saneyoshi. 2009. "Dinosaur Footprints from the Upper Cretaceous of Mongolia." *Geological Quarterly* 53(4): 449–60.

Kegel, W. 1965. Lineament-Tektonik In Nordost-Brasilien. Geologische Rundschau, Bd. 54: 1240–60.

Kellner, A. W. A. 1996. "First Early Cretaceous Theropod Dinosaur from Brazil with Comments on Spinosauridae." Neues Jahrbuch Paläont. Abh. 199: 151–66.

———. 1999. "Short Note on a New Dinosaur (Theropoda, Coelurosauria from the Santana Formation (Romualdo Member, Albian), Northeastern Brazil." *Boletim do Museu Nacional, Geologia* 49: 1–8.

Kellner, A. W. A., S. A. K. Azevedo, E. B. Machado, L. B. Carvalho, and D. D. R. Henriques. 2011. "A New Dinosaur (Theropoda, Spinosauridae) from the Cretaceous (Cenomanian) Alcântara Formation, Cajual Island, Brazil." *Anais da Academia Brasileira de Ciências*, 83: 99–108.

Kellner, A. W. A., and D. A. Campos. 2000. "Brief Review of Dinosaur Studies and Perspectives in Brazil." *Anais da Academia Brasileira de Ciências* 72(4): 509–38.

Kershaw, P., and B. Wagstaff. 2001. "The Southern Conifer Family Araucariaceae: History, Status, and Value for Paleoenvironmental Reconstruction." *Annu. Rev. Ecol. Syst.* 32:397–414.

Kim, J. Y., K. S. Kim, M. G. Lockley, and J. S. Seung. 2010. "Dinosaur Skin Impressions from the Cretaceous of Korea: New Insights into Modes of Preservation." *Palaeogeography, Palaeoclimatology, Palaeoecology* 293: 167–74.

Kim, K. S., M. G. Lockley, J. D. Lim, S. M. Bae, and A. Romilio. 2020. Trackway evidence for large bipedal crocodylomorphs from the Cretaceous of Korea. *Scientific Reports* 10: 8680, 10:8680, https://doi.org/10.1038/s41598-020-66008-7.

Kim, K. S., M. G. Lockley, J. D. Lim, and L. Xing. 2019. Exquisitely-preserved, high-definition skin traces in diminutive theropod tracks from the Cretaceous of Korea. *Scientific Reports* 9: 2039, https://doi.org/10.1038/s41598-019-38633-4.

Kozu, S., A. Sardsud, D. Saesaengseerung, C. Pothichaiya, S. Agematsu, and K. Sashida. 2017. "Dinosaur Footprint Assemblage from the Lower Cretaceous Khok Kruat Formation, Khorat Group, Northeastern Thailand." *Geoscience Frontiers* 8, 6 (2017): 1479–93. https://doi.org/10.1016/j.gsf.2017.02.003.

Kuban, G. J. 1991a. "Color Distinctions and Other Curious Features of Dinosaur Tracks Near Glen Rose, Texas." In *Dinosaur Tracks and Traces*, edited by D. D. Gillette and M. G. Lockley, 427–40. New York: Cambridge University Press.

———. 1991b. "Elongate Dinosaur Tracks." In *Dinosaur Tracks and Traces*, edited by D. D. Gillette and M. G. Lockley, 57–72. New York: Cambridge University Press.

Lallensack, J. N., T. Engler, and H. J. Barthel. 2019. Shape Variability in Tridactyl Dinosaur Footprints: The Significance of Size and Function. *Palaeontology* 2019, 1–26.

Lallensack, J. N., A. H. van Heteren, and O. Wings. 2016. "Geometric Morphometric Analysis of Intratrackway Variability: A Case Study on Theropod and Ornithopod Dinosaur Trackways from Münchehagen (Lower Cretaceous, Germany)." *PeerJ* 4 (2016):e2059. https://doi.org/10.7717/peerj.2059.

Lapparent, A. F. 1962. *Footprints of Dinosaurs in the Lower Cretaceous of Vest-Spitsbergen-Svalbard.* Arbok: Norsk Polarinstitutt 1960, 14–21.

Lapparent, A.-F., and C. Montenat. 1967. "Les empreintes de pas de reptiles de l'infralias du Veillon (Vendée)." Mémoire de la Société géologique de France, (n.s.) 46(107), 43 p., 13 pl.

Lee, A. H., A. K. Huttenlocker, K. Padian, and H. N. Woodward. 2013. "Analysis of Growth Rates." In *Bone Histology of Fossil Tetrapods: Advancing Methods, Analysis, and Interpretation,* edited by K. Padian and E. T. Lamm, 217–51. Berkeley: University of California Press.

Lee, A. H., and S. Werning. 2008. "Sexual Maturity in Growing Dinosaurs Does Not Fit Reptilian Growth Models." *Proceedings of the National Academy of Sciences USA* 105: 582–87.

Lehman, T. M. 1990. "The Ceratopsian Subfamily Chasmosaurinae: Sexual Dimorphism and Systematics." In *Dinosaur Systematics: Perspectives and Approaches,* edited by P. J. Currie and K. Carpenter, 211–29. New York: Cambridge University Press.

Leonardi, G. 1975. "On the Trackways of the Recent South American Lizard Tupinambis teguixin (Linnaeus, 1758), Lacertilia, Teiidae." *Anais Academia brasileira de Ciências,* 47(Suplemento): 301–10.

———. 1977. "Two Simple Instruments for Ichnological Research, Principally in the Field of Vertebrates." *Dusenia* 10(3): 185–88.

———. 1979a. "Nota Preliminar Sobre Seis Pistas de Dinossauros Ornithischia da Bacia do Rio do Peixe (Cretáceo Inferior) em Sousa, Paraíba, Brasil." *Anais da Academia Brasileira de Ciências* 51(3): 501–16.

———. 1979b. "New Archosaurian Trackways from the Rio do Peixe Basin, Paraíba, Brasil." *Annali dell'Università di Ferrara,* N.S., S. IX 5(14): 239–49.

———. 1979c. "Four Years of Vertebrate Ichnology in Brasil." *Ichnology Newsletter,* Menlo Park 10: 25–26.

———. 1979d. Um Glossário comparado da Icnologia de Vertebrados em português e uma história desta ciência no Brasil. Universidade Estadual de Ponta Grossa. Imprensa Universitária, Cadernos Universitários, 17, 55 pp.

———. 1980a. "*Isochirotherium* sp.: Pista de um gigantesco Tecodonte na Formaçao Antenor Navarro (Triássico), Sousa, Paraíba, Brasil." *Revista Brasileira de Geociências* 10(4): 186–90.

———. 1980b. Dez novas pistas de Dinossauros (Theropoda Marsh, 1881) na Bacia do Rio do Peixe, Paraíba, Brasil; 1: 243–248 in 1er Congreso Latinoamericano de Paleontología, 1, Actas, Buenos Aires, 1978, Argentina.

———. 1980c. "Vertebrate Ichnology in Brasil and Italy." *Ichnology Newsletter* 11: 10.

———. 1980d. "Dados icnológicos sobre a raridade de espécimes jovens em populações de dinossauros do Brasil." *Anais da Academia Brasileira de Ciências* 52(3): 647.

———. 1980e. Ornithischian trackways of the Corda Formation (Jurassic), Goiás, Brasil; 1: 215–22 in Congreso Latinoamericano de Paleontología, 1, Actas, Buenos Aires, 1978, Argentina.

———. 1981a. "Novo Ichnogênero de Tetrápode Mesozóico da Formaçao Botucatu, Araraquara, SP." *Anais da Academia Brasileira de Ciências* 53(4): 793–805.

———. 1981b. As localidades com rastros fósseis de Tetrápodes na América Latina; 2: 929–40 in Congresso Latino-Americano de Paleontología, 2, Anais, Porto Alegre, 1981, Brazil.

———. 1981c. "Ichnological Data on the Rarity of Young in North East Brasil Dinosaurian Populations." *Anais da Academia Brasileira de Ciências* 53(2): 345–46.

———. 1981d. "Novo Ichnogênero de Tetrápode Mesozóico da Formação Botucatu, Araraquara, SP." *Anais da Academia Brasileira de Ciências,* 53(4): 793–805.

———. 1981e. "News." *Ichnology Newsletter* 12: 12.

———. 1982. "News." *Ichnology Newsletter* 13: 14.

———. 1984a. "Ichnological Data on the Rarity of Young in North East Brasil Dinosaurian Populations." *Anais da Academia Brasileira de Ciências* 53(2): 345–46.

———. 1984b. Le impronte fossili di dinosauri. In Bonaparte, J. F., E. H. Colbert, P. J. Currie, A/J de Ricqlès, Z. Keilan-Jaworowska, G. Leonardi, F. Morello, P. Taquet (eds.). Sulle orme dei dinosauri. Venezia-Mestre, Erizzo, 1984. (Esplorazioni e ricerche, IX), 335 p.: 161–86.

———. 1984c. "Rastros de um mundo perdido." *Ciência Hoje* 2(15): 48–60.

———. 1984d. "Notes at the Margin of an Ichnological Expedition to Riachão (Maranhão, Brasil)." *Dusenia* 14(1): 23–27.

———. 1985a. "Vale dos dinossauros: uma janela na noite dos tempos." *Revista brasileira de Tecnologia* 16(1): 23–28.

———. 1985b. "Mais pegadas de dinossauros na Paraíba." *Ciência Hoje* 3(16): 94.

———. 1985c. Breve inventário das icnofaunas dinossaurianas da América do Sul e considerações estatísticas sobre sua distribuição; p. 44 in Congresso Brasileiro de Paleontologia, 9, Resumo das Comunicações, Fortaleza, Ceará, Brazil.

———, ed. 1987a. Glossary and Manual of Tetrapod Footprint Palaeoichnology. Brasília, DNPM (Serviço Geológico do Brasil), 1987. 117 p., 20 tav., 20 pages of tables.

———. 1987b. Pegadas de dinossauros (Carnosauria, Coelurosauria, Iguanodontidae) na Formação Piranhas da Bacia do Rio do Peixe, Sousa, Paraíba, Brasil; 1: 337–51, 3 plates in Congresso Brasileiro de Paleontologia, 10, Anais, Rio de Janeiro, 1987. Sociedade Brasileira de Paleontologia, Rio de Janeiro, Brazil.

———. 1987c. "News." *Ichnology Newsletter* 16: 15–16.

———. 1988. "News." *Ichnology Newsletter* 17: 20.

———. 1989a. "Inventory and Statistics of the South American Dinosaurian Ichnofauna and Its Paleobiological Interpretation." In *Dinosaur Tracks and Traces,* edited by D. D. Gillette and M. G. Lockley, 165–78. New York: Cambridge University Press.

———. 1989b. "News." *Ichnology Newsletter* 18: 26–27.

———. 1992. "Sulle prime impronte fossili del Paraguay." *Paleocronache* 1992(1): 66–67.

———. 1994. Annotated Atlas of South America Tetrapod Footprints (Devonian to Holocene) with an appendix on Mexico and Central America. Companhia de Pesquisa de Recursos Minerais, Brasília, 1994. 248 pp., 35 plates, Brasil.

———. 1996. Le piste di dinosauri dei Lavini di Marco (Rovereto, TN, Italia) e alcune questioni generali sull'icnologia dei tetrapodi. Atti Accademia Roveretana degli Agiati, a.246, 1996, ser. VII, vol.VI, B: 65–104.

———. 1997. "Problemática actual de las icnitas de dinosaurios." *Revista de la Sociedad geológica de España* 10(3–4): 341–53.

———. 1998. "Dinosaurs of Italy." *Records of the Western Australian Museum,* Supplement 57: 405–06.

———. 2000. I dinosauri d'Italia e delle aree adiacenti; pp. 275–295 in G. Leonardi and P. Mietto (eds.), Dinosauri in Italia. Le orme giurassiche dei Lavini di Marco (Trentino) e gli altri resti fossili italiani. Accademia Editoriale, Pisa-Roma, 2000. Italy.

———. 2008. Trinta e três anos à procura de pegadas fósseis nas bacias brasileiras; p. 811–13 in Congresso Brasileiro de Geologia, 44, Anais, Curitiba, Paraná, Brazil.

———. 2008. "Vertebrate Ichnology in Italy." *Studi Trentini di Scienze Naturali Acta Geologica* 83 (2008): 213–21.

———. 2011. "What Do the Dinosaur Tracks of the Rio do Peixe Basins (Paraíba, Brasil) Point At?" In *Paleontologia: Cenários de Vida*, edited by I. S. Carvalho, N. K. Srivastava, O. Strohschoen, and C. C. Lana, 669–80. Rio de Janeiro: Editora Interciência.

———. 2015. Quaderno di bordo, esplorazioni nelle "Valli dei Dinosauri"; 567 pp. In L. Campanelli (ed.), Pietre fossili. Maestri muti. Ariccia, Aracne, 2015, Italy.

Leonardi, G., and M. Avanzini. 1994. "Dinosauri in Italia." *Le Scienze—Quaderni* 76: 69–81.

Leonardi, G., and G. Borgomanero. 1981. "Sobre uma possível ocorrência de Ornithischia na Formação Santana, Chapada do Araripe." *Revista brasileira de Geociências* 11(1): 1–4.

Leonardi, G., and I. S. Carvalho. 1999. Jazigo icnofossilífero do Ouro—Araraquara (SP) [Ichnosite of Ouro - Araraquara, São Paulo state, Brazil]; 13 pp. in C. Schobbenhaus, D. A. Campos, E. T. Queiroz, M. Winge and M. Berbert-Born (eds.), Sítios Geológicos e Paleontológicos do Brasil. http://sigep.cprm.gov.br/sitio079/sitio079english.htm.

———. 2000. As pegadas de dinossauros das bacias Rio do Peixe, PB. [The Dinosaur Footprints from Rio do Peixe Basins, Paraíba State, Northeastern Brasil]; 15 pp. in C. Schobbenhaus, D. A. Campos, E. T. Queiroz, M. Winge and M. Berbert-Born (eds.), Sítios Geológicos e Paleontológicos do Brasil. h http://sigep.cprm.gov.br/sitio079/sitio079english.htm.

———. 2002. Icnofósseis da Bacia do Rio do Peixe, PB. O mais marcante registro de pegadas de dinossauros do Brasil; pp. 101–11 in C. Schobbenhaus, D. A. Campos, E. T. Queiroz, M. Winge and M. Berbert-Born (eds.), *Sítios geológicos e paleontológicos do Brasil. Brasília*. Brasil, Departamento Nacional de Produção Mineral, 2002, Brazil.

———. 2020a. Vertebrate trace fossils: the Congo's *Brasilichnium* mammaloid fossil footprints. *Italian Journal of Geociences*, 140(1), doi.org/10.3301/IJG.2020.24.

———. 2020b. Review of the early mammal *Brasilichnium* and *Brasilichnium*-like tracks from South America. *Journal of South American Earth Sciences*, 10.1016/j.jsames.2020.102940.

Leonardi, G., I. S. Carvalho, and M. Fernandes. 2007a. The desert ichnofauna from Botucatu Formation (Upper Jurassic–Lower Cretaceous), Brasil; p. 292 in Congresso Brasileiro de Paleontologia, 20, Anais de Resumos, Sociedade Brasileira de Paleontologia, Búzios, Rio de Janeiro, Brazil.

———. 2007b. The desert ichnofauna from Botucatu Formation (Upper Jurassic–Lower Cretaceous), Brazil; 1: 379–391 in I.S. Carvalho, R.C.T. Cassab, C. Schwanke, M.A. Carvalho, A.C.S. Fernandes, M.A.C. Rodrigues, M.S.S. Carvalho, M. Arai, M.E.Q. Oliveira (eds.), Paleontologia: Cenários de Vida. Interciência, 2007, Rio de Janeiro, Brazil.

Leonardi, G., C. S. Ferreira, C. S., and F. L. S. Cunha. 1979. "Evidências do Mesozóico na região dos Rios Guamá e Capim, PA. O registro das pegadas fósseis de répteis (Prov. Triássico-Jurássico) no arenito aflorante do Rio Guamá (Projeto ABC/FINEP)." *Anais Academia Brasileira Ciências* 51(2): 360.

Leonardi, G., and L. C. Godoy. 1980. Novas Pistas de Tetrápodes da Formação Botucatu no Estado de São Paulo; 5: 3080–89 in Congresso Brasileiro de Geologia, 31, Anais, Camboriú, Santa Catarina, Brazil.

Leonardi, G. and M. Lanzinger. 1992. "Dinosauri nel Trentino: venticinque piste fossili nel Liassico di Rovereto (Trento, Italia)." *Paleocronache* 1992(1): 13–24.

Leonardi, G. and S. Leghissa. 1990. Una pista di Sauropode scoperta nei calcari cenomaniani dell'Istria. [Milano], Centro di Cultura Giuliano Dalmata, 1990.

Leonardi, G., C. V. Lima, and F. H. L. Oliveira. 1987a. Os dados numéricos relativos às pistas (e suas pegadas) das Icnofaunas dinossaurianas do Cretáceo inferior da Paraíba, e sua interpretação estatística. I - Parâmetros das pistas; 1: 377–94 in Congresso Brasileiro de Paleontologia, 10, Anais, Rio de Janeiro, Sociedade Brasileira de Paleontologia, Rio de Janeiro, Brazil.

———. 1987b. Os dados numéricos relativos às pistas (e suas pegadas) das icnofaunas dinossaurianas do Cretáceo Inferior da Paraíba, e sua interpretação estatística. II - Parâmetros das pegadas; 1: 395–417 in Congresso Brasileiro de Paleontologia, 10, Anais, Rio de Janeiro, Sociedade Brasileira de Paleontologia, Rio de Janeiro, Brazil.

———. 1987c. Os dados numéricos relativos às pistas (e suas pegadas) das icnofaunas dinossaurianas do Cretáceo Inferior da Paraíba, e sua interpretação estatística. III – Estudo estatístico; 1: 419–444 in Congresso Brasileiro De Paleontologia, 10, Anais, Rio de Janeiro, Sociedade Brasileira de Paleontologia, Rio de Janeiro, Brazil.

Leonardi, G., and M. G. Lockley. 1995. "A Proposal to Abandon the Ichnogenus *Coelurosaurichnus* Huene, 1941—A Junior Synonym of *Grallator* E. Hitchcock, 1858." *Journal of Vertebrate Paleontology* 15(3): 40A.

Leonardi, G., and P. Mietto, eds. 2000. Dinosauri in Italia. Le orme giurassiche dei Lavini di Marco (Trentino) e gli altri resti fossili italiani. Accademia Editoriale, Pisa-Roma.

Leonardi, G., and G. C. B. Muniz. 1985. Observações icnológicas (Invertebrados e Vertebrados) no Cretáceo continental do Ceará (Brasil), com menção a moluscos dulçaquícolas; p. 45 in Congresso Brasileiro de Paleontologia, 9, Fortaleza, Sociedade Brasileira de Paleontologia, Resumo das Comunicações, Fortaleza, Ceará, Brazil.

Leonardi, G., and F. H. L. Oliveira.1990. "A Revision of the Triassic and Jurassic Tetrapod Footprints of Argentina and a New Approach on the Age and Meaning of the Botucatu Formation Footprints (Brazil)." *Revista Brasileira de Geociências* 20(1–4): 216–29.

Leonardi, G., and M. F. C. F. Santos. 2006. "New Dinosaur Tracksites from the Sousa Lower Cretaceous Basin (Paraíba, Brasil)." *Studi Trentini di Scienze Naturali, Acta Geologica* 81(2004): 5–21.

Leonardi, G., and W. A. S. Sarjeant. 1986. "Footprints Representing a New Mesozoic Vertebrate Fauna from Brazil." *Modern Geology*, 10: 73–84.

Leonardi, G., and M. Spezzamonte. 1994. "New Tracksites (Dinosauria: Theropoda and Ornithopoda) from the Lower Cretaceous of the Ceará, Brasil." *Studi Trentini di Scienze Naturali—Acta Geologica* 69(1992): 61–70.

Leonardi, G., and G. Teruzzi. 1993. "Prima segnalazione di uno scheletro fossile di Dinosauro (Theropoda, Coelurosauria) in Italia (Cretacico di Pietraroia, Benevento)." *Paleocronache* 1993(1): 7–14.

Li, R., M. G. Lockley, M. Matsukawa, and M. Liu. 2015. "Important Dinosaur-Dominated Footprint Assemblages from the Lower Cretaceous Tianjialou Formation at the Houzuoshan Dinosaur Park, Junan County, Shandong Province, China." *Cretaceous Research* 52(2015): 83–100.

Li, R., M. G. Lockley, M. Matsukawa, K. Wang, and M. Liu. 2011. "An Unusual Theropod Track Assemblage from the Cretaceous of the Zhucheng Area, Shandong Province, China." *Cretaceous Research* 32(2011): 422–32.

Lima, M. R. 1983. "Paleoclimatic Reconstruction of the Brazilian Cretaceous Based on Palynology Data." *Revista Brasileira de Geociências* 13: 223–28.

——. 1990. Estudo palinológico de sedimentos da Bacia de Icó, Cretáceo do Estado do Ceará, Brasil. Boletim do Instituto de Geociências-USP, Série Científica 21: 35–43.

Lilienstern, H. R., von. 1939. "Fährten und Spuren im Chirotheriumsandstein von Südthuringen." Fortschr. Geol. Palaeont 12(40): 293–387.

Lima, M. R., and M. P. C. A. Coelho. 1987. Estudo palinológico da sondagem estratigráfica de Lagoa do Forno, Bacia do Rio do Peixe, Cretáceo do Nordeste do Brasil. Boletim do Instituto de Geociências–USP, Série Científica 18: 67–83.

Lima Filho, M. F. 1991. Evolução tectono-sedimentar da Bacia do Rio do Peixe-PB. Pernambuco. Dissertação de Mestrado, Universidade Federal de Pernambuco, Centro de Tecnologia [Unpublished].

Lima Filho, M. F., J. M. Mabesoone, and M. S. S. Viana. 1999. Late Mesozoic history of sedimentary basins in NE Brasilian Borborema Province before the final separation of South America and Africa 1: Tectonic-sedimentary evolution; pp. 605–11 in Simpósio Sobre o Cretáceo do Brasil, 5, Boletim, UNESP Rio Claro, Brazil.

Lins, F. A. P. L. 1987. Geofísica aplicada ao estudo do arcabouço tectônico das bacias sedimentares entre as bacias Potiguar e Rio do Peixe. Dissertação de Mestrado, Universidade Federal de Pernambuco, Centro de Tecnologia [Unpublished].

Lockley, M. G. 1987. Dinosaur Footprints from the Dakota Group of Eastern Colorado. Mountain Geologist 24(4): 107–22.

——. 1991. Tracking Dinosaurs: A New Look at an Ancient World. Cambridge, MA: Cambridge University Press.

——. 2000. "An Amended Description of the Theropod Footprint Bueckeburgichnus maximus Kuhn 1958, and Its Bearing on the Megalosaur Tracks Debate." Ichnos 7: 217–25.

——. 2007. A tale of two ichnologies: the different goals and potentials of invertebrates and vertebrate (tetrapod) ichnotaxonomy and how they relate to ichnofacies analysis. Ichnos 14 (1–2): 39–57.

Lockley, M. G., K. Cart, J. Foster, and S. G. Lucas. 2017. "Early Jurassic Batrachopus-Rich Track Assemblages from Interdune Deposits in the Wingate Sandstone, Dolores Valley, Colorado, USA." Palaeo 491 (2018): 185–95. https://doi.org/10.1016/j.palaeo .2017.12.008.

Lockley, M. G., K. Cart, J. Martin, R. Prunty, K. Houck, K. Hups, J.-D. Lim, et al. 2014a. "A Bonanza of New Tetrapod Tracksites from the Cretaceous Dakota Group, Western Colorado: Implications for Paleoecology." In Fossil Footprints of Western North America, edited by M. G. Lockley and S. G. Lucas. New Mexico Museum of Natural History and Science, Bulletin 62: 393–410.

Lockley, M. G., and K. Conrad. 1991. "The Paleoenvironmental Context, Preservation and Paleoecological Significance of Dinosaur Tracksites in the Western USA." In Dinosaur Tracks and Traces, edited by D. D. Gillette and M. G. Lockley, 121–34. New York: Cambridge University Press.

Lockley, M. G., K. Houck, S.-Y. Yang, M. Matsukawa, and S.-K. Lim. 2006. "Dinosaur-Dominated Footprint Assemblages from the Cretaceous Jindong Formation, Hallyo Haesang National Park area, Goseong County, South Korea: Evidence and Implications." Cretaceous Research 27(1): 70–101.

Lockley, M. G., and A. P. Hunt. 1995. Dinosaur Tracks and Other Fossil Footprints of the Western United States. New York: Columbia University Press.

Lockley, M. G., A. P. Hunt, and C. Meyer. 1994. "Vertebrate Tracks and the Ichnofacies Concept: Implications for Paleoecology and Palichnostratigraphy." In The Paleobiology of Trace Fossils, edited by S. Donovan, 241–268. Chichester, UK: Wiley.

Lockley, M. G., J. Y. Kim, K. Kim, S. H. Kim, M. Matsukawa, R. Li, J. Li, and S.-Y. Yang. 2008. Minisauripus—the Track of a Diminutive Dinosaur from the Cretaceous of China and South Korea: Implications for Stratigraphic Correlation and Theropod Foot Morphodynamics." Cretaceous Research 29(1): 115–30.

Lockley, M. G., M. Matsukawa, and L. Jianjun. 2003a. "Crouching Theropods in Taxonomic Jungles: Ichnological and Ichnotaxonomica Investigations of Footprints with Metatarsal and Ischial Impressions." Ichnos 10: 169–77.

Lockley, M. G., M. Matsukawa, and I. Obata. 1989. "Dinosaur Tracks and Radial Cracks: Unusual Footprint Features." Bulletin of the National Science Museum, Tokyo, Series C, 15:151–60.

Lockley, M. G., and C. Meyer. 1999. Dinosaur Tracks and Other Fossil Footprints of Europe. New York: Columbia University Press.

Lockley, M. G., G. Nadon, and P. J. Currie. 2003b. "A Diverse Dinosaur-Bird Footprint Assemblage from the Lance Formation, Upper Cretaceous, Eastern Wyoming: Implications for Ichnotaxonomy." Ichnos 11: 229–49.

Lockley, M. G., and A. Rice. 1980. "Did 'Brontosaurus' Ever Swim Out to Sea? Evidence from Brontosaur Trackways and Other Dinosaur Footprints." Ichnos 1: 81–90.

Lockley, M. G., A. S. Schulp, C. A. Meyer, G. Leonardi, and D. K. Mamani. 2002. "Titanosaurid Trackways from the Upper Cretaceous of Bolivia: Evidence for Large Manus, Wide-Gauge Locomotion and Gregarious Behaviour." Cretaceous Research 23(3): 383–400.

Lockley, M. G., and L. Xing. 2015. "Flattened Fossil Footprints: Implications for Paleobiology." Palaeogeography, Palaeoclimatology, Palaeoecology 426: 85–94.

Lockley, M.G., L. Xing, J. A. F. Lockwood, and S. Pond. 2014b. "A Review of Large Cretaceous Ornithopod Tracks, with Special Reference to Their Ichnotaxonomy." Biological Journal of the Linnean Society 113: 721–36.

Lopes, R. F., C. R. A. Candeiro, and S. de Valais. 2019. Geoconservation of the paleontological heritage of the geosite of dinosaur footprints (sauropods) in the locality of São Domingos, municipality of Itaguatins, state of Tocantins, Brazil. Environmental Earth Sciences 78, 707. 8 pp. https://doi.org/10.1007 /s12665-019-8722-1.

Lull, R. S. 1904. "Fossil Footprints of the Jura-Trias of North America." Memoirs of the Boston Society of Natural History 5: 420–22.

——. 1953. "Triassic Life of the Connecticut Valley." Bulletin, State Geological and Natural History Survey of Connecticut, 81.

Lužar-Oberiter, B., B. Kordić, and A. Mezga. 2017. "Digital Modelling of the Late Albian Solaris Dinosaur Tracksite (Istria, Croatia)." Palaios, 32(12):739–49. https://doi.org/10.2110/palo .2017.034.

Mabesoone, J. M. 1972. Sedimentos do Grupo Rio do Peixe (Paraíba); 1: 236 in Congresso Brasileiro de Geologia, 26, Boletim, Belém, 1972. Sociedade Brasileira de Geologia, Belém, Pará, Brazil.

——. 1975. Desenvolvimento paleoclimático do Nordeste Brasileiro; 5: 75–94 in Simpósio de Geologia do Nordeste, 7, Atas, Fortaleza, Sociedade Brasileira de Geologia, Núcleo Nordeste, Brazil.

——. 1994. "Sedimentary Basins of Northeast Brasil." Federal University Pernambuco, Geology Department, Special Publication 2.

Mabesoone, J. M., and V. A. Campanha. 1973/1974. "Caracterização estratigráfica dos grupos Rio do Peixe e Iguatu." Estudos Sedimentológicos, Natal 3/4: 21–41.

Mabesoone, J. M., and A. Campos e Silva. 1972. "Formação Moura: depósito correlativo do interior nordestino." *Estudos Sedimentológicos, Natal* 2:35–41.

Mabesoone, J. M., P. J. Lima, and E. M. D. Ferreira. 1979. Depósitos de cones aluviais antigos, ilustrados pelas formações Quixoá e Antenor Navarro (Nordeste do Brasil); 7: 225–35 in Simpósio de Geologia do Nordeste, 9, Anais, Recife, Sociedade Brasileira de Geologia/Núcleo Nordeste, Brazil.

Mabesoone, J. M., M. S. S. Viana, and V. H. Neumann. 2000. "Late Jurassic to Mid-Cretaceous Lacustrine Sequences in the Araripe-Potiguar Depression of Northeastern Brasil." In *Lake Basins through Space and Time*, edited by E. H. Gierlowski-Kordesch and K. R. Kelts. *AAPG Studies in Geology* 46: 197–208.

Machado, D. L., L. K. Dehira, C. D. R. Carneiro, and F. F. M. Almeida.1990. "Reconstruções paleoambientais do Juro-Cretáceo no Nordeste Oriental Brasileiro." *Revista Brasileira de Geociências* 19: 470–85.

Maiorino, L., A. K. Farke, T. Kotsakis, and P. Piras. 2015. "Re-Evaluating Sexual Dimorphism in Protoceratops andrewsi (Neoceratopsia, Protoceratopsidae)." *PLoS ONE* 10(5): e0126464.

Maisey, J. G. "Northeastern Brazil: Out of Africa? [Nordeste do Brasil: Fora da África?]." 2011. In *Paleontologia: Cenários de Vida. Editora Interciência*, edited by I. S. Carvalho, N. K. Srivastava, O. Strohschoen Jr. and C. C. Lana, 515–29. 4: 669–80. Rio de Janeiro.

Marchetti, L. 2019. "Can Undertracks Show Higher Morphologic Quality than Surface Tracks? Remarks on Large Amphibian Tracks from the Early Permian of France." *Journal of Iberian Geology* 45: 353–363.

Marchetti, L., M. Belvedere, S. Voigt, H. Klein, D. Castanera, I. Díaz-Martínez, D. Marty, et al.. 2019. "Defining the Morphological Quality of Fossil Footprints. Problems and Principles of Preservation in Tetrapod Ichnology with Examples from the Palaeozoic to the Present." *Earth-Science Reviews* 193: 109–145

Marsh, O. C. 1899. "Footprints of Jurassic dinosaurs." *Am. J. Sci.* (ser. 4)7: 227–32.

Martin, A. J. 2016. "A Close Look at Victoria's First Known Dinosaur Tracks." *Memoirs of Museum Victoria* 74: 63–71.

Marty, D. 2005. Sedimentology and taphonomy of dinosaur track-bearing Plattenkalke (Kimmeridgian, Canton Jura, Switzerland). Zitteliana B26: 20.

———. 2012. Formation, taphonomy, and preservation of vertebrate tracks; p. 37 in A. Richter and M. Reich (eds.), Dinosaur Tracks 2011. An International Symposium. Obernkirchen, April 14–17, 2011. Göttingen, Göttingen Universitätsdrucke, 2012, Germany.

Marty, D., A. Strasser, and C. A. Meyer. 2009. "Formation and Taphonomy of Human Footprints in Microbial Mats of Present-Day Tidal-flat Environments: Implications for the Study of Fossil Footprints." *Ichnos*, 16: 127–42.

Mateus, O., M. Marzola, A. S. Schulp, L. L. Jacobs, M. J. Polcyn, V. Pervov, A. O. Gonçalves, and M. L. Morais. 2017. "Angolan Ichnosite in a Diamond Mine Shows the Presence of a Large Terrestrial Mammaliamorph, a Crocodylomorph, and Sauropod Dinosaurs in the Early Cretaceous of Africa." *Palaeogeography, Palaeoclimatology, Palaeoecology* 471(2017): 220–32.

Matos, R. M. D. 1992. "The Northeast Brazilian Rift System." *Tectonics* 11: 766–91.

Matthews, N., T. Noble, and B. Breithaupt. 2016. "Close-Range Photogrammetry for 3-D Ichnology: The Basics of Photogrammetric Ichnology." In *Dinosaur Tracks: The Next Steps*, edited P. L. Falkingham, D. Marty, and A. Richter, 28–55. Bloomington: Indiana University Press.

Maury, C. J. 1930. O Cretáceo da Paraíba do Norte. Rio de Janeiro. DNPM/DGM, Monografia.

———. 1934. Fossil Invertebrata from Northeastern Brazil. Bulletin of American Museum of Natural History, 67: 123–79; in Fundação Guimarães Duque, 1982 Coleção Mossoroense, 194, pp. 52–58.

McCrea, R. T. 2000. "Vertebrate Palaeoichnology of the Lower Cretaceous (Lower Albian) Gates Formation of Alberta." MSc thesis, University of Saskatchewan, Saskatchewan, Canada.

McCrea, R. T., L. G. Buckley, J. O. Farlow, M. G. Lockley, P. J. Currie, N. A. Matthews, and S. G. Pemberton. 2014. "A 'Terror of Tyrannosaurs': The First Trackways of Tyrannosaurids and Evidence of Gregariousness and Pathology in Tyrannosauridae." *PLoS ONE* 9(7): e103613. https://doi.org/10.1371/journal.pone.0103613.

McCrea, R. T., M. G. Lockley, and C. Meyer. 2001. "The Global Distribution of Purported Ankylosaur Track Occurrences." In *The Armored Dinosaurs*, edited by K. C. Karpenter, 413–454. Bloomington: University of Indiana Press.

Medeiros, M. A. 2006. "Large Theropod Teeth from the Eocenomanian of Northeastern Brazil and the Occurrence of Spinosauridae." *Revista Brasileira de Paleontologia* 9(3): 333–38.

Medeiros, M. A. M., J. C. Della Favera, M. A. F. Reis, T. L. Vieira e Silva, E. B. Oliveira, R. A. Biassusi, and R. G. Silveira. 2007. "O Laser Scanner e a Paleontologia em 3D." *Anuário do Instituto de Geociências* 30(1): 94–100.

Melchor, R. N., D. L. Rivarola, A. M. Umazano, M. N. Moyano, F. R. Mendoza Belmontes. 2019. "Elusive Cretaceous Gondwanan Theropods: The Footprint Evidence from Central Argentina." *Cretaceous Research* 97: 125–42, https://doi.org/10.1016/j.cretres.2019.01.004.

Mendonça Filho, J. G., I. S. Carvalho, and D. A. Azevedo. 2006. "Aspectos Geoquímicos do Óleo da Bacia de Sousa (Cretáceo Inferior), Nordeste do Brasil: Contexto Geológico." *Geociências* 25(1): 91–98.

Menezes, M. N., H. I. Araújo-Júnior, P. F. Dal' Bó, and M. A. A. Medeiros. 2019. Integrating Ichnology and Paleopedology in the Analysis of Albian Alluvial Plains of the Parnaíba Basin, Brazil. *Cretaceous Research* 96: 210–226.

Meyer, C. A., M. G. Lockley, G. Leonardi, and F. Anaya. 1999. "Late Cretaceous Vertebrate Ichnofacies of Bolivia—Facts and Implications." *Journal of Vertebrate Paleontology*, 19 (suppl. to #3): 63A.

Meyer, C. A., D. Marty, and M. Belvedere. 2018. "Titanosaur Trackways from the Late Cretaceous El Molino Formation of Bolivia (Cal Orck'o, Sucre)." *Annales Societatis Geologorum Poloniae* 88: 223–241, doi: https://doi.org/10.14241/asgp.2018.014.

Meyer, C. A., R. Menegat, G. Garcia, A. Alem, and M. Jaldin. 2019. "New Dinosaur Tracks from the Late Cretaceous El Molino Formation of Toro Toro (Dep. Potosi, Bolivia)." *Hallesches Jahrbuch für Geowissenschaften*, Beiheft 46, 63.

Milàn, J., D. B. Loope, and R. G. Bromley. 2008. Crouching Theropod and *Navahopus* Sauropodomorph Tracks from the Early Jurassic Navajo Sandstone of USA. *Acta Palaeontologica Polonica* 53 (2): 197–205.

Milner, A. R. C. and M. G. Lockley. 2016. "Dinosaur Swim Track Assemblages: Characteristics, Contexts, and Ichnofacies Implications." In *Dinosaur Tracks: The Next Steps*, edited by P. L. Falkingham, D. Marty, and A. Richter, 152–180, chap. 10. Bloomington: Indiana University Press.

Milner, A. R. C., M. G. Lockley, and J. I. Kirkland. 2006. "A Large Collection of Well Preserved Theropod Dinosaur Swim Tracks from the Lower Jurassic Moenave Formation, St. George, Utah." In *The Triassic-Jurassic Terrestrial Transition*, edited by J. D. Harris, S. G. Lucas, J. A. Spielmann, M. G. Lockley, A. R. C. Milner, and J. I. Kirkland. New Mexico Museum of Natural History and Science, Bulletin 37: 315–28.

Molnar, R. E. 1980. "An Ankylosaur (Ornithischia: Reptilia) from the Lower Cretaceous of Southern Queensland." *Memoirs of the Queensland Museum* 20: 77–87.

Molnar, R. E., and J. O. Farlow. 1990. "Carnosaur Paleobiology." In *The Dinosauria*, edited by D. B. Weishampel, P. Dodson, and H. Osmólska, 210–24. Berkeley: University of California Press.

Moraes, L. J. 1924. Serras e montanhas do Nordeste; pp. 43–58 in Inspectoria de Obras Contra As Seccas. Geologia. Rio de Janeiro. Ministério da Viação e Obras Publicas. (Série I. D. Publ. 58). 2nd ed. Coleção Mossoroense, 35(1). Fundação Guimarães Duque, Rio Grande do Norte, Brazil.

Moratalla, J. J. 2012. The Lower Cretaceous Dinosaur Movements Through the Lacustrine System of the Cameros Basin (Spain) Written in Their Tracks; pp. 46 in: A. Richter and M. Reich (eds.), Dinosaur Tracks 2011. An International Symposium. Obernkirchen, April 14–17, 2011. Göttingen, Göttingen Universitätsdrucke, 2012, Germany.

Moratalla, J. J., J. Marugán-Lobón, H. Martín-Abad, E. Cuesta, and A. D. Buscalioni. 2017. A New Trackway Possibly Made by a Trotting Theropod at the Las Hoyas Fossil Site (Early Cretaceous, Cuenca Province, Spain): Identification, Bio-dynamics, and Palaeoenvironmental Implications. *Palaeontologia Electronica* 20.3.59A: 1–14. https://doi.org/10.26879/770.

Moratalla, J. J., J. L. Sanz, and S. Jiménez. 1988a. "Multivariate Analysis on Lower Cretaceous Dinosaur Footprints: Discrimination between Ornithopods and Theropods." *Geobios* 21: 395–408.

Moratalla, J. J., J. L. Sanz, S. Jiménez, and M. G. Lockley. 1992. "A Quadrupedal Ornithopod Trackway from the Lower Cretaceous of La Rioja (Spain): Inferences on Gait and Hand Structure." *Journal of Vertebrate Paleontology* 12(2): 150–57.

Moratalla, J. J., J. L. Sanz, I. Melero, and S. Jiménez. 1988b. Yacimientos Paleoicnológicos de La Rioja (Huellas de dinosaurios). Gobierno de La Rioja, Logroño, 1988.

Moreau, J.-D., V. Trincal, D. André, L. Baret, A. Jacquet, and M. Wienin. 2018. "Underground Dinosaur Tracksite inside a Karst of Southern France: Early Jurassic Tridactyl Traces from the Dolomitic Formation of the Malaval Cave (Lozère)." *International Journal of Speleology*, 47 (1): 29–42. https://doi.org/10.5038/1827-806X.47.1.2149.

Moreno, K., and M. J. Benton. 2005. "Occurrence of Sauropod Dinosaur Tracks in the Upper Jurassic of Chile (Redescription of *Iguanodonichnus frenki*)." *Journal of South American Earth Sciences* 20(2005): 253–57.

Moreno, K., S. De Valais, N. Blanco, A. J. Tomlinson, J. Jacay, and J. O. Calvo. 2011. "Large Theropod Dinosaur Footprint Associations in Western Gondwana: Behavioural and Palaeogeographic Implications." *Acta Palaeontologica Polonica* 57(1): 73–83.

Muniz, G. C. B. 1985. *Cochlichnus sousensis*, icnoespécie da Formação Sousa, Grupo Rio do Peixe, no Estado da Paraíba. Coletânea de Trabalhos Paleontológicos. DNPM, Brasília, Brazil, pp. 239–42.

Mussa, D. 2001. "Os gêneros *Dadoxylon* Endlicher 1847 e *Araucarioxylon* Kraus 1870." *Revista Brasileira de Paleontologia* 2: 115–16.

Navarrete, R., C. L. Liesa, D. Castanera, A. R. Soria, J. P. Rodríguez-López, and J. I. Canudo. 2014. "A Thick Tethyan Multi-bed Tsunami Deposit Preserving a Dinosaur Megatracksite within a Coastal Lagoon (Barremian, Eastern Spain)." *Sedimentary Geology* 313: 105–27.

Nicosia, U., M. Avanzini, C. Barbera, M. A. Conti, F. Dalla Vecchia, C. Dal Sasso, P. Gianola, et al. 2005. I vertebrati continentali del Paleozoico e Mesozoico. In: Bonfiglio L. (ed.). Paleontologia dei Vertebrati in Italia. Memorie del Museo Civico di Storia Naturale di Verona, 2. Serie, Sezione Scienze della Terra, 6.

Nicosia, U., M. Marino, N. Mariotti, C. Muraro, S. Panigutti, F.M. Petti, E. Sacchi. 1999a. "The Late Cretaceous Dinosaur Tracksite near Altamura (Bari, Southern Italy). I—Geological Framework." *Geologica Romana* 35: 231–236.

———. 1999b. "The Late Cretaceous Dinosaur Tracksite near Altamura (Bari, Southern Italy), II—*Apulosauripus federicianus* New Ichnogen. and New Ichnosp." *Geologica Romana* 35: 237–247.

O'Connor, P. M., and L. P. A. M. Claessens. 2005. "Basic Avian Pulmonary Design and Flow-through Ventilation in Non-avian Theropod Dinosaurs." *Nature* 436: 253–56.

Oliveira, A. I., and O. H. Leonardos. 1943. *Geologia do Brasil*. 2nd ed. Série Didática 2, Ministério da Agricultura. Serviço de Informação Agrícola.

Ősi, A., J. Pálfy, L. Makádi, Z. Szentesi, P. Gulyás, M. Rabi, G. Botfalvai, and K. Hips. 2011. "Hettangian (Early Jurassic) Dinosaur Tracksites from the Mecsek Mountains, Hungary." *Ichnos* 18: 79–94.

Ostrom, J. H. 1980. "The Evidence for Endothermy in Dinosaurs." In *A Cold Look at the Warm-Blooded Dinosaurs*, edited by D. K. Thomas and E. C. A. Olson, 15–54. Boulder, CO: Westview Press.

Padian, K., and J. R. Horner. 2010. "The Evolution of 'Bizarre Structures' in Dinosaurs: Biomechanics, Sexual Selection, Social Selection or Species Recognition?" *Journal of Zoology* 11(2010): 1–15.

Padian, K., J. R. Horner, and A. Ricqles. 2004. "Growth in Small Dinosaurs and Pterosaurs: The Evolution of Archosaurian Growth Strategies." *Journal of Vertebrate Paleontology* 24(3): 555–71.

Paik, I. S., M. Huh, K. H. Park, K. G. Hwang, K. S. Kim, and H. J. Kim. 2006. "Yeosu Dinosaur Track Sites of Korea: The Youngest Dinosaur Track Records in Asia." *Journal of Asian Earth Sciences* 28(4–6): 457–68.

Paik, I. S., H. J. Kim, and Y. I. Lee. 2001. "Dinosaur Track-Bearing Deposits in the Cretaceous Jindong Formation, Korea: Occurrence, Palaeoenvironments and Preservation." *Cretaceous Research* 22(1):79–92.

Paik, I. S., H. J. Kim, H. Lee and S. Kim. 2017. "A Large and Distinct Skin Impression on the Cast of a Sauropod Dinosaur Footprint from Early Cretaceous Floodplain Deposits, Korea." *Scientific Reports* 7: 16339. https://doi.org/10.1038/s41598-017-16576-y.

Paik, I. S., H. J. Kim, S. G. Baek, and Y. K. Seo. 2020. "New evidence for truly gregarious behavior of ornithopods and solitary hunting by a theropod." *Episodes*, https://doi.org/10.18814/epiiugs/2020/020069

Paul, G. S. 1987a. "Predation in the Meat-Eating Dinosaurs." In *Symposium of Mesozoic Terrestrial Ecosystems*, edited by P. J. Currie and E. H. Koster, 171–76. Drumheller, Alberta, Canada: Short Papers Tyrrell Museum Palaeontol.

———. 1987b. "The Science and Art of Restoring the Life Appearance of Dinosaurs and Their Relatives." In *Dinosaurs Past and Present*, vol. II, edited by S. J. Czerkas and E. C. Olson, 5–49. Los Angeles County: Natural History Museum.

Peabody, F. 1948. "Reptile and Amphibian Trackways from the Lower Triassic Moenkopi Formation of Arizona and Utah." *Bull. Dep. Geol. Sci.* 27(8): 205–468.

Pérez-Lorente, F. 2001. Paleoicnología. Los dinosaurios y sus huellas en La Rioja. Apuntes para los cursos y campos de trabajo de verano. Gobierno de La Rioja (ed.), La Rioja, 227 pp.

———. 2002. La distribución de yacimientos y de tipos de huellas de dinosaurio en la Cuenca de Cameros (La Rioja, Burgos, Soria, España). Zubía 14(2002): 191–210.

———, ed. 2003. Dinosaurios y otros reptiles mesozoicos en España. Fundación Patrimonio Paleontológico de La Rioja: Instituto de Estudios Riojanos, Universidad de La Rioja, Logroño, 444 pp.

———. 2015. *Dinosaur Footprints and Trackways of La Rioja.* Bloomington: Indiana University Press.

Petri, S. 1983. "Brazilian Cretaceous Paleoclimates: Evidence from Clay-Minerals, Sedimentary Structures and Palynomorphs." *Revista Brasileira de Geociências* 13(4): 215–22.

———. 1998. "Paleoclimas da era Mesozóica no Brasil—evidências paleontológicas e sedimentológicas." *Revista da Universidade de Guarulhos* 6: 22–38.

Petti, F. M., M. Antonelli, P. Citton, N., Mariotti, M. Petruzzelli, J. Pignatti, S. D'Orazi Porchetti, et al. 2020a. "Cretaceous Tetrapod Tracks from Italy: A Treasure Trove of Exceptional Biodiversity." *Journal Of Mediterranean Earth Sciences* 12: 167–191.

Petti, F. M., M. Avanzini, M. Antonelli, M. Bernardi, G. Leonardi, R. Manni, P. Mietto, et al. 2020b. "Jurassic Tetrapod Tracks from Italy: A Training Ground for Generations of Researchers." *Journal of Mediterranean Earth Sciences* 12: 137–65.

Petti, F. M., M. Petruzzelli, J. Conti, L. Spalluto, A. Wagensommer, M. Lamendola, R. Francioso, G. Montrone, L. Sabato, and M. Tropeano. 2018. "The Use of Aerial and Close-Range Photogrammetry in the Study of Dinosaur Tracksites: Lower Cretaceous (upper Aptian/lower Albian) Molfetta Ichnosite (Apulia, Southern Italy)." *Palaeontologia Electronica* 21.3.3T 1–18. https://doi.org/10.26879/845.

Platt B. F., C. A. Suarez, S. K. Boss, M. Williamson, J. Cothren, and J. A. C. Kvamme. 2018. "LIDAR-Based Characterization and Conservation of the First Theropod Dinosaur Trackways from Arkansas, USA." *PLoS ONE* 13(1):e0190527. https://doi .org/10.1371/journal.pone.0190527.

Ponte, F. C. 1992. Origem e evolução das pequenas bacias cretácicas do interior do Nordeste do Brasil; pp. 55–58 in Simpósio sobre as Bacias Cretácicas Brasileiras, 2, Resumos expandidos , Rio Claro, São Paulo, UNESP, Brazil.

Ponte, F. C., R. Dino, M. Arai, and A. C. Silva-Telles. 1990. Estratigrafia comparada das bacias sedimentares mesozóicas do Interior do Nordeste do Brasil. Uma síntese; pp. 70–73 in Simpósio das Bacias Cretácicas Brasileiras, 1, Boletim de Resumos, Rio Claro, 1990. UNESP/IGCE, Rio Claro, Brasil.

Pontzer, H., V. Allen, and J. R. Hutchinson. 2009. "Biomechanics of Running Indicates Endothermy in Bipedal Dinosaurs." *PLoS ONE* 4(11): 1–9 p. e7783. https://doi.org/10.1371/journal.pone .0007783.

Popoff, M. 1988. "Du Gondwana à l'Atlantique sud: les conexions du fossé de la Bénoué avec bassins du Nord-Est brésilien jusqu'à l'ouverture du golfe de Guinée au Crétacé inférieur." *Journal of African Earth Sciences* 7: 409–31.

Price, L. I. 1959. "Sobre um crocodilídeo notossuquio do Cretácico Brasileiro." *Boletim da Divisão de Geologia e Mineralogia Rio de Janeiro*, 118: 1–55.

———. 1961. "Sobre os dinossáurios do Brasil." *Anais da Academia Brasileira de Ciências*, 33(3–4): 28–29.

Prince, N. K., and M. G. Lockley. 1991. "The Sedimentology of the Purgatoire Tracksite Region, Morrison Formation of Southeastern Colorado." In *Dinosaur Tracks and Traces*, edited by D. D. Gillette and M. G. Lockley, 155–63. New York: Cambridge University Press.

Rand, H. M.1984. Reconhecimento gravimétrico da Bacia do Rio do Peixe; pp. 42–47 in Simpósio de Geologia do Nordeste, 11,

Anais, Natal, Sociedade Brasileira de Geologia, Natal, Rio Grande do Norte, Brazil.

Ratcliffe, B. C., and J. A. Fagerstrom. 1980. "Invertebrate Lebensspuren of Holocene Floodplains: Their Morphology, Origin and Paleoecological Significance." *Journal of Paleontology* 54(3): 614–30.

Razzolini, N. L., B. Vila, D. Castanera, P. L. Falkingham, J. L. Barco, J. I. Canudo, P. L. Manniing, and A. Galobart. 2014. "Intra-Trackway Morphological Variations Due to Substrate Consistency: The El Frontal Dinosaur Tracksite (Lower Cretaceous, Spain)." *PLoS ONE* 9(4): e93708. https://doi.org/10.1371 /journal.pone.0093708.

Razzolini, N. L., B. Vila, I. Díaz-Martínez, Ph. L. Manning, and A. Galobart. 2016. "Pes Shape Variation in an Ornithopod Dinosaur Trackway (Lower Cretaceous, NW Spain): New Evidence of an Antalgic Gait in the Fossil Track Record." *Cretaceous Research* 58(2016): 125–34.

Regali, M. S. P. 1990. Biocronoestratigrafia e paleoambiente do Eocretáceo das bacias do Araripe (CE) e Rio do Peixe (PB), NE-Brasil; pp. 163–72 in Simpósio Sobre a Bacia do Araripe e Bacias Interiores do Nordeste, 1, Atas, Crato. D.A. Campos, M.S.S. Vianna, P.M. Brito and G. Beurlen (eds.), Crato, Ceará, Brazil.

Reid, R. E. H. 1997. "Dinosaurian Physiology: The Case for 'Intermediate' Dinosaurs." In *The Complete Dinosaur*, edited by J. O. Farlow and M. K. Brett-Surman, 449–73. Bloomington: Indiana University Press.

Reineck, H. E., and I. B. Singh. 1986. *Depositional Sedimentary Environments*, 2nd edition. Berlin: Springer-Verlag.

Rich, T. H., and P. V. Rich. 1989. "Polar Dinosaurs and Biotas of the Early Cretaceous of Southeastern Australia." *National Geographic Research* 5: 15–53.

Richter, A., and A. Böhme. 2016. "Too Many Tracks-Preliminary Description and Interpretation of the Diverse and Heavily Dinoturbated Early Cretaceous 'Chicken Yard' Ichnoassemblage (Obernkirchen Tracksite, Northern Germany)." *Dinosaur Tracks: The Next Steps*, edited by P. L. Falkingham, D. Marty, and A. Richter, 334–57, chap. 17. Bloomington: Indiana University Press.

Richter, A., and M. Reich, eds. Dinosaur Tracks 2011. An International Symposium. Obernkirchen, April 14–17, 2011. Göttingen, Göttingen Universitätsdrucke, 2012, Germany.

Ricqles. A. J. 1969. L'histologie osseuse envisage comme indicateur de la physiologie termique chez les tetrapods fossiles. C. R. Acad. Sci. Paris, 268: 782–85.

———. 1974. "Evolution of Endothermy: Histological Evidence." *Evol. Theory*, 1: 51–80.

———. 1984. Paleoistologia dei dinosauri. In: Bonaparte, J. F., E. H. Colbert, P. J. Currie, A/J de Ricqlès, Z. Keilan-Jaworowska, G. Leonardi, F. Morello, P. Taquet (eds.). Sulle orme dei dinosauri. Venezia-Mestre, Erizzo, 1984. (Esplorazioni e ricerche, IX): 161–86.

Roesner, E. H., C. C. Lana, A. L. Hérissé, and J. H. G. Melo. 2011. Bacia do Rio do Peixe (PB): novos resultados biocronoestratigráficos e paleoambientais; pp. 135–41 in I. S. Carvalho, N. K. Srivastava, O. Strohschoen Jr., and C. C. Lana, eds., Paleontologia: Cenários de Vida. Editora Interciência, Rio de Janeiro, Brazil.

Romano, M., and P. Citton. 2020. "Tetrapod Ichnology in Italy: The State of the Art." Guest editorial. *Journal of Mediterranean Earth Sciences* 12(2020): 1-265. https://doi.org/10.3304/jmes.2020 .17068.

Romano, M., P. Citton, and M. Avanzini. 2018. "A Review of the Concepts of 'Axony' and Their Bearing on Tetrapod Ichnology.

Historical Biology 32(5): 611–619 https://doi.org/10.1080/08912963.2018.1516766.

Romano, M., and M. A. Whyte. 2015. "Could Stegosaurs Swim? Suggestive Evidence from the Middle Jurassic Tracksite of the Cleveland Basin, Yorkshire, UK." *Proceedings of the Yorkshire Geological Society* 60(3): 227–33.

Romilio, A., J. M. Hacker, R. Zlot, G. Poropat, M. Bosse and S. W. Salisbury. 2017. "A Multidisciplinary Approach to Digital Mapping of Dinosaurian Tracksites in the Lower Cretaceous (Valanginian–Barremian) Broome Sandstone of the Dampier Peninsula, Western Australia." *PeerJ* (March 21, 2017): 26. https://doi.org/10.7717/peerj.3013.

Romilio, A., and S. W. Salisbury. 2012. Re-interpretation of the dinosaur track-maker identities and tracksite scenario at Lark Quarry, of the Mid-Cretaceous (late Albian–Cenomanian) Winton Formation, central-western Queensland, Australia; pp. 53 in A. Richter and M. Reich (eds.), Dinosaur Tracks 2011. An International Symposium. Obernkirchen, April 14–17, 2011. Göttingen, Göttingen Universitätsdrucke, 2012, Germany.

Rozadilla, S., F. L. Agnolin, F. E. Novas, A. M. Aranciaga Rolando, M. J. Motta, J. M. Lirio, and M. P. Isasi. 2016. "A New Ornithopod (Dinosauria, Ornithischia) from the Upper Cretaceous of Antarctica and Its Palaeobiogeographical Implications." *Cretaceous Research* 57(2016): 311–24.

Ruben, J. A., W. J. Hillenius, N. R. Geist, A. Leitch, T. D. Jones, Philip J. Currie, John R. Horner and George Espe III. 1996. "The Metabolic Status of Some Late Cretaceous Dinosaurs." *Science* 273: 1204–07.

Salgado L., and J. F. Bonaparte. 1991. "Un nuevo saurópodo Dicraeosauridae, *Amargasaurus cazaui* gen. et sp. nov., de la formación La Amarga, Neocomiano de la provincia de Neuquén, Argentina." *Ameghiniana* 28(3–4): 333–46.

Salgado, L., I. S. Carvalho, and A. Garrido. 2006. "*Zapalasaurus bonapartei*, un Nuevo dinosaurio saurópodo de la formacion La Amarga (Cretácico Inferior) noroeste de Patagonia, provincial de Neuquén, Argentina." *Geobios* 39(2006): 695–707.

Salisbury, S. W., A. Romilio, M. C. Herne, R. T. Tucker, and J. P. Nair. 2017. "The Dinosaurian Ichnofauna of the Lower Cretaceous (Valanginian–Barremian) Broome Sandstone of the Walmadany Area (James Price Point), Dampier Peninsula, Western Australia." Society of Vertebrate Paleontology Memoir 16. *Journal of Vertebrate Paleontology* 36(6, Supplement). https://doi.org/10.1080/02724634.2016.1269539.

Sampson, S. D., and M. J. Ryan. 1997. "Variation." In *Encyclopedia of Dinosaurs*, edited by P. J. Currie and K. Padian, 773–80. San Diego: Academic Press.

Sant'Anna, A. J. 2009. Investigação geoquímica orgânica na Bacia de Sousa—Paraíba—Correlação óleo-rocha geradora. Programa de Pós-Graduação em Geologia, Universidade Federal do Rio de Janeiro. Dissertação de Mestrado.

Santi, G. and U. Nicosia. 2008. "The Ichnofacies Concept in Vertebrate Ichnology." *Studi Trentini di Scienze Naturali, Acta Geologica* 83: 223–229.

Santos, C. L. A., L. Vale, M. F. C. F. Santos, G. Leonardi, M. D. Wanderley, and A. C. Brito. 1985. Técnicas de moldagem em pegadas de répteis na Bacia do Rio do Peixe, Sousa, Paraíba, Brasil; p. 47 in Congresso Brasileiro de Paleontologia, 9, Resumos, Fortaleza, 1985. Sociedade Brasileira de Paleontologia, Fortaleza, Ceará, Brazil.

Santos, E. J., and B. B. Brito Neves. 1984. Província Borborema; 123–86 in F. F. M. Almeida and Y. Hasui (coord.). O Pré-Cambriano no Brasil. Editora Edgard Blücher, São Paulo, Brazil.

Santos, M. F. C. S., and C. L. A. Santos. 1987a. Sobre a ocorrência de pegadas e pistas de dinossauros na localidade de Engenho Novo, Antenor Navarro, Paraíba (Grupo Rio do Peixe, Cretáceo Inferior); 1: 353–66 in Congresso Brasileiro de Paleontologia, 10, Anais, Rio de Janeiro, 1987, Brazil

———. 1987b. Novas pegadas de dinossauros retiradas de uma cerca de pedras no sítio Cabra Assada, Antenor Navarro, Paraíba (Grupo Rio do Peixe, Cretáceo Inferior); 1: 367–76 in Congresso Brasileiro de Paleontologia, 10, Anais, Rio de Janeiro, 1987, Brazil.

———. 1989. Alguns parâmetros relativos às pegadas de dinossauros em Várzea dos Ramos, município de Sousa, Paraíba; 1: 373–80 in Congresso Brasileiro de Paleontologia, 11, Anais, Sociedade Brasileira de Paleontologia, Curitiba, 1989, Paraná, Brazil.

Santos, V. F., P. M. Callapez, and N. P. C. Rodrigues. 2013. "Dinosaur Footprints from the Lower Cretaceous of the Algarve Basin (Portugal): New Data on the Ornithopod Palaeoecology and Palaeobiogeography of the Iberian Peninsula. *Cretaceous Research*, 40(2013): 158–69.

Santos, W. F. S. 2014. "Sítios Paleontológicos, Estratégias de Geoconservação e Geoturismo na Bacia de Sousa (Paraíba): potencial da área para se tornar um Geoparque." DSc Thesis, Programa de Pós-Graduação em Geologia, Rio de Janeiro Federal University, Geoscience Institute [Unpublished].

Santos, W. F. S., I. S. Carvalho, and J. B. Brilha. 2019. "Public Understanding on Geoconservation Strategies at the Passagem das Pedras Geosite, Paraíba (Brazil): Contribution to the Rio do Peixe Geopark Proposal." *Geoheritage* 11: 2065–2077, https://doi.org/10.1007/s12371-019-00420-y.

Santos, W. F. S., I. S. Carvalho, J. B. Brilha, and G. Leonardi. 2015. "Inventory and Assessment of Palaeontological Sites in the Sousa Basin (Paraíba, Brazil): Preliminary Study to Evaluate the Potential of the Area to Become a Geopark." *Geoheritage* 8 (2016): 315–32. http://doi.org/10.1007/s12371-015-0165-9.

Sarjeant, W. A. S., J. B. Delair, and M. G. Lockley. 1998. "The Footprints of Iguanodon: A History and Taxonomic Study." *Ichnos* 6:183–202.

Schanz, T., M. Datcheva, H. Haase, and D. Marty. 2016. "Analysis of Desiccation Cracks Patterns for Quantitative Interpretation on Fossil Tracks." In *Dinosaur Tracks: The Next Steps*, edited by P. L. Falkingham, D. Marty, and A. Richter, 366–79, chap. 19. Bloomington: Indiana University Press.

Schobbenhaus, C., D. A. Campos, E. T. Queiroz, M. Winge, and M. L. C. Berbert-Born, eds. 2002. Sítios Geológicos e Paleontológicos do Brasil. D.N.P./C.P.R.M./SIGEP, Brasília, Brazil, 2002, 540 pp.

Schweitzer, M. H., J. L. Wittmeyer, and J. R. Horner. 2005. "Gender-Specific Reproductive Tissue in Ratites and *Tyrannosaurus rex*." *Science* 308: 1456–60.

Sciscio, L., E. M. Bordy, M. Reid, and M. Abrahams. 2016. Sedimentology and Ichnology of the Mafube Dinosaur Track Site (Lower Jurassic, eastern Free State, South Africa): A Report on Footprint Preservation and Palaeoenvironment. *PeerJ* 4:e2285; doi 10.7717/peerj.2285.

Seebacher, F. 2003. "Dinosaur Body Temperatures: The Occurrence of Endothermy and Ectothermy." *Paleobiology* 29: 105–22.

Segura, M., F. Barroso-Barcenilla, M. Berrocal-Casero, D. Castanera, J. F. García-Hidalgo, and V. F. Santos. 2016. "A New Cenomanian Vertebrate Tracksite at Tamajón (Guadalajara, Spain): Palaeoichnology and Palaeoenvironmental implications." *Cretaceous Research* 57(2016): 1–11.

Sénant, J., and M. Popoff. 1989. Les bassins du Rio do Peixe (NE Brésil): extension intracontinentale Crétacée et réactivation des grands cisaillements ductiles Pan-Africains. C. R. Acad. Sci. Paris, 308, série 11, pp. 1613–19.

Sereno, P. C., A. L. Beck, D. B. Dutheil, B. Gado, H. C. E. Larsson, G. H. Lyon, J. D. Marcot, et al. 1998. "A Long-Snouted Predatory Dinosaur from Africa and the Evolution of Spinosaurids." *Science* 282: 1298–1302.

Sereno, P. C., R. N. Martínez, J. A. Wilson, D. J. Varricchio, O. A. Alcober, *et alii.* 2008. "Evidence for Avian Intrathoracic Air Sacs in a New Predatory Dinosaur from Argentina." *PLoS ONE* 3(9): e3303. https://doi.org/10.1371/journal.pone.0003303.

Schult, M. F., and J. O. Farlow. 1992. "Vertebrate Trace Fossils." In *Trace Fossils*, edited by C. G. Maples and R. R. West, 34–63. Paleontological Society Short Course in Paleontology.

Shillito, A. P., and N. S. Davies. 2019. "Dinosaur-Landscape Interactions at a Diverse Early Cretaceous Tracksite (Lee Ness Sandstone, Ashdown Formation, southern England)." *Palaeogeography, Palaeoclimatology, Palaeoecology* 514: 593–612.

Silva Filho, R. P. 2009. Análise de fácies da Formação Sousa (Cretáceo Inferior), bacia de Sousa (PB), e seu contexto em um sistema petrolífero. Monografia de Conclusão de Curso. Universidade Federal do Rio de Janeiro. Departamento de Geologia [Unpublished].

Siqueira, L. M. P., M. A. R. Polck, A. C. G. Hauch, C. A. Silva, F. B. Chaves, I. T. Yamamoto, J. P. Araujo, et al. 2011. "Sítios Paleontológicos das Bacias do Rio do Peixe: Georreferenciamento, Diagnóstico de Vulnerabilidade e Medidas de Proteção." *Anuário do Instituto de Geociências* 34(1): 9–21.

Skelton, P., ed. 2003. *The Cretaceous World.* Cambridge, UK: Cambridge University Press.

Slater, S. M., C. H. Wellman, M. Romano, and V. Vajda. 2017. "Dinosaur-Plant Interactions within a Middle Jurassic Ecosystem—Palynology of the Burniston Bay Dinosaur Footprint Locality, Yorkshire, UK." *Palaeobio Palaeoenv* 98(2018): 139–51. https://doi.org/10.1007/s12549-017-0309-9.

Sopper, R. H. 1923. "Geologia e suprimento d'água subterrâneo no Rio Grande do Norte e Paraíba." *Inspetoria de Obras Contra as Seccas* 26: 1–93.

Soto-Acuña, S., R. A. Otero, D. Rubilar-Rogers, and A. O. Vargas. 2015. "Arcosaurios no avianos de Chile." *Publicación Ocasional del Museo Nacional de Historia Natural, Chile* 63: 209–63.

Sousa, A. J., I. S. Carvalho, and E. P. Ferreira. 2019. "Non-marine Ostracod Biostratigraphy of Cretaceous Rift Lake Deposits (Sousa Basin, Brazil): Paleogeographical Implications and Correlation with Gondwanic Basins." *Journal of South American Earth Sciences* 96 (2019). https://doi.org/10.1016/j.jsames.2019.102345.

Souto, P. R. F., and M. A. Fernandes. 2015. "Fossilized Excreta Associated to Dinosaurs in Brazil." *Journal of South American Earth Sciences* 57(2015): 32–38.

Spotilla, J. R., P. W. Lommen, G. S. Bakken, and D. M. Gates. 1973. "A Mathematical Model for Body Temperature of Large Reptiles: Implications for Dinosaur Endothermy. *American Naturalist* 107: 391–404.

Squires, R. L., and D. M. Advocate. 1984. "Meniscate Burrows from Miocene Lacustrine-Fluvial Deposits, Diligencia Formation, Orocopia Mountains, Southern Califórnia." *Journal of Paleontology* 58(2): 593–97.

Srivastava, N. K., and I. S. Carvalho. 2004. "Bacias do Rio do Peixe." *Fundação Paleontológica Phoenix* 71: 1–4.

Stewart, W. N. 1983. *Paleobotany and the Evolution of Plants.* Cambridge, UK: Cambridge University Press.

Strevell, C. N. 1932. Dinosauropodes. Salt Lake City: Deseret News.

Suárez Riglos, M. 1995. "Huellas de dinosaurios en Sucre." *Associación Sucrenes de Ecología Anuario* 95, 44–48.

Sues, H. D., E. Frey, D. M. Martill, and D. M. Scott. 2002. "*Irritator challengeri*, a Spinosaurid (Dinosauria: Theropoda) from the Lower Cretaceous of Brazil." *Journal of Vertebrate Paleontology* 22(3): 535–47.

Taquet, P. 1976. Géologie et Paléontologie du gisement de Gadoufaoua (Aptien du Niger). Cahiers de Paléontologie. CNRS, Paris, 1976.

Tasch, P. 1987. *Fossil Conchostraca of the Southern Hemisphere and Continental Drift. Paleontology, Biostratigraphy and Dispersal.* Colorado: Geological Society of America.

Teilhard de Chardin, P., and C. C. Young. 1929. "On Some Traces of Vertebrate Life in the Jurassic and Triassic Beds of Shansi and Shensi." *Geological Society of China Bulletin* 8: 131–35.

Thompson, D. W. 1942. *On Growth and Form.* Dover reprint of 1942, 2nd ed. (1st ed., 1917). Cambridge, UK: Cambridge University Press.

Thulborn, R. A. 1989. "The Gaits of Dinosaurs." In *Dinosaur Tracks and Traces*, edited by D. D. Gillette and M. G. Lockley, 39–50. New York: Cambridge University Press.

———. 1990. *Dinosaurs Tracks.* London: Chapman and Hall.

———. 2012. "Impact of Sauropod Dinosaurs on Lagoonal Substrates in the Broome Sandstone (Lower Cretaceous), Western Australia." *PLoS ONE* 7(5): e36208. doi:10.1371/journal.pone.0036208.

Thulborn, R. A., and M. Wade. 1984. "Dinosaur Trackways in the Winton Formation (Mid-Cretaceous) of Queensland." *Memoirs of the Queensland Museum* 21(2): 413–517.

Tinoco, I. M., and I. Katoo. 1975. Conchostráceos da Formação Sousa, Bacia do Rio do Peixe, Estado da Paraíba; pp. 135–47 in Simpósio de Geologia, 7, Atas, Fortaleza, 1975. Sociedade Brasileira de Geologia/Núcleo Nordeste, Fortaleza, Ceará, Brazil.

Tinoco, I. M., and J. M. Mabesoone. 1975. Observações paleoecológicas sobre as bacias mesozóicas de Iguatu, Rio do Peixe e Intermediárias; pp. 95–107 in Simpósio de Geologia, 7, Atas, Fortaleza, 1975. Sociedade Brasileira de Geologia/Núcleo Nordeste, Fortaleza, Ceará, Brazil.

Trompette, R., M. Egydio-Silva, A. Tommasi, A. Vauchez, and A. Uhlein. 1993. "Amalgamação do Gondwana Ocidental no Panafricano-Brasiliano e o papel da geometria do Cráton do São Francisco na arquitetura da Faixa Ribeira." *Revista Brasileira Geociências* 23: 187–93.

Tsukiji, Y., Y. Azuma, F. Shiraishi, M. Shibata, and Y. Noda. 2017. "New Ornithopod Footprints from the Lower Cretaceous Kitadani Formation, Fukui, Japan: Ichnotaxonomical Implications." *Cretaceous Research* 84(2018): 501–14. https://doi.org/10.1016/j.cretres.2017.12.011.

Valença, L. M. M., V. H. Neumann, and J. M. Mabesoone. 2003. "An Overview on Callovian-Cenomanian Intracratonic Basins of Northeast Brazil: Onshore Stratigraphic Record of the Opening of the Southern Atlantic." *Geologica Acta*1(3): 261–75.

Vasconcelos, E. C. 1980. Estudo faciológico da Formação Sousa. Dissertação de Mestrado. Universidade Federal de Pernambuco, Centro de Tecnologia, Recife [Unpublished].

Vialov, O. S. 1988. "On the Classification of Dinosaurian Traces." *Ezhegodnik Vsesoyuznogo Paleontologicheskogo Obshchestva* 31: 322–25.

Viana, M. S. S., M. F. Lima Filho, and I. S. Carvalho. 1993. Borborema Megatracksite: uma base para correlação dos "arenitos inferiores" das bacias intracontinentais do Nordeste do Brasil; in Simpósio de Geologia do Nordeste, Sociedade Brasileira de Geologia/Núcleo Nordeste, Boletim 13: 23–25.

Vila, B., D. Castanera, J. Marmi, J. I. Canudo, and A. Galobart. 2015. "Crocodile Swim Tracks from the Latest Cretaceous of Europe." *Lethaia*, 48(2): 256–66.

Wade, M., 1989. "The Stance of Dinosaurs and the Cossack Dancer Syndrome." In *Dinosaur Tracks and Traces*, edited by

D. D. Gillette and M. G. Lockley, 73–82. New York: Cambridge University Press.

Wagensommer, A., M. Latiano, G. Leroux, G. Cassano, S.d'Orazi Porchetti, 2012. "New Dinosaur Tracksites from the Middle Jurassic of Madagascar: Ichnotaxonomical, Behavioural and Palaeoenvironmental Implications." *Palaeontology* 55(1): 109–126.

Weishampel, D. B., P. Dodson, and H. Osmólska, eds. 2004. *The Dinosauria*. Berkeley: University of California Press.

Whyte, M. A., and M. Romano. 2001. "A Dinosaur Ichnocoenosis from the Middle Jurassic of Yorkshire, UK." *Ichnos*, 8: 233–34.

Wings, O., G. N. Lallensack, and H. Mallison. 2016. "The Early Cretaceous Dinosaur Trackways in Münchehagen (Lower Saxony, Germany): 3-D Photogrammetry as Basis for Geometric Morphometric Analysis of Shape Variation and Evaluation of Material Loss during Excavation." In *Dinosaur Tracks: The Next Steps*, edited by P. L. Falkingham, D. Marty, and A. Richter, 56–70, chap. 3. Bloomington: Indiana University Press.

Woodward, H. N., T. H. Rich, A. Chinsamy, and P. Vickers-Rich. 2011. "Growth Dynamics of Australia's Polar Dinosaurs." *PLoS ONE* 6(8): e23339. https://doi.org/10.1371/journal.pone.0023339.

Xing, L., L. G. Buckley, M. G. Lockley, R. T. McCrea, and Y. Tang. 2017a. "Lower Cretaceous Avian Tracks from Jiangsu Province, China: A First Chinese Report for Ichnogenus *Goseongornipes* (Ignotornidae)." *Cretaceous Research* 84(2018): 571–77. https://doi.org/10.1016/j.cretres.2017.12.016.

Xing, L., D. Li, P. L. Falkingham, M. G. Lockley, M. J. Benton, H. Klein, J. Zhang, H. Ran, H., W. S. Persons IV, and H. Dai. 2016a. "Digit-Only Sauropod Pes Trackways from China — Evidence of Swimming or a Preservational Phenomenon?" *Scientific Reports* 6, (2016). https://doi.org/10.1038/srep21138.

Xing, L., M. G. Lockley, N. Hu, G. Li, G. Tong, M. Matsukawa, H. Klein, Y. Ye, J. Zhang, and W. S. Persons IV. 2016b. "Saurischian Track Assemblages from the Lower Cretaceous Shenhuangshan Formation in the Yuanma Basin, Southern China." *Cretaceous Research* 65(2016): 1–9.

Xing , L., M. G. Lockley, K. S. Kim, H. Klein, M. Matsukawa, J. D. Lim, W. S. Persons IV, and X. Xu. 2017b. "Mid-Cretaceous Dinosaur Track Assemblage from the Tongfosi Formation of China: Comparison with the Track Assemblage of South Korea." *Cretaceous Research* 74(2017): 155–64. https://doi.org/10.1016/j.cretres.2017.02.019.

Xing, L., M.G. Lockley, H. Klein, R. Zeng, S. Cai, X. Luo, and C. Li. 2018. Theropod Assemblages and a New Ichnotaxon *Gigandipus chiappei* ichnosp. nov. from the Jiaguan Formation, Lower Cretaceous of Guizhou Province, China, Geoscience Frontiers (2018), doi: 10.1016/ j.gsf.2017.12.012.

Xing, L., M. G. Lockley, D. Marty, H. Klein, G. Yang, J. Zhang, G. Peng, et al. 2016c. "A Diverse Saurischian (Theropod-Sauropod) Dominated Footprint Assemblage from the Lower Cretaceous Jiaguan Formation in the Sichuan Basin, Southwestern China: A New Ornithischian Ichnotaxon, Pterosaur Tracks and an Unusual Sauropod Walking Pattern." *Cretaceous Research* 60 (2016): 176–93.

Xing, L., M. G. Lockley, D. Marty, L. Piñuela, H. Klein, J. Zhang and W. S. Persons IV. 2015a. "Re-description of the Partially Collapsed Early Cretaceous Zhaojue Dinosaur Tracksite (Sichuan Province, China) by Using Previously Registered Video Coverage." *Cretaceous Research* 52(2015): 138–52.

Xing, L., M. G. Lockley, D. Marty, J. Zhang, Y. Wang, H. Klein, R. T. McCrea, et al. 2015b. "An Ornithopod-Dominated Tracksite from the Lower Cretaceous Jiaguan Formation (Barremian–Albian) of Qijiang, South-Central China: New

Discoveries, Ichnotaxonomy, Preservation and Palaeoecology." *PLoS ONE* 10(10): e0141059. https://doi.org/10.1371/journal.pone.0141059.

Xing, L., M. G. Lockley, A. Romilio, H. Klein, J. Zhang, H. Chen, J. Zhang, M. E. Burns, and X. Wang. 2017c. "Diverse Sauropod-Theropod-Dominated Track Assemblage from the Lower Cretaceous Dasheng Group of Eastern China: Testing the Use of Drones in Footprint Documentation." *Cretaceous Research* 84(2018): 588–99. https://doi.org/10.1016/j.cretres.2017.12.012.

Xing, L., M. G. Lockley, K. Tong, G. Peng, A. Romilio, H. Klein, W.S. Persons IV, Y. Ye, and Jiang, S. 2019. "A Diversified Dinosaur Track Assemblage from the Lower Cretaceous Xiaoba Formation of Sichuan Province, China: Implications for Ichnological Database and Census Studies." *Cretaceous Research* 96:120–134, https://doi.org/10.1016/j.cretres.2018.11.026.

Xing, L., M. G. Lockley, G. Yang, J. Cao, R. T. McCrea, H. Klein, J. Zhang, W. S. Persons IV, and H. Dai. 2016d. "A Diversified Vertebrate Ichnite Fauna from the Feitianshan Formation (Lower Cretaceous) of Southwestern Sichuan, China." *Cretaceous Research* 57(2016): 79–89.

Xing, L., M. G. Lockley, J. Zhang, H. Klein, D. Marty, G. Peng, Y. Ye, R. T. McCrea, W. F. Persons IV, and T. Xu. 2015c. "The Longest Theropod Trackway from East Asia, and a Diverse Sauropod-, Theropod-, and Ornithopod-Track Assemblage from the Lower Cretaceous Jiaguan Formation, Southwest China." *Cretaceous Research* 56(2015): 345–62.

Xing, L., M. G. Lockley, J. Zhang, H. Klein, W. S. Persons IV, and H. Dai. 2014a. "Diverse Sauropod-, Theropod-, and Ornithopod-Track Assemblages and a New Ichnotaxon *Siamopodus xui* ichnosp. nov. from the Feitianshan Formation, Lower Cretaceous of Sichuan Province, Southwest China." *Palaeogeography, Palaeoclimatology, Palaeoecology* 414: 79–97.

Xing, L., M. G. Lockley, J. Zhang, A. R. C. Milner, H. Klein, D. Li, W. S. IV Persons, and J. Ebi. 2013. "A New Early Cretaceous Dinosaur track Assemblage and the First Definite Non-avian Theropod Swim Trackway from China." *Chinese Science Bulletin* 58(19): 2370–78.

Xing, L., D. Marty, K. Wang, M. G. Lockley, S. Chen, X. Xu, Y. Liuf, et al. 2015d. "An Unusual Sauropod Turning Trackway from the Early Cretaceous of Shandong Province, China." *Palaeogeography, Palaeoclimatology, Palaeoecology* 437(2015) 74–84.

Xing, L., G. Peng, M. G. Lockley, Y. Ye, H. Klein, J. Zhang, and W. S. Persons IV. 2015e. "Early Cretaceous Sauropod and Ornithopod Trackways from a Stream Course in Sichuan Basin, Southwest China; in R. M. Sullivan and S. G. Lucas (eds.), Fossil Record 4." *New Mexico Museum of Natural History and Science Bulletin* 68: 319–25.

Xing, L.-D., M. G. Lockley, H. Klein, J.-P. Zhang, T. Wang, W.S. Persons, and Z.-M. Dong. 2015f. A tetrapod footprint assemblage with possible swim traces from the Jurassic-Cretaceous boundary, Anning Formation, Konglongshan, Yunnan, China. Palaeoworld, 2015: http://dx.doi.org/10.1016/j.palwor.2015.06.002

Xing, L.-D., G. Nied´zwiedzki, M. G. Lockley, J.-P. Zhang, X.-F. Cai, W. S. P. Iv, and Y. Ye. 2014b. "*Asianopodus*-Type Footprints from the Hekou Group of Honggu District, Lanzhou City, Gansu, China and the 'Heel' of Large Theropod Tracks." *Palaeoworld*, 23(3–4): 304–13.

Zheng, X.-T., H.-L. You, X. Xu, and Z.-M. Dong. 2009. "An Early Cretaceous Heterodontosaurid Dinosaur with Filamentous Integumentary Structures." *Nature* 458: 333–36.

Appendix A: Glossary of Brazilian Geographical Names and Terms (Paraíba)

Açude reservoir
Aroeira kind of northeastern Brazil tree (Anacardiaceae)
Baixio sand bar
Barragem dam
Brejo humid and fertile meadow
Caatinga a kind of semiarid environment in the northeast of Brazil
Caiçara stockade or palisade for cattle or sheep
Canto coarse gravel; also, corner
Corredor a road with barbed-wire fence on both sides
Córrego creek
Cupim termite
Curral corral
Engenho cane sugar factory
Fazenda farm
Ilha (fluvial) island
Inverno (= **winter**) rain season
Jangada raft
Juazeirinho small Juazeiro (northeastern Brazil tree, Ramnaceae)
Lagoa lake
Lajão paved ford
Lajedo rocky natural pavement
Letreiro inscription, namely, Indio petroglyphs
Mãe-d'Água spring, source
Matadouro slaughterhouse
Morada dwelling (place)
Passagem ford
Pedregulho gravel
Piau name of several species of fluvial fishes (Caraciformes)
Pocinho small residual pool (see below, poço)
Poço residual pool on a riverbed during the drought season (NE)
Poço do Motor Pool of the Motor-Pump
Rancho small farm
Riacho, riachão creek
Rio river
Saguim marmoset
Serrote hill range
Sertão wilderness, semiarid region in Northeast Brazil
Sitio small farm
Sousa the name of the town means "downriver"
Tanque (at Sousa) very small reservoir caved in the rock near the river
Várzea fertile and flat land alongside a watercourse
Vau ford
Volta meandering of a river
Zoador a place where waters whisper; small rapid

Appendix B: Dates of the Discovery of the Tracksites and Their Discoverers

1924	Passagem das Pedras	Luciano Jacques de Moraes
1977	Poço do motor	Giuseppe Leonardi
1977	Serrote do Letreiro	Giuseppe Leonardi, with Robson Araujo Marques
1977	Pedregulho	Giuseppe Leonardi, with Luiz Carlos da Silva Gomes
1977	Piedade	Giuseppe Leonardi, with Luiz Carlos da Silva Gomes
1977	Juazeirinho	Giuseppe Leonardi, with Luiz Carlos da Silva Gomes
1979	Piau-Caiçara	Giuseppe Leonardi, with Robson Araujo Marques
1979	Curral Velho	Giuseppe Leonardi
1979	Mãe d'Água	Giuseppe Leonardi
1979	Matadouro	Giuseppe Leonardi
1979	Serrote do Pimenta	Giuseppe Leonardi
1979	Riacho do Cazé	Giuseppe Leonardi
1984	Zoador	Giuseppe Leonardi
1984	Poço da Volta	Giuseppe Leonardi
1984	Barragem do Domício	Giuseppe Leonardi
1984	Engenho Novo	Giuseppe Leonardi
1984	Aroeira	Giuseppe Leonardi
1984	Cabra Assada	Giuseppe Leonardi, with M. de Fátima C. F. dos Santos and Claude Luiz Aguilar Santos
1985	Várzea dos Ramos	Wilton V. Barbosa
1987	Arapuã	G. Leonardi and Ismar de Souza Carvalho
1987	Baleia	G. Leonardi and Ismar de Souza Carvalho
1987	Pocinho	Ismar de Souza Carvalho
1987	Grotão (Pombal)	G. Leonardi and Ismar de Souza Carvalho
1988	Piau II	Giuseppe Leonardi
1988	Sitio Saguim	Giuseppe Leonardi
1988?	Abreu	Antonio Leme
1993	Fazenda Paraíso	Sérgio A. K. Azevedo
2003	Floresta dos Borba	G. Leonardi, Luiz Carlos Gomes, and Robson A. Marques
2003	Lagoa do Forno I	Giuseppe Leonardi
2003	Lagoa do Forno II	Giuseppe Leonardi
2003	Várzea dos Ramos II	Giuseppe Leonardi
2003	Várzea dos Ramos III	Giuseppe Leonardi
2004	Araçá de Cima	Luiz Carlos Gomes
2005	Tapera	Hebert B. N. Campos
2005	Riachão do Oliveira	Giuseppe Leonardi
2008	Baixio do Padre	Giuseppe Leonardi and M. de Fátima C. F. dos Santos

Appendix C: List of the Codes and Localities

Code	Municipality	Locality	Formation
ANAC	São João do Rio do Peixe	Araçá de Cima	Sousa
ANAR	São João do Rio do Peixe	Aroeira	Sousa
ANBD	São João do Rio do Peixe	Barragem do Domício	Sousa
ANCA	São João do Rio do Peixe	Cabra Assada	Sousa-Rio Piranhas
ANEN	São João do Rio do Peixe	Fazenda Engenho Novo	Sousa
ANJU	São João do Rio do Peixe	Sítio Juazeirinho	Sousa
ANPV	São João do Rio do Peixe	Poço da Volta	Sousa
ANZO	São João do Rio do Peixe	Zoador	Sousa
APTA	Aparecida	Tapera	Sousa
APVR II	Aparecida	Várzea dos Ramos II	Sousa
APVR III	Aparecida	Várzea dos Ramos III	Sousa
POGR	Pombal	Grotão	Antenor Navarro
SOAB	Sousa	Abreu	Sousa
SOBP	Sousa	Baixio do Padre	Sousa
SOCA	Sousa	Fazenda Caiçara-Piau	Sousa
SOCV	Sousa	Curral Velho	Rio Piranhas
SOES	Sousa	Fazenda Estreito, (Serrote do Pimenta)	Antenor Navarro
SOFB	Sousa	Floresta dos Borba	Antenor Navarro
SOFP	Sousa	Fazenda Paraíso	Rio Piranhas
SOLF I	Sousa	Lagoa do Forno I	Sousa-Rio Piranhas
SOLF II	Sousa	Lagoa do Forno II	Sousa-Rio Piranhas
SOMA	Sousa	Matadouro Municipal	Sousa
SOMD	Sousa	Mãe d'Água	Rio Piranhas
SOPE	Sousa	Fazenda Pedregulho	Sousa
SOPI	Sousa	Fazenda Piedade	Sousa
SOPU	Sousa	Piau II	Sousa
SOPM	Sousa	Poço do Motor	Sousa
SOPP	Sousa	Passagem das Pedras	Sousa
SORC	Sousa	Riacho do Cazé	Antenor Navarro
SORO	Sousa	Riachão dos Oliveira	Antenor Navarro
SOSA	Sousa	Sítio Saguim	Sousa
SOSL	Sousa	Serrote do Letreiro	Antenor Navarro
SOVR (APVR I)	Aparecida	Várzea dos Ramos	Sousa
UIAR	Uiraúna	Arapuã	Antenor Navarro
UIBA	Uiraúna	Baleia	Antenor Navarro
UIPO	Uiraúna	Poçinho	Antenor Navarro

Index

Italics denote an illustration or a table.

GIUSEPPE LEONARDI is an Associate Senior Researcher in the Department of Geology at Federal University of Rio de Janeiro and Associate Curator of the Museo della Scienza-MUSE of Trento (Italy). He is the author of *Annotated Atlas of South America Tetrapod Footprints* and of *Glossary and Manual of Tetrapod Footprint Paleoichnology*.

ISMAR DE SOUZA CARVALHO is a Professor in the Department of Geology at Federal University of Rio de Janeiro and a Researcher in the Geosciences Center at University of Coimbra, the Brazilian National Research Council (CNPq), and the Rio de Janeiro Scientist Program (FAPERJ).